80,-

Astrophysik

'19

**Space Sciences Series of ISSI**

Volume 70

More information about this series at http://www.springer.com/series/6592

Saskia Foerster · Luis Guanter ·
Teodolina Lopez · José Moreno ·
Michael Rast · Michael E. Schaepman
Editors

# Exploring the Earth System with Imaging Spectroscopy

Previously published in *Surveys in Geophysics*, Volume 40, Issue 3, 2019

Springer

*Editors*
Saskia Foerster
GFZ German Research Centre
for Geosciences
Potsdam, Germany

Luis Guanter
GFZ German Research Centre
for Geosciences
Potsdam, Germany

Teodolina Lopez
International Space Science Institute
Bern, Switzerland

José Moreno
University of Valencia
Valencia, Spain

Michael Rast
ESA/ESRIN
Frascati, Italy

Michael E. Schaepman
University of Zürich
Zürich, Switzerland

ISSN 1385-7525
Space Sciences Series of ISSI
ISBN 978-3-030-24909-0



Cover illustration: Schematic representation of future hyperspectral satellites observing the Earth’s surface (© ESA/NASA, modified GFZ Potsdam/Theres Kuester).

This Springer imprint is published by the registered company Springer Nature Switzerland AG
The registered company address is: Gewerbestrasse 11, 6330 Cham, Switzerland

# Contents

Surveys in Geophysics (2019) 40:297–301
https://doi.org/10.1007/s10712-019-09516-0

# Guest Editorial: International Space Science Institute (ISSI) Workshop on Space-Borne Imaging Spectroscopy for Exploring the Earth's Ecosystems

**Saskia Foerster[1] · Luis Guanter[1] · Teodolina Lopez[2] · José Moreno[3] · Michael Rast[4] · Michael E. Schaepman[5]**

Received: 1 February 2019 / Accepted: 6 February 2019 / Published online: 25 March 2019


Earth's terrestrial and aquatic ecosystems are undergoing accelerated changes as a consequence of climate change, inadequate land use and many other processes, including feedback mechanisms. In the coming decades, the world's increasing population and living standards will prompt a further increase in our demand for global food and water resources. Likewise, the increasing demand of the world's growing economies for resources such as energy, raw materials, water and land will remain a major driver of global change and environmental degradation. Furthermore, extreme weather conditions, hazards and pollution endanger human health, property and the environment. Our ability to address these increasingly urgent challenges very much depends on the improved detection, quantification and understanding of the relevant processes. A solid scientific foundation is required for informed decision making aimed at sustainable resources management.

During recent decades, imaging spectroscopy has proven to be a valuable source of information for quantifying and modelling Earth surface processes (Goetz et al. 1985; Green et al. 1998; Schaepman et al. 2009). As opposed to multispectral imaging systems, imaging spectrometers with narrow and contiguous spectral bands provide quantitative 'fingerprints' of those materials having diagnostic absorption features which can be observed by means of dispersive (prism or diffraction grating) optical systems. Since the 1980s, airborne imaging spectrometers such as AVIRIS (Vane et al. 1993), APEX (Schaepman et al. 2015) and others have been shown to provide accurate and quantitative information for many application fields. These range from mineral exploration, soil mapping, vegetation monitoring, mapping of pollution and hazardous

✉ Saskia Foerster
saskia.foerster@gfz-potsdam.de

1 Helmholtz Centre Potsdam, GFZ German Research Centre for Geosciences, Telegrafenberg, 14473 Potsdam, Germany

2 International Space Science Institute (ISSI), Hallerstrasse 6, 3012 Bern, Switzerland

3 Image Processing Laboratory (IPL), Parc Científic, Universitat de València, C/ Catedrático José Beltrán 2, 46980 Paterna, València, Spain

4 ESA-ESRIN, Largo Galileo Galilei 1, 00044 Frascati, Italy

5 Remote Sensing Laboratories (RSL), Department of Geography, University of Zurich, Winterthurerstrasse 190, 8057 Zurich, Switzerland

materials, inland and coastal water monitoring to urban applications. Airborne imaging spectrometers have demonstrated a solid science baseline, enabling us to scale these applications to space missions. Following on from early space-borne systems such as the Hyperion imaging spectrometer aboard NASA's Earth observing mission EO-1 and ESA's compact high-resolution imaging spectrometer CHRIS aboard the PROBA satellite, there are currently various plans to deploy imaging spectrometers in space, such as the approved Italian PRISMA (Loizzo et al. 2018) and German EnMAP (Guanter et al. 2015) missions. Further, there are a number of concepts under development, such as the US surface biology and geology SBG (Green 2018a), the Israeli/Italian SHALOM (Feingersh and Ben-Dor 2015) and the ESA Copernicus CHIME (Nieke and Rast 2018) concepts. Furthermore, the International Space Station (ISS) has become home to several spectrometers targeting the Earth's surface. Examples include the hyperspectral DESIS mission by DLR and Teledyne Brown Engineering (Müller et al. 2018). The Japanese hyperspectral mission HISUI (Matsunaga et al. 2018) and the US Earth surface mineral dust source investigation mission EMIT (Green 2018b) will be deployed on the ISS in the 2020/21 timeframe.

Even though the full solar reflected spectrum from the visible to the short wavelength infrared optical range (VSWIR) has to be contiguously covered on space missions in order to address all the aforementioned scientific and operational applications, the number of visible and near-infrared systems (VNIR, suitable for many vegetation and water applications) that can be accommodated aboard small satellites and Cubesats is increasing. For these, a coarser spatial resolution and narrower swath can be compensated for by a larger number of satellites that can be deployed at a comparably low budget. On the other hand, the so-called reference missions, mainly provided by space agencies, tend to be VSWIR systems with ~ tens m spatial sampling. These typically incorporate a spectral sampling interval between 5 and 10 nm and desired temporal revisit times of the order of 10–20 days.

The advent of new technologies, for instance in the area of detector development, but also in terms of satellite formations and small satellites, gives a new direction into which imaging spectroscopy from space, aircraft and possibly high-altitude platforms can be expected to evolve. In the long run, hyperspectral systems, owing to their diagnostic capacity, are expected to supersede multispectral systems. One of the challenging aspects of space-borne hyperspectral remote sensing remains the massive amount of data generated and resulting in large three-dimensional data cubes (with spectral, spatial and temporal dimensions) provided by those spectrometers. These easily outrun data rates of conventional multispectral systems, by a large factor. However, recent developments in machine learning, data compression and data analytics open new doors for the handling and analysis of huge amounts of imaging spectroscopy data.

With the growing availability of high-quality data and advances in data analysis and handling techniques, it is expected that—in the future—data from space-borne imaging spectrometers will significantly improve our knowledge of the complex processes and feedback mechanisms of the Earth system and will support many aspects of environmental management. While most satellite missions target the measurement of one or a few variables, imaging spectroscopy can provide a large range of measurements in parallel. Hence, a set of carefully balanced requirements of measured variables, and consequently of mission objectives, is needed. Therefore, a challenging task in the definition of future imaging spectroscopy missions lies in the identification and consolidation of the user and observational requirements to support environmental management and policies in various fields, including agriculture and forestry, raw materials, water resources, hazards and pollution and biodiversity.

The papers presented in this volume are the outcome of a Workshop on 'Exploring the Earth's Ecosystems on a Global Scale: Requirements, Capabilities and Directions in Space-borne Imaging Spectroscopy' held at the International Space Science Institute (ISSI) in Bern, Switzerland, from 21 to 25 November 2016. The overall objective of the Workshop was to provide a better and more comprehensive understanding of the observational requirements for a range of major application fields, of the capacity of space-borne imaging spectroscopy for observing terrestrial and aquatic ecosystems as well as of potential new fields where additional geo-biophysical observations and long-term monitoring from space will make a significant contribution. Furthermore, the Workshop with its international experts in the field of imaging spectroscopy provided an excellent podium on which to discuss new operational applications, as well as the requirements for future missions and associated ground segments. The Workshop discussion was organized along the main fields of vegetation biochemistry and functioning, ecosystem structure and composition, biodiversity, inland and coastal waters, ice and snow, soils, geology and minerals, natural and man-made hazards, integrated Earth system applications, agriculture, forestry and urban environments, modelling and simulation, data processing and retrievals, the synergistic use of spectroscopy and other observational approaches, in situ measurements, validation schemes and spectral libraries, science traceability and uncertainty estimates.

The papers in this volume reflect and summarize the presentations and discussions. They comprise an overview on imaging spectroscopy and its mission legacy (Rast and Painter 2019) followed by a review on the retrieval of atmospheric parameters and surface reflectance from VSWIR imaging spectroscopy data (Thompson et al. 2019). The majority of papers focus on specific application fields of space-borne imaging spectroscopy describing the state-of-the-art situation, current gaps and user and observational requirements, in addition to challenges and future directions. These application fields include soils (Chabrillat et al. 2019), inland and coastal waters (Giardino et al. 2019), natural and anthropogenic hazards (Ong et al. 2019), urban environments (van der Linden et al. 2019), vegetation biochemistry and functioning (Gamon et al. 2019), agriculture (Hank et al. 2019) and forest ecosystems (Hill et al. 2019). The volume is completed by papers on specific across-application topics of radiative transfer models, retrieval methods and the emulation of spectral databases for vegetation observation (Verrelst et al. 2019), upscaling imaging spectroscopy observations (Malenovsky et al. 2019) and the synergistic use of imaging spectroscopy along with other remote sensing observational approaches (Guanter et al. 2019).

To conclude, a topic which started as a purely experimental and research-oriented observing technology on airborne platforms increasingly assumes the character of an operational observing capacity as we move towards space-borne imaging spectroscopy missions with global frequent mapping capabilities. It can be expected that future repeated coverage of the Earth's surface with space-borne imaging spectroscopy data will bring a major advance in both the quantity and depth of information that future models and service developments can be based upon. Thus, this volume provides a broad overview of user and observational requirements, the cases for the science, the capabilities of space-borne imaging spectroscopy technology and promising avenues into a hyperspectral future.

**Acknowledgements** We would like to thank the International Space Science Institute (ISSI), the European Space Agency (ESA) and all contributors to the Workshop and to this issue, the referees, the Editor in Chief, Michael Rycroft, and the staff of Springer Nature involved in its production. The first author would like to acknowledge the EnMAP mission scientific preparation program funded under the DLR Space Administration with resources from the German Federal Ministry of Economic Affairs and Energy. The contribution of the last author is supported by the University of Zurich Research Priority Program on Global Change and Biodiversity (URPP GCB).

## References


Chabrillat et al. (2019) Surv Geophys (in review)

Feingersh T, Ben-Dor E (2015) SHALOM—a commercial hyperspectral space mission. In: Qian S (ed) Optical payloads for space missions. https://doi.org/10.1002/9781118945179.ch11

Gamon et al. (2019) Surv Geophys (in review)

Giardino C, Brando VE, Gege P, Pinnel N, Hochberg E, Knaeps E, Reusen I, Doerffer R, Bresciani M, Braga F, Foerster S, Champollion N, Dekker A (2019) Imaging spectrometry of inland and coastal waters: state of the art, achievements and perspectives. Surv Geophys. https://doi.org/10.1007/s10712-018-9476-0

Goetz AF, Vane G, Solomon JE, Rock BN (1985) Imaging spectrometry for earth remote sensing. Science 228(4704):1147–1153. https://doi.org/10.1126/science.228.4704.1147

Green RO (2018a) Global VSWIR imaging spectroscopy and the 2017 Decadal Survey. In: IGARSS 2018–2018 IEEE International Geoscience and Remote Sensing Symposium, Valencia, pp 183–185. https://doi.org/10.1109/igarss.2018.8518744

Green RO (2018b). The earth surface mineral dust source investigation (EMIT). https://hyspiri.jpl.nasa.gov/downloads/2018_Workshop/day1/13_HyspIRI_EMIT_Overview_20180815b.pdf. Last accessed 6 Jan 2019

Green RO, Eastwood ML, Sarture CM, Chrien TG, Aronsson M, Chippendale BJ, Faust JA, Pavri BE, Chovit CJ, Solis M, Olah MR (1998) Imaging spectroscopy and the airborne visible/infrared imaging spectrometer. Remote Sens Environ 65:227–248. https://doi.org/10.1016/S0034-4257(98)00064-9

Guanter L, Kaufmann H, Segl K, Foerster S, Rogaß C, Chabrillat S, Küster T, Hollstein A, Rossner G, Chlebek C, Straif C, Fischer S, Schrader S, Storch T, Heiden U, Mueller A, Bachmann M, Mühle H, Müller R, Habermeyer M, Ohndorf A, Hill J, Buddenbaum H, Hostert P, van der Linden S, Leitão PJ, Rabe A, Doerffer R, Krasemann H, Xi H, Mauser W, Hank T, Locherer M, Rast M, Staenz K, Sang B (2015) The EnMAP spaceborne imaging spectroscopy mission for earth observation. Remote Sens 7(7):8830–8857. https://doi.org/10.3390/rs70708830

Guanter L, Brell M, Chan JC-W, Giardino C, Gomez-Dans J, Mielke C, Morsdorf F, Segl K, Yokoya N (2019) Synergies of spaceborne imaging spectroscopy with other remote sensing approaches. Surv Geophys. https://doi.org/10.1007/s10712-018-9485-z

Hank TB, Berger K, Bach H, Clevers JGPW, Gitelson A, Zarco-Tejada P, Mauser W (2019) Spaceborne imaging spectroscopy for sustainable agriculture: contributions and challenges. Surv Geophys. https://doi.org/10.1007/s10712-018-9492-0

Hill, Buddenbaum and Townsend (2019) Surv Geophys (in review)

Loizzo R, Guarini R, Longo F, Scopa T, Formaro R, Facchinetti C and Varacalli G (2018) Prisma: the Italian hyperspectral mission. In: IGARSS 2018–2018 IEEE International Geoscience and Remote Sensing Symposium, Valencia, pp 175–178. https://doi.org/10.1109/igarss.2018.8518512

Malenovsky et al. (2019) Surv Geophys (in review)

Matsunaga T, Iwasaki A, Tsuchida S, Iwao K, Tanii J, Nakamura R, Yamamoto H, Kato S, Obata K, Kashimura O, Tanii J, Mouri K, Tachikawa T (2018) HISUI Status Toward FY2019 Launch. In: IGARSS 2018–2018 IEEE International Geoscience and Remote Sensing Symposium, Valencia, pp 160-163. https://doi.org/10.1109/igarss.2018.8518639

Müller R, Bachmann M, Alonso K, Carmona E, Cerra D, de Ios Reyes R, Gerasch B, Krawczyk H, Ziel V, Heiden U, Krutz D (2018) Processing, validation and quality control of spaceborne imaging spectroscopy data from desis mission on the ISS. In: IGARSS 2018–2018 IEEE International Geoscience and Remote Sensing Symposium, Valencia, pp 189–191. https://doi.org/10.1109/igarss.2018.8519572

Nieke J, Rast M (2018) Towards the Copernicus Hyperspectral Imaging Mission For The Environment (CHIME). In: IGARSS 2018–2018 IEEE International Geoscience and Remote Sensing Symposium, Valencia, pp 157–159. https://doi.org/10.1109/igarss.2018.8518384

Ong et al. (2019) Surv Geophys (in review)

Rast and Painter (2019) Surv Geophys (in review)

Schaepman ME, Ustin SL, Plaza AJ, Painter TH, Verrelst J, Liang S (2009) Earth system science related imaging spectroscopy—an assessment. Remote Sens Environ Supplement 1:A123–S137. https://doi.org/10.1016/j.rse.2009.03.001

Schaepman ME, Jehle M, Hueni A, D'Odorico P, Damm A, Weyermann J, Schneider FD, Laurent V, Popp C, Seidel FC, Lenhard K, Gege P, Küchler C, Brazile J, Kohler P, De Vos L, Meuleman K, Meynart R, Schläpfer D, Kneubühler M, Itten KI (2015) Advanced radiometry measurements and Earth science applications with the Airborne Prism Experiment (APEX). Remote Sens Environ 158:207–219. https://doi.org/10.1016/j.rse.2014.11.014

Thompson DR, Guanter L, Berk A, Gao B-C, Richter R, Schläpfer D, Thome KJ (2019) Retrieval of atmospheric parameters and surface reflectance from visible and shortwave infrared imaging spectroscopy data. Surv Geophys. https://doi.org/10.1007/s10712-018-9488-9

van der Linden S, Okujeni A, Canters F, Degerickx J, Heiden U, Hostert P, Priem F, Somers B, Thiel F (2019) Imaging spectroscopy of urban environments. Surv Geophys. https://doi.org/10.1007/s10712-018-9486-y

Vane G, Green RO, Chrien TG, Enmark HT, Hansen EG, Porter WM (1993) The airborne visible/infrared imaging spectrometer (AVIRIS). Remote Sens Environ 44:127–143. https://doi.org/10.1016/0034-4257(93)90012-M

Verrelst J, Malenovský Z, Van der Tol C, Camps-Valls G, Gastellu-Etchegorry J-P, Lewis P, North P, Moreno J (2019) Quantifying vegetation biophysical variables from imaging spectroscopy data: a review on retrieval methods. Surv Geophys 1:1. https://doi.org/10.1007/s10712-018-9478-y

Surveys in Geophysics (2019) 40:303–331
https://doi.org/10.1007/s10712-019-09517-z

# Earth Observation Imaging Spectroscopy for Terrestrial Systems: An Overview of Its History, Techniques, and Applications of Its Missions

Michael Rast[1] · Thomas H. Painter[2]

Received: 12 June 2018 / Accepted: 8 February 2019 / Published online: 2 March 2019



## Abstract

Imaging spectroscopy in the visible-to-shortwave infrared wavelength range (VSWIR), or nowadays more commonly known as 'hyperspectral imaging', for terrestrial Earth Observation remote sensing, dates back to the early 1980s when its development started with mainly airborne demonstrations. From its initial use as a research tool, imaging spectroscopy encompassing the VSWIR spectral range has gradually evolved towards operational and commercial applications. Today, it is one of the fastest growing research areas in remote sensing owing to its diagnostic power by means of discrete spectral bands that are contiguously sampled over the spectral range with which a target is observed. The main principles of imaging spectroscopy rely on the exploitation of light dispersion technologies to split the incoming light through a telescope before being projected onto detector arrays. The light dispersion can be achieved by using prism or diffractive grating optical systems, perpetually aiming for improved performances in terms of efficiency, straylight rejection, and polarization sensitivity. The sensor technique has been first used in airborne imaging spectroscopy since the early 1980s and later in spaceborne hyperspectral missions from the end of the 1990s onwards. Currently, several hyperspectral spaceborne systems are under development and in preparation to be launched within the next few years. Through hyperspectral remote sensing, physical, chemical, and biological components of the observed matter can be separated and resolved thus providing a spectral 'fingerprint'. The analyses of the spectral absorptions often give rise to quantitative retrievals of components of the observed target. The derived information is vital for the generation of a wide variety of new quantitative products and services in the domain of agriculture, food security, raw materials, soils, biodiversity, environmental degradation and hazards, inland and coastal waters, snow hydrology and forestry. Many of these are relevant to various international policies and conventions. Originally developed as a powerful detection and analysis tool for applications predominantly related to planetary exploration and non-renewable resources, imaging spectroscopy now covers many disciplines in atmospheric, terrestrial vegetation, cryosphere, and marine research and application fields. There is an increasing number of visible/near-infrared (VNIR) imaging spectrometers emerging also as small payloads on small satellites and cubesats, built and launched by small-medium enterprises. These are targeted to address commercial applications mainly in agriculture, resources and environmental management, and hazard observations.


Extended author information available on the last page of the article

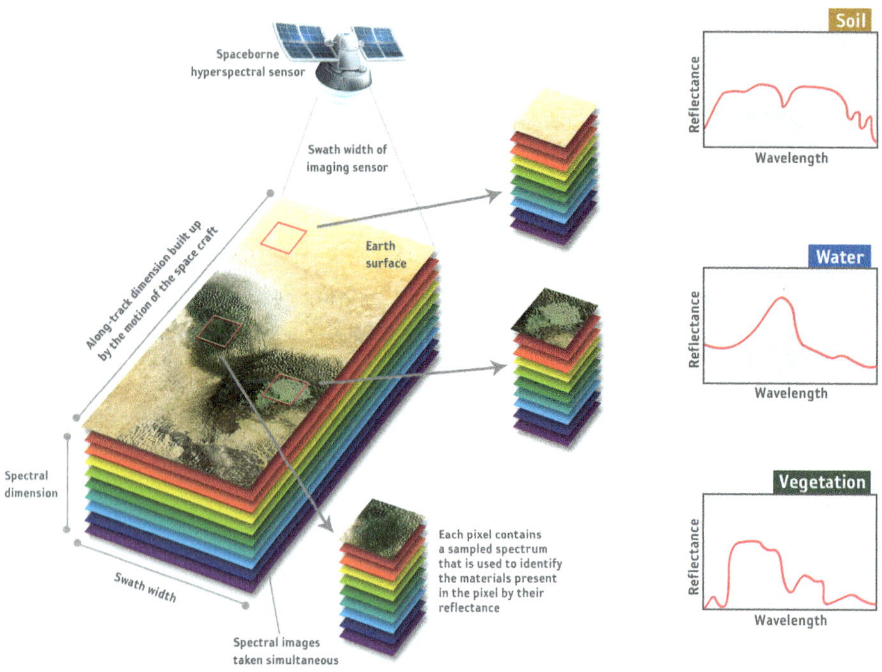


**Fig. 1** Imaging spectroscopy—principle of observation (ESA Earth Observation Graphic Bureau)

**Keywords** Earth observation · Satellite remote sensing · Imaging spectroscopy · Terrestrial ecosystems

## 1 Introduction

Imaging spectroscopy entails quantitative sensing of the target in the optical range of the electromagnetic spectrum (visible through shortwave infrared) with contiguous, gapless spectral sampling from the visible to the shortwave infrared portion of the electromagnetic spectrum. With this complete spectrally resolved sampling of reflected radiation, imaging spectroscopy is the asymptote concept of passive optical remote sensing, with the only direction forward being improvements in resolutions (spectral, spatial, and radiometric). With a much larger number of spectral sampling intervals or so-called spectral bands than is commonly applied with multispectral remote sensing systems, such as those onboard the US Landsat series or the European Copernicus Sentinel-2 satellites (Drusch et al. 2012), imaging spectroscopy today enables the markedly greater diversity of geophysical, geo-biophysical, and geo-biochemical observation and monitoring of the surface (see Fig. 1). An imaging spectrometer can be defined as an instrument that acquires images in dispersed multiple contiguous spectral bands with high spectral resolution (Goetz et al. 1985; Vane and Goetz 1988; Van der Meer and De Jong 2011).

With increasingly well-established spectroscopic techniques, optical remotely sensed imaging spectroscopy can deliver significant enhancement in quantitative value-added products. This will support the generation of a wide variety of new products and services in the domain of agriculture, food security, water resources, raw material resources, soils, biodiversity, environmental degradation and hazards, inland and coastal waters, urban environments and forestry. Such products are relevant to various EU policies that are currently not being met or can be significantly improved, and also to the downstream private sector.

Hyperspectral imaging has already supported or been used for a large range of applications (to name a few: Asner et al. 2017; Chang 2007; Giardino et al. 2018; Green et al. 2006; Hochberg et al. 2015; King et al. 2014; Ong et al. 2003; Painter et al. 2016; Roberts et al. 2012; Swayze et al. 2000; Ustin et al. 2004). Based on successful airborne deployments over the last three decades and the past satellite missions along with preparatory activities of some national demonstrative satellite missions, hyperspectral imaging from satellite is now ready for operational use.

In general, the application potential of a hyperspectral mission is a direct result of adding an increased number of narrow spectral (contiguous) bands with a relatively high signal-to-noise ratio (SNR) to the conventional passive optical multispectral remote sensing missions, such as Sentinel-2 and Landsat, thereby allowing direct and indirect identification of target compositions and quantities (see Fig. 2).

With hyperspectral remote sensing physical, chemical and biological components of the observed matter can be separated and resolved thus providing a spectral 'fingerprint'. The analyses of the spectral absorptions also often give rise to quantitative retrievals of components of the observed target. Key environmental information can be derived and quantified from the measured spectral signal, e.g., directly through distinct absorption features or indirectly, through inversion, assimilation, spectral un-mixing, and/or (de-) correlation

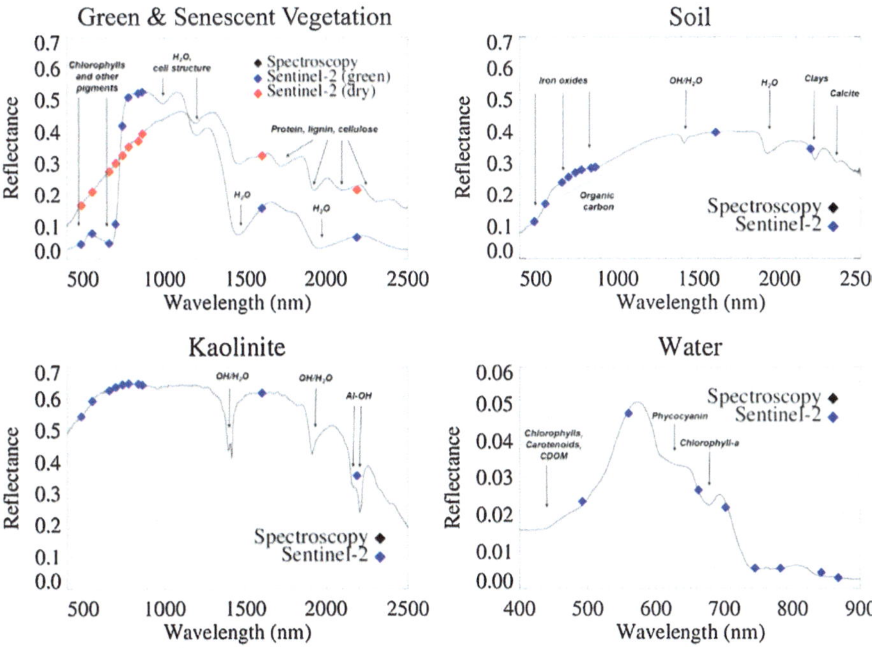


**Fig. 2** Reflectance spectra for different Earth surface materials at high spectral resolution and resampled to the spectral response of the multispectral instrument onboard Sentinel-2 (courtesy L. Guanter, GFZ—Helmholtz Centre Potsdam)

techniques. Potential supported policies comprise the UN Sustainable Development Goals (SDGs) and the European Commission Common Agricultural Policy (CAP).

The objective of this paper is to provide an overview of (terrestrial) imaging spectroscopy for geo-scientific applications, more specifically, about its principles (Sect. 2), instrument concepts (Sect. 3), past, current, and planned airborne and spaceborne imaging spectrometers (Sect. 4), the main applications fields (Sect. 5), and future observational requirements (Sect. 6).

## 2 Principles of Imaging Spectroscopy

Spectroscopy is defined as the study of light as a function of wavelength that has been emitted, reflected, or scattered from a solid, liquid, or gas. In remote sensing, the quantity most used is (surface) reflectance (expressed as a percentage). Spectroradiometry is the associated technology for measuring the power of optical radiation in narrow, contiguous wavelength intervals. The quantities measured are usually expressed as spectral irradiance (commonly measured in W $m^{-2}$ $nm^{-1}$) and spectral radiance (commonly measured in W $sr^{-1}$ $m^{-2}$ $nm^{-1}$) (Schaepman 2009).

Remote, direct identification of surface materials on a picture element (pixel) basis can be accomplished by proper sampling of absorption features in the reflectance spectrum following atmospheric correction of the obtained at-sensor radiance values. The airborne and spaceborne sensors are capable of acquiring images simultaneously in 10 s to 100 s of contiguous bands (for instance the US Airborne Visible/InfraRed Imaging Spectrometer—AVIRIS—acquires 224 bands, 210 remaining after removing the overlaps between spectrometers). The ability to acquire laboratory-like spectra remotely has become a major advance in remote sensing capability (Goetz et al. 1985).

The objective of imaging spectroscopy is to measure quantitatively the components of the Earth system from calibrated spectra acquired as images for scientific research and applications. The radiance of reflected solar electromagnetic energy from the Earth's surface is dispersed in a spectrometer and used to form contiguous, inherently registered spectral images of the scene over the VSWIR spectral range. Each picture element in the scene has associated with it sufficient information for the reconstruction of a complete reflectance spectrum. The value of this technique is that it allows the diagnostic narrow-band spectral features that are present in most terrestrial materials to be used to uniquely identify those materials (Vane and Goetz 1988). These spectral features are typically 20–40 nm in width at the half-band depth (Hunt 1980).

Spectral imaging systems that acquire data in contiguous 10-nm-wide spectral bands can, therefore, produce data with sufficient resolution for the direct identification of those materials with diagnostic features. Conversely, multispectral sensors, with broader spectral bands can not resolve these features because their spectral bandwidths are often as broad as 100–200 nm. Moreover, in multispectral systems the spectral bands are non-contiguously dispersed over the sensor's spectral range (Vane and Goetz 1988).

Starting in the early 1970s, the power of spectroscopic mineral analyses, where charge transfer bands in the ultraviolet and electronic transitions at longer wavelengths of minerals were analysed and diagnostic spectral features appeared as results of either electronic or vibrational processes, was mainly applied in the laboratory (Rast 1991). In the early 1980s, the capacity of spectroscopy analysis was transferred onto airborne imaging systems for the observation of terrestrial and aquatic ecosystems (Vane et al. 1984).

# 3 Instrument Concepts and Calibration

Imaging spectrometers usually consist of four major elements: A front optic that collects the signal emerging from the observed target/scene; a dispersive element that splits the incoming signal into narrow spectral intervals; an optical component focusing the dispersed signal; and a detector assembly, which is located in the focal plane.

## 3.1 Instrument Concepts

Three general types of imaging spectrometer systems are used for observations from aircraft and satellite (personal correspondence with U. Del Bello, ESA-ESTEC, Del Bello et al. 2003; Kunkel et al. 2000; Schaepman 2009; Eismann 2012; Rast et al. 2004):

### 3.1.1 Prism or Grating Spectrometers

In a dispersive imaging spectrometer, two types of dispersive elements are used for the spectral separation of the incoming signal: prism or grating. In prism-based imaging spectrometers, the incoming signal/light is projected onto a prism, resulting in angular dispersion of the incident light into different wavelength components. The thickness of the glass prism and its wedge shape contribute to aberrations in such instruments. To minimize aberrations over the field, imaging spectrometers sometimes employ a combination of prisms with curved surfaces.

In grating-based imaging spectrometers, the spectral dispersion can be varied by modifying the grating design. Blazed grating are used to increase the diffraction efficiency over a limited spectral interval.

Drawbacks of gratings are the emergence of multi-order diffraction, which is mitigated by filters. There are different types of gratings, i.e. convex gratings, volume phase holographic gratings, and spherical transmission gratings.

### 3.1.2 Fourier Transform Interferometers

Fourier transform interferometers split the incoming radiation into two beams, introducing a controlled phase shift, and recombining them generating interference and introducing sinusoidal variations in spectral transmission, depending on the wavelength and the optical path-length difference in the interferometer. The light intensity is modulated by the path difference of the two beams, and the amplitude of the signal is sampled at an appropriate sampling rate during the acquisition. Subsequently, a Fourier transform converts the amplitude-modulated signal into a frequency spectrum. Not limited by a dispersing component, Fourier transform interferometers have a high spectral resolution and a high optical throughput and spectral accuracy. However, the sampling scheme of Fourier transform interferometers over an extended time period for each sample of a scene along-track in the absence of an entrance slit is of disadvantage and necessitates intensive raw image data processing.

### 3.1.3 Narrow-Band Adaptive Filters

There are various types of filter imaging spectrometers, which are strictly speaking not falling under the definition of a spectrometer, since the incoming light is not dispersed or 'broken up' into its wavelength components, but through filtering into the discrete wavelengths desired:

- A linear variable filter or wedged filter is a band-pass filter in which the thickness of the filter coating varies linearly along its length so that the spectral content transmitted through it varies in that direction resulting in principle in an unlimited number of different filters. The number of separate spectral channels depends solely on the number of elements in the detector array. The transmission characteristic of a linear variable filter varies with the incidence angle of the incoming light and usually has a moderate spectral resolution of a few tens of nanometres.
- An acousto-optic tunable filter is a pass-band transmission filter, which exploits the acousto-optic interaction inside an anisotropic medium. It consists of a birefringent crystal, onto which a piezoelectric transducer is bonded. The acoustic waves that are generated by means of the transducer create a periodic modulation of the index of refraction in the crystal, and therefore a moving phase grating (or diffraction grating). For a given acoustic frequency, only a limited optical wavelength band can satisfy the phase matching condition and can, as a consequence, be diffracted. The frequency of the excitation source is varied to change the period of the grating formed and thus select the required transmission wavelength. The advantage of acousto-optic tunable filters is the short response time (typically microseconds) and the consequent fast spectral band selection. For a large number of spectral bands, however, the measurement time significantly increases and the optical throughput is generally low.
- A Fabry–Perot interferometer is the basis of tuned etalon spectral imagers. The etalon, composed of two plane parallel partially reflecting surfaces with a precise spacing between them, is the core element of the Fabry–Perot interferometer. It produces very narrow bands and filter characteristics, and central wavelength and bandwidth can be tuned by varying the separation between the parallel surfaces or by electrically controlling the refractive index of the etalon material. It has a limited field of view owing to the light transmission being dependent on the angle of incidence and its sensitivity to temperature variation. Moreover, the operating wavelength range is rather limited.

### 3.2 Calibration

In order to be able to carry out a quantitative assessment of imaging spectroscopy data and to validate geo-biophysical observations, the measurement values obtained must be accurate in both a relative and absolute sense. Two major factors influence the representativeness of these data, namely the atmospheric influence and the properties of the sensors system including its carrier (Thome 2001).

Absolute and relative radiometric calibration of a hyperspectral imaging system can in principle be provided by repeated observations of well-documented and calibrated ground targets. The spectrometers flown in space so far indicate, however, that onboard calibration is highly desirable, since it is independent from the correction of atmospheric and on-ground morphology effects. Moreover, onboard calibration provides accurate monitoring of the system degradation, which is specifically important for spaceborne systems susceptible to degrading influences such as hard radiation (Guanter et al. 2006).

Radiometric calibration is commonly provided using reflectance-calibrated sun diffusers deployed in the optical path of the system. More than one diffuser is often implemented, in order to employ one less frequently deployed diffuser to check for, e.g., effects of Sun exposure on the more frequently used radiance diffuser's reflectance (Delwart et al. 2004).

Dark current calibration is often provided involving the system's shutter. The drift of the dark current between calibrations is often monitored using non-illuminated or covered lines on the spectrometer's detector. Non-illuminated detector areas may also be used to assess stray light in comparison with pre-flight predictions (Rast et al. 1999).

Spectral calibration requirements often include in-flight characterization of wavelengths imaged on detector elements using onboard spectral sources, such as a set of lasers, illuminated diffusers, or doped diffusers. However, sufficiently accurate wavelength calibration can be achieved using atmospheric absorption features (and possibly some deeper solar Fraunhofer lines).

In addition to the various onboard calibration devices, such as solar diffusers, lamps or LEDs, lunar observations can be employed to monitor reflective solar band on-orbit calibration stability and to specifically independently track the performance of solar diffuser calibration. This approach has successfully been used for the calibration of MODIS on the Terra and Aqua satellites, for the Visible Infrared Imaging Radiometer Suite (VIIRS) on the Suomi National Polar-orbiting Partnership (S-NPP) (Xiong et al. 2016) and for Hyperion on EO-1 (Kieffer et al. 2002). The Moon is considered to be a very stable radiometric reference in the VIS/NIR spectral regions (Kieffer 1997), being viewed by the satellite sensors mostly through a dedicated port in the spacecraft regularly and at very similar phase angles, which can be achieved by in-orbit spacecraft manoeuvres (Sun and Xiong 2011).

## 4 Past, Current and Future Hyperspectral Airborne and Spaceborne Systems

### 4.1 Airborne Imaging Spectrometers

Most imaging spectroscopy over the last 30 years has been based on airborne spectrometers covering the visible and near-infrared (VNIR) and, in many cases, also the shortwave infrared (SWIR) spectral ranges (~400–1000 nm and 1000–2500 nm, respectively). Since the 1980s, developments in VNIR and SWIR spectrometers were initiated by the US Airborne Imaging Spectrometer AIS-I (Vane and Goetz 1985), built by JPL, which flew from 1982 to 1985 and the Canadian Fluorescence Line Imager (FLI) by Moniteq Ltd., CDN (Gower et al. 1992), which flew from 1983 to 1990. In particular, the Airborne Visible/Infrared Imaging Spectrometer (AVIRIS), designed by NASA JPL in California and operated under NASA contract since 1986, has been a 'trailblazer' for airborne imaging spectroscopy since the late 1980s in a large number of airborne deployments and field campaigns.

The most widely used in the last few years have been the Australian HyMAP (Cocks et al. 1998), the Compact Airborne Spectrographic Imager (CASI) by ITRES Research Ltd., CDN (Gower et al. 1992) and the Airborne Prism Experiment (APEX), by University of Zürich, CH and VITO, BE (Schaepman et al. 2015). Building upon AVIRIS, the Airborne Visible Infrared Imaging Spectrometer—Next Generation (AVIRIS NG) has been developed to eventually replace the AVIRIS instrument with an improved radiometric performance and a higher spectral sampling interval of 5 nm. One of the major assets of AVIRIS, as well as AVIRIS NG, is that the sensors can be flown on lower flying aircraft (e.g., Twin Otter) but also on high-altitude airborne platforms (e.g., ER-2).

In addition, other well-known sensor systems, such as the US built Digital Airborne Imaging Spectrometer (DAIS) by GER Corp. USA (Chang et al. 1993), the German Reflective Optics System Imaging Spectrometer (ROSIS) by German industry and research institutions (Gege et al. 1998), the Finnish Airborne Imaging Spectrometer for Applications (AISA) by SPECIM (Makisara et al. 1993), the US Hyperspectral Digital Imagery Collection Experiment (HYDICE) (Rickard et al. 1993), the Italian Multispectral Infrared Visible Imaging Spectrometer (MIVIS) (Bianchi et al. 1996), as well as the Airborne Hyperspectral Scanner (AHS) (Fernández-Renau et al. 2005), both by Deadalus Inc., US, and the NASA Portable Remote Imaging Spectrometer (PRISM) (Mouroulis et al. 2014), were used in many airborne deployments for research, for commercial applications and as test-beds for spaceborne imaging spectrometer concepts studied worldwide.

In the early 2000s, the coupling of VNIR/VSWIR imaging spectrometers with scanning lidar became more prevalent, as for instance in the Carnegie Airborne Observatory (CAO), which provides an in-flight fusion of a VSWIR imaging spectrometer data with a scanning waveform LiDAR for ecosystem research (Asner et al. 2007), the National Ecological Observatory Network (NEON) on the Airborne Observation Platform (AOP) with again a VSWIR imaging spectrometer, a scanning small-footprint waveform LiDAR for 3-D canopy structure measurements and a high-resolution airborne digital camera (Kampe et al. 2010) and the Airborne Snow Observatory (ASO), a coupled visible/near-infrared imaging spectrometer and scanning LiDAR, combined with distributed snow modelling, developed for the measurement of snow spectral albedo/broadband albedo and snow depth/Snow Water Equivalent (Painter et al. 2016).

From the end of the 1990s, the encouraging results and advances made by means of this observation technique led to the launch of the first generation of spaceborne imaging spectrometers with mainly HYPERION on the US—Earth Observing-1 (EO1) satellite (2000), the European CHRIS (Compact High-Resolution Imaging Spectrometer) on PROBA-1 (2001), or in dedicated application domains MERIS on Envisat (2002), and MOS on IRS-P3 (1996). Today, the US AVIRIS NG and the European APEX count to the best performing and calibrated airborne imaging spectrometers. They have besides fulfilling their scientific application campaign objectives also provided the test-bed for mission requirements analyses for future spaceborne Imaging Spectrometer concepts, such as the US HIRIS and HyspIRI and the European LSPIM and SPECTRA.

In addition to the aforementioned VNIR and SWIR spectrometers, various hyperspectral mid- to thermal Infrared sensors in the 2–14 μm spectral range have been developed. These include the Spatially Enhanced Broadband Array Spectrograph System (SEBASS), intended to explore the utility of hyperspectral infrared sensors for remotely identifying solids, liquids, gases, and chemical vapours in the 2 to 14 μm spectral region (Hackwell et al. 1996), the airborne thermal infrared imaging spectrometer, 'Mako', with 128 bands in the thermal infrared covering 7.8 to 13.4 μm (Hall et al. 2011) and the Hyperspectral Thermal Emission Spectrometer (HyTES), developed as part of the activities associated with the Hyperspectral Infrared Imager (HyspIRI) of the NASA's Tier 2 Decadal Survey Missions for Earth science (Hook et al. 2013).

More recently, the use of Unmanned Airborne Vehicles (UAVs) has been increasing and it is to be expected that these will constitute a significant part of airborne imaging spectroscopy, in particular as the instruments are increasingly more compact and comparably lightweight (Adão et al. 2017).

## 4.2 Spaceborne Imaging Spectrometers

Over the last two decades, several true dispersive element Imaging spectrometer missions have been flown in space (Staenz et al. 2013), of which the most prominent ones include:

**MOS:** The Multispectral Optoelectronic Scanner (MOS) was an experimental imaging pushbroom spectrometer for VNIR/SWIR range observations, developed by DLR (German Aerospace Center). The objective was to monitor the Earth's surface (surface–atmosphere interaction, ocean colour, phytoplankton, regional and global distributions of man-made aerosols and their links to gaseous admixtures, spectral and spatial cloudiness characteristics, etc.) in the VNIR/SWIR region of 400–1600 nm. It featured 18 spectral bands, four of which in the NIR had 1.4 nm, 13 bands in the VNIR with 10 nm and one band in the SWIR with 100 nm. The sensor had a 200-km swath and was launched on the Indian satellite IRSP-3 in March 1996. Its operation ended in March 2004 (Zimmermann and Neumann 2000).

**Hyperion:** Launched onboard EO-1 in November 2000, the pushbroom imaging spectrometer Hyperion was built by Northrup Grumman (formerly TRW), USA, as a technology demonstration mission with an estimated 1-year lifespan and completed over 16 years of successful operations. Hyperion featured a grating imaging spectrometer encompassing the 400–2500 nm spectral range with a spectral resolution of 10 nm in VNIR and SWIR at 7.5 km swath width and a spatial resolution of 30 m. It was operated as the heritage orbital spectrometer for future global platforms, including the proposed NASA Hyperspectral Infrared Imager (HyspIRI) concept and the forthcoming German satellite, EnMAP until the spacecraft was decommissioned in March 2017 (Pearlman et al. 2003; Middleton et al. 2013).

**CHRIS**: The Compact High-Resolution Imaging Spectrometer (CHRIS) is a hyperspectral instrument developed and built by former Sira Electro-Optics Ltd., later SSTL, UK. With the objective to collect Hemispherical Conical Reflectance Factor data (as sample to estimate Bidirectional Reflectance Distribution Function) for a better understanding of surface spectral reflectance, CHRIS was launched in October 2001 onboard the European Space Agency (ESA) 'On-Board Autonomy' mission PROBA. It is operating until today. The initial technology objective was to explore the capabilities of imaging spectrometers on agile small satellite platforms. CHRIS' full programmability of its spectral bands provides 63 spectral bands with 34 m resolution, or 19 bands with 18 m resolution over a 13-km swath, or 37 bands with 18 m resolution over a 6.5-km swath in the VNIR range (400–1050 nm) (Cutter 2006).

**MERIS:** MERIS was a pushbroom wide-field imaging spectrometer launched onboard the ESA Envisat satellite in March 2002 to observe the ocean, land surfaces and clouds with high spectral resolution. Developed and built by former Aerospatiale (F), MERIS had individual gain settings to optimize the dynamic range of each of its fifteen spectral bands, which were fully programmable in width and position over its spectral range of 390–1040 nm. The total swath was covered by five identical modules (cameras) each having a 14° FOV, with 0.4° overlap between adjacent cameras resulting in a total swath of 1150 km. MERIS had a dual spatial resolution of 300 and 1200 m. After onboard digital conversion, the data were corrected in real-time with the calibration coefficients to produce raw image data in 15 spectral bands (Rast et al. 1999). MERIS operations ended in April 2012 with a loss of the communication link with the Envisat satellite.

**HICO:** The Hyperspectral Imager for the Coastal Ocean (HICO) was developed by the Naval Research laboratory, USA, and implemented on the International Space

Station (ISS) in September 2009. It was the first spaceborne imaging spectrometer designed to acquire data over coastal regions at 90 m spatial resolution over the 380 to 960 nm spectral range with a 5.7 nm spectral resolution. As a technical demonstrator, HICO was designed to collect one 50 × 200 km scene per orbit. The regions to be collected were determined weekly by a scheduling team. During its operation, which ended in September 2014, HICO collected over 10,000 scenes from around the world (Corson et al. 2008).

So far, **Hyperion** onboard NASA's Earth Observing-1 (EO-1) spacecraft, which was operated until 2017, and the Compact High-Resolution Imaging Spectrometer (**CHRIS**) on ESA's Proba-1 microsatellite have been the main spaceborne providers of space-based hyperspectral data for terrestrial ecosystem observations over the last few decades. Both missions have exceeded their planned one-year lifetime by far and have despite of some shortcomings in sampling capacity and some of the performances been very valuable sources of publicly available spaceborne imaging spectrometer data.

In May 2018, China launched the meteorological satellite 'Gao Fen-5'. Among the six payloads on Gaofen-5 is the Advanced Hyperspectral Imager (AHSI), a VSWIR spectrometer, with 400–2500 nm spectral range, a spectral resolution of 5 nm in the VNIR and 10 nm in the SWIR, approximately 300 spectral bands, a spatial resolution of 30 m and 60 km swath width (personal communication with A. Skidmore, Univ. Twente and Andy Zmuda, ESA-ESRIN). At the time of this paper's submission, unfortunately no official publication reference for this mission could be obtained (Fig. 3).

The most recently launched spaceborne spectrometer is the DLR Earth Sensing Imaging Spectrometer (**DESIS**) for coastal zone/aquaculture, land use/cover and forestry monitoring launched in June, and operating since August 2018. DESIS is a VNIR spectrometer built by DLR to be deployed on the International Space Station ISS, with a 30 m resolution over a 30-km swath and 235 spectral bands, each 2.55 nm wide in the spectral range between 400 and 1000 nm (Müller et al. 2016).

Looking towards the future, with a launch planned in early 2019, one of the upcoming terrestrial hyperspectral imaging space-based mission in Europe will be the Italian **PRISMA** (PRecursore IperSpettrale della Missione Applicativa). PRISMA will be a pre-operational system and a technology demonstrator, which will focus with its acquisition on the Euro-Mediterranean Regions. The system will cover a 400–2500 nm

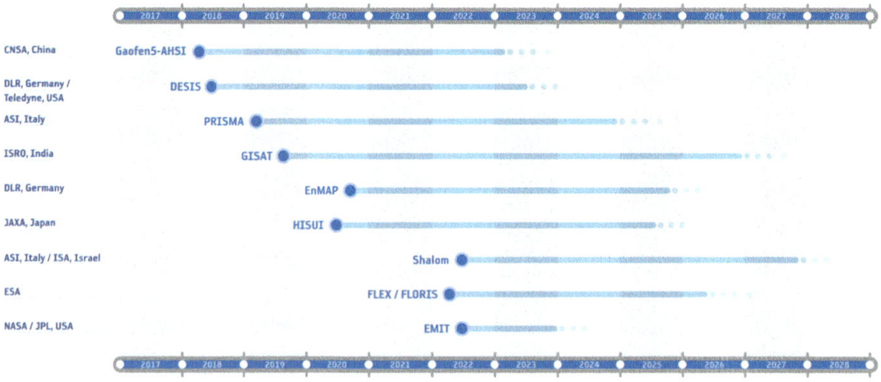


**Fig. 3** Recently launched and planned/approved spaceborne Imaging Spectrometers (ESA Earth Observation Graphic Bureau)

spectral range with 237 spectral bands at a variable spectral bandwidth between 6 and 12 nm, and a 30 m spatial resolution over a 30-km swath (Pignatti et al. 2013).

Also planned for launch in 2019 is the Indian geostationary Imaging Spectrometer **GISAT**, which is expected to provide near real-time images over India sector-wise every 5 min and the entire Indian landmass covered every 30 min at 50 m to 1.5 km spatial resolution, depending on the spectral bands used. Application objectives include disaster monitoring, natural hazards and any short-term events (Misra 2017).

As another imaging spectrometer to be mounted on the ISS, the Hyperspectral Imager Suite (**HISUI**), developed by Japanese Ministry of Economy, Trade, and Industry (METI) will be launched in early 2020. HISUI consists of a Hyperspectral and a Multispectral Imager. The Hyperspectral Imager has 185 bands in the 400–2500 nm region with 30 m spatial resolution and 30 km swath. Cross-calibration with DESIS is planned (Matsunaga et al. 2017).

Subsequently, towards the end of 2020, the German Environmental Mapping and Analysis Program (**EnMAP**) is expected to be launched. It will cover a 30-km-wide area in the across-track direction with a ground sampling distance of 30 m over its 400 to 2500 nm spectral range, with 244 spectral bands and a spectral sampling interval varying from 6.5 to 10 nm. Although primarily science-driven, the EnMAP mission will also address applications beyond scientific issues (Guanter et al. 2015).

PRISMA and EnMAP will co-exist with other comparable and complementary missions, such the Italian-Israeli Spaceborne Hyperspectral Applicative Land and Ocean Mission **SHALOM**, which is planned for launch in 2022 with a focus on operational and commercial use. With 10 m spatial resolution over a 30-km swath, the mission covers the spectral range between 400 and 2500 nm with 275 spectral bands, focusing on ecosystems primarily in the coastal zone (Feingersh and Ben-Dor 2015).

In the framework of its Earth Observation Envelope Programme, in 2015 ESA selected the Fluorescence Explorer **FLEX** as its eighth Earth Explorer Mission for launch in 2022. FLEX carries a fluorescence spectrometer (**FLORIS**) to map vegetation fluorescence for quantifying photosynthetic activity, leading to better insight into plant health and stress. FLORIS acquires data in the 500–780 nm spectral range with a high spectral resolution of ~ 0.3 nm around the Oxygen absorption bands. Other spectral areas with less pronounced absorption features will be measured at medium spectral resolution between 0.5 and 3 nm. FLORIS will provide imagery with about 300 m spatial resolution over a swath width of at least 105 km (Kraft et al. 2017).

As one of the two NASA's Earth Venture Instruments, the Earth surface Mineral dust source InvesTigation (**EMIT**) (Green 2018) will use a VSWIR imaging spectrometer mounted to the exterior of the ISS to determine the mineral composition of natural sources that produce dust aerosols around the world. By measuring in detail which minerals make up the dust, EMIT, which is slated for launch to the ISS in 2022, will help to answer the essential question of whether this type of aerosol warms or cools the atmosphere through constraint of regional and global climate modelling.

In early 2018, the US National Academies of Sciences, Engineering, and Medicine released the 2017 Earth Science Decadal Survey, providing a distillation of the broader Earth Science community's recommendations for required measurements in the coming decade (National Academies of Sciences, Engineering, and Medicine 2018). Among these is a measurement concept, entitled Surface Biology and Geology (SBG), that will use a pointable VSWIR imaging spectrometer in polar orbit to address cutting edge science in ecosystems, the cryosphere, the hydrosphere, and solid Earth. This concept replaces the Hyperspectral Infrared Imager (HyspIRI) that has been in intensive study over the last decade (Green et al. 2013).

Since 2017, an evolution of the Copernicus Space Component is foreseen for the mid-2020s by the European Commission (EC), in order to meet priority user needs not addressed by the existing Earth observation infrastructure from space. The EC together with ESA is studying six Copernicus high-priority mission candidates. One of them is the Copernicus Hyperspectral Imaging Mission for the Environment (CHIME) with the goal to provide routine hyperspectral observations through the Copernicus Programme in support of EU- and related policies for the management of natural resources, assets, and benefits. This unique VIS-SWIR spectroscopy-based observational capability will in particular support new and enhanced services for food security, agriculture, and raw materials. This includes sustainable agricultural and biodiversity management, soil properties characterization, sustainable mining practices, and environment preservation (CHIME MAG 2018).

Furthermore, the French HYPXIM (Carrere et al. 2013), the China Commercial Remote-sensing Satellite System (CCRSS) and the planned Imaging Spectrometer on the Indian Cartosat-3 are among others additional Imaging Spectrometer concepts under development.

An overview of the most recently launched, planned and approved terrestrial spaceborne imaging spectrometer missions is shown in Fig. 3.

## 5 Hyperspectral Applications

Through the large spectral range that imaging spectrometers cover, there is a wide field of geo-scientific applications encompassing the atmosphere, biosphere, geosphere, hydrosphere, and cryosphere. Some of the main application fields, to which imaging spectroscopy can make major contributions are briefly addressed in the following sections. There are several papers in this Special Issue on 'Space-borne Imaging Spectroscopy for exploring the Earth's Ecosystems' by the International Space Science Institute (ISSI) addressing these applications in detail.

### 5.1 Vegetation

The properties of plant canopy structure are highly important for ecosystem and vegetation functioning studies. They affect most processes that couple plants to their environment, such as radiation interception and hence photosynthesis and evapotranspiration (Asner et al. 2009) and also provide useful information for the detection of water and nitrogen contents and related stress in plants (Field et al. 1992). One of the key variables in this context is the Leaf Area Index (LAI), controlling many biological processes in plant canopies and characterizing the structure and the functioning of vegetation, with implications in the prediction of plant growth and productivity estimations. Benefits and improvements in the estimation of LAI by hyperspectral data, as compared to multispectral data, are expected in the reduction of soil influence and atmospheric effects, the distinction from competing influences of pigments (e.g., chlorophyll) on canopy reflectance and the reduction of the saturation of vegetation indices for high LAI values (Lee et al. 2004).

Owing to the importance of pigments for photosynthetic function, variations in pigments content can provide information on the physiological state of leaves or plants. Generally, when plants are under stress or during leaf senescence, the content of, and proportions between, individual pigments (chlorophylls and carotenoids) changes correspondingly (Chapin 1991). Leaf chlorophyll density, i.e. the amount of chlorophyll a and b per unit leaf area, is sensitive to soil nitrogen availability, making it probably the most effective biophysical indicator of nitrogen deficiency (Schepers et al. 1996). The fact that the spectral absorbance properties of chlorophyll are apparent in the reflectance spectra of leaves offers the opportunity of using hyperspectral measurements for quantifying chlorophyll concentration in plants (Zarco-Tejada et al. 2001; Colombo et al. 2008). In addition to chlorophyll, hyperspectral data allow the estimations of other pigments, such as carotenoids and anthocyanin (Ustin et al. 2004), which reveal information on photosynthetic and plant stress.

Studies have demonstrated that a high spectral resolution, i.e. the use of narrow bands, is essential to discriminate leaf biochemical components. Nitrogen has long been recognized as the limiting nutrient for plant growth (Heimann and Reichstein. 2008), and thus its monitoring has been a long-term objective for imaging spectroscopy. NASA's Accelerated Canopy Chemistry Program (ACCP 1994) was established in 1991–1992 to estimate nitrogen and lignin concentrations in vegetation from for example AVIRIS (Martin and Aber 1997; Kokaly et al. 2009). Several papers have since reported the detection of nitrogen from imaging spectroscopy (Feng et al. 2008; Kokaly et al. 2009; Schlerf et al. 2010). The detection of other biochemical constituents, which benefit from hyperspectral imaging include the distinction of Carotenoid content from Chlorophyll (content) (Asner et al. 2012; Asner and Martin 2016), and plant functional types (Schweiger et al. 2016), since broadband indices are unable to exploit the subtle spectral features of these pigments. A comprehensive review of the applications of imaging spectroscopy for vegetation can be found in Ustin et al. 2004.

Information on light use efficiency and the concept of fraction of photosynthetically active radiation (PAR) absorbed for vegetation photosynthesis can also be addressed with improved accuracy through the use of hyperspectral data, for example by using the Photochemical Reflectance Index (PRI) (Gamon et al. 1992). The PRI has been proposed to detect the relative downregulation of photosynthesis and has also shown good potential in early water stress detection (Gamon et al. 1997; Suárez et al. 2009). Moreover, hyperspectral data will determine improvement of plant classification, exploring automated techniques incorporating expert knowledge on biophysical variables dynamics (e-GEOS 2018). In addition to plant functioning and structure, imaging spectroscopy has also been used for mapping invasive species (Somers and Asner; Ustin et al. 2004), forest degradation and drought impacts with Hyperion (Asner et al. 2009), and vegetation and land cover mapping such as the vegetation map over California using AVIRIS data (Roberts et al. 1998).

### 5.2 Agriculture and Food Security

In the context of the main elements of food security (i.e. availability, access, utilization, and stability), spectroscopic techniques can be used to accurately identifies crops (Thenkabail et al. 2000; Miglani et al. 2008), grasslands species (Schmidt and Skidmore 2010; Thenkabail et al. 2004), and to exploit spectral absorption features from leaf pigments (e.g., Chapin 1991; Schepers et al. 1996; Ustin et al. 2004), water content (Gao and Goetz 1995; Serrano et al. 2000; Ustin et al. 1998) and biochemical compounds (Feng et al. 2008; Asner et al. 2012; Asner and Martin 2016; Schweiger et al. 2016) in order to characterize plant functioning and type. Disease detection benefits greatly from extra information hyperspectral imaging may provide, since the spectral signal of plants changes if affected by fungi or bacterial damage (West et al. 2003).

For example, Shafri and Hamdan (2009) used AISA to map disease infection in oil palm plantations based on information from vegetation indices and the red edge spectral region, and Apan et al. (2003) used EO-1 Hyperion imagery to detect orange rust disease in sugarcanes. Sankaran et al. (2010) provide a comprehensive review of hyperspectral imaging for plant disease detection. In addition, hyperspectral imaging may support a better soil management to increase crop production and yield (Fig. 4). This leads to two wide areas of application for hyperspectral techniques regarding agriculture and food security:

i. Vegetation—productivity and functioning, are central to food security applications. Hyperspectral observations are needed for the monitoring of plant productivity and are based on the exploitation of spectral indicators of green biomass, pigment pools, and photosynthetic functioning (Kokaly et al. 2009). Of notable importance is the assessment of stress in crops and ecosystems using hyperspectral remote sensing, for example the detection of drought stressed plants which are characterized by early and accelerated leaf senescence (Munné-Bosch and Alegre 2004). These applications would benefit from a high temporal and spectral resolution to monitor vegetation growth and early-stress detection. Spatial resolution should ideally be in accordance with Sentinel-2 (10–20 m), which would allow the mapping of small agricultural units, but an upper limit up to 30 m is acceptable for many agricultural regions and applications.

Vegetation—crop types as well as vegetation structure are recognized as Essential Biodiversity Variables (EBVs) (Pereira et al. 2013). Hyperspectral applications in biodiversity and forestry take advantage of the high potential of hyperspectral information to separate

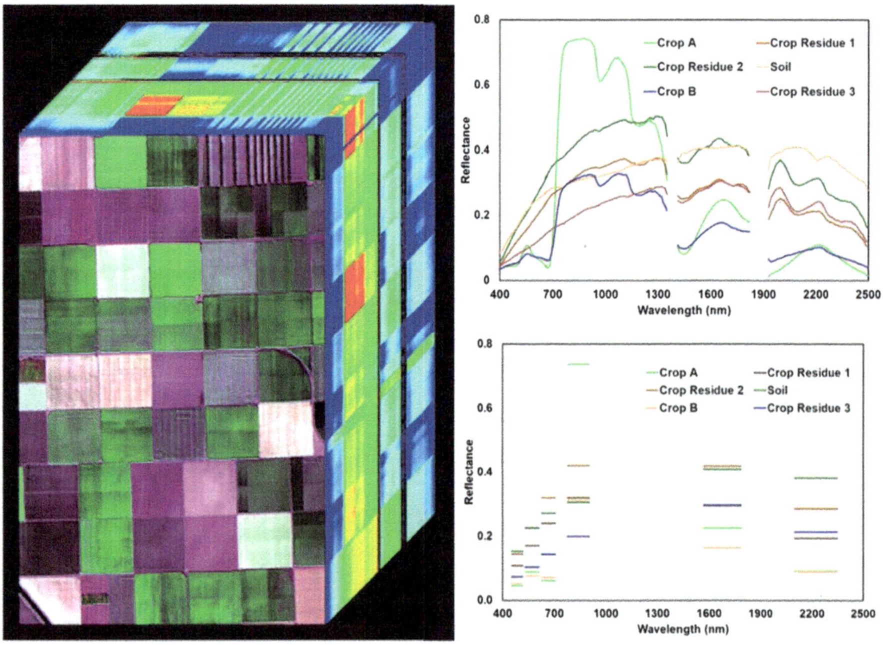


**Fig. 4** Imaging spectrometer data set from the San Joaquin Valley, California, highlighting the addition spectral content available for a larger set of agriculture and food security products. (Courtesy R.O. Green, JPL)

vegetation classes and to retrieve critical plant functional traits (Schweiger et al. 2016). The highest spatial resolution possible would be required for the monitoring of ecosystem composition in the most heterogeneous areas (e.g., for forests, for monitoring of protected habitats and for the monitoring of species occurrence). Also, the detection of invasive species for land management activities would highly benefit from the highest spatial resolution possible (ideally in the range of 20 m). On the other hand, a relatively high temporal resolution would still be necessary to track seasonal changes in plant structure and traits, which is also relevant to biodiversity applications.

ii. Soils—the potential of hyperspectral data for mapping and monitoring soil properties relies mostly on direct chromophores that are soil mineralogy (clay, carbonates and iron oxides) (Ben-Dor et al. 2009; Swayze et al. 2000, 2014) and organic carbon and on soil properties that are highly correlated with those parameters (e.g., Cation Exchange Capacity is highly correlated with clay content and type) (Gomez et al. 2008; Lagacherie et al. 2008). Current multispectral sensors cannot extract mineralogical composition, and so hyperspectral imagery is crucial for the derivation of fundamental soil properties linked to food security such as organic carbon content, clay and water topsoil content, soil degradation stages, and soil quality status. In the case of soil mineralogical composition, high spatial resolution is preferable to high temporal resolution; in the case of soil organic carbon, high spatial resolution would be required for the derivation of soil organic content baseline inventories, but high temporal resolution would also be important for a monitoring system optimized to quantify changes especially during the vegetation-free period.

In the case of indirect soil properties, high spectral resolution and knowledge of the relationship between all soil properties are important. This can be extracted from Global Soil Spectral Libraries that are now very popular and widely available for potential users (Rossel et al. 2016).

While the multitemporal domain is the necessary requirement for food availability and production, specific information extracted at regular intervals by a spaceborne hyperspectral imager will provide a unique opportunity to extend the analysis also for food nutrition. In this context, soil quality indices can be derived from combining several important properties, as demonstrated by (Paz-Kagan et al. 2015).

### 5.3 Raw Materials

The potential to map mineral resources and other raw materials is one of the most compelling cases for imaging spectroscopy compared to multispectral systems (King et al. 2014; Swayze et al. 2014). Currently, no multispectral system is able to resolve mineralogical compositional information. In general, geological applications benefit the most from high spatial resolution, where many minerals and raw materials present absorption features. Mineral mapping and geological applications typically occur over relatively stable areas, so temporal resolution is not considered as a limiting factor.

Mining applications such as resource management will benefit from moderate temporal resolution and high spatial resolution. Mine environmental monitoring including mine waste rehabilitation, dust contamination, and other surface pollution is an application requiring a high spectral resolution and a moderate temporal resolution of several months, but a higher spatial resolution would be preferred over high temporal resolution. The specific case where high temporal resolution is required is for emergency responses in the case of an accidental release such as a tailings dam breach. Here, a high revisit time and real-time product delivery are required.

With a long history of development, hyperspectral imaging for mineral resources applications has matured and has become one of the most advanced applications. It is now at a stage where regional scale maps are being produced for baseline geological mapping and mineral exploration. Such data help to better define prospective ore bodies and focus exploration, thereby also reducing the environmental impact. Additionally, acquired with appropriate calibration data, the same data are also useful for baseline environmental monitoring in raw material exploration areas.

With the increasing use of hyperspectral sensing for mineral mapping, semi-operational software packages have become available and are readily used for the identification of surface minerals materials, e.g., from the USGS (Clark et al. 2003) and from the GFZ (Mielke et al. 2016), which have proven their suitability for the use with spaceborne imaging spectroscopy data, e.g., from Hyperion and the planned EnMAP mission (see Fig. 5).

## 5.4 Hydrology/Cryosphere

Water is arguably the most critical resource on planet Earth. Within the water cycle, the cryosphere contains vast volumes of freshwater, whereby changes in snow cover, mountain glaciers, and ice sheets have first order feedbacks on regional to global climatic change, water resources, and sea level rise (IPCC 2013; Gardner et al. 2013). Monitoring of the snow cover is of importance for hydrology in high latitude or alpine areas, while in low

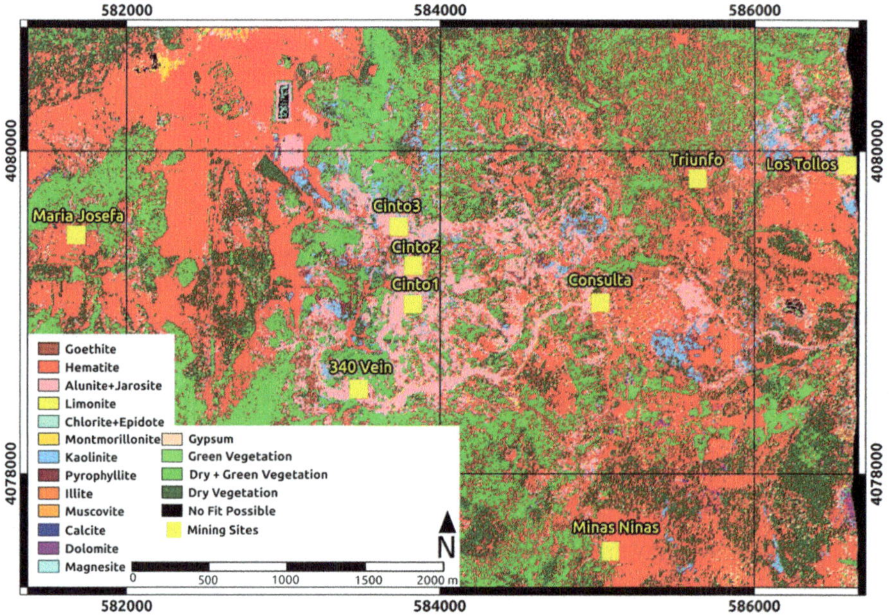


**Fig. 5** Automated mineral classification (EnGeoMAP) of a high sulphidation, epithermal gold deposit, in Rodalquilar, Spain, calculated from airborne HyMAP data, used as simulation input to produce the EnMAP data (Mielke et al. 2016)

mountain ranges significant snow dynamics occur as well. Due to temperature alteration, the snow pack accumulates and melts off several times during a winter season.

New hyperspectral products lead to deeper knowledge of snow physical properties that can be used to constrain regional to global scale climate and water cycle modelling. In addition to the fractional snow cover (Painter et al. 2003), also snow grain size, radiative forcing by light-absorbing impurities in snow (Painter et al. 2013), snow biological constituents (Painter et al. 2001), and snow surface liquid water content (Green et al. 2006) can simultaneously be retrieved. An example showing the combined impacts of grain size and dust on snow albedo is shown in Fig. 6. From the composite of these, snow albedo can accurately be mapped and the controls on its deviation from clean and dry properties well understood (Painter et al. 2013). Likewise, these properties can be retrieved from imaging spectrometer data for glacier ice properties for further constraint on mountain glaciers (Naegeli et al. 2015). A comprehensive review of the use of hyperspectral imaging for snow and ice can be found in Dozier et al. (2009).

The NASA Airborne Snow Observatory (ASO) described above, along with snow depth and snow water equivalent maps entire snow covered mountain basins every 1–2 weeks

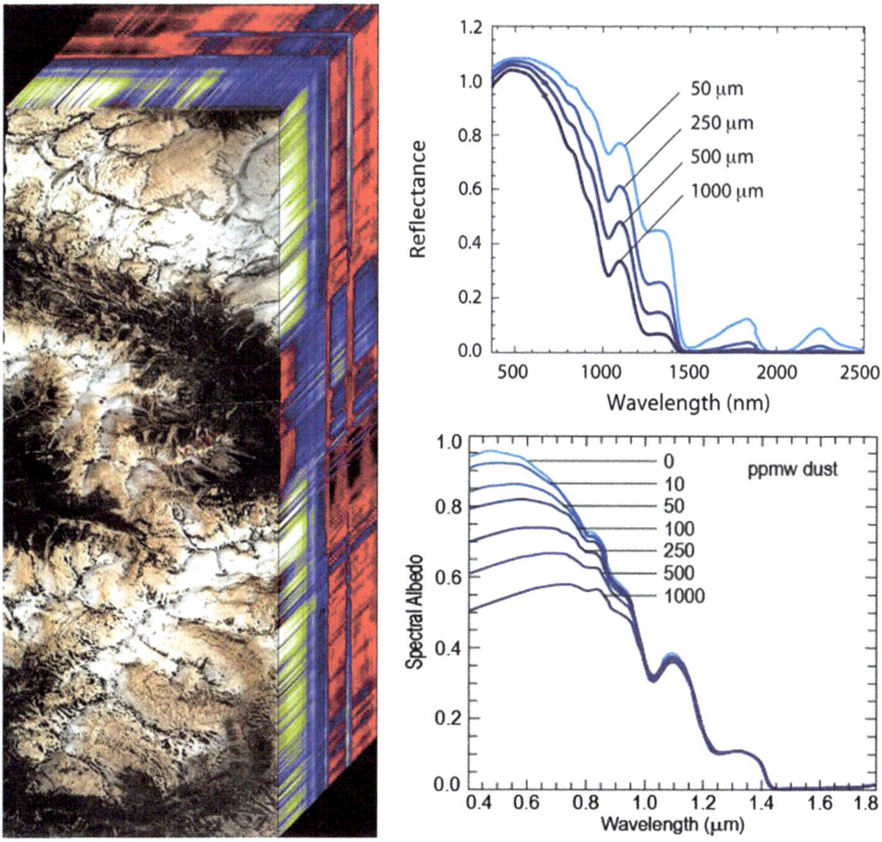


**Fig. 6** AVIRIS imaging spectrometer measurements of dust-impacted snow in the Colorado Rocky Mountains (left). VSWIR spectroscopic leverage used to retrieve grain size (right top) and dust albedo impact (right bottom) to constrain radiative forces and model melting rates and water resource runoff (Painter et al. 2013)

with these properties at 3 m spatial resolution and provides the first coupled information at temporal scales required to constrain physically based hydrologic models (Painter et al. 2016; Henn et al. 2018; Hedrick et al. 2018). These models, constrained by the ASO measurements, are now being implemented for operational runoff forecasting and water management in the states of California and Colorado, USA.

In the US 2017 Earth Science Decadal Survey, the Surface Biology and Geology measurement concept as a spaceborne VSWIR imaging spectrometer was indicated as a designated measurement that must be implemented to fill critical science voids. Among the most important measurements that motivated this concept as a Designated measurement was the Hydrology panel's objective to: '*Quantify rates of snow accumulation, snowmelt, ice melt, and sublimation from snow and ice worldwide at scales driven by topographic variability*', describing the need to understand the controls on snow albedo (that are only available from the imaging spectroscopy retrievals described above) and, snow- and ice melt.

### 5.5 Coastal and Inland Waters

Inland and coastal water ecosystems are vital for human consumption, irrigation, sanitation, transportation, recreation, and industry. Earth observation has been used for decades to acquire information from local to global scales of aquatic ecosystems. These systems encompass gradients of clear to turbid and oligotrophic to hypertrophic productive waters and with varying contents of macrophytes, macro-algae, benthic micro-algae or corals. To observe such optical complexity, imaging spectroscopy provides more accurate assessments than multispectral sensors starting from turbidity and transparency, chlorophyll, suspended matter and coloured dissolved organic matter concentration up to more sophisticated products, such as particle size distributions, phytoplankton types and pigments, harmful algal blooms, estimating water depth and mapping heterogeneous substrates and cover types, as well as different types of sea grass (see Fig. 7). Submerged aquatic vegetation is one of the biological quality elements used in monitoring the ecological status of surface waters within the process recommended by the European Water Framework Directive, where the abundance of aquatic plants is needed to define the deviation from the ecological conditions of reference (Hestir et al. 2015).

Since 2000, spaceborne hyperspectral sensors (e.g., HYPERION, CHRIS, HICO) have shown increasing capabilities in inland, estuarine, and coastal water applications. Water quality-related applications, like the retrieval of chlorophyll content and secondary pigments as proxies for harmful algae blooms, the identification of phytoplankton types, or the benthic habitat mapping, strongly benefit from hyperspectral information (Hestir et al. 2015; Giardino et al. 2018). Those conclusions are supported by the findings of a recent Committee on Earth Observation Satellites (CEOS) study towards the assessment of potential hyperspectral mission concepts for aquatic ecosystems (Dekker et al. 2018).

### 5.6 Environmental Degradation and Hazards

Natural and anthropogenic hazards have the potential to impact on all aspects of society including its economy, its habitats health and well-being, and the environment. Diagnostic information to inform decisions is critical for hazard management, whether as an emergency response for routine monitoring or assessments of potential risks. Hyperspectral imaging has unique contributions to make via the ability to provide some key quantitative diagnostic information.

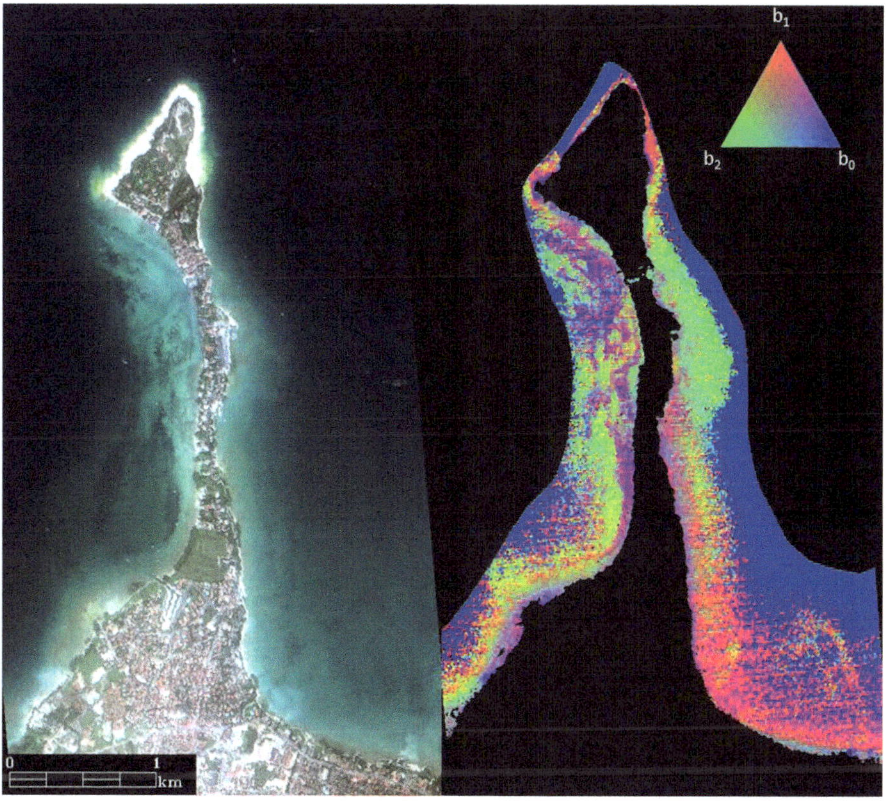


**Fig. 7** Airborne Multispectral IR and Visible Imaging Spectrometer (MIVIS) data Lake Garda, Italy, 15.7.2015. Left: Pseudo-true-colour composite. Right: Submerged vegetation to 7 m depth, containing main bottom cover classes of sand (b0, blue), submersed benthic vegetation (b1, red) such as *Chara sp.*, and submersed tall vegetation (b2, green) such as *Vallisneria spiralis*; black areas are masked optically deep-water areas and land (Courtesy C. Giardino, IREA-CNR)

Examples where imaging spectroscopy has been used for mapping environmental degradation and hazards other than those provided for the raw materials are steadily growing.

Hyperspectral images were acquired following the 11th September 2001 World Trade Centre (WTC) attack, and the data were used for mapping thermal hotspots and potentially hazardous dust (Clark et al. 2005); see Fig. 8.

Hyperspectral imaging was used to estimate the distribution and volume of the oil slick on the surface of the ocean and the impact of oil on the terrestrial ecosystem (Kokaly et al. 2013) from the largest oil spill in US history in the Gulf of Mexico. Spaceborne hyperspectral imaging from EO1 Hyperion also mapped a very large accidental gas release (Thompson et al. 2016). Pioneering work from Chabrillat et al. (2002) demonstrated the potential of hyperspectral imaging for detecting and mapping swelling clays with a study along the Front Range Urban corridor near Denver, Colorado, USA. Chudnovsky et al. (2009) documented the use of spaceborne hyperspectral imaging data acquired by NASA EO-1 Hyperion sensor for determining the composition of the dust at the Bodélé Depression of Northern Chad, considered to be one of the world's largest sources of atmospheric mineral dust. Further, imaging spectrometer measurements have in the recent past been used for detailed

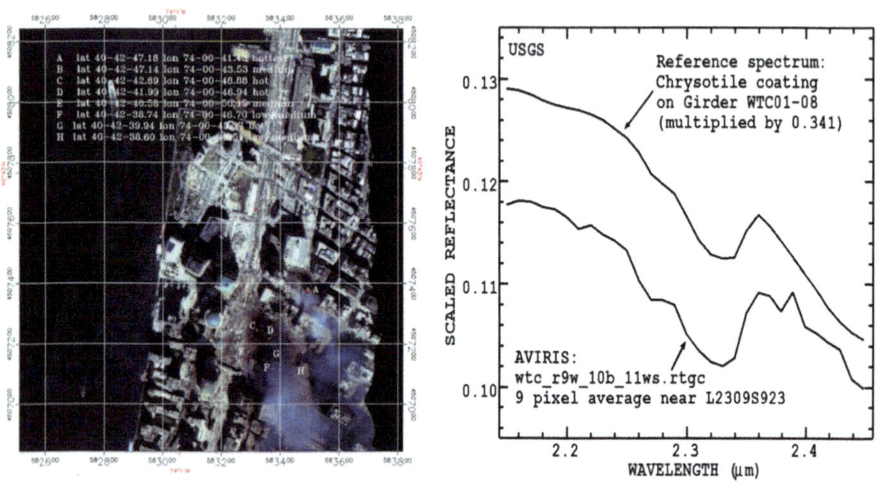


**Fig. 8** AVIRIS measurements of the WTC site on the 16th of September 2001. Hot spot fire temperatures were calculated from AVIRIS spectra and delivered to ground teams (left). Surface debris composition includes Chrysotile. Mineral asbestos was mapped from AVIRIS spectroscopy by the USGS (Clark et al. 2005)

mapping of fire fuel (Roberts et al. 2003), temperature (Green 1996; Dennison 2006; Veraverbeke et al. 2018), burn severity (Kokaly et al. 2007), and recovery (Riaño et al. 2002).

## 6 Observational Requirements

Observational requirements of a spaceborne imaging spectrometer mission are driven by the mission's scientific and/or operational objectives and are based on experience, state-of-the-art technology and results of previous hyperspectral airborne and experimental spaceborne systems. On a rather generic level, overarching observational requirements comprise:

- Spectral resolution: Contiguous spectral coverage between 400 and 2500 nm with a spectral sampling interval at FWHM of less or equal 10 nm is aspired to by land-observing sensors, driven by the need for atmospheric characterization and reflectance retrieval, as well as some geophysical, geo-biophysical, and geochemical retrievals for environmental measurements and agricultural applications. For ocean-observing sensors focusing on 'Ocean Colour' variables often, a spectral sampling around 5 nm is required. For radiative forcing by impurities in snow, 20 nm is the generally accepted requirement. Ultimately, a sensor with a semi-static spectral resolution that can accommodate the science requirements is more cost-effective than varying the spectral resolution across the spectrum.
- Spatial resolution: While airborne imaging spectrometers typically have a spatial sampling of $\leq 1$ m (depending on the flight altitude), spaceborne land observation imaging spectrometers have a spatial resolution in the range of tens of metres, i.e. 20–60 m to resolve the scales of spatial variability of the observed constituent. For Ocean Colour imaging spectrometers, the spatial sampling requirements range typically between 100 and 300 m.

- Temporal resolution: The cadence of clear-sky acquisitions varies according to the science or application question. For snow and ice, an average of one clear acquisition per week is required, whereas for ocean observation a cadence of the order of about 3 days is required. For agricultural applications, repeat observations every 5–10 days are needed, for biodiversity and biogeochemical measures seasonal clear acquisitions will suffice but generally require a broader spatial coverage.

The above sensor requirements for future terrestrial imaging spectrometers were subject to and outcome of the discussion at the 'conclusion session' of the Workshop: 'Exploring the Earth's ecosystems on a global scale', in November 2016 at the International Space Science Institute in Bern, Switzerland.

Desired temporal and spatial resolutions stretch the technology and the data rates for a spaceborne imaging spectrometer. In future, the requirements for terrestrial and oceanic science might be achievable under likely budget availability, but for the next decade or two, a feasible strategy is to exploit the combination of imaging spectroscopy with the variety of multispectral sensors already operating or in the international launch queues. Specifically, imaging spectroscopy can play a valuable role in calibrating multispectral sensors and validating interpretations made from them. In this vein, compelling opportunities are seen in synthesizing observations from imaging spectrometers, the Landsat and Sentinel-2 missions, the global daily observations from sensors such as VIIRS and the MetOP satellite series, and the very frequent observations from geostationary missions such as GOES-16 and 17, Himawari-8 and 9, Fengyun-4, and Meteosat-9 through 11 (presentation J. Dozier at ISSI Workshop: 'Exploring the Earth's ecosystems on a global scale', November 2016).

## 7 Conclusions, Outlook, and Perspectives

Owing to well-established spectroscopic techniques, optical hyperspectral remote sensing has the potential to deliver significant enhancement in quantitative value-added retrievals, as compared to multispectral remote sensing, thus supporting the generation of a wide variety of new products and services in the domain of agriculture, food security, raw materials, soils, biodiversity, environmental degradation and hazards, snow, water resources, inland and coastal waters, and forestry.

Corresponding variables have been derived from the observed spectra, e.g., directly through distinct absorption features or indirectly, through inversion of physically based models, assimilation, spectral un-mixing, and/or (de-) correlation techniques. Based on successful airborne deployments and some satellite missions along with preparatory activities, imaging spectroscopy from satellite is now ready for operational use.

The potential of airborne and spaceborne imaging spectrometers has intensively been researched to monitor land and water surfaces and the atmosphere to provide valuable quantitative information for the better understanding of a large number of environmental and ecosystem processes. These applications include vegetation monitoring and agriculture, geology and soils, the mapping of snow properties and coastal and inland waters.

Over the last three decades, remotely sensed imaging spectroscopy has steadily evolved and by now has become one of the most powerful diagnostic and quantitative retrieval techniques from both air- and spacecraft. Indeed, spectroscopy from space represents the asymptote for optical remote sensing—that is, the pinnacle of optical RS—spectral and spatial resolutions will continue to improve, but the concept of imaging spectroscopy will continue for as long as the Sun is being used as an illumination source. Advances in technology facilitate improvement in the spectral sampling and the radiometric performance of imaging spectroscopy systems, thus yielding higher signal-to-noise ratios. Starting from airborne in the late 1980s and also spaceborne technology demonstrators from the early 2000s, imaging spectrometers are meanwhile viewed also as potential operational sensing systems, such as the Airborne Snow Observatory's spectrometer use in this operational programme for snow water resources. One of the remaining challenges is to deploy systems with high radiometric performance in the shortwave infrared where many diagnostic absorption features of vegetation, soils and minerals are to be found. In parallel, compact imaging spectrometers encompassing the visible–near-infrared range have been deployed into space and are increasing in number rapidly.

The superiority of imaging spectrometers over multispectral imaging systems with a limited number of spectral bands gives rise to the expectation that in the future multispectral systems will be superseded by imaging spectrometers in Earth observation remote sensing of terrestrial as well as aquatic ecosystems.

Evolution in technology also enables building increasingly compact systems and deploying them into space. It can hence be expected that smaller VNIR imaging spectrometers together with some VSWIR systems will form monitoring constellations in future.

With the advent of Big Data, Artificial Intelligence, Machine Learning and Data Analytics, it can be expected that the handling of large data volumes such as those acquired by imaging spectrometers will be less of a challenge in future to the Earth observation science and operational user community, and the retrievals will thus be poised to straightforwardly inform climate modelling, hydrologic modelling, and resource management and operations.

**Acknowledgements** The paper is an outcome of a Workshop on Requirements, capabilities, and directions in Spaceborne Imaging Spectroscopy held at the International Space Science Institute (ISSI) in Bern, Switzerland, in November 2016. Part of this work was performed at the Jet Propulsion Laboratory, California Institute of Technology, under contract with NASA. The authors acknowledge the support of J. Adams (ESA-ESRIN), U. del Bello (ESA-ESTEC), C. Giardino (IREA-CNR), R.O. Green (JPL), L. Guanter (GFZ), S. Förster (GFZ), and C. Mielke (GFZ).

## References

ACCP (1994) "Accelerated canopy chemistry program final report to NASA-EOS-IWG". In: Aber J (ed), Washington, DC http://daac.ornl.gov/ACCP/accp.html

Adão T, Hruška J, Pádua L, Bessa J, Peres E, Morais R, Sousa JJ (2017) Hyperspectral imaging: a review on UAV-based sensors, data processing and applications for agriculture and forestry. Remote Sens 9(11):1110

Apan A, Held A, Phinn S, Markley J (2003) Detecting sugarcane 'orange rust' disease using EO-1 hyperion hyperspectral imagery. Int J Remote Sens 25(2):489–498. https://doi.org/10.1080/01431160310001618031

Asner GP, Martin RE (2016) Convergent elevation trends in canopy chemical traits of tropical forests. Glob Change Biol 22:2216–2227. https://doi.org/10.1111/gcb.13164

Asner GP, Knapp DE, Kennedy-Bowdoin T, Jones MO, Martin RE, Boardman JW, Field CB (2007) Carnegie airborne observatory: in-flight fusion of hyperspectral imaging and waveform light detection and ranging for three-dimensional studies of ecosystems. J Appl Remote Sens 1(1):013536. https://doi.org/10.1117/1.2794018

Asner GP, Knapp DE, Balaji A, Páez-Acosta G (2009) Automated mapping of tropical deforestation and forest degradation: CLASlite. J Appl Remote Sens 3:033543. https://doi.org/10.1117/1.3223675

Asner GP, Knapp DE, Boardman J, Green RO, Kennedy-Bowdoin T, Eastwood M, Martin RE, Anderson C, Field CB (2012) Carnegie airborne observatory-2: increasing science data dimensionality via high-fidelity multi-sensor fusion. Remote Sens Environ 1(124):454–465. https://doi.org/10.1016/j.rse.2012.06.012

Asner GP, Martin RE, Knapp DE, Tupayachi R, Anderson CB, Sinca F, Vaughn NR, Llactayo W (2017) Airborne laser-guided imaging spectroscopy to map forest trait diversity and guide conservation. Science 355:385–389. https://doi.org/10.1126/science.aaj1987

Ben-Dor E, Chabrillat S, Demattê J, Taylor G, Hill J, Whiting M, Sommer S (2009) Using imaging spectroscopy to study soil properties. Remote Sens Environ 113:S38–S55

Bianchi R, Cavalli RM, Fiumi L, Marino CM, Pignatti S (1996) Airborne imaging spectrometry: a new approach to environmental problems. In: Proceeding of the XVIII ISPRS, pp128–132

Carrere V, Briottet X, Jacquemoud S, Marion R, Bourguignon A, Cham M, Dumont M, Minghelli-Roman A, Weber C, Lefevrefonollosa M-J, Mandea M (2013) "HYPXIM: a second generation high spatial resolution hyperspectral satellite for dual applications," 2013 5th Workshop on Hyperspectral Image and Signal Processing: Evolution in Remote Sensing (WHISPERS). Gainesville, FL 2013:1–4. https://doi.org/10.1109/WHISPERS.2013.8080685

Chabrillat S, Goetz AF, Krosley L, Olsen HW (2002) Use of hyperspectral images in the identification and mapping of expansive clay soils and the role of spatial resolution. Remote Sens Environ 82(2):431–445. https://doi.org/10.1016/S0034-4257(02)00060-3

Chang SH, Westfield MJ, Lehmann F, Oertel D, Richter R (1993) 79-channel airborne imaging spectrometer. In: Vane G (ed) Imaging spectrometry of the terrestrial environment, vol 1937. International Society for Optics and Photonics, Bellingham, pp 164–173. https://doi.org/10.1117/12.157053

Chang CI (2007) Hyperspectral data exploitation: theory and applications. Wiley, Hoboken

Chapin FS (1991) Integrated responses of plants to stress. Bioscience 41:29–36. https://doi.org/10.2307/1311538

CHIME mission advisory group (MAG) (2018) "Copernicus hyperspectral imaging mission requirements document" ESA-EOPSM-CHIM-MRD-3216, version 1.2

Chudnovsky A, Ben-Dor E, Kostinski A, Koren I (2009) "Mineral content analysis of atmospheric dust using hyperspectral information from space." Geophys Res Lett 36(15). https://doi.org/10.1029/2009GL037922

Clark RN, Swayze GA, Livo KE, Kokaly RF, Sutley SJ, Dalton JB, McDougal RR, Gent CA (2003) "Imaging spectroscopy: earth and planetary remote sensing with the USGS tetracorder and expert systems." J Geophys Res Planets 108(E12). https://doi.org/10.1029/2002JE001847

Clark RN, Swayze GA, Hoefen TM, Green RO, Livo KE, Meeker GP, Sutley SJ, Plumlee GS, Pavri B, Sarture C (2005) Environmental mapping of the world trade center area with imaging spectroscopy after the September 11, 2001 attack. In: Vane G, Green RO, Chrien TG, Enmark HT, Hansen EG, Porter WM (eds) The airborne visible/infrared imaging spectrometer mapping. ACS Publications, Washington. https://doi.org/10.1021/bk-2006-0919.ch004

Cocks T, Jenssen R, Stewart A, Wilson I, Shields T (1998) The hymap airborne hyperspectral sensor: the system, calibration and performance. In: EARSEL workshop on imaging spectroscopy

Colombo R, Meroni M, Marchesi A, Busetto L, Rossini M, Giardino C, Panigada C (2008) Estimation of leaf and canopy water content in poplar plantations by means of hyperspectral indices and inverse modeling. Remote Sens Environ 112(1820–1834):20. https://doi.org/10.1016/j.rse.2007.09.005

Corson MR, DR Korwan, RL Lucke, WA Snyder, CO Davis (2008) The hyperspectral imager for the coastal ocean (HICO) on the international space station. In: Proceedings of international geoscience and remote sensing symposium (IGARSS'08). 4, pp I-101–I-104

Cutter MA (2006) The PROBA-1/CHRIS hyperspectral mission—five years since launch. In: Proceedings of the 4S symposium: small satellite systems and services, Chia Laguna Sardinia, Italy, Sept. 25–29, 2006, ESA SP-618

Dekker A, Pinnel N, Gege P, Briottet X, Peters S, Turpie K, Sterckx S, Costa M, Giardino C, Brando V (2018). Feasibility study of an Aquatic ecosystem earth observing system." CEOS Report copyright CSIRO, Australia, http://ceos.org/document_management/Publications/CEOS_FS-Aquatic-Ecosystem-EO-System_v2.5_low-res_April2018.pdf

Del Bello U, Bezy JL, Fuchs J, Rast M (2003) System definition of the ESA earth explorer spectra mission. In: Lurie JB, Aten ML, Weber K (eds) Sensors, systems, and next-generation satellites VI, vol 4881. International Society for Optics and Photonics, Bellingham, pp 1–12. https://doi.org/10.1117/12.463003

Delwart S, Bourg L, Huot JP (2004) MERIS 1st year: early calibration results. In: Meynart R, Neeck S, Shimoda H, Habib S (eds) Sensors, systems, and next-generation satellites vii, vol 5234. International Society for Optics and Photonics, Bellingham, pp 379–391. https://doi.org/10.1117/12.508485

Dennison PE (2006) Fire detection in imaging spectrometer data using atmospheric carbon dioxide absorption. Int J Remote Sens 27(14):3049–3055. https://doi.org/10.1080/01431160600660871

Dozier J, Green RO, Nolin AW, Painter TH (2009) Interpretation of snow properties from imaging spectrometry. Remote Sens Environ 113:S25–S37. https://doi.org/10.1016/j.rse.2007.07.029

Drusch M, Del Bello U, Carlier S, Colin O, Fernandez V, Gascon F, Hoersch B, Isola C, Laberinti P, Martimort P, Meygret A (2012) Sentinel-2: ESA'S optical high-resolution mission for GMES operational services. Remote Sens Environ 120:25–36. https://doi.org/10.1016/j.rse.2011.11.026

e-GEOS (2018) Hyperspectral imaging mission concepts, ESA contract for ESA-ESRIN, e-GEOS-HYP-ES-0055, January 2018

Eismann MT (2012) Hyperspectral remote sensing. SPIE Press Book, Bellingham

Feingersh T, Ben-Dor E (2015) SHALOM–A commercial hyperspectral space mission. In: Qian SE (ed) Optical payloads for space missions. Wiley, Hoboken

Feng W, Zhu Y, Yao X, Tian YC, Cao WX (2008) Monitoring leaf nitrogen status with hyperspectral reflectance in wheat. Eur J Agron 28:394–404. https://doi.org/10.1016/j.eja.2007.11.005

Fernández-Renau A, Gómez JA, de Miguel E (2005) The INTA AHS system. In: Meynart R, Neeck S, Shimoda H, Habib S (eds) Sensors, systems, and next-generation satellites IX, vol 5978. International Society for Optics and Photonics, Bellingham. https://doi.org/10.1117/12.629440

Field CB, Chapin FS, Matson PA, Mooney HA (1992) Responses of the terrestrial ecosystems to changing atmosphere. Annu Rev Ecolol Syst 23:201–235

Gamon JA, Peñuelas J, Field CB (1992) A narrow-waveband spectral index that tracks diurnal changes in photosynthetic efficiency. Remote Sens Environ 41:35–44. https://doi.org/10.1016/0034-4257(92)90059-S

Gamon J, Serrano L, Surfus J (1997) The photochemical reflectance index: an optical indicator of photosynthetic radiation use efficiency across species, functional types, and nutrient levels. Oecologia 112:492–501

Gao BC, Goetz AFH (1995) Retrieval of equivalent water thickness and information related to biochemical components of vegetation canopies from AVIRIS data. Remote Sens Environ 52(3):155–162. https://doi.org/10.1016/0034-4257(95)00039-4

Gardner AS, Moholdt G, Cogley JG, Wouters B, Arendt AA, Wahr J, Berthier E, Hock R, Pfeffer WT, Kaser G, Ligtenberg SRM, Bolch T, Sharp MJ, Hagen JO, van den Broeke MR, Paul F (2013) A reconciled estimate of glacier contributions to sea level rise: 2003 to 2009. Science 340:852–857. https://doi.org/10.1126/science.1234532

Gege P, Beran D, Mooshuber W, Schulz J, van der Piepen H (1998) System analysis and performance of the new version of the imaging spectrometer ROSIS, In: Proc 1st EARSEL workshop on imaging spectroscopy, Zurich, pp 29–35

Giardino C, Brando VE, Gege P, Pinnel N, Hochberg E, Knaeps E, Reusen I, Doerffer R, Bresciani M, Braga F, Foerster S, Champollion N, Dekker A (2018) Imaging spectrometry of inland and coastal waters: state-of-the-art, achievements and perspectives. Surv Geophys. https://doi.org/10.1007/s10712-018-9476-0

Goetz AF, Vane G, Solomon JE, Rock BN (1985) Imaging spectrometry for earth remote sensing. Science 228(4704):1147–1153. https://doi.org/10.1126/science.228.4704.1147

Gomez C, Lagacherie P, Coulouma G (2008) Continuum removal versus PLSR method for clay and calcium carbonate content estimation from laboratory and airborne hyperspectral measurements. Geoderma 148(2):141–148. https://doi.org/10.1016/j.geoderma.2008.09.016

Gower JFR, Borstad GA, Anger CD, Edel HR (1992) CCD-based imaging spectroscopy for remote sensing: the FLI and CASI programs. Can J Remote Sens 18(4):199–208. https://doi.org/10.1080/07038992.1992.10855325

Green RO (1996) Estimation of biomass fire temperature and areal extent from calibrated AVIRIS spectra. http://hdl.handle.net/2014/25024

Green RO (2018) The earth surface mineral dust source investigation (EMIT) https://hyspiri.jpl.nasa.gov/downloads/2018_Workshop/day1/13_HyspIRI_EMIT_Overview_20180815b.pdf

Green RO, Painter TH, Roberts DA, Dozier J (2006) Measuring the three phases of water in a melting snow environment with an imaging spectrometer in the solar reflected spectrum. Water Resour Res

Green RO, Hook SJ, Middleton E, Turner W, Ungar S, Knox R (2013) The HyspIRI decadal survey mission: update on the mission concept and prepatory airborne science campaign. In: Proceedings of the international geoscience and remote sensing symposium (IGARSS'13), Melbourne, Australia, p 4

Guanter L, Richter R, Moreno J (2006) Spectral calibration of hyperspectral imagery using atmospheric absorption features. Appl Opt 45(10):2360–2370. https://doi.org/10.1364/AO.45.002360

Guanter L, Kaufmann H, Segl K, Förster S, Rogaß C, Chabrillat S, Küster T, Hollstein A, Rossner G, Chlebek C, Straif C, Fischer S, Schrader S, Storch T, Heiden U, Mueller A, Bachmann M, Mühle H, Müller R, Habermeyer M, Ohndorf A, Hill J, Buddenbaum H, Hostert P, van der Linden S, Leitão PJ, Rabe A, Doerffer R, Krasemann H, Xi H, Mauser W, Hank T, Locherer M, Rast M, Staenz K, Sang

B (2015) The EnMAP spaceborne imaging spectroscopy mission for earth observation. Remote Sens 7(7):8830–8857. https://doi.org/10.3390/rs70708830
Hackwell JA, Warren DW, Bongiovi RP, Hansel SJ, Hayhurst TL, Mabry DJ, Sivjee MG, Skinner JW (1996) LWIR/MWIR imaging hyperspectral sensor for airborne and ground-based remote sensing. In: Lurie JB, Aten ML, Weber K (eds) Imaging spectrometry II, vol 2819. International Society for Optics and Photonics, Bellingham, pp 102–108. https://doi.org/10.1117/12.258057
Hall JL, Boucher RH, Gutierrez DJ, Hansel SJ, Kasper BP, Keim ER, Moreno NM, Polak ML, Sivjee MG, Tratt DM, Warren DW (2011) First flights of a new airborne thermal infrared imaging spectrometer with high area coverage. Infrared technology and applications XXXVII (Vol. 8012, p. 801203). International Society for Optics and Photonics. https://doi.org/10.1117/12.884865
Hedrick AR, Marks D, Havens S, Robertson M, Johnson M, Sandusky M, Marshall HP, Kormos PR, Bormann KJ, Painter TH (2018) Direct insertion of NASA airborne snow observatory-derived snow depth time series into the iSnobal energy balance snow model. Water Resour Res 54(10):8045–63
Heimann M, Reichstein M (2008) Terrestrial ecosystem carbon dynamics and climate feedbacks. Nature 451(7176):289–292. https://doi.org/10.1038/nature06591
Henn B, Painter TH, Bormann KJ, McGurk B, Flint AL, Flint LE, White V, Lundquist JD (2018) High-elevation evapotranspiration estimates during drought: using streamflow and NASA airborne snow observatory SWE observations to close the upper tuolumne river basin water balance. Water Resour Res 54(2):746–66
Hestir EL, Brando VE, Bresciani M, Giardino C, Matta E, Villa P, Dekker AG (2015) Measuring freshwater aquatic ecosystems: the need for a hyperspectral global mapping satellite mission. Remote Sens Environ 167:181–195. https://doi.org/10.1016/j.rse.2015.05.023
Hochberg EJ, Roberts DA, Dennison PE, Hulley GC (2015) Special issue on the hyperspectral infrared imager (HyspIRI): emerging science in terrestrial and aquatic ecology, radiation balance and hazards. Remote Sens Environ 167:1–5. https://doi.org/10.1016/j.rse.2015.06.011
Hook SJ, Johnson WR, Abrams MJ (2013) NASA's hyperspectral thermal emission spectrometer (HyTES). In: Thermal infrared remote sensing. Springer, Dordrecht. 93–115 https://doi.org/10.1007/978-94-007-6639-6_5
Hunt GR (1980) Electromagnetic radiation: the communication link in remote sensing. Remote Sens Geol: 5–45
IPCC (2013): Summary for Policymakers. In: Climate Change 2013: The Physical Science Basis. Contribution of Working Group I to the Fifth Assessment Report of the Intergovernmental Panel on Climate Change [Stocker, T.F., D. Qin, G.-K. Plattner, M. Tignor, S. K. Allen, J. Boschung, A. Nauels, Y. Xia, V. Bex and P.M. Midgley (eds.)]. Cambridge University Press, Cambridge, United Kingdom and New York, NY, USA
Kampe TU, Johnson BR, Kuester MA, Keller M (2010) NEON: the first continental-scale ecological observatory with airborne remote sensing of vegetation canopy biochemistry and structure. J Appl Remote Sens 4(1):043510. https://doi.org/10.1117/1.3361375
Kieffer HH (1997) Photometric stability of the lunar surface. Icarus 130(2):323–327. https://doi.org/10.1006/icar.1997.5822
Kieffer HH, Jarecke PJ, Pearlman J (2002) Initial lunar calibration observations by the EO-1 hyperion imaging spectrometer. In: Imaging spectrometry VII 2002 Jan 17 (Vol. 4480, pp. 247-259). International Society for Optics and Photonics. https://doi.org/10.1117/12.453347
King TVV, Berger BR, Johnson MR (2014) Characterization of potential mineralization in Afghanistan: four permissive areas identified using imaging spectroscopy data, USGS Open-File Report 2014-1071, https://doi.org/10.3133/ofr20141071. https://doi.org/10.3133/ofr20141071
Kokaly RF, Rockwell BW, Haire SL, King TV (2007) Characterization of post-fire surface cover, soils, and burn severity at the Cerro Grande Fire, New Mexico, using hyperspectral and multispectral remote sensing. Remote Sens Environ 106(3):305–325. https://doi.org/10.1016/j.rse.2006.08.006
Kokaly RF, Asner GP, Ollinger SV, Martin ME, Wessman CA (2009) Characterizing canopy biochemistry from imaging spectroscopy and its application to ecosystem studies. Remote Sensing of Environment. 113:S78–S91. https://doi.org/10.1016/j.rse.2008.10.018
Kokaly RF, Couvillion BR, Holloway JM, Roberts DA, Ustin SL, Peterson SH, Khanna S, Piazza SC (2013) Spectroscopic remote sensing of the distribution and persistence of oil from the deepwater horizon spill in Barataria Bay marshes. Remote Sens Environ 129:210–230. https://doi.org/10.1016/j.rse.2012.10.028
Kraft S, Del Bello U, Harnisch B, Bouvet M, Drusch M, Bézy, JL (2017) Fluorescence imaging spectrometer concepts for the earth explorer mission candidate FLEX. In: International conference on space optics—ICSO 2012 International Society for Optics and Photonics. 10564, p 105641 https://doi.org/10.1117/12.2309086

Kunkel B, Harms J, Kummer U, Schmidt E, Del Bello U, Harnisch B, Meynart R (2000) Hyperspectral imager survey and developments for scientific and operational land processes monitoring applications. In: Observing land from space: science, customers and technology. Springer, Dordrecht. pp 303–327 https://doi.org/10.1007/0-306-48124-3_30

Lagacherie P, Baret F, Feret J-B, Netto JM, Robbez-Masson JM (2008) Estimation of soil clay and calcium carbonate using laboratory, field and airborne hyperspectral measurements. Remote Sens Environ 112(3):825–835. https://doi.org/10.1016/j.rse.2007.06.014

Lee KS, Cohen WB, Kennedy RE, Maiersperger TK, Gower ST (2004) Hyperspectral versus multispectral data for estimating leaf area index in four different biomes. Remote Sens Environ 91:508–520. https://doi.org/10.1016/j.rse.2004.04.010

Makisara K, Meinander M, Rantasuo M, Okkonen J, Aikio M, Sipola K (1993) Airborne imaging spectrometer for applications (AISA). In: Geoscience and Remote Sensing Symposium, IGARSS'93. Better understanding of earth environment, International (pp 479–481). IEEE. https://doi.org/10.1109/igarss.1993.322291

Martin ME, Aber JD (1997) "High spectral resolution remote sensing of forest canopy lignin, nitrogen, and ecosystem processes." Ecology applications 7(2):431–443. https://doi.org/10.1890/1051-0761(1997)007%5b0431:HSRRSO%5d2.0.CO;2

Matsunaga T, Iwasaki A, Tsuchida S, Iwao K, Tanii J, Kashimura O, Nakamura R, Yamamoto H, Kato S, Obata K, Mouri K (2017) Current status of hyperspectral imager suite (HISUI) onboard international space station (ISS). In: Geoscience and remote sensing symposium (IGARSS), pp 443–446 https://doi.org/10.1109/igarss.2017.8126989

Middleton EM, Ungar SG, Mandl DJ, Ong L, Frye SW, Campbell PE, Landis DR, Young JP, Pollack NH (2013) The earth observing one (EO-1) satellite mission: over a decade in space. IEEE J Sel Top ApplEarth Obs Remote Sens 6(2):243–256. https://doi.org/10.1109/JSTARS.2013.2249496

Mielke C, Rogass C, Boesche N, Segl K, Altenberger U (2016) EnGeoMAP 2.0—automated hyperspectral mineral identification for the german EnMAP space mission. Remote Sens 8(2):127. https://doi.org/10.3390/rs8020127

Miglani SS, Pandey RR, Parihar JS (2008) Evaluation of EO-1 hyperion data for agricultural application. J Indian Soc Remote Sens 36:255–266. https://doi.org/10.1007/s12524-008-0026-y

Misra T (2017) Indian remote sensing sensor system: current and future perspective. In: Proceedings of the national academy of sciences, India section A: physical sciences, 87(4): pp 473–486. https://doi.org/10.1007/s40010-017-0429-7

Mouroulis P, Van Gorp B, Green RO, Dierssen H, Wilson DW, Eastwood M, Boardman J, Gao B-C, Cohen D, Franklin B, Loya F, Lundeen S, Mazer A, McCubbin I, Randall D, Richardson B, Rodriguez JI, Sarture C, Urquiza E, Vargas R, White V, Yee K (2014) The Portable Remote Imaging Spectrometer (PRISM) coastal ocean sensor: design, characteristics and first flight results, Appl. Opt. 53(7) 1363–1380 http://dx.doi/org/10.1364/AO.99.099999

Müller R, Avbelj J, Carmona E, Gerasch B, Graham L, Günther B, Heiden U, Kerr G, Knodt U, Krutz D, Krawczyk H (2016) The new hyperspectral sensor DESIS on the multi-payload platform MUSES installed on the ISS. The international archives of the photogrammetry. Remote Sens Spatial Inf Sci, 41, 461–467. https://doi.org/10.5194/isprsarchives-xli-b1-461-2016

Munné-Bosch S, Alegre L (2004) Die and let live: leaf senescence contributes to plant survival under drought stress. Funct Plant Biol 31(3):203–216. https://doi.org/10.1071/FP03236

Naegeli K, Damm A, Huss M, Schaepman M, Hoelzle M (2015) Imaging spectroscopy to assess the composition of ice surface materials and their impact on glacier mass balance. Remote Sens Environ 168:388–402

National Academies of Sciences, and Medicine (2018) "Thriving on our changing planet: a decadal strategy for earth observation from space," Washington, DC

Ong CC, Cudahy TJ, Caccetta MS, Piggott MS (2003) Deriving quantitative dust measurements related to iron ore handling from airborne hyperspectral data. Mining Technol 112(3):158–163. https://doi.org/10.1179/037178403225003555

Painter TH, Duval B, Thomas WH, Mendez M, Heintzelman S, Dozier J (2001) Detection and quantification of snow algae with an airborne imaging spectrometer. Appl Environ Microbiol 67(11):5267–5272. https://doi.org/10.1128/AEM.67.11.5267-5272.2001

Painter TH, Dozier J, Roberts DA, Davis RE, Green RO (2003) Retrieval of subpixel snow-covered area and grain size from imaging spectrometer data. Remote Sens Environ 85(1):64–77. https://doi.org/10.1016/S0034-4257(02)00187-6

Painter TH, Seidel F, Bryant AC, Skiles SM, Rittger K (2013) Imaging spectroscopy of albedo and radiative forcing by light absorbing impurities in mountain snow. J Geophys Res Atmos 118(17):9511–9523. https://doi.org/10.1002/jgrd.50520

Painter TH, Berisford DF, Boardman JW, Bormann KJ, Deems JS, Gehrke F, Hedrick A, Joyce M, Laidlaw R, Marks D, Mattmann C (2016) The airborne snow observatory: fusion of scanning lidar, imaging spectrometer, and physically-based modeling for mapping snow water equivalent and snow albedo. Remote Sens Environ 31(184):139–152. https://doi.org/10.1016/j.rse.2016.06.018

Paz-Kagan T, Zaady E, Salbach C, Schmidt A, Lausch A, Zacharias S, Notesco G, Ben-Dor E, Karnieli A (2015) Mapping the spectral soil quality index (SSQI) using airborne imaging spectroscopy. Remote Sens Environ 7(11):15748–15781. https://doi.org/10.3390/rs71115748

Pearlman J, Barry PS, Segal C, Shepanski J, Beiso D, Carman SL (2003) Hyperion, a space-based imaging spectrometer. IEEE Trans Geosci Remote Sens 41(6):1160–1173. https://doi.org/10.1109/TGRS.2003.815018

Pereira HM, Ferrier S, Walters M, Geller GN, Jongman R, Scholes RJ, Bruford MW, Brummitt N, Butchart S, Cardoso A (2013) Essential biodiversity variables. Science 339(6117):277–278. https://doi.org/10.1126/science.1229931

Pignatti S, Acito N, Amato U, Casa R, de Bonis R, Diani M, Laneve G, Matteoli S, Palombo A, Pascucci S, Romano F, Santini F, Simoniello T, Ananasso C, Corsini G, Cuomo V (2013) "The PRISMA Hyperspectral Mission: Science Activities and Opportunities for Agriculture and Land Monitoring," Proceedings of the International Geoscience and Remote Sensing Symposium (IGARSS'13), Melbourne, Australia, 4 pages https://doi.org/10.1109/igarss.2013.6723850

Rast M (1991) Imaging spectroscopy: fundamentals and considerations for the application of spaceborne systems. ESA SP 1144:144p

Rast M, Bézy JL, Bruzzi S (1999) The ESA medium resolution imaging spectrometer MERIS- review of the instrument and it's mission. Int J Remote Sens 20(9):1681–1702. https://doi.org/10.1080/014311699212416

Rast M, Baret F, Hurk B, Knorr W, Mauser W, Menenti M, Miller J, Moreno J, Schaepman ME, Verstraete M (2004) SPECTRA-surface processes and ecosystem changes through response analysis (http://esamultimedia.esa.int/docs/SP_1279_2_SPECTRA.pdf)

Riaño D, Chuvieco E, Ustin S, Zomer R, Dennison P, Roberts D, Salas J (2002) Assessment of vegetation regeneration after fire through multitemporal analysis of AVIRIS images in the Santa Monica Mountains. Remote Sens Environ 79(1):60–71. https://doi.org/10.1016/S0034-4257(01)00239-5

Rickard LJ, Basedow RW, Zalewski EF, Silverglate PR, Landers M (1993) HYDICE: An airborne system for hyperspectral imaging. In: Imaging Spectrometry of the Terrestrial Environment (Vol. 1937, pp. 173-180). International Society for Optics and Photonics. https://doi.org/10.1117/12.157055

Roberts DA, Gardner M, Church R, Ustin S, Scheer G, Green RO (1998) Mapping chaparral in the Santa Monica Mountains using multiple endmember spectral mixture models. Remote Sens Environ 65:267–279. https://doi.org/10.1016/S0034-4257(98)00037-6

Roberts DA, Dennison PE, Gardner ME, Hetzel Y, Ustin SL, Lee CT (2003) Evaluation of the potential of Hyperion for fire danger assessment by comparison to the Airborne Visible/Infrared Imaging Spectrometer. IEEE Trans Geosci Remote Sens 41(6):1297–1310. https://doi.org/10.1109/TGRS.2003.812904

Roberts DA, Quattrochi DA, Hulley GC, Hook SJ, Green RO (2012) Synergies between VSWIR and TIR data for the urban environment: an evaluation of the potential for the hyperspectral infrared imager (hyspiri) decadal survey mission. Remote Sens Environ 15(117):83–101. https://doi.org/10.1016/j.rse.2011.07.021

Rossel RV, Behrens T, Ben-Dor E, Brown D, Demattê J, Shepherd K, Shi Z, Stenberg B, Stevens A, Adamchuk V (2016) A global spectral library to characterize the world's soil. Earth Sci Rev 155:198–230. https://doi.org/10.1016/j.earscirev.2016.01.012

Sankaran S, Mishra A, Ehsani R, Davis C (2010) A review of advanced techniques for detecting plant diseases. Comput Electron Agric 72(1):1–13. https://doi.org/10.1016/j.compag.2010.02.007

Schaepman ME (2009) Imaging spectrometers. SAGE Handb Remote Sens 18:166

Schaepman ME, Jehle M, Hueni A, D'Odorico P, Damm A, Weyermann J, Schneider FD, Laurent V, Popp C, Seidel FC, Lenhard K (2015) Advanced radiometry measurements and Earth science applications with the airborne prism experiment (APEX). Remote Sens Environ 158:207–219. https://doi.org/10.1016/j.rse.2014.11.014

Schepers S, Blackmer TM, Wilhelm WW, Resende M (1996) Transmittance and reflectance measurements of corn leaves from plants with different nitrogen and water supply. J Plant Physiol 148:523–529. https://doi.org/10.1016/S0176-1617(96)80071-X

Schlerf M, Atzberger C, Hill J, Buddenbaum H, Werner W, Schüler G (2010) Retrieval of chlorophyll and nitrogen in Norway spruce (Picea abies L. Karst.) using imaging spectroscopy. Int J Appl Earth Obs Geoinf 12(1):17–26

Schmidt KS, Skidmore AK (2010) Exploring spectral discrimination of grass species in African rangelands. Int J Remote Sens 22(17):3421–3434. https://doi.org/10.1080/01431160152609245
Schweiger AK, Schütz M, Risch AC, Kneubühler M, Haller R, Schaepman ME (2016) How to predict plant functional types using imaging spectroscopy: linking vegetation community traits, plant functional types and spectral response. Methods Ecol Evolut https://doi.org/10.1111/2041-210X.12642
Serrano L, Ustin SL, Roberts DA, Gamon JA, Peñuelas J (2000) Deriving water content of chaparral vegetation from AVIRIS data. Remote Sens Environ 74(3):570–581. https://doi.org/10.1016/S0034-4257(00)00147-4
Shafri HZM, Hamdan N (2009) Hyperspectral imagery for mapping disease infection in oil palm plantation using vegetation indices and red edge techniques. Am J Appl Sci 6(6):1031–1035
Staenz K, Mueller A, Heiden U (2013) Overview of terrestrial imaging spectroscopy missions. IEEE Int Geosci Remote Sens Symp—IGARSS 2013:3502–3505. https://doi.org/10.1109/IGARSS.2013.6723584
Suárez L, Zarco-Tejada PJ, Berni JA, González-Dugo V, Fereres E (2009) Modelling PRI for water stress detection using radiative transfer models. Remote Sens Environ 113:730–744. https://doi.org/10.1016/j.rse.2008.12.001
Sun J, Xiong X (2011) Solar and lunar observation planning for Earth-observing sensor, Proc SPIE, 8176 https://doi.org/10.1117/12.897751
Swayze GA, Smith K, Clark R, Sutley S, Pearson R, Rust G, Vance J, Hageman P, Briggs P, Meier A, Singleton M, Roth S (2000) Using imaging spectroscopy to map acidic mine waste. Environ Sci Technol 34(1):47–54. https://doi.org/10.1021/es990046w
Swayze GA, Clark RN, Goetz AF, Livo K, Breit G, Sutley S, Kruse F, Snee L, Lowers H, Post J, Stoffregen R, Ashley R (2014) Mapping advanced argillic alteration at Cuprite, Nevada using imaging spectroscopy. Econ Geol 109(5):1179–1221. https://doi.org/10.2113/econgeo.109.5.1179
Thenkabail PS, Smith RB, De Pauw E (2000) Hyperspectral vegetation indices and their relationship with agricultural crop characteristics. Remote Sens Environ 71:158–182. https://doi.org/10.1016/S0034-4257(99)00067-X
Thenkabail PS, Enclona EA, Ashton MS, Van Der Meer B (2004) Accuracy assessments of hyperspectral waveband performance for vegetation analysis applications. Remote Sens Environ 91:354–376. https://doi.org/10.1016/j.rse.2004.03.013
Thome KJ (2001) Absolute radiometric calibration of Landsat 7 ETM + using the reflectance-based method. Remote Sens Environ 78(1–2):27–38. https://doi.org/10.1016/S0034-4257(01)00247-4
Thompson D, Thorpe A, Frankenberg C, Green R, Duren R, Guanter L, Hollstein A, Middleton E, Ong L, Ungar S (2016) Space-based remote imaging spectroscopy of the Aliso Canyon CH4 superemitter. Geophys Res Lett 43(12):6571–6578
Ustin SL, Roberts DA, Jacquemoud S, Pinzón J, Gardner M, Scheer G et al (1998) Estimating canopy water content of chaparral shrubs using optical methods. Remote Sens Environ 65(3):280–291. https://doi.org/10.1016/S0034-4257(98)00038-8
Ustin SL, Roberts DA, Gamon JA, Asner GP, Green RO (2004). "Using imaging spectroscopy to study ecosystem processes and properties." BioScience 54(6): 523–534. https://doi.org/10.1641/0006-3568(2004)054%5b0523:UISTSE%5d2.0.CO;2
Van der Meer FD, De Jong SM (eds) (2011) Imaging spectrometry: basic principles and prospective applications (Vol. 4). Springer Science & Business Media
Vane G, Goetz AFH (1985) Introduction to the proceedings of the airborne imaging spectrometer (AIS) data analysis workshop, JPL Publication 85–41, Vane G. & Goetz A.F.H. eds. pp 1–21
Vane G, Goetz AF (1988) Terrestrial imaging spectroscopy. Remote Sens Environ 24(1):1–29. https://doi.org/10.1016/0034-4257(88)90003-X
Vane G, Goetz AFH, Wellman JB (1984) Airborne imaging spectrometer: a new tool for remote sensing. IEEE Trans Geosci Remote Sens 6:546–549. https://doi.org/10.1109/TGRS.1984.6499168
Veraverbeke S, Dennison P, Gitas I, Hulley G, Kalashnikova O, Katagis T, Kuai L, Meng R, Roberts D, Stavros N (2018) Hyperspectral remote sensing of fire: state-of-the-art and future perspectives. Remote Sens Environ 216:105–121. https://doi.org/10.1016/j.rse.2018.06.020
West JS, Bravo C, Oberti R, Lemaire D, Moshou D, McCartney HA (2003) The potential of optical canopy measurement for targeted control of field crop diseases. Annu Rev Phytopathol 41:593–614. https://doi.org/10.1146/annurev.phyto.41.121702.103726
Xiong X, Sun J, Fulbright J, Wang Z, Butler JJ (2016) Lunar calibration and performance for S-NPP VIIRS reflective solar bands. IEEE Trans Geosci Remote Sens 54(2):1052–1061. https://doi.org/10.1109/TGRS.2015.2473665

Zarco-Tejada PJ, Miller JR, Mohammed GH, Noland TL, Sampson PH (2001) Vegetation stress detection through chlorophyll a+$b$ estimation and fluorescence effects on hyperspectral imagery. J Environ Qual 31(5):1433–1441. https://doi.org/10.2134/jeq2002.1433

Zimmermann G, Neumann A (2000) The imaging spectrometer experiment mos on ipr-p3-three years of experience. J Spacecr Technol 10(1):1–9



## Affiliations

**Michael Rast[1] · Thomas H. Painter[2]**

✉ Michael Rast
michael.rast@esa.int

Thomas H. Painter
thomas.painter@jpl.nasa.gov

1 European Space Agency—ESRIN, Largo Galileo Galilei 1, 00044 Frascati, Italy

2 NASA-Jet Propulsion Laboratory, 4800 Oak Grove Drive, Pasadena, CA 91109, USA

Surveys in Geophysics (2019) 40:333–360
https://doi.org/10.1007/s10712-018-9488-9

# Retrieval of Atmospheric Parameters and Surface Reflectance from Visible and Shortwave Infrared Imaging Spectroscopy Data

David R. Thompson[1] · Luis Guanter[2] · Alexander Berk[3] · Bo-Cai Gao[4] · Rudolf Richter[5] · Daniel Schläpfer[6] · Kurtis J. Thome[7]

Received: 5 October 2017 / Accepted: 7 July 2018 / Published online: 24 July 2018


## Abstract
Remote imaging spectroscopy in the 0.4–2.5-μm visible and shortwave infrared (VSWIR) range captures the majority of solar-reflected energy and enables a wide range of earth surface studies. This spectral range is also influenced by atmospheric effects including absorption from atmospheric gases and aerosols, Rayleigh scattering, and particle scattering. Globally consistent surface measurements must compensate for these atmospheric effects. This article reviews the physical and mathematical foundations of modern VSWIR atmospheric retrieval, focusing on imaging spectrometers. We assess sensitivity of the retrieval to errors in atmospheric state estimation. Finally, we describe some promising avenues of future research to support the next generation of orbital imaging spectrometers.

**Keywords** Imaging spectroscopy · Atmospheric correction · Hyperspectral imaging · Surface reflectance

## 1 Introduction and Background

The 0.4–2.5-μm visible shortwave infrared (VSWIR) range captures the majority of solar-reflected energy, enabling diverse spectroscopic studies of the Earth's surface. Applications span aquatic and terrestrial ecology (Mumby et al. 2004; Ustin et al. 2004),

✉ David R. Thompson
David.R.Thompson@jpl.nasa.gov

[1] Jet Propulsion Laboratory, California Institute of Technology, 4800 Oak Grove Dr., Pasadena, CA 91109, USA

[2] GFZ German Research Centre for Geosciences, Helmholtz Centre Potsdam, Telegrafenberg, Building A 17, Room 20.22, 14473 Potsdam, Germany

[3] Spectral Sciences, Inc., 4 Fourth Ave., Burlington, MA 01803, USA

[4] Naval Research Laboratory, 4555 Overlook Ave., Washington, DC 20375, USA

[5] German Aerospace Center (DLR), Münchener Straße 20, 82234 Weßling, Germany

[6] ReSe Applications LLC, Langeggweg 3, 9500 Wil, Switzerland

[7] NASA/Goddard Space Flight Center, 8800 Greenbelt Rd, Greenbelt, MD 20771, USA

urban and agricultural studies (Herold et al. 2004), geology, hydrology, natural hazards remediation (Clark et al. 2003), and many more. Such investigations must estimate spectral *reflectance*—roughly speaking, the fraction of incident light that is reflected from the land or water surface (Schaepman-Strub et al. 2006). This involves first calibrating the measurement from raw instrument digital numbers (DN) to spectral radiance, and then inverting atmospheric influence to estimate surface properties (Fig. 1).

This article considers atmospheric inversion for *orbital VSWIR imaging spectrometers.* Future orbital missions will enable global studies comparing different regions and continents to understand the Earth system. Such studies have challenges commensurate with their value: estimating surface reflectance without biases across varied surface and imaging conditions requires consistent, accurate atmospheric correction. Moreover, global studies will lack in situ validation data for most scenes, so it will be important to understand any uncertainties that influence downstream analysis. We will investigate the challenge from first principles, surveying algorithms used in contemporary atmospheric correction practice. We aim to determine: first, the most significant uncertainties in global-scale atmospheric correction, and their effect on the resulting reflectance; and second, the best opportunities for future research to improve atmospheric correction accuracy in support of global missions.

The most important uncertainties depend on the specific instrument and application. Here, we address them in a general sense with an idealized instrument similar to many existing and planned VSWIR spectrometers: a spectral range of 0.4–2.5 μm sampled at 0.01-μm intervals, a ground sampling distance (GSD) of 30 m, and noise equivalent change in radiance (NEdL) on the order of 0.01 $\mu W\ cm^{-1}\ sr^{-1}\ nm^{-1}$. This is sufficient to reveal general sensitivity trends, and the analytical framework is generic if precise predictions for a specific instrument are needed.

We begin by reviewing the radiative transfer background and atmospheric processes relevant to mapping spectrometers in the VSWIR range. We describe retrieval algorithms including estimation of aerosol and gas interference as well as correction for surface effects such as bidirectional reflectance and topography. Finally, we describe common uncertainties in calibration and atmosphere and simulate propagation of errors using a state-of-the-art atmospheric radiative transfer code. This permits a sensitivity analysis tracing imperfect atmospheric retrievals to errors in downstream analyses. We close by summarizing our findings and outlining future research that could address remaining challenges.

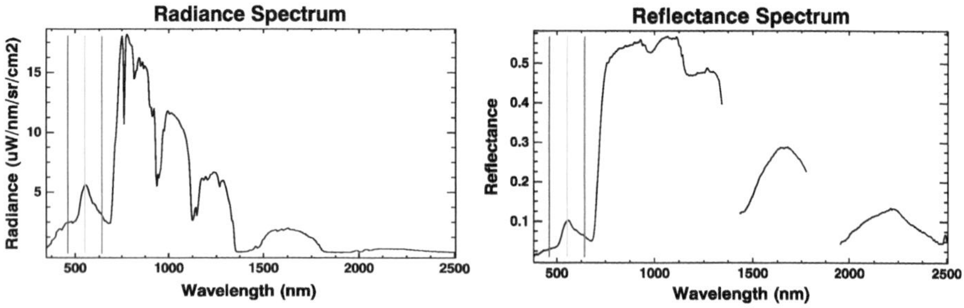


**Fig. 1** Typical analysis progresses from radiance at the sensor (left) to atmospherically corrected reflectance (right). The plot shows individual spectra acquired by NASA's Next Generation Airborne Visible Infrared Imaging Spectrometer (AVIRIS-NG) instrument. Adapted from Thompson et al. (2018a). Red, green, and blue lines indicate visible wavelengths

## 2 Principles of Radiative Transfer for VSWIR Imaging Spectroscopy

Global atmospheric correction begins with the physics of radiative transfer. This is not always necessary in airborne investigations, where physically motivated atmospheric correction is just one of several options (Gao et al. 2009). Alternatives include *empirical* approaches like scene averaging (Kruse 1988), flat fielding (Roberts et al. 1986), quick atmospheric correction (QUAC, Bernstein et al. 2005), and cloud shadow methods (Reinersman et al. 1998). Empirical methods are useful in isolated cases, but rely on specific characteristics of the scene—such as the presence of clouds, a spatially homogeneous atmosphere, or a known distribution of spectra—which precludes scaling them to global missions. Empirical line methods (Conel 1987; Thompson et al. 2016a) fit scene-specific corrections to match in situ reference spectra obtained by field teams. These relationships can be highly accurate over small geographic regions but also fail on a global scale. We focus here on physically motivated corrections based on radiative transfer models, which have superior stability and analytical interpretability for cross-regional studies.

Physically based methods for the retrieval of atmospheric parameters and surface reflectance rely on mathematical models of the spectral radiance reflected by the Earth's surface and atmosphere at the top of atmosphere (TOA). They typically decompose this measurement into a sum of radiative contributions from different processes, such as the photons scattered by the atmosphere into the sensor line of sight or those multiply scattered between the atmosphere and the surface. Tanré et al. (1981) proposed one useful and accurate formulation of the TOA radiance, implemented in the second simulation of the satellite signal in the solar spectrum (6S) code (Vermote et al. 1997b). It is part of the operative Moderate Resolution Imaging Spectroradiometer (MODIS) atmospheric correction algorithm (Vermote et al. 1997a) and the ATmospheric REMoval (ATREM) atmospheric correction code for VSWIR imaging spectrometers (Gao et al. 1993). In this formulation, the intrinsic directional properties of surface reflectance are rigorously coupled with the downwelling direct and diffuse irradiance fluxes, whereas gas absorption is decoupled from the total transmittance for practical reasons and included in the formalism as a multiplicative factor. The upward TOA radiance is expressed as a sum of five terms: (1) the photons reflected by the atmosphere before reaching the surface; (2) the photons directly transmitted from the Sun to the target and directly reflected back to the sensor; (3) the photons scattered by the atmosphere, then reflected by the target and directly transmitted to the sensor; (4) the photons directly transmitted to the target but scattered by the atmosphere on their way to the sensor, and (5) the photons having at least two interactions with the atmosphere and one with the target surface. The observed spectral reflectance at the TOA ($\rho^*_{\mathrm{obs}}$), resulting from the normalization of the TOA radiance $L_{\mathrm{obs}}$ by the incoming extra-terrestrial solar irradiance $\mu_0 F_0$, can be written:

$$
\begin{aligned}
\rho^*_{\mathrm{toa}}(\mu_0,\mu_\mathrm{v},\phi) = \frac{\pi L_{\mathrm{obs}}}{\mu_0 F_0} = T_\mathrm{g}(\mu_0,\mu_\mathrm{v})\Big[\rho_\mathrm{c}(\mu_0,\mu_\mathrm{v},\phi) \\
+ e^{\frac{-\tau}{\mu_0}} e^{\frac{-\tau}{\mu_\mathrm{v}}} \rho_\mathrm{s}(\mu_0,\mu_\mathrm{v},\phi) \\
+ e^{\frac{-\tau}{\mu_\mathrm{v}}} t_\mathrm{d}(\mu_0)\rho_{\mathrm{hd}} + e^{\frac{-\tau}{\mu_0}} t_\mathrm{d}(\mu_\mathrm{v})\rho_{\mathrm{dh}} \\
+ t_\mathrm{d}(\mu_0)t_\mathrm{d}(\mu_\mathrm{v})\rho_{\mathrm{hh}} \\
+ \frac{\left[e^{\frac{-\tau}{\mu_0}} + t_\mathrm{d}(\mu_0)\right]\left[e^{\frac{-\tau}{\mu_\mathrm{v}}} + t_\mathrm{d}(\mu_\mathrm{v})\right] S_\mathrm{c}(\rho_{\mathrm{hh}})^2}{1 - S_\mathrm{c}\rho_{\mathrm{hh}}}\Big]
\end{aligned}
\tag{1}
$$

Here $\mu_0$ is the cosine of the solar zenith angle, $\mu_v$ is the cosine of the view zenith angle, $\varphi$ is the relative azimuth angle between the illumination and the observation, and $T_g$ is the transmittance due to gas absorption. There are also several terms related to the influence of atmospheric molecular and Rayleigh scattering, after the gas absorption contribution has been factored out: $\rho_c$ the intrinsic atmospheric reflectance, also called atmospheric path reflectance; $S_c$ the spherical albedo, the reflectance of the atmosphere for isotropic light entering it from the surface; $e^{-\tau/\mu_0}$ and $t_d(\mu_0)$ are the downward direct and diffuse transmittances of the atmosphere along the path of the incoming solar beam; $e^{-\tau/\mu_v}$ and $t_d(\mu_v)$ are the upward transmittances of the atmosphere in the viewing direction. The other terms describe surface properties: $\rho_s$ is the surface reflectance; $\rho_{hd}$, $\rho_{dh}$, and $\rho_{hh}$ are the surface hemispherical-directional, directional-hemispherical, and hemispherical-hemispherical reflectances, respectively.

These latter terms are also called coupling terms, as they couple atmospheric radiative transfer with the surface reflectance (United States National Bureau of Standards and Nicodemus 1977; Schaepman-Strub et al. 2006). For the remote sensing situation, they are defined as:

$$\rho_{hd} = \frac{\int_0^{2\pi}\int_0^1 \mu L_\downarrow(\mu_0, \mu, \phi')\,\rho_s(\mu_0, \mu_v, \phi' - \phi)\mathrm{d}\mu\mathrm{d}\phi'}{\int_0^{2\pi}\int_0^1 \mu L_\downarrow(\mu_0, \mu, \phi')\mathrm{d}\mu\mathrm{d}\phi'} \tag{2}$$

$$\rho_{dh}(\mu_0, \mu_v, \phi) = \rho_{hd}(\mu_v, \mu_v, \phi) \tag{3}$$

$$\rho_{hh} = \rho_{dh}(\mu_0, \mu_v, \phi) \simeq \frac{\int_0^1\int_0^{2\pi}\int_0^1 \rho_s(\mu, \mu', \phi)\mu\mu'\mathrm{d}\mu'\mathrm{d}\mu\mathrm{d}\phi}{\int_0^1\int_0^{2\pi}\int_0^1 \mu\mu'\mathrm{d}\mu'\mathrm{d}\mu\mathrm{d}\phi} \tag{4}$$

where $L_\downarrow(\mu_0, \mu, \varphi')$ is the downward diffuse irradiance from the Sun at $\mu_0$. Equations 2 and 3 can be computed exactly from the downward radiation, whereas Eq. 4 (involving at least two interactions between the atmosphere and the surface) is typically approximated by taking $\rho_{hh}$ equal to the hemispherical albedo of the target. Multiple scattering tends to be azimuthally independent, so this approximation replaces a double integration by a single one. It is justified by the limited impact on the total signal of this last contribution.

Although Eq. 1 gives a rigorous formulation of the radiative coupling between the atmosphere and the surface, there is no closed form solution to invert surface reflectance since $\rho_s$ is embedded in the integrals accounting for the coupling terms in Eqs. 2, 3, and 4. This can be inverted by an iterative optimization using an initial guess for $\rho_s$ (Hu et al. 1999; Vermote et al. 1997a), but accurate guesses are seldom possible for single-view (as opposed to multi-angular) remote sensing. Assume that a Lambertian reflectance is often more practical for operational atmospheric correction. No real surface shows perfect Lambertian behavior, but many can be represented with this approximation to a reasonable extent. Under the Lambertian assumption,

$$\rho_{hd} = \rho_{dh} = \rho_{hh} = \rho_s \tag{5}$$

Equation 1 becomes:

$$\rho^*_{toa} = T_g\left[\rho_a + \frac{T_d T_u \rho_s}{1 - S\rho_s}\right] \tag{6}$$

in which the angular and spectral dependencies of the different terms are omitted and $T_d$ and $T_u$ represent the total downward and upward transmittance (for direct + diffuse radiation), respectively. The atmospheric optical parameters and $\rho_s$ are decoupled in Eq. 6, so a simple expression for $\rho_s$, which is now a Lambertian-equivalent surface reflectance, can be derived analytically:

$$\hat{\rho}_s = \frac{\rho^*_{toa}/T_g - \rho_c}{T_c + S_c(\rho^*_{toa}/T_g - \rho_c)} \tag{7}$$

Some care is required with the gas absorption $T_g$. Gas absorption features manifest at spectral resolution much higher than the typical VSWIR imaging spectrometer. To simulate a radiance measurement, it is typically necessary to calculate these terms at a very high spectral resolution prior to sampling by the instrument response function. Figure 2 illustrates a typical instantiation of $T_t$, the transmittance due to atmospheric gases, after sampling by the instrument response function. Annotations attribute features to different atmospheric constituents. An alternative to calculating $T_g$ at high spectral resolution is to treat each band as a distribution of magnitudes with different populations, an approach known as the correlated $k$ method (Fu and Liou 1992). The remaining terms are spectrally smooth and can be calculated at a coarser resolution. This decoupling is accurate in situations where scattering is weak.

The approach of decoupling gas absorption from gas and particle scattering may not be accurate in wavelengths with strong gaseous absorptions, or where scattering is strong, but the simplifications in the equations and the subsequent saving in computation time make useful efficient radiative transfer calculations in the VSWIR spectral range. A similar formulation which does not assume decoupling between atmospheric absorption and scattering can be written simply as:

$$\rho^*_{toa} = \frac{L_{obs}\pi}{\mu_o F_o} = \rho_a + \frac{T\rho_s}{1 - S\rho_s} \tag{8}$$

This defines $T$ to be the transmittance including both scattering and gas absorption components along the Sun–sensor path. Similarly, $\rho_a$ and $S$ represent the atmospheric reflectance and spherical albedo incorporating the coupled effects of gas absorption. The corresponding target reflectance estimate is:

$$\hat{\rho}_s = \frac{1}{T/\left(\rho^*_{toa} - \rho_a\right) + S} \tag{9}$$

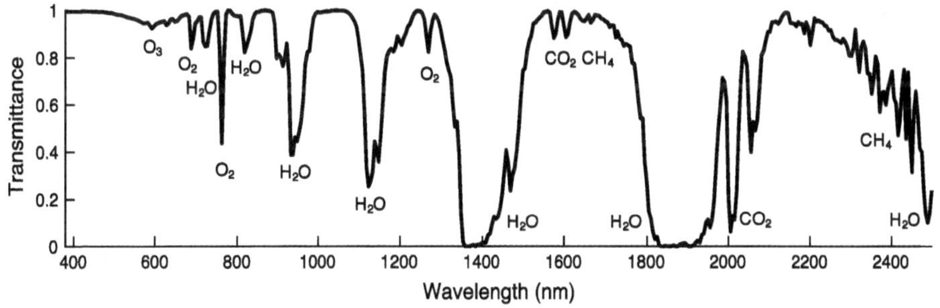


**Fig. 2** Atmospheric transmittance due to gas absorption, $T_g$, as a function of wavelength

Note that in the coupled model, $S$, $T$, and $\rho_a$ include spectrally sharp gas absorption features and must be calculated at high spectral resolution or with correlated $k$.

In practice, the quantities described thus far are calculated using radiative transfer codes such as 6S (Vermote et al. 1997b) or DISORT (Stamnes et al. 1988), the core of MODTRAN (Berk et al. 2016). These solvers propagate the solar flux at the top of atmosphere along the proper geometric path through an altitude-stratified atmospheric model populated with appropriate gas and particle abundances. The solver calculates diffuse and direct fluxes at each layer, accounting for absorption and scattering. Any of the above quantities that cannot be directly calculated in this fashion can be found from multiple runs of the radiative transfer code for a handful of different constant surface albedo values. For example, $\rho_s = 0$ provides the path radiance, and two additional albedos (e.g., 0.5 and 1.0) give a system of equations with closed form solutions for $T$ and S.

## 3 Retrieval of Atmospheric Parameters and Surface Reflectance

The atmosphere changes on relatively short spatial and temporal timescales, so some constituents must be estimated from the measurement data. Contemporary codes differ in their parameterizations and inversion methodology. This section describes approaches for retrieving atmospheric state variables such as the pressure altitude (Thompson et al. 2015a, 2018a), water vapor (Gao and Goetz 1990; Schläpfer et al. 1998), and aerosol parameters (Gao et al. 2000; Guanter et al. 2007). We also show methods for post-processing reflectance data to account for surface Bidirectional Reflectance (Schläpfer et al. 2015) and other effects.

### 3.1 Pressure Altitude

The dominant source of spatial variability in the atmosphere is the vertical depth of the column. *Pressure altitude* describes the effective surface pressure experienced by photons transecting the atmosphere along the instrument's optical path from Sun to ground to sensor. The pressure altitude is mostly a function of surface topography and is generally close to the altitude of the surface under nominal barometric conditions. The pressure altitude influences absorption and Rayleigh scattering by well-mixed gases, so it is important for both the transmittance and path radiance terms of Eq. 1. Most atmospheric correction routines either obtain pressure altitude directly from a digital elevation model (DEM) or use a single presumed elevation everywhere when the scene is flat. However, it is also possible to retrieve the pressure altitude from the data itself, which may be a more accurate estimate of the optical path under the precise scattering conditions of the acquisition.

Most strategies for retrieving pressure altitude rely on the depth of the oxygen A band feature. The continuum interpolated band ratio or CIBR (Green et al. 1989; Bruegge et al. 1990) measures the depth of the absorption feature relative to a linear continuum. One can easily estimate the continuum by interpolating linearly across the feature from the extremes. Green et al. (1993) and Thompson et al. (2015a) use the CIBR approach for pressure altitude estimation, improving accuracy over scenes with topographic variability. Oxygen band depths are also influenced by aerosol scattering (via the additive path radiance term), so this retrieval method tends to work best under strong aerosol constraints.

### 3.2 Column Water Vapor

Absorption by water vapor is another major source of spatial variability in measured radiances. It is common to use near-infrared water vapor absorption bands at 820, 940, or 1130 nm for columnar water vapor retrieval (Fig. 3). The optimal spectral region depends on the sensor characteristics and the absolute humidity of the atmosphere. While high-quality, full spectral range imaging spectrometers offer flexibility in choosing the spectral range, the choice is limited to the 820/940 nm range for visible and near infrared (VIS-NIR) spectrometers and may even be restricted further to 820 nm by calibration accuracy. The 820 nm range, on the other hand, bears the problem that it shows lower absorption. This may be critical when processing imagery with low water vapor content. Two common methods for water vapor retrieval are differential absorption methods using relative absorption strength (Schläpfer et al. 1998) and curve fitting techniques (Gao and Goetz 1990).

In the differential absorption approach, the water vapor amount is related to relative depth of the absorption feature as calculated by a two- or three-channel band ratio (Bruegge et al. 1990). However, this relation varies depending on the additive effect of aerosol scattering. The atmospheric pre-corrected differential absorption (APDA, Schläpfer et al. 1998) technique uses a terrain dependent correction of the atmospheric path radiance term before the apparent transmittance values are related to total columnar water vapor. A radiative transfer code like MODTRAN is used to calculate the spectral path radiance and to find an exponential conversion function between transmittance and water vapor amount for the atmospheric conditions of a specific scene. The APDA technique reduces the errors due to background reflectance variations significantly compared to differential absorption techniques based on top of atmosphere (TOA) radiance. Validation of the results in comparison with water vapor contents derived from AERONET has shown an RMSE accuracy of 0.12-cm column water vapor (Makarau et al. 2017).

Spectral curve fitting is an alternative that leverages the full spectroscopic measurement, matching the shape of the $H_2O$ absorption feature (Gao and Goetz 1990; Thompson et al. 2015a) over a spectral interval bracketing it. For example, Gao and Goetz (1990) calculate a water vapor absorption with a full radiative transfer model and then

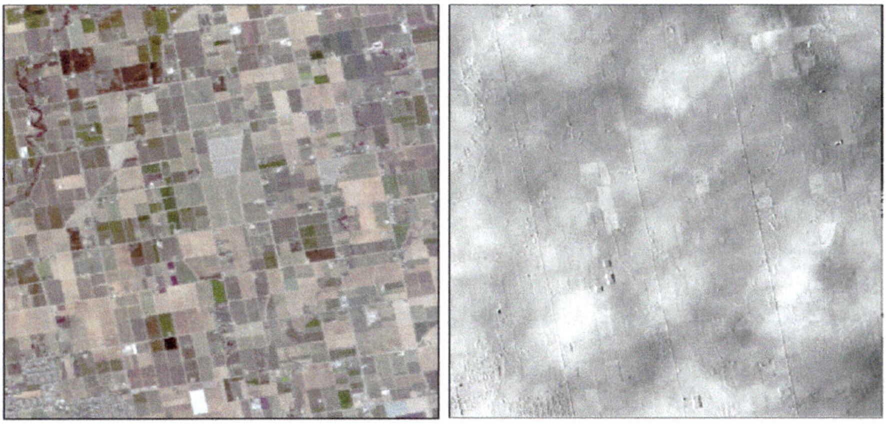

**Fig. 3** (left) An image from the Airborne Visible Infrared Imaging Spectrometer (AVIRIS-C), visible channels. (right) Corresponding map of atmospheric water vapor retrieved using the spectrum fitting technique in Thompson et al. (2015a)

estimate a linear scaling that best fits the absorption feature in TOA spectra. To avoid indeterminacy, most such methods model surface reflectance in this short interval as a linear slope. The coefficients can be retrieved as free parameters in a joint optimization or estimated directly by interpolating across the top of the absorption feature.

In reality, the reflectance is never a perfect linear continuum; more complex surface reflectance shapes can distort both ratios and spectral fits. Liquid water and ice are common offenders, since they are pervasive in snow and tree canopies and have absorption features overlapping with those of vapor. To remedy this, Gao and Goetzt (1995) jointly fit liquid and vapor, treating them as Beer–Lambert absorptions that are log-linear in the path lengths. A similar assumption holds (approximately) for the continuum over short spectral intervals, allowing all parameters to be fit simultaneously as a linear equation mapping path length and reflectance parameters to the negative log TOA reflectance. Thompson et al. (2015a) adopted a similar approach while constraining the retrieved parameters to be nonnegative. In both cases, the water vapor parameter is not perfectly log-linear at typical imaging spectrometer resolutions due to under-resolved spectrally fine structure. This necessitates an initial guess derived from band ratio methods to determine an "effective" absorption coefficient. Green et al. (2006) used a nonlinear fit to match the vapor feature more accurately at the cost of additional computational complexity. They pre-calculate transmittance lookup tables at many different water vapor concentrations and match measured radiances using nonlinear least squares optimization. In both linear and nonlinear cases, fitting atmosphere and surface simultaneously improved accuracy *vis-a-vis* a linear reflectance continuum.

Figure 4 shows an example of such approaches applied to a mosaic by the "Classic" Airborne Visible InfraRed Imaging Spectrometer (AVIRIS-C) over Yosemite National Park on May 2013. Color channels indicate water in different phases: Blue represents water vapor, which fills topographic lows; green represents liquid water at the surface, which is present mainly in tree canopies; and red represents ice, which is present as melting snow at high elevations and appears yellow due to coincidence with the liquid phase.

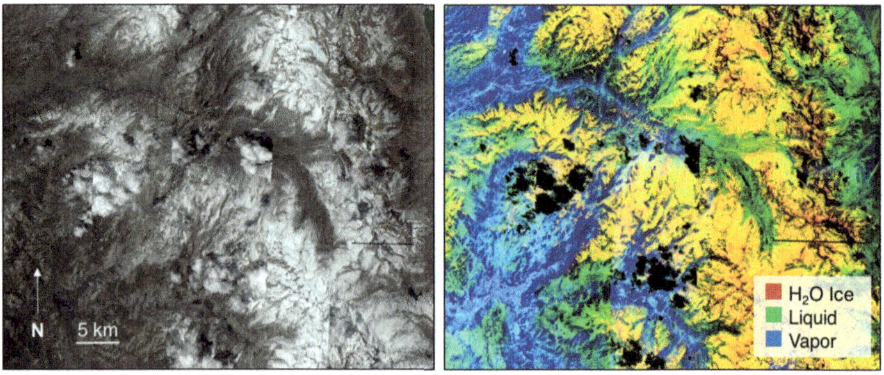


**Fig. 4** Three phases of water apparent in an AVIRIS-C mosaic of flightlines over Yosemite National Park. (left) Mosaic in visible wavelengths. (right) Abundance of water phases. Blue represents $H_2O$ vapor, green is liquid water in vegetation canopies, and red represents ice. Yellow areas are a combination of liquid and ice, i.e., melting snow. Black areas have no data, due mainly to clouds and cloud shadows

### 3.3 Ozone

Apart from oxygen and water vapor, other gases play a role in the visible shortwave infrared interval. Methane, carbon monoxide, nitrous oxide, and carbon dioxide all have absorption features in this range (Gao et al. 1993). However, most of these gas concentrations are sufficiently predictable that dynamic estimation is not usually needed to determine surface reflectance. For example, accurate VSWIR surface reflectance retrieval must account for ozone in the 450–800-nm spectral region, which has a peak absorption (minimum transmittance) around 600 nm. Typically, ozone column information (in Dobson Units, or DU) is drawn from external sources, e.g., ozone-measuring instruments such as TOMS, GOME, OMI or numerical weather predictions, e.g., the National Center for Environmental Prediction (NCEP) or the European Center for Medium-Range Weather Forecasts (ECMWF). Contrary to aerosol and water vapor variations, ozone concentrations usually change slowly on a monthly scale; typical variations are 50–100 Dobson Units (DU), typical spatial changes are on a grid of 1°–5°, i.e., about 100–500 km (Loyola et al. 2009). If the ozone column is known, then the ozone gas transmittance can be calculated with a knowledge of the spectral absorption coefficient and solar/view geometry (Yamamoto et al. 2008). Another approach for ozone correction is based on lookup tables derived from radiative transfer calculations (Richter et al. 2014) covering the relevant solar and view geometries and a range of ozone columns (e.g., 200–600 Dobson Units).

### 3.4 Atmospheric Aerosols and Haze

Unlike water vapor features that show distinctive spectral shapes over narrow intervals, atmospheric aerosols and haze induce spectrally smooth perturbations that can be difficult to disentangle from legitimate changes in surface reflectance. There are many methods to estimate the aerosol scattering and haze distribution from imaging spectroscopy data, and no universal standard has yet emerged. One may distinguish between methods that rely on known reflectance spectra from reference surfaces, statistical approaches, climatology-driven approaches, and radiative transfer approaches.

Reference surface methods exploit known reflectance properties for one or more locations in the scene, optimizing aerosol optical properties to match the target. Historically investigators have used a wide range of reference surface types. One of the first involved dark objects (Tanré et al. 1983). The result can be strongly affected by the surface reflectance characteristics since not all scenes contain good quality dark spectra. The dark dense vegetation approach (DDV) is an established method that exploits the reliable reflectance shape and magnitude of dense vegetation in visible wavelengths (Kaufman and Sendra 1988; Kaufman et al. 1997). Because vegetation is pervasive over much of Earth's surface, this method has been largely successful and has been incorporated in automatic atmospheric compensation routines. However, it fails when pure vegetation pixels are not present. Guanter et al. (2008) overcame this challenge with a mixture model that captures partially vegetated spectra. They use combinations of library spectra to span a subspace of legitimate reflectances. Other measures are necessary for aquatic scenes (Gao et al. 2000) where investigators can exploit the known shape of water-leaving reflectance in infrared wavelengths. If the water is deep and not too turbid, the infrared reflectance should be zero after accounting for spectrally smooth effects of aerosols and sun glint. When no nearby vegetation or deep clear water is available, satellite reference information from MODIS or similar instruments is another alternative (Hagolle et al. 2015).

Statistical methods are an efficient way to deal with high aerosol load and haze situations. One example employs the haze optimized transform (HOT, Zhang et al. 2002). Another uses the haze thickness map, or HTM (Makarau et al. 2014). This statistical dehazing can be performed on at-sensor radiance data as an optional pre-processing to the physical atmospheric compensation. The HTM method is an advanced locally adaptive dark-object subtraction algorithm which uses statistical methods to derive the spatial haze distribution. It treats the aerosol contribution as an additive path radiance term raising the "zero point" of the radiance spectrum. Because this term varies over the image, a local subwindow search for dark pixels estimates the haze contribution separately at each location and band in the scene.

Despite progress on these fronts, aerosol correction remains a challenging atmospheric correction problem in the general case. Methods that work well in isolated test cases may fail if their assumptions are violated, such as in scenes lacking dark pixels or aerosols with unexpected optical absorption properties. Aerosol correction can also suffer under very hazy conditions where particle scattering modifies the optical path length, changing the shape of gas absorption features and contaminating other gas retrieval and correction. To address these shortcomings, recent research has focused increasingly on methods that incorporate rigorous radiative transfer modeling. One is an iterative approach to derive aerosol amounts over dark objects and shaded areas (c). The atmospheric compensation method itself is used for the aerosol amount inversion by evaluating the brightness of corrected cast shadow areas in comparison with directly illuminated regions. A precondition to this method is the presence of shaded pixels, which are most abundant in images with spatial sampling finer than 10 m. Another radiative transfer approach uses sophisticated surface models to estimate the joint probability of different aerosol/surface combinations given the remote measurement; Bayesian formalisms can incorporate prior climatology and surface expectations with remote data to determine the full posterior distribution (Frouin and Pelletier 2015).

### 3.5 Cirrus

A unique challenge of high altitude and orbital observations is frequent contamination by thin cirrus clouds. Fortunately, the VSWIR interval permits effective correction algorithms (Makarau et al. 2016). Through analysis of high-altitude AVIRIS-C data, Gao and Kaufman demonstrated that narrow channels within the 1.38- and 1.88-μm atmospheric band centers are very effective in detecting thin cirrus clouds (Gao and Kaufman 1995). In the absence of cirrus clouds, channels near 1.38 μm receive little scattered solar radiance by the surface and the low-level water clouds because of the total absorption of solar radiation by atmospheric water vapor above them. When cirrus clouds are present, these channels receive scattered solar radiance by the cirrus clouds, which contrast well on the nearly black background. However, the 1.38-μm channels are slightly affected by absorption from water vapor located above and within thin cirrus clouds. These effects must be removed in order to use the 1.38-μm signal for quantitative cirrus correction in other channels (Gao et al. 1998, 2002).

For thin cirrus, one can model the cloud as a homogeneous layer located above a "virtual surface," which includes the effects of Rayleigh scattering and land or ocean surface reflection and scattering. Accounting for transmission through the cirrus cloud and multiple reflections between cloud and the "virtual surface," we have the following relationship (similar to what is done for aerosol correction, e.g., Fraser and Kaufman 1985):

$$\rho^*_{\mathrm{toa}} = \rho_{\mathrm{i}} + \frac{T_{\mathrm{i}}\rho_{\mathrm{v}}}{1 - S_{\mathrm{i}}\rho_{\mathrm{v}}} \tag{10}$$

where $\rho_{\mathrm{i}}$ is the reflectance of the cirrus (ice) cloud, $T_{\mathrm{i}}$ is the two-way transmission (direct + diffuse) through the cloud, $\rho_{\mathrm{v}}$ is the reflectance of the "virtual surface," and $S_{\mathrm{i}}$ is the cloud-base reflectance of upward radiation. Because $S_{\mathrm{i}}\rho_{\mathrm{v}} \ll 1$ for thin cirrus, the equation can be simplified to:

$$\rho^*_{\mathrm{toa}} = \rho_{\mathrm{i}} + T_{\mathrm{i}}\rho_{\mathrm{v}} \tag{11}$$

In order to remove the cirrus effect from the satellite measurement $\rho_{\mathrm{obs}}$, namely to derive $\rho_{\mathrm{v}}$, cirrus reflectance $\rho_{\mathrm{i}}$ and transmittance $T_{\mathrm{i}}$ must be known. Through analysis of AVIRIS-C data, Gao et al. (1998) found that cirrus reflectance $\rho_{\mathrm{i}}$ for channels between 0.4 and 1.0 μm was linearly related to $\rho_{\mathrm{i}}$ at 1.375 μm, i.e.,

$$\rho_i = \rho_i(1.375\,\mu\mathrm{m})/\mathrm{K}, \quad 0.4 < \lambda < 1.0\,\mu\mathrm{m} \tag{12}$$

where $K$ is an empirical parameter derived from data. It is essentially the 1.375-μm channel transmittance for water vapor above and within cirrus clouds. The weak ice absorption at 1.375 μm also slightly reduces the $K$ value. Algebraic substitution gives:

$$T_i\rho_v = \rho^*_{toa} - \rho_i(1.375\,\mu\mathrm{m})/\mathrm{K}, \quad 0.4 < \lambda < 1.0\,\mu\mathrm{m} \tag{13}$$

The image of $T_{\mathrm{i}}\rho_{\mathrm{v}}$ is referred to as the "cirrus-path-radiance-corrected" image. It is similar to the "virtual surface" reflectance image of $\rho_{\mathrm{v}}$ due to high transmittance of thin cirrus clouds ($T_i$ is usually greater than 0.9).

In order to use the 1.38-micron channels for quantitative correction of thin cirrus effects in other channels, one must account for the absorption by water vapor above and within cirrus clouds. The water vapor transmittance factor, $K$, can be obtained from a scatter plot between the 1.38-μm channel and a channel in atmospheric window region, such as 1.25 μm. The minimum value of $\rho_{\mathrm{obs}}$ at 1.25 μm for each possible value of 1.38 μm is a slope that estimates $K$ (Gao et al. 1998). An empirical algorithm for the derivation of cirrus reflectances has already been implemented for processing MODIS data (Gao et al. 2002). This would also estimate thin cirrus reflectances in the 0.4–1.0 μm region using the empirically derived $K$ value in the equation above, permitting cirrus correction in wavelengths shorter than 1 μm.

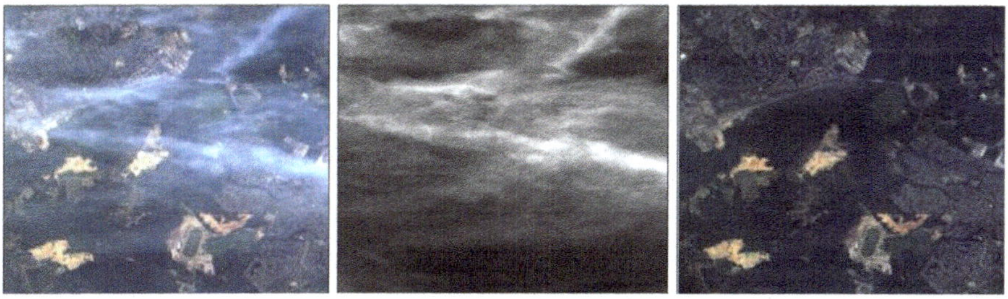

**Fig. 5** Left: An AVIRIS-C red/green/blue (RGB) image acquired over Bowie, Maryland in July 1997. Center: The 1.38-μm image. Right: the cirrus-corrected RGB image

Figure 5 shows examples of cirrus detection and corrections with a narrow channel located within the strong water vapor absorption region near 1.38 μm. The left panel shows visible wavelengths of an AVIRIS-C image acquired over Bowie, Maryland in the summer of 1997. Surface features appear blurred because of the presence of thin cirrus clouds. The center panel is the 1.38-μm channel image of the same scene. Only the upper level thin cirrus clouds are seen. The surface features seen in the left panel disappear completely in the center because of absorption of water vapor beneath cirrus. The right panel is the cirrus-corrected RGB image. The corrections were made using the procedures outlined above. Comparing left and right panels illustrates that removing thin cirrus improves contrast in the surface features.

### 3.6 Bidirectional Reflectance Distribution Function (BRDF) Retrieval and Compensation

The output of atmospheric compensation is typically described as bottom of atmosphere (BOA) reflectance and is also often referred to as hemispherical conical reflectance factor (HCRF) or hemispherical directional reflectance factor (HDRF) (Schaepman-Strub et al. 2006). This represents the fractional reflectance of the hemispherical incident illumination onto the surface, sampled in the direction of the sensor. However, as noted in the original coupled BRDF-atmosphere formulation of Eq. 1, the reflectance is never perfectly isotropic. In fact, the sensor direction is just one sample of a richer Bidirectional reflectance distribution function, or BRDF, which describes how reflectance changes for different incidence and observation angles. The BRDF captures specular properties like glint and can be used at larger ground sampling distances to represent geometric effects from surface faceting and cast shadows.

According to United States National Bureau of Standards and Nicodemus (1977), the HCRF and HDRF are defined with respect to a complete, full-sky illumination field which is not the natural remote sensing situation. Accounting for location-specific illumination effects such as obscuration can help retrieve values close to a physical HDRF as intermediate output of atmospheric compensation. This involves correcting for topographic illumination and incidence BRDF effects. Still stronger bidirectional reflectance effects occur in the observation direction for off-nadir viewing. If the application desires an observation-independent quantity, one can transform the HDRF to an equivalent nadir-perspective measurement by accounting for the BRDF of each pixel.

The BRDF Effects Correction method BREFCOR (Schläpfer et al. 2015) uses the Ross-Thick-Li-Sparse (RTLS) reciprocal BRDF model. The model is tuned to the surface cover properties based on a continuous BRDF cover characterization index (BCI). This allows a surface cover dependent but yet continuous correction of the HDRF to bihemispherical reflectance (BHR), i.e., to the spectral albedo. The method is of generic nature and has been successfully applied to 4-band photogrammetric imagery, to multispectral space borne multi-angle imagery, and to airborne imaging spectroscopy data (Schläpfer et al. 2015).

### 3.7 Topographic Corrections

A presentation of atmospheric correction techniques would be incomplete without at least mentioning the problem of topographic correction. Surface topography affects imaging spectroscopy in several ways. At a coarse level, the surface elevation defines the bottom of the atmospheric column, influencing the amount of scattering and absorption by gases and aerosols; this is the subject of Sect. 3.1 above. This section deals with subtler effects related to the influence of local slope and topography on the incident light field. An adequate discussion of the available methods is out of scope for this work, so we restrict our treatment to the main challenges and some state-of-the-art solutions. In common to all topographic correction methods is the correction of the varying illumination field for each pixel based on the local description of the topography. This correction may be achieved by physical modeling (Teillet et al. 1982; Sandmeier and Itten 1997; Li et al. 2015). Or by empirical and semi-empirical approaches (Kobayashi and Sanga-Ngoie 2008; Soenen et al. 2005; Riano et al. 2003), commercially available codes for the inversion of atmospheric parameters of imaging spectrometers are ACORN (www.aigllc.com) ATCOR (www.rese-apps.com), FLAASH (www.harrisgeospatial.com), but only ATCOR includes an option for topographic compensation (Richter 1998).

In rugged terrain, a Digital Elevation/Surface Model (DEM/DSM) defines the elevation map for the imaging spectrometer scene. The elevation map is resampled to obtain the elevation on a per-pixel level, and then the slope and aspect maps have to be calculated. Critical issues are the adequate spatial resolution of the DEM (at least comparable to the pixel size of the scene) and the accuracy of the DEM itself. For steep terrain but also for high spatial resolution data, the maps of sky view factor (portion of visible sky dome per pixel) and topographic shadow (self and cast shadow from surrounding obstacles) are required. In shadow regions, the direct solar irradiance is zero, and even the diffuse solar flux (due to scattered radiation) is reduced according to a *sky view factor*. This makes an atmospheric inversion extremely difficult, especially for low-radiometric-resolution instruments (8–10 bit/pixel). Even for a 12–16-bit dynamic range, the correction of shadow areas is difficult causing over- and under-correction artifacts. A shadow fraction parameter can be introduced based on blue/green and blue/red radiance ratios to better estimate the diffuse illumination caused by scattering (Schläpfer et al. 2013). This shadow fraction improves the illumination map calculated from the DEM. Finding accurate DEM/DSMs is difficult for very high-resolution ($< 10$ m) data; at the time of this writing, freely available global DEMs have a spatial resolution up to 30 m (e.g., ASTER 2018, SRTM 2018). Better DEMs and DSMs may become available in the future, but their better resolution will not always be sufficient to describe the underlying physics in the case of forests or vegetation canopies.

Even for moderately steep terrain (pixel slope values of 10°–20°) spatially varying topographic effects often dominate over atmospheric effects. Despite the illumination effects, directional effects of surface reflectance play a strong role, i.e., even for nadir-looking sensors any terrain changes of surface slope and orientation cause significant changes of the recorded radiance (Shepherd and Dymond 2003; Richter et al. 2009). In addition to the adjacency effect, one often considers the radiation reflected from facing surfaces, such as mountains or buildings within the line-of-sight of each pixel. In the ideal case, the incidence BRDF for each pixel is required. However, per-pixel knowledge of the incidence BRDF (distinct from the "observer BRDF" of Sect. 3.6) is scarce for very high-resolution imagery. Li et al. (2015) approximated this problem by spatially resampling incidence BRDF functions calculated from MODIS scenes, i.e., using 500-m incidence BRDF data on a 30-m raster. This leaves the effect of observation angle uncorrected. For the compensation of BRDF effects, one may use a global incidence BRDF model derived from MODIS data or one can try to fit the best-suited model to the given data and observation angles (Jensen et al. 2018).

### 3.8 Spectral Polishing

As the spectral resolution of imaging spectrometers has gradually improved with time, spectral artifacts have become more and more visible in the atmospherically corrected data. There are two common reasons for these high-frequency spikes (Schläpfer and Richter 2011). First, *systematic* deviations may be caused by inappropriate radiometric standards, by spectral miscalibration, by systematic errors in the description of the atmospheric absorption or from uncertainties in the solar reference spectrum. Non-systematic (statistical) variations may be caused by the intrinsic variation of the atmosphere and the sun, the statistical photon shot noise, or due to readout noise in the detector electronics. Investigators often correct these artifacts with system- and situation-specific techniques known colloquially as *spectral polishing* (Boardman 1998; Gao and Liu 2013). The term can be defined as the removal of statistical noise and calibration artifacts in the spectral domain from atmospherically corrected imaging spectroscopy surface reflectance data.

For instruments at resolutions in a range of 10–30 nm, systematic errors are most prominent. One can correct such errors by searching for systematic deviations from average of each detected pixel, using flat field or empirical line analysis. Such corrections implicitly correct solar spectrum uncertainties which are relevant at the scale of 5–10 nm (Gao et al. 2009). In certain aquatic applications, uncertainties in the solar spectrum, and the instruments' sampling thereof, persist because they distort radiances rather than reflectances—if the path radiance is a large fraction of the measured signal, then solar-related systematic errors will not be multiplicative after the conversion to reflectance. In such cases, it is possible to optimize the solar spectrum directly based on spectrally smooth reference targets (Thompson et al. 2015c).

With the advance of high-resolution pushbroom-type imaging spectrometry systems, statistical variations can also be prominent. Investigators often use statistical strategies to correct these deviations. This exploits the fact that for many applications in the VSWIR range, surface reflectance signatures do not show important variability at resolutions below 10 nm. Such applications include general water quality studies, hydrology studies of snow and ice, basic mineralogical analyses and some terrestrial and aquatic ecosystem investigations. Here, the criterion of a "good" spectrum at resolutions below 10 nm is its smoothness in the spectral space. Filtering may be done by straight forward smoothing algorithm, Savitzky–Golay filters, derivative filters, or Fourier transform filtering. Schläpfer and Richter (2011) find that derivative filters or Savitzky–Golay filters perform best in reconstructing a spectrum affected by statistical noise. Other applications, such as mapping of subtle mineralogical distinctions, canopy chemistry or species discrimination (Kokaly et al. 2009), or discrimination and detection of anthropogenic materials, benefit from higher spectral resolution. Investigators are often cautious about deploying statistical polishing in these domains.

## 4 Uncertainties in Atmospheric Correction

This section describes uncertainties in VSWIR atmospheric correction for global imaging spectrometers. We begin by reviewing prior studies and some established strategies for field validation. Then, a simulation demonstrates the sensitivity of typical aquatic and terrestrial ecosystem analyses to atmospheric state estimation errors.

### 4.1 Empirical Uncertainty Analysis

Countless prior field studies have compared in situ reflectance measurements against remote airborne retrievals; these give an empirical picture of typical errors. Most field studies find that, excluding BRDF post-processing or topographic treatment, different algorithmic packages typically provide broadly similar results with some minor differences in the atmospheric state retrieval (Kruse 2004). Errors hinge mainly on local atmospheric conditions, manifesting as slight slope differences or offsets related to scattering profiles with sharper residuals around atmospheric absorptions. Current methods often achieve error better than 1–2% in absolute reflectance for aquatic environments, where the surface variability is a small fraction of total radiance (Thompson et al. 2015b). Such accuracy is also regularly achieved for uniform terrestrial validation sites under clear skies.

Figure 6 shows one example from a validation experiment over Ivanpah Playa conducted on March 23, 2017. Here, NASA's Next Generation Airborne Visible Infrared Imaging Spectrometer (AVIRIS-NG) overflew a bright playa onboard a high-altitude ER-2 aircraft at 19.8 km, which is above the majority of atmospheric absorption and scattering and therefore comparable to the orbital case. Simultaneous reflectance spectra were acquired with in situ instruments deployed by a ground team. The right panel compares the in situ data to the remote retrieval. Our atmospheric correction calculates $T$, $S$, and $\rho_a$ coefficients for different atmospheric states using MODTRAN with a correlated k method. We then estimated atmospheric water vapor using nonlinear least squares fitting across 940- and 1140-nm intervals. The result is within 1–2% absolute reflectance across most of the spectrum. In more unconstrained terrestrial cases, current methods commonly achieve errors within 2–5% absolute reflectance against reference spectra (Richter and Schlaepfer 2002; Matthew et al. 2002; Kruse 2004; Thompson et al. 2015a). Critically, atmospheric correction can still recover relative reflectance shapes—particularly over short intervals—with higher fidelity, permitting accurate recognition and mapping of spectral signatures.

Other studies evaluate atmospheric impacts on downstream products. Makarau et al. (2017) compared retrieved $H_2O$ vapor profiles to AERONET ground station retrievals, showing RMSE accuracies of 0.12-cm column water vapor. Palacios et al. (2015) compared different reflectance retrievals for open water phytoplankton studies. While measurements often agreed to within 1–2% absolute reflectance, accurate scattering compensation proved critical; subtle slope differences relate to blue-end distortions caused poor statistical

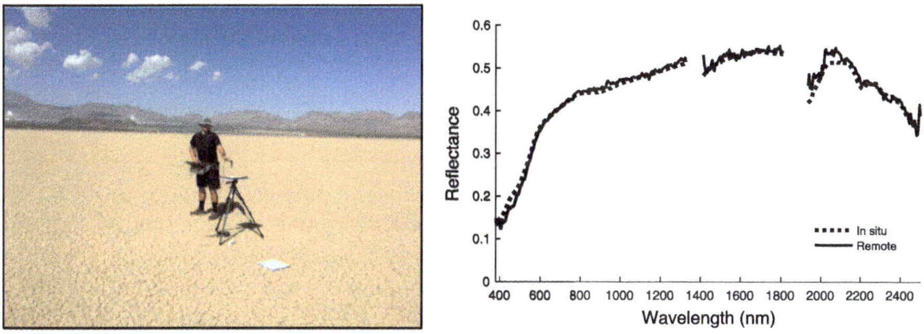


**Fig. 6** In situ validation example from Ivanpah Playa by AVIRIS-NG, 23 March 2017, acquired from an ER-2 aircraft at 19.8 km altitude with 39.2 degrees solar elevation. Left: field procedure. Right: comparison of spectra. No spectral polishing has been applied

correlations to chlorophyll-a retrievals for all but one of the attempted atmospheric correction strategies. Other studies demonstrate the importance of BRDFs; Weyermann et al. (2014) evaluated vegetation indices under uncorrected surface anisotropy, finding considerable variability in small-scale retrievals of band ratios. Bachmann et al. (2015) showed how both radiance and reflectance errors can influence vegetation indices measured by the EnMAP instrument.

### 4.2 Simulation Method

Here, we explore sensitivities in a simple, controlled simulation. Atmospheric correction uncertainties are influenced by factors ranging from calibration to imaging conditions to the atmosphere itself. For generality, we will limit our study in a few important respects. First, we use a generic instrument specification: a spectral range of 0.38–2.5 μm sampled at 0.01-μm intervals, a ground sampling distance (GSD) of 30 m, and noise equivalent change in radiance (NEdL) on the order of 0.01 $\mu W\ cm^{-1}\ sr^{-1}\ nm^{-1}$. This is typical for existing airborne instruments and anticipated orbital spectrometers. We will also assume appropriate corrections for cross-track non-uniformity of sensor look angle (for imaging geometry) and spectral response (for matching atmospheric features). Such corrections are well understood, and any discrepancies would be highly instrument-specific. Third, we analyze spectra independently, abstracting away from domain-specific scene structure. This simplifies the statistical interpretation but precludes a treatment of adjacency effects in heterogeneous environments like urban settings. Finally, we focus on the influence of atmospheric parameter errors on an idealized, uncoupled surface—one that is locally flat and uniform Lambertian. This is optimistic for natural scenes. However, departures and discrepancies in adjacency and BRDF are case-specific and challenging to simulate. Moreover, effects of locally non-uniform scene content become less relevant at larger pixel sizes of orbital instruments where one analyzes aggregated 30-m areas. As a consequence of these limitations, the reader should consider the resulting uncertainties as a lower bound and most appropriate for favorable surfaces and illumination at 30-m sampling. However, in this context it provides a controlled experiment to compare the sensitivity of typical investigations to the most critical atmospheric variables.

We begin with a set of reference reflectances representing two typical, representative application domains with relevant features across a wide range of wavelengths (Fig. 9). The first is a terrestrial ecosystem investigation using unmixing to determine the fractional composition of different vegetated and non-vegetated surfaces as in prior studies of forest health, invasive species, and draught. We form a reference surface using component spectra drawn from AVIRIS-NG airborne data: a green vegetation spectrum, Non-Photosynthetic Vegetation (NPV), and soil, each in 30% fractions. This leaves an additional 10% mixing fraction for a zero spectrum representing photometric shade. We combined these proportionally in a linear mixture model to form the "ideal" reflectance. The left panel of Fig. 7 shows the mixture model, with alternative spectra illustrating perturbations in each component. The second test case is a shallow water aquatic remote sensing scenario, relevant to inland or coastal bathymetry and water quality estimation challenges. We form a model spectrum using the HOPE parameterization of the water column (Lee et al. 1998, 1999), a model with five free parameters representing different particle concentrations and depth. We use typical values of particle backscatter, phytoplankton, and gelbstoff, and a depth of 3 m. The bottom substrate is a sand spectrum extracted from PRISM images of beach sand.

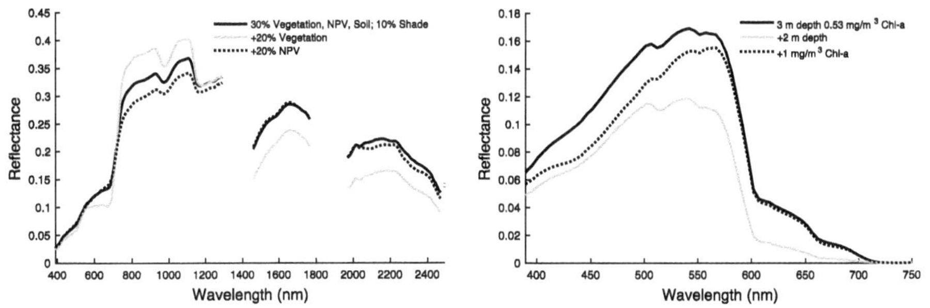


**Fig. 7** Example spectra used in the sensitivity study: terrestrial (left) and aquatic (right). The dark lines show the reference cases, while the lighter lines show how different model parameters affect the spectrum shape

The right panel of Fig. 7 illustrates the result with different perturbations of the reference case.

We perform atmospheric modeling with the MODTRAN 6.0 radiative transfer code (Berk et al. 2016a, b). We begin with a *nominal* reference atmosphere, using the MODTRAN temperate midlatitude summer model, with 1.5-g $cm^{-2}$ column water vapor, and thin rural aerosols providing 25 km of ground-level visibility. We combine this with the reference reflectances to simulate a noisy radiance measurement for a typical 30° solar zenith and nadir-viewing observation. We then simulate alternative observations under perturbed atmospheres with changes in two key atmospheric parameters: aerosol scattering, including total optical depth and a categorical type representing different optical properties; and atmospheric column water vapor. We use atmospheric correction coefficients ($S$, $T$, and $\rho_a$) from the nominal case to estimate the reflectance spectra under the perturbed atmosphere, to evaluate the effect of misestimating atmospheric constituents.

Finally, we retrieve geophysical variables from the resulting reflectance spectra using a handful of different techniques designed to span the relevant wavelengths and algorithm options. For terrestrial spectra we estimate NPV and green vegetation fractions by inverting the linear mixture model with the original reference spectra. We also report the estimated green fraction. In the aquatic case, we estimate bottom albedo, bathymetric depth, and chlorophyll-a concentrations by inverting the HOPE model. All numerical model inversions optimize the least squares spectrum fit of the model to the atmospherically corrected observation using a Levenberg–Marquardt solution with finite difference derivative calculations. We exclude saturated channels inside deep water vapor features at 1380 and 1850 nm. As an alternative to these spectrum fitting methods, we also consider band ratio products based on arithmetic with just a few wavelengths. For terrestrial spectra, we calculate the Cellulose Absorption Index, or CAI, as a proxy for the NPV fraction (Nagler et al. 2003), and the Normalized Difference Vegetation Index, or NDVI, as a proxy for vegetation fraction (Carlson and Ripley 1997). In aquatic spectra, we calculate the 440-nm/670-nm band ratio which indicates relative changes in chlorophyll-a (Cannizzaro and Carder 2006).

### 4.3 Simulation Results

Figure 8 illustrates how different perturbations of the estimated atmospheric state influence reflectance retrievals. The left and right panels show terrestrial and aquatic cases,

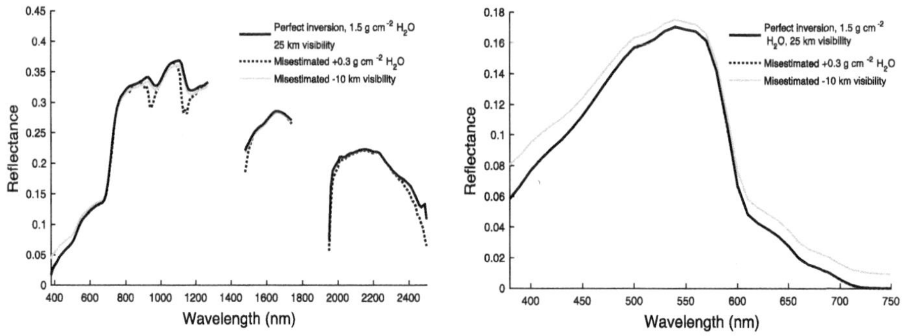


**Fig. 8** Sensitivity of reflectance spectra to different atmospheric state estimation errors

respectively. Errors in columnar $H_2O$ show as localized residuals near water features at 940 and 1140 nm and bend the spectrum shape more generally near the deep water vapor absorptions. This slope change is most apparent in the 2000–2500-nm region that is affected by sporadic unresolved water lines. The aquatic spectra are nearly insensitive to $H_2O$ vapor. In contrast, aerosol errors modify the slope of the spectrum across the entire interval, with a difference that is most notable in short wavelengths. This suggests that many aquatic applications (such as chlorophyll retrievals in the 400–550 nm range) are highly impacted. Vegetation studies may or may not be influenced, depending on their use of visible wavelength features and their sensitivity to magnitude of the near-infrared reflectance.

Figures 9 and 10 quantify this intuition. They show how accuracy in the $H_2O$ and aerosol depth estimates propagate to the derived target quantities. Figure 9 shows error propagation from estimated $H_2O$. The aquatic applications are insensitive, while terrestrial applications show a range of different effects: calculations using NIR channels shorter than 840 nm, such as NDVI, are relatively immune; spectrum fitting degrades gracefully with increasing $H_2O$ error; and the CAI degrades rapidly. AERONET comparisons demonstrate noise equivalent $H_2O$ sensitivities of 0.12 cm (Makarau et al. 2017). Extrapolating this to our reference atmosphere induces a < 3% noise equivalent change in the resulting fractions. Insofar as $H_2O$ retrieval uncertainty is due to systematic modeling effects that are not visible as obvious residuals, the actual impact on the spectrum shape, and the resulting influence on derived fractions, could be smaller in practice.

Figure 10 shows the influence of atmospheric aerosol optical depth (AOD). Aquatic domains are highly sensitive to aerosols, a fact long recognized in the ocean color remote sensing community (Palacios et al. 2015). The phytoplankton algorithms, which rely on short wavelengths, are most affected. In contrast, terrestrial applications are less sensitive, with AOD errors of 0.1 producing no more than 5% change in derived variables. This is probably optimistic since it assumes the optical properties of the aerosol particles are well-characterized, when in practice it is difficult to know their size distribution and absorption properties. Large areas of the tropics have high aerosol optical depth, and global missions must consider these effects.

This brief study is no substitute for an investigation-specific simulation, but it underscores several key aspects of atmospheric correction. First, different domains and algorithms vary widely in their sensitivity to atmospheric state. Some algorithms like spectral unmixing can be skewed by narrowband residuals. These should be inoculated whenever

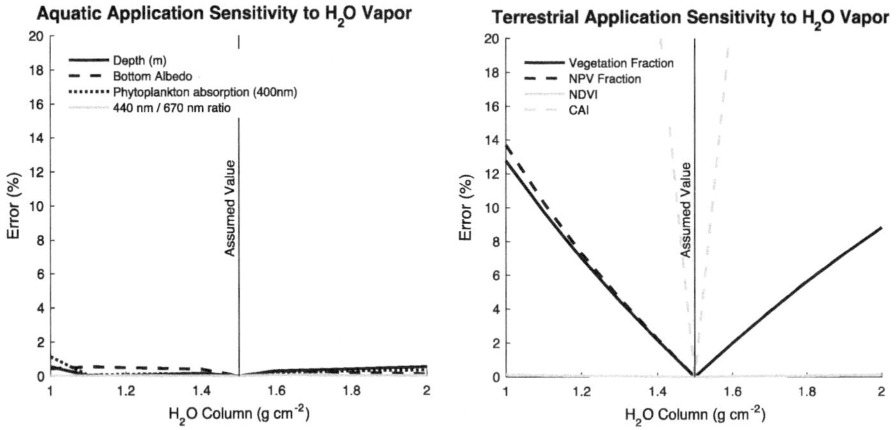


**Fig. 9** Sensitivity of reference applications to misestimating atmospheric water vapor. Errors show the percent departure from the undistorted value of each estimated variable. Aquatic traits are the bottom albedo, phytoplankton absorption, and the band ratio of 440 and 670 nm that is often used as a proxy for chlorophyll-a. Terrestrial traits are the total fractional coverage of vegetation and Non-Photosynthetic Vegetation (NPV), the Normalized Difference Vegetation Index (NDVI), and the Cellulose Absorption Index (CAI)

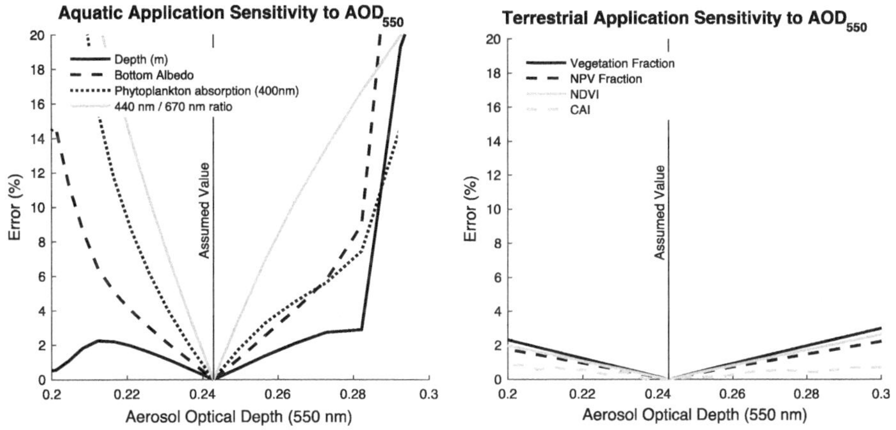


**Fig. 10** Sensitivity of reference applications to misestimating aerosol optical depth. Errors show the percent departure from the undistorted value of each estimated variable

possible by incorporating the reflectance measurement error in an appropriate objective function. It is also widely understood that visible wavelength studies such as aquatic science are highly susceptible to aerosol interference. This also holds to a lesser degree for terrestrial vegetation due to the value of visible wavelengths for characterizing plant pigmentation and productivity. Due to the wide range of aerosol types and AODs that may appear in on global scales, estimating these parameters accurately will be important to achieve consistent results.

# 5 Atmospheric Studies

While imaging spectrometers are primarily used to study the Earth's surface, certain atmospheric measurements have intrinsic scientific value. Solar-reflected imaging spectrometers typically operate at much higher spatial resolutions than purpose-built atmospheric sounding instruments, so they reveal unique fine-scale phenomena such as individual sources of greenhouse gas emissions or small cloud structures. Future orbital instruments will extend the reach of these investigations.

## 5.1 Monitoring of Greenhouse Gas Sources and Urban Air Quality

Most trace gases like $CH_4$ and $CO_2$ are predictable enough for accurate reflectance estimation. However, investigators can estimate these concentrations still more precisely to support specific campaign objectives. Recently, new investigators have pursued detection and mapping of point source greenhouse gas emitters (Frankenberg et al. 2016). Imaging spectrometers are suited for this purpose since they cannot only measure the total magnitude of an emissions plume, but also map its extent and—if spatial resolution is sufficient—trace it to specific source locations on the Earth's surface. In the case of $CH_4$, such techniques could capture events such as spontaneous leaks as well as standard losses from fossil fuel transport and processing or biogenic emissions from animal husbandry. In many industries, a small fraction of facilities (the so-called *superemitters*) are responsible for a disproportionate fraction of total $CH_4$ emissions. Cataloguing these superemitters offers an efficient path to reduce anthropogenic radiative forcing—and often reduces lost revenue for the extractive industry responsible. Thompson et al. (2016a) demonstrated detection of a large $CH_4$ plume from orbit using the Hyperion imaging spectrometer, suggesting that future orbital missions may be capable of detecting the strongest $CH_4$ superemitters.

Band ratios provide insufficient sensitivity to map local variability in greenhouse gases. Instead, investigations typically use the entire spectral measurement with techniques based on differential optical absorption spectroscopy, or DOAS (Thorpe et al. 2017). DOAS has also been used to monitor non-greenhouse trace gases such as $NO_2$, which are important for urban air quality (Tack et al. 2017, Popp et al. 2012). A popular alternative is the matched filter (Thompson et al. 2015b). It uses the fact that for optically thin enhancements, the Beer–Lambert absorption reduces in first-order Taylor expansion to a linear function of optical path length. This allows fast, accurate retrievals of local enhancements relative to the background using a linear projection operator. To avoid confusion by other variability in the scene, it is common to first characterize the statistics of the local radiance background. This enables a whitening normalization to remove the effects of substrate and other incidental atmospheric variability. Figure 11 shows an example from the AVIRIS-NG instrument, acquired in October 2016. A $CO_2$ plume from a powerplant is visible in the enhancement map at right.

## 5.2 Cloud Properties

Many existing orbital instruments can characterize cloud optical properties, such as thermodynamic phase, particle size, and optical depth, at kilometer scales. These properties relate to clouds' lifetimes, evolution, and radiative forcing potential. Imaging spectroscopy observations can provide complementary information on these properties at sub-kilometer spatial resolutions, revealing fine-scale variability that could help parameterize cloud

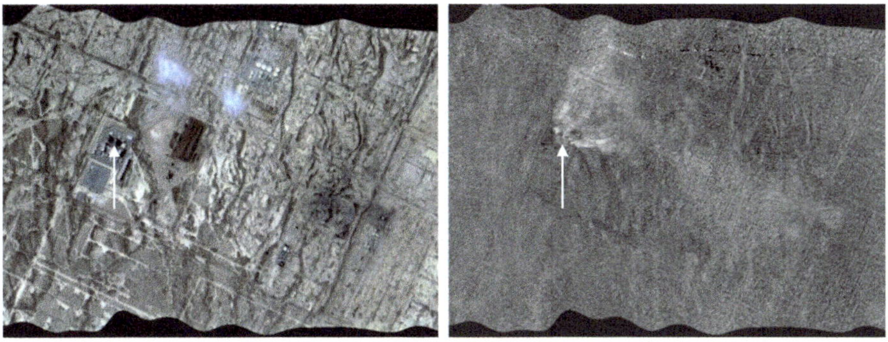

**Fig. 11** AVIRIS-NG image from 29 October 2016. Left: visible RGB image. Right: enhancement of $CO_2$ absorption via the Jacobian-based, columnwise matched filter approach of Thompson et al. (2015b). Arrows indicate a facility emitting a large $CO_2$ plume

formation and evolution models for global climate models (Tan and Storelvmo 2016). The VSWIR interval reveals cloud-top particle size and optical depth, as well as thermodynamic phase (the partitioning of clouds into liquid and ice particles). A recent validation experiment demonstrated that joint spectrum fitting of the three water phases, described above in the context of atmospheric correction, could also retrieve cloud thermodynamic phase with high accuracy (Thompson et al. 2016b). Spectroscopic curve fitting can be less vulnerable to ambiguities in the retrieval than multi-band techniques using band ratios—particularly for mixed cloud cases. The validation experiment observed the same cloud locations within 10 min from both an airborne imaging spectrometer and an in situ aircraft platform. The in situ platform ingested cloud particles through an external port; microscopic high-speed imaging of this airstream provided an authoritative direct classification.

Figure 12 shows an example of co-located spectra. The inset images show microscopic cloud particles: Rough shapes are ice, large annuli are supercooled water droplets, and small points are standard liquid cloud particles. The campaign demonstrated that the relative liquid and ice thicknesses retrieved from remote imaging spectroscopy at 1.4–1.8 μm were highly correlated with the clouds' actual liquid and ice content. Future orbital instruments could perform similar retrievals over wide spatiotemporal intervals—a new measurement modality to monitor cloud thermodynamic phase as climate changes (Thompson et al. 2018a).

## 5.3 Cross-Track Uniformity and Spectral Response

Sharp atmospheric features enable in-flight characterization of the instrument spectral response. This is useful to validate ground spectral calibration, to update calibrations over time, and to estimate response properties under realistic illumination and environmental conditions. One important artifact for some imaging spectrometers is the spectral "smile," also known as spectral non-uniformity or spectral aberration (Mouroulis et al. 2000). It is a consequence of optical aberrations causing the spectrometer entrance slit, representing the across-track swath, to be projected as a curve on the rectilinear detector array. This effect can be estimated in-flight from atmospheric absorption features (Gao et al. 2004, Guanter et al. 2006). These techniques are commonly used operationally, such as the study of Thompson et al. (2015b) which characterized the cross-track deviation in a linear shift through the wavelength change in the oxygen A band. The HATCH code was developed to correct imaging

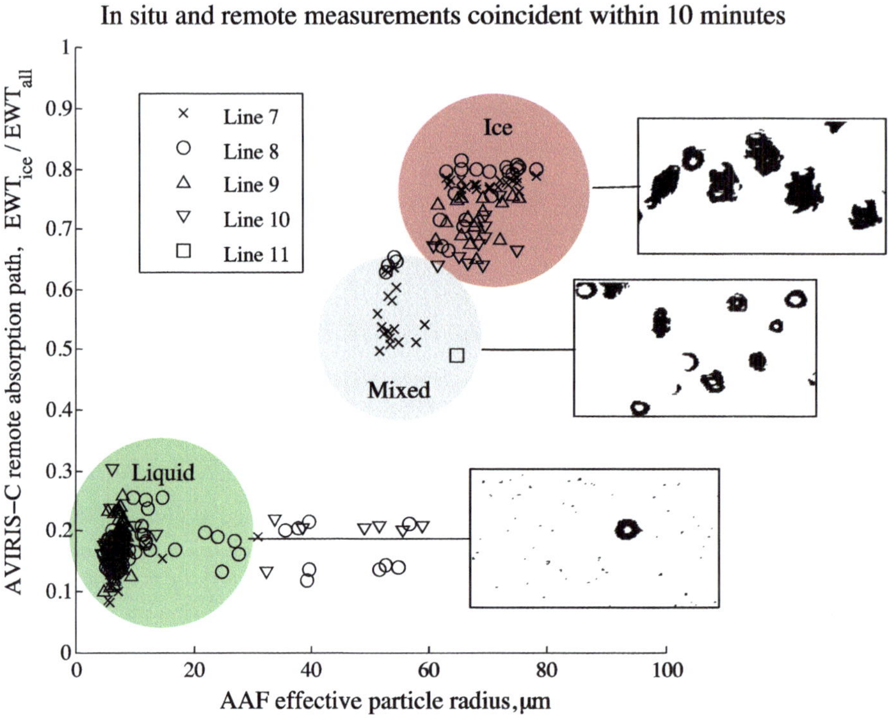


**Fig. 12** Cloud particles measured at identical locations, less than 10 min apart by in situ high-speed imaging and remote imaging spectroscopy. The remotely estimated water fraction of ice ($EWT_{ice}/EWT_{all}$) indicates cloud particle phase. Each point in the plot is a spectrum from one of five flightlines ("lines 7–11"). Image inserts show direct microscopic images of cloud particles obtained in situ, confirming the remote classification; Atmospheric radiation measurement Aerial Facility (AAF). Adapted from Thompson et al. (2016b)

spectrometer data for the non-uniformity (Goetz et al. 2003). An optimized version of spectral non-uniformity compensation with respect to execution time and accuracy was published for operational processing (Richter et al. 2011), where the wavelength shift in across-track direction is modeled as a polynomial function; this feature is an integral part of the atmospheric correction. A similar approach is also implemented in the FLAASH code (Perkins et al. 2012). Atmospheric features can reveal other subtler aspects of the spectral response. Prior research has used sharp features like the oxygen A band to estimate the full width at half maximum (Kuhlmann et al. 2016). Recent studies have estimated more subtle tail deviations of the spectral response function from the ideal Gaussian line shape (Thompson et al. 2018b). Such features affect the response three to four orders of magnitude below the peak, but can still influence sensitive retrievals such as surface pressure estimates.

## 6 Challenges and Future Directions

In many respects, model-based atmospheric correction for VSWIR imaging spectrometers is a mature technology, with decades of regular operational use on airborne instruments. Under favorable conditions, investigations regularly achieve reflectance accuracy to within a 2–5% or better, as judged by many independent validation campaigns. However, the general atmospheric correction problem is far from solved. The best accuracies reported in prior validation studies belie significant planning, investigator field work and analyst attention to the atmospheric correction process. They may not represent the average-case result for a fully automated production system, which must handle global observations over many different atmospheres and challenging weather conditions. Global imaging spectroscopy will require sustained attention to atmospheric correction accuracy and algorithm development in order to meet the promise of improved access to data, reduce institutional "barriers to entry" for imaging spectroscopy products, and enable bias-free science investigations spanning many regions and climatologies. In addition, new and sensitive applications such as fine-scale ecosystem process mapping will require very high accuracy and consistency in atmospheric correction. Such applications rely on subtler spectral signatures, raising accuracy standards.

Several topics warrant additional attention for global spectroscopic studies. Arguably, trace gas estimation is now accurate enough that it is not a significant contributor to error in retrieved reflectance. In contrast, consistent aerosol estimation remains a challenge. Not only are the perturbations of aerosols spectrally smooth, and therefore more difficult to disentangle from surface reflectance effects, but they are also highly variable with respect to aerosol optical properties. Additionally, many current methods for aerosol estimation rely on shaded pixels that may not be apparent at orbital 30 m ground sampling distances. Applications like shallow water coastal studies and vegetation studies can be sensitive to slope changes in the aerosol-sensitive visible wavelengths. Fortunately, these applications also offer fairly strong constraints on reasonable surface reflectance shapes, enabling statistical recognition of aerosol contaminated data. This points toward the utility of future aerosol estimation methods that combine surface and atmosphere models, such as the Bayesian approach of Frouin and Pelletier (2015), or the *optimal estimation* approach used in atmospheric sounding instruments (Rodgers 2000). Optimal estimation was recently demonstrated with imaging spectrometer flight data, showing the feasibility of this approach (Thompson et al. 2018c). These unified inversions of surface and atmosphere hold several potential advantages: They could add statistical rigor, seamlessly incorporating domain knowledge or climatology in the form of statistical priors; they could combine surface and atmospheric information across the VSWIR interval, potentially resolving the problem of indeterminacy in smooth perturbations; and they could fit a joint solution encompassing multiple atmospheric correction steps that have conventionally been handled with independent sequential operations.

Another fruitful topic for further research is the characterization and propagation of uncertainties through the atmospheric correction process. Most current downstream algorithms rely on spectrum fitting and matching presuming uniform errors (and often, uncorrelated errors) across wavelengths. In practice, atmospheric correction uncertainties manifest more strongly in some channels than others and may induce spurious correlations. Reporting such errors would permit downstream algorithms to incorporate them to improve resilience against pathological atmospheric effects. Atmospheric correction should ideally respect the errors and correlations in the input data—the underlying instrument radiances.

Finally, there has been increasing recognition in the potential for VSWIR atmospheric science investigations by imaging spectrometers. The ability to image atmospheric content at sub-kilometer spatial resolutions opens new applications including monitoring anthropogenic greenhouse gases, cloud studies, and process studies of ecosystem gas exchange and evapotranspiration. Despite the long history of success to date in atmospheric correction,

VSWIR atmospheric studies remain an exciting, relevant, and challenging domain for ongoing study.

**Acknowledgements** We acknowledge the critical support and facilitation of the International Space Science Institute (ISSI), Bern, Switzerland. A portion of this research was performed at the Jet Propulsion Laboratory, California Institute of Technology. AVIRIS-C and AVIRIS-NG are supported by National Aeronautics and Space Administration Earth Science, Science Mission Directorate. U.S. Federal Government support acknowledged. 

## References

ASTER (2018) The advanced spaceborne thermal emission and reflection radiometer global digital elevation map. https://asterweb.jpl.nasa.gov/gdem.asp. Last access 17 May 2018

Bachmann M, Makarau A, Segl K, Richter R (2015) Estimating the influence of spectral and radiometric calibration uncertainties on EnMAP data products—examples for ground reflectance retrieval and vegetation indices. Remote Sens 7(8):10689–10714

Berk A et al (2016a) Algorithm theoretic basis document (ATBD) for next generation MODTRAN®. Spectral Sciences Inc, Burlington

Berk A, van den Bosch J, Hawes F, Perkins T, Conforti PF, Anderson GP, Kennett RG, Acharya PK (2016b). MODTRAN®6.0.0 user's manual (revision 5). Spectral Sciences, Inc., Burlington. SSI-TR-685

Bernstein LS, Adler-Golden SM, Sundberg RL, Levine RY, Perkins TC, Berk A, Ratkowski AJ, Felde G, Hoke ML (2005) Validation of the QUick Atmospheric Correction (QUAC) algorithm for VNIR-SWIR multi-and hyperspectral imagery. In: Defense and security. International Society for Optics and Photonics, pp 668–678

Boardman JW (1998) Post-ATREM polishing of AVIRIS apparent reflectance data using EFFORT: a lesson in accuracy versus precision. In: Summaries of the 7th JPL airborne earth science workshop, JPL Publication 97–21, 1:53

Bruegge CJ, Conel JE, Margolis JS, Green RO, Toon GC, Carrere V, Holm RG, Hoover G (1990) In-situ atmospheric water-vapor retrieval in support of AVIRIS validation. In: Proceedings of SPIE - The International Society for Optical Engineering, vol 1298

Cannizzaro JP, Carder KL (2006) Estimating chlorophyll a concentrations from remote-sensing reflectance in optically shallow waters. Remote Sens Environ 101(2006):13–24

Carlson TN, Ripley DA (1997) On the relation between NDVI, fractional vegetation cover, and leaf area index. Remote Sens Environ 62(3):241–252

Clark RN, Swayze GA, Livo KE, Kokaly RF, Sutley SJ, Dalton JB, McDougal RR, Gent CA (2003) Imaging spectroscopy: earth and planetary remote sensing with the USGS Tetracorder and expert systems. J Geophys Res 108:5131. https://doi.org/10.1029/2002JE001847,E12

Conel JE, Green RO, Vane G, Bruegge CJ, Alley RE (1987) AIS-2 radiometry and a comparison of methods for the recovery of ground reflectance. In: Vane G (ed) Proceedings of the 3rd airborne imaging spectrometer data analysis workshop, JPL Publ. volume 87–30 Jet Propulsion Laboratory, Pasadena, CA, pp 18–47

Frankenberg C, Thorpe A, Thompson DR, Hulley G, Kort E, Vance N, Borchard J, Krings T, Gerilowski K, Sweeney C, Conley S, Bue B, Aubrey A, Hook S, Green RO (2016) Airborne methane remote measurements reveal heavy-tail flux distribution in Four Corners region. Proc Natl Acade Sci 113(35):9734–9739

Fraser RS, Kaufman YJ (1985) The relative importance of aerosol scattering and absorption in remote sensing. IEEE J Geosci Remote Sens GE-23:525–633

Frouin Robert, Pelletier Bruno (2015) Bayesian methodology for inverting satellite ocean-color data. Remote Sens Environ 159:332–360

Fu Q, Liou KN (1992) On the correlated k-distribution method for radiative transfer in nonhomogeneous atmospheres. J Atmos Sci 49(22):2139–2156

Gao B-C, Goetz AF (1990) Column atmospheric water vapor and vegetation liquid water retrievals from airborne imaging spectrometer data. J Geophys Res Atmos 95(D4):3549–3564

Gao B-C, Goetzt AF (1995) Retrieval of equivalent water thickness and information related to biochemical components of vegetation canopies from AVIRIS data. Remote Sens Environ 52(3):155–162

Gao B-C, Kaufman YJ (1995) Selection of the 1.375-μm MODIS channel for remote sensing of cirrus clouds and stratospheric aerosols from space. J Atmos Sci 52:4231–4237
Gao B-C, Liu M (2013) A fast smoothing algorithm for post-processing of surface reflectance spectra retrieved from airborne imaging spectrometer data. Sensors 13:13879–13891. https://doi.org/10.3390/s131013879
Gao B-C, Heidebrecht KB, Goetz AFH (1993) Derivation of scaled surface reflectances from AVIRIS data. Remote Sens Environ 44:165–178
Gao B-C et al (1998) Correction of thin cirrus path radiance in the 0.4–1.0 μm spectral region using the sensitive 1.375-μm cirrus detecting channel. J Geophys Res 103:32169–32176
Gao B-C, Montes MJ, Ahmad Z, Davis CO (2000) Atmospheric correction algorithm for hyperspectral remote sensing of ocean color from space. Appl Opt 39(6):887–896
Gao B-C, Yang P, Han W, Li R-R, Wiscombe WJ (2002) An algorithm using visible and 1.38-micron channels to retrieve cirrus cloud reflectances from aircraft and satellite data. IEEE Trans. Geosci. Remote Sensing 40:1659–1668
Gao B-C, Montes MJ, Davis CO (2004) Refinement of wavelength calibrations of hyperspectral imaging data using a spectrum-matching technique. Remote Sens Environ 90:424–433
Gao B-C, Montes MJ, Davis CO, Goetz AF (2009) Atmospheric correction algorithms for hyperspectral remote sensing data of land and ocean. Remote Sens Environ 113:S17–S24
Goetz AFH, Kindel BC, Ferri M, Qu Z (2003) HATCH: results from simulated radiances, AVIRIS and Hyperion. IEEE TGRS 41:1215–1222
Green RO, Carrere V, Conel JE (1989) Measurement of atmospheric water vapor using the Airborne Visible/Infrared Imaging Spectrometer. In: Proceedings of the ASPRS conference on image processing, Reno, NV
Green RO, Conel JE, Roberts DA (1993) Estimation of aerosol optical depth, pressure elevation, water vapor, and calculation of apparent surface reflectance from radiance measured by the airborne visible/infrared imaging spectrometer (AVIRIS) using a radiative transfer code. Optical Engineering and Photonics in Aerospace Sensing. International Society for Optics and Photonics. pp 2–11
Green RO, Painter TH, Roberts DA, Dozier J (2006) Measuring the expressed abundance of the three phases of water with an imaging spectrometer over melting snow. Water Resour Res 42(10):W10402
Guanter L, Richter R, Moreno J (2006) Spectral calibration of hyperspectral imagery using atmospheric absorption features. Appl Opt 45:2360–2370
Guanter L, Estellés V, Moreno J (2007) Spectral calibration and atmospheric correction of ultra-fine spectral and spatial resolution remote sensing data. Application to CASI-1500 data. Remote Sens Environ 109(1):54–65
Guanter L, Gómez-Chova L, Moreno J (2008) Coupled retrieval of aerosol optical thickness, columnar water vapor and surface reflectance maps from ENVISAT/MERIS data over land. Remote Sens Environ 112(6):2898–2913
Hagolle O, Huc M, Villa Pascual D, Dedieu G (2015) A multi-temporal and multi-spectral method to estimate aerosol optical thickness over land, for the atmospheric correction of FormoSat-2, LandSat, VENμS and Sentinel-2 images. Remote Sens 7(3):2668–2691
Herold M, Roberts DA, Gardner ME, Dennison PE (2004) Spectrometry for urban area remote sensing—development and analysis of a spectral library from 350 to 2400 nm. Remote Sens Environ 91(3):304–319
Hu B, Lucht W, Strahler AH (1999) The interrrelationship of atmospheric correction of reflectances and surface BRDF retrieval: a sensitivity study. IEEE Trans Geosci Remote Sens 37:724–738
Jensen DJ, Simard M, Cavanaugh KC, Thompson DR (2018) Imaging spectroscopy BRDF correction for mapping Louisiana's coastal ecosystems. IEEE Trans Geosci Remote Sens 56(3):1739–1748
Kaufman YJ, Sendra C (1988) Algorithm for automatic atmospheric corrections to visible and near-IR satellite imagery. Int J Remote Sens 9(8):1357–1381
Kaufman YJ, Wald A, Remer LA, Gao B-C, Li RR, Flynn L (1997) The MODIS 2.1-μm channel—correlation with visible reflectance for use in remote sensing of aerosol. IEEE Trans Geosci Remote Sens 35:1286–1298
Kobayashi S, Sanga-Ngoie K (2008) The integrated radiometric correction of optical remote sensing imageries. Int J Remote Sens 29:5957–5985
Kokaly RF, Asner GP, Ollinger SV, Martin ME, Wessman CA (2009) Characterizing canopy biochemistry from imaging spectroscopy and its application to ecosystem studies. Remote Sens Environ 113:S78–S91
Kruse FA (1988) Use of airborne imaging spectrometer data to map minerals associated with hydrothermally altered rocks in the northern Grapevine Mountains, Nevada and California. Remote Sens Environ 24:31–51

Kruse FA (2004) Comparison of ATREM, ACORN, and FLAASH atmospheric corrections using low-altitude AVIRIS data of Boulder, CO. In: Summaries of 13th JPL Airborne Geoscience Workshop, Jet Propulsion Laboratory, Pasadena, CA

Kuhlmann G, Hueni A, Damm A, Brunner D (2016) An algorithm for in-flight spectral calibration of imaging spectrometers. Remote Sens 8:1017

Lee Z, Carder KL, Mobley CD, Steward RG, Patch JS (1998) Hyperspectral remote sensing for shallow waters. I. A semianalytical model. Appl Opt 37(27):6329–6338

Lee Z, Carder KL, Mobley CD, Steward RG, Patch JS (1999) Hyperspectral remote sensing for shallow waters: 2. Deriving bottom depths and water properties by optimization. Appl Opt 38(18):3831–3843

Li A, Wang Q, Bian J, Lei G (2015) An improved physics-based model for topographic correction of Landsat TM images. Remote Sens 7:6296–6319

Loyola DG, Coldewey-Egbers RM, Dameris M, Gamy H, Stenke A, Van Roozendael M, Lerot C, Balis D, Koukouli M (2009) Global long-term-monitoring of the ozone layer—a prerequisite for predictions. Int J Remote Sens 30:4295–4318

Makarau A, Richter R, Müller R, Reinartz P (2014) Haze detection and removal in remotely sensed multi-spectral imagery. IEEE TGRS 52:5895–5905

Makarau A, Richter R, Schläpfer D, Reinartz P (2016) Combined haze and cirrus removal for mulispectral imagery. IEEE GRSL 13:379–383

Makarau A, Richter R, Schläpfer D, Reinartz P (2017) APDA water vapor retrieval validation for Sentinel-2 imagery. IEEE GRSL 14:227–231

Matthew MW, Adler-Golden SM, Berk A, Felde G, Anderson GP, Gorodetzky D, Paswaters S, Shippert M (2002) Atmospheric correction of spectral imagery: evaluation of the FLAASH algorithm with AVIRIS data. In: Proceedings of the 31st applied imagery pattern recognition workshop. IEEE, pp 157–163

Mouroulis P, Green RO, Chrien TG (2000) Design of pushbroom imaging spectrometers for optimum recovery of spectroscopic and spatial information. Appl Opt 39:2210–2220

Mumby PJ, Skirving W, Strong AE, Hardy JT, LeDrew EF, Hochberg EJ et al (2004) Remote sensing of coral reefs and their physical environment. Mar Pollut Bull 48(3):219–228

Nagler PL, Inoue Y, Glenn EP, Russ AL, Daughtry CST (2003) Cellulose absorption index (CAI) to quantify mixed soil–plant litter scenes. Remote Sens Environ 87(2):310–325

Palacios SL, Kudela RM, Guild LS, Negrey KH, Torres-Perez J, Broughton J (2015) Remote sensing of phytoplankton functional types in the coastal ocean from the HyspIRI Preparatory Flight Campaign. Remote Sens Environ 167:269–280

Perkins T, Adler-Golden S, Matthew MW, Berk A, Bernstein LS, Lee J, Fox M (2012) Speed and accuracy improvements in FLAASH atmospheric correction of hyperspectral imagery. Opt Eng 51:111707–111708

Popp C, Brunner D, Damm A, Van Roozendael M, Fayt C, Buchmann B (2012) High-resolution NO2 remote sensing from the Airborne Prism EXperiment (APEX) imaging spectrometer. Atmos Meas Tech 5(9):2211–2225

Reinersman PN, Carder KL, Chen RF (1998) Satellite-sensor calibration verification with the cloud-shadow method. Appl Opt 37:5541–5549

Riano D, Chuvieco E, Salas J, Aguado I (2003) Assessment of different topographic corrections in Landsat-TM data for mapping vegetation types. IEEE TGRS 41:1056–1061

Richter R (1998) Correction of satellite imagery over mountainous terrain. Appl Opt 37:4004–4015

Richter R, Schlaepfer D (2002) Geo-atmospheric processing of airborne imaging spectrometry data, Part 2: atmospheric/topographic correction. Int J Remote Sens 23(13):2631–2649

Richter R, Kellenberger T, Kaufmann H (2009) Comparison of topographic correction methods. Remote Sens 1:184–196

Richter R, Schläpfer D, Müller A (2011) Operational atmospheric correction for imaging spectrometers accounting for the smile effect. IEEE TGRS 49:1772–1780

Richter R, Heege T, Kiselev V, Schläpfer D (2014) Correction of ozone influence on TOA radiance. Int J Remote Sens 35:8044–8056

Roberts DA, Yamaguchi Y, Lyon R (1986) Comparison of various techniques for calibration of AIS data. In: Vane G, Goetz AFH (eds) Proceedings of the 2nd airborne imaging spectrometer data analysis workshop, JPL Publication 86-35, 21–30, Jet Propulsion Lab, Pasadena, CA

Roberts DA, Yamaguchi Y, Lyon R (1986) Comparison of various techniques for calibration of AIS data. In: Vane G, Goetz AFH (eds) Proceedings of the 2nd Airborne imaging spectrometer data analysis workshop, JPL Publication, vol 86–35, Jet Propulsion Lab, Pasadena, CA, pp 21–30

Rodgers CD (2000) Inverse methods for atmospheric sounding: theory and practice. World Scientific, Singapore

Sandmeier S, Itten KI (1997) A physically-based model to correct atmospheric and illumination effects in optical satellite data of rugged terrain. IEEE TGRS 35:708–717
Schaepman-Strub G, Schaepman ME, Painter TH, Dangel S, Martonchik JV (2006) Reflectance quantities in optical remote sensing—definitions and case studies. Remote Sens Environ 103(1):27–42
Schläpfer D, Richter R (2011) Spectral polishing of high resolution imaging spectroscopy data. In: Proceedings of the 7th SIG-IS workshop on imaging spectroscopy, Edinburgh, UK, p 7. http://www.daniel-schlaepfer.ch/pdf/Schlaepfer_IS2011_polish.pdf
Schläpfer D, Borel CC, Keller J, Itten KI (1998) Atmospheric precorrected differential absorption technique to retrieve columnar water vapor. Remote Sens Environ 65(3):353–366
Schläpfer D, Richter R, Damm A (2013) Correction of shadowing in imaging spectroscopy data by quantification of the proportion of diffuse illumination. In: Presented at the 8th EARSeL SIG-IS Workshop on Imaging Spectroscopy, Nantes, FR, pp 10. http://www.daniel-schlaepfer.ch/pdf/Schlaepfer_Earsel2013_Shadow.pdf
Schläpfer D, Richter R, Feingersh T (2015) Operational BRDF effects correction for wide-field-of-view optical scanners (BREFCOR). IEEE Trans Geosci Remote Sens 53(4):1855–1864
Shepherd JD, Dymond JR (2003) Correcting satellite imagery for the variance of reflectance and illumination with topography. Int J Remote Sens 24:3503–3514
Soenen SA, Peddle DR, Coburn CA (2005) SCS + C: a modified sun-canopy-sensor topographic correction in forested terrain2. IEEE TGRS 43:2148–2159
SRTM (2018) Shuttle radar topography mission digital elevation model. https://lta.cr.usgs.gov/SRTM1Arc. Last access 17 May 2018
Stamnes K, Tsay SC, Wiscombe W, Jayaweera K (1988) Numerically stable algorithm for discrete-ordinate-method radiative transfer in multiple scattering and emitting layered media. Appl Opt 27(12):2502–2509
Tack F, Merlaud A, Iordache M-D, Danckaert T, Yu H, Fayt C, Meuleman K, Deutsch F, Fierens F, Van Roozendael M (2017) High-resolution mapping of the $NO_2$ spatial distribution over Belgian urban areas based on airborne APEX remote sensing. Atmos Meas Tech 10:1665–1688
Tan I, Storelvmo T (2016) Sensitivity study on the influence of cloud microphysical parameters on mixed-phase cloud thermodynamic phase partitioning in CAM5. J Atmos Sci 73(2):709–728
Tanré D, Herman M, Deschamps PY (1981) Influence of the background contribution upon space measurements of ground reflectances. Appl Opt 20:3676–3684
Tanré D, Herman M, Deschamps PY (1983) Influence of the atmosphere on space measurements of directional properties. Appl Opt 22(5):733–741
Teillet PM, Guindon B, Goodenough DG (1982) On the slope-aspect correction of multispectral scanner data. Can J Remote Sens 8:84–106
Thompson DR, Gao BC, Green RO, Roberts DA, Dennison PE, Lundeen SR (2015a) Atmospheric correction for global mapping spectroscopy: ATREM advances for the HyspIRI preparatory campaign. Remote Sens Environ 167:64–77
Thompson DR, Leifer I, Bovensmann H, Eastwood M, Fladeland M, Frankenberg C, Gerilowski K, Green RO, Kratwurst S, Krings T, Luna B, Thorpe A (2015b) Real-time remote detection and measurement for airborne imaging spectroscopy: a case study with methane. Atmos Meas Tech 8:4383–4397
Thompson DR, Siedel F, Gao B-C, Gierach M, Kudela R, Green RO, Mouroulis P (2015c) Optimizing solar irradiance for coastal spectroscopy. Geophys Res Lett 42:4116–4123
Thompson DR, Thorpe AK, Frankenberg C, Green RO, Duren R, Hollstein A, Guanter L, Middleton E, Ong L, Ungar S (2016a) Orbital measurement of the Aliso Canyon $CH_4$ super-emitter. Geophys Res Lett 43:6571–6578
Thompson DR, McCubbin I, Gao B-C, Green RO, Matthews AA, Mei F, Meyer K, Platnick S, Schmid B, Tomlinson J, Wilcox E (2016b) Measuring cloud thermodynamic phase with shortwave infrared imaging spectroscopy. J Geophys Res Atmos 121(15):9174–9190
Thompson DR, Boardman JW, Eastwood ML, Green RO, Haag JM, Mouroulis P, Van Gorp BE (2018a) Imaging spectrometer stray spectral response: in-flight characterization, correction, and validation. Remote Sens Environ 204:850–860
Thompson DR, Kahn BH, Green RO, Chien SA, Middleton EM, Tran DQ (2018b) Global spectroscopic survey of cloud thermodynamic phase at high spatial resolution, 2005–2015. Atmos Meas Tech 11:1019–1030. https://doi.org/10.5194/amt-11-1019-2018
Thompson DR, Natraj V, Green RO, Helmlinger MC, Gao B-C, Eastwood ML (2018c) Optimal estimation for imaging spectrometer atmospheric correction. Remote Sens Environ. https://doi.org/10.1016/j.rse.2018.07.003
Thorpe AK, Frankenberg C, Thompson DR, Duren RM, Aubrey AD, Bue BD, Green RO, Gerilowski K, Krings T, Borchardt J, Kort EA, Sweeney C, Conley S, Roberts DA, Dennison PE (2017) Airborne

DOAS retrievals of methane, carbon dioxide, and water vapor concentrations at high spatial resolution: application to AVIRIS-NG. Atmos Meas Tech 10(10):3833–3850
United States National Bureau of Standards, Nicodemus FE (1977) Geometrical considerations and nomenclature for reflectance, vol 160. US Department of Commerce, National Bureau of Standards
Ustin SL, Roberts DA, Gamon JA, Asner GP, Green RO (2004) Using imaging spectroscopy to study ecosystem processes and properties. Bioscience 54(6):523–534
Vermote EF, El-Saleous N, Justice CO, Kaufman YJ, Privette JL, Remer L, Roger JC, Tanré D (1997a) Atmospheric correction of visible to middle infrared EOS-MODIS data over land surface: background, operational algorithm and validation. J Geophys Res 102:17131–17141
Vermote EF, Tanré D, Deuzé JL, Herman M, Morcrette JJ (1997b) Second simulation of the satellite signal in the solar spectrum, 6S: an overview. IEEE Trans Geosci Remote Sens 35:675–686
Weyermann J, Damm A, Kneubuhler M, Schaepman ME (2014) Correction of reflectance anisotropy effects of vegetation on airborne spectroscopy data and derived products. IEEE Trans Geosci Remote Sens 52(1):616–627
Yamamoto H, Tsuchida S, Yoshioka H (2008) A study on ASTER/MODIS radiometric and atmospheric correction. IGARSS. https://doi.org/10.1109/IGARSS.2008.4779982
Zhang Y, Guindon B, Cihlar J (2002) An image transform to characterize and compensate for spatial variations in thin cloud contamination of Landsat images. Remote Sens Environ 82(2–3):173–187

Surveys in Geophysics (2019) 40:361–399
https://doi.org/10.1007/s10712-019-09524-0

# Imaging Spectroscopy for Soil Mapping and Monitoring

S. Chabrillat[1] · E. Ben-Dor[2] · J. Cierniewski[3] · C. Gomez[4] · T. Schmid[5] · B. van Wesemael[6]

Received: 16 February 2018 / Accepted: 28 February 2019 / Published online: 20 March 2019


**Abstract**

There is a renewed awareness of the finite nature of the world's soil resources, growing concern about soil security and significant uncertainties about the carrying capacity of the planet. Regular assessments of soil conditions from local through to global scales are requested, and there is a clear demand for accurate, up-to-date and spatially referenced soil information by the modelling scientific community, farmers and land users, and policy- and decision-makers. Soil and imaging spectroscopy, based on visible–near-infrared and short-wave infrared (400–2500 nm) spectral reflectance, has been shown to be a proven method for the quantitative prediction of key soil surface properties. With the upcoming launch of the next generation of hyperspectral satellite sensors in the next years, a high potential to meet the demand for global soil mapping and monitoring is appearing. In this paper, we briefly review the basic concepts of soil spectroscopy with a special attention to the effects of soil roughness on reflectance and then provide a review of state of the art, achievements and perspectives in soil mapping and monitoring based on imaging spectroscopy from air- and spaceborne sensors. Selected application cases are presented for the modelling of soil organic carbon, mineralogical composition, topsoil water content and characterization of soil crust, soil erosion and soil degradation stages based on airborne and simulated spaceborne imaging spectroscopy data. Further, current challenges, gaps and new directions toward enhanced soil properties modelling are presented. Overall, this paper highlights the potential and limitations of multiscale imaging spectroscopy nowadays for soil mapping and monitoring, and capabilities and requirements of upcoming spaceborne sensors as support for a more informed and sustainable use of our world's soil resources.

**Keywords** Soil mapping and monitoring · Imaging spectroscopy · Hyperspectral · Soil organic carbon · Soil mineralogical composition · Surface roughness · Soil moisture · Vegetation cover · Spaceborne instruments

✉ S. Chabrillat
sabine.chabrillat@gfz-potsdam.de

Extended author information available on the last page of the article

## 1 Introduction

"All natural resources… are soil or derivatives of soil. Farms, ranges, crops, and livestock, forests, irrigation water and even water power resolve themselves into questions of soil. Soil is therefore the basic natural resource" said Aldo Leopold in 1921 "Erosion and Prosperity" (Meine and Knight 1999). Nearly all of the food, fuel and fibres used by humans are produced on soil. Soil is also essential for water and ecosystem health. It is second only to the oceans as a global carbon sink, with an important role in the potential slowing of climate change. Several soil functions depend on a multitude of soil organisms, which makes soil an important part of our biodiversity. Nowadays, in the face of rapidly growing population, there is a renewed awareness of the finite nature of the world's soil resources (Hartemink and McBratney 2008), growing concern about soil security (FAO and ITPS 2015) and significant uncertainties about the carrying capacity of the planet (i.e. the number of people that the Earth can support (UNEP 2012)). Hartemink (2008) further acknowledged "Soils are back on the global agenda". This growing concern has been answered with a growing number of soil policies and regulations around the world concerned with, e.g., increasing soil degradation and loss of organic carbon in topsoils and aiming at more soil management and soil protection such as the EU Soil Thematic Strategy and Soil Framework Directive. The European Commission recognized that soil resources in many parts of Europe are being over-exploited, degraded and irreversibly lost due to inappropriate land management practices, industrial activities and land-use changes that lead to soil sealing, contamination, erosion and loss of organic carbon (JRC 2012). Soil scientists are being challenged to provide assessments of soil conditions from local to global scales (Grunwald et al. 2011; Arrouays et al. 2017). However, only a few countries have the necessary survey and monitoring programs to meet these new needs and existing global data sets are out of date. For example, the state-of-the-art Harmonized World Soil Database (FAO 2012) providing up-to-date information on world soil resources at approximately 1 km scale (30 arc-second database) was last updated in 2013 and recognizes that the reliability of the information contained in the database is variable. A particular issue is the clear demand for a new regional to global coverage with accurate, up-to-date and spatially referenced soil information as expressed by the scientific community, farmers and land users, and policy- and decision-makers (EC 2006).

In this regard, optical remote sensing observations and in particular reflectance spectroscopy at the remote sensing scale, referred to as imaging spectroscopy (IS), or hyperspectral imaging, have been shown to be powerful techniques for the quantitative determination and modelling of a range of soil properties. These soil properties include topsoil mineralogical composition such as soil organic carbon (SOC) content, textural composition, iron or carbonate content, etc., and physical attributes (e.g., Ben-Dor et al. 2009). The attractiveness of imaging spectroscopy is that measurements are rapid and estimates of soil properties are inexpensive compared to conventional soil analyses, as it exploits the information carried out by the visible and near-infrared (Vis–NIR: 400–1100 nm) and shortwave infrared (SWIR: 1100–2500 nm) part of the electromagnetic spectrum (Goetz et al. 1985). IS has been used since more than 20 years in various soil applications such as evaluation and monitoring of soil quality and soil function (e.g., soil moisture and carbon storage), soil fertility and soil threats (e.g., acidification and erosion) and soil pedogenesis (i.e. soil formation and evolution). Further, soil degradation (salinity, erosion and deposition), soil mapping and classification, soil genesis and formation, soil contamination and soil hazards (swelling soils) are also important soil science issues nowadays examined with IS, enlarging the soil spectroscopy into the spatial domain from mainly airborne platforms (e.g., see review of Ben-Dor et al. 2018). Research on quantitative soil spectroscopy for the prediction of soil properties has largely benefited from technological and methodological developments over the past decades. The availability of new high signal-to-noise ratio airborne hyperspectral sensors allowed the delivery, at remote sensing scale, of laboratory-like reflectance data. Simultaneously, developments in multivariate statistics and chemometrics opened significant new possibilities toward soil spectral modelling and quantitative analyses of the physical and biochemical composition of the Earth's soil based on spectral reflectance.

In the upcoming future, a large availability of high signal-to-noise ratio satellite imaging spectrometers is expected. Several target missions having high spatial resolution and limited coverage per day are soon to be launched such as the Italian PRISMA (PRecursore IperSpettrale della Missione Applicativa) (Loizzo et al. 2018), launch 2019, 30 m; the German EnMAP (Environmental Mapping and Analysis Program) (Guanter et al. 2015), launch 2020, 30 m; the Japanese HISUI (Hyperspectral Imager Suite) (Matsunaga et al. 2018), to be put on the International Space Station in 2019, 30 m; and the Italy–Israeli SHALOM (Spaceborne Hyperspectral Applicative Land and Ocean Mission) (Ben-Dor et al. 2014), launch 2024, possibly 10 m. Among these missions, the open data policy of the German EnMAP mission is worth noting. Furthermore, several global mapping missions are planned or under study for the upcoming future such as the NASA-SBG (Surface Biology and Geology), former HyspIRI mission (Lee et al. 2015; Green 2018), HySpex2 proposed as Earth Explorer Mission ESA (Briottet et al. 2017) and the Sentinel-10/CHIME satellite proposed as ESA candidate mission (Copernicus Hyperspectral Imaging Mission for the Environment) (Rast et al. 2019). These satellites with their variable spatial coverage and different ground sampling distances will represent for soil scientists a major step toward global soil mapping and monitoring as a response to the need for accurate, up-to-date information on the world's state of soils.

Nevertheless, to be able to answer that demand and to reach the full potential of imaging spectroscopy from orbital utilization for soil mapping, challenges have been identified and, for example, linked to limitations in reference data availability (e.g., global standardized soil spectral libraries databases) and in methodological approaches and tools adequate to process the spectral data into practical soil model solutions that are globally applicable (e.g., Ben-Dor et al. 2018). An area of active research nowadays is thus on the demonstration of the potential and limitations of hyperspectral imagery for soil mapping and monitoring from airborne to spaceborne scale and on the development of enhanced databases and methods to be ready for upcoming hyperspectral satellite launches with adapted strategies and tools for the computation and delivery of global soil maps.

In this frame, we present in this paper a timely review of state of the art, challenges and limitations of imaging spectroscopy for soil applications, for young and senior research scholars, undergraduate and graduate students in Earth sciences and remote sensing, soil scientists, institutional and industrial soil entities that are concerned with the future use of remote sensing data for soil mapping and monitoring. This paper is divided into three sections. First, a review of basic concepts of soil spectroscopy for soil properties' determination is presented, including a thorough review of the effect of soil roughness on reflectance. Then, a review of the applications of imaging spectroscopy for the mapping of soil properties is presented, including selected application cases demonstrating the potential and limitations of imaging spectroscopy for the mapping of SOC, common soil properties, soil moisture, soil crust, soil erosion and degradation. Finally, a review of current challenges and gaps related to the use of air- to spaceborne imaging spectroscopy data is presented and discussed in view of future avenues of research, perspectives and user requirements for upcoming satellites.

# 2 Theory and Concepts

## 2.1 Principles of Soil Spectroscopy for Soil Properties Identification and Modelling

The soil reflectance spectrum ($\rho$) is a collection of values obtained at every spectral band ($\lambda$) from the ratio of radiance ($E$) and irradiance ($L$) fluxes across most of the spectral region of the solar emittance function. The reflectance values are traditionally described, from a practical standpoint, by a relative ratio against a perfect reflector spectrum measured at the same geometry and position of the soils (Palmer 1982; Baumgardner et al. 1985; Jackson et al. 1987). The reflectance information is used to identify material by the nature of the reflectance spectrum it provides. The more wavelengths involved in the measurement scheme, the more information can be obtained. The nature of the spectrum is composed of absorption features of chemical constituents ("peaks") (e.g., absorption of OH of water molecules) and overall spectral shape of the physical properties ("albedo") (e.g., particle size) (Ben-Dor and Banin 1995a, b).

In soils, the optical activity of chemical chromophores (i.e. those parts of a molecule responsible for its colour) is due to vibration overtones and combination modes of functional groups at the molecular level across the SWIR spectral region and to electronic transitions in atoms across the Vis–NIR spectral regions at specific wavelengths. A compressive description on the exact soil chromophore and its electromagnetic activity can be found in, e.g., Ben-Dor et al. (1999), Stenberg et al. (2015), Demattê et al. (2015). The physical chromophores are due to scattering effects based on particle size and shape distribution in the material. The water molecule influences the absorption features at specific wavelengths that are the results of overtone and combination modes from the IR region as well as the results of the physical effects that scatter the light in a way that the spectrum shape and base line are changing. Figure 1 provides a typical soil spectrum with the direct known chromophores. It can be seen that there are physical (baseline height and spectral shape) and chemical (absorption) features. In the Vis–NIR region, electronic transitions are dominated mainly in iron oxides with also organic matter (OM) that refer to the slope of the spectrum part (probably related to soil OM structure). In the SWIR region, the water

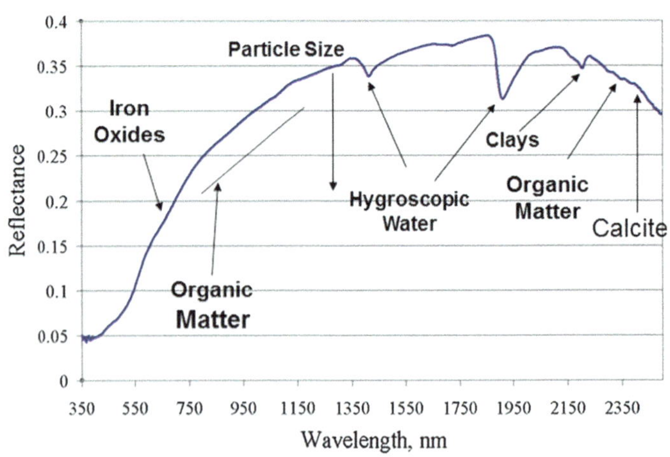


**Fig. 1** A soil spectrum (Haploxeralf) that represents the major chromophores in soils (after Ben-Dor et al. 2008)

molecules in hygroscopic water play a major role in 1.4 and 1.9 μm where clay minerals and calcite are at around 2.2 and 2.3 μm, respectively. The wavelengths may vary based on the structure of the minerals and their crystal shape and purity. All possible absorption features and their quantum mechanisms in the Earth's crust minerals were summarized, e.g., in Ben-Dor et al. (1999).

Soil is a complex system that is extremely variable in physical structure and chemical composition both temporally and spatially. Soil spectroscopy, although being complex, can cluster several soil properties with a single measurement that can be extracted based on radiative transfer models (e.g., Hapke 1981) or empirical data-mining (chemometric) approach (Ben-Dor and Banin 1995b). The prediction of a soil spectrum from physical bases is quite difficult (Liang and Townshend 1996). Each soil property $S_i$ has to respect the following rules to be successfully predicted by soil spectroscopy: Rule (1.1) the soil property $S_i$ has a specific spectral signature due to a chemical or physical structure (Ben-Dor et al. 2002) or Rule (1.2) the soil property $S_i$ is correlated with a soil property $S_j$ having a specific spectral signature due to an associated chemical or physical structure (Ben-Dor et al. 2002); and additionally Rule (2) the soil property $S_i$ has to have a quite high amount of variability (Gomez et al. 2012a, b). Until now, and from our knowledge, the limits of the quantification of a soil property that has got a low variability within the scene have not been explored in detail as this minimum of variability may depend on (1) the studied soil property and (2) the link with other soil properties on the soil samples. This last rule implies that the predictability of a primary soil property depends on the soil diversity of the study site. For example, Ben-Dor et al. (2002) had already stressed that the soil is a complex matrix and that the spectral features of one component (e.g., OM) can be hidden or slightly shifted by another component [e.g., iron (hydr)oxides]. Nevertheless, separating the spectral information from the different soil attributes is possible as already shown in previous studies. Laboratory soil spectroscopy, linked with statistical analyses, has, for example, been used since, e.g., the earlier works of Demattê et al. (2004) over soils from São Paulo State, Brazil, for the derivation of soil survey maps and soil classes. It was the first paper applying soil spectroscopy for pedological mapping. The authors developed a spectral reflectance-based methodology that was able to evaluate soil types and soil tillage systems.

Although there is a strong relationship between the soil chromophores as observed in the spectral domain and the chemical/physical characteristics of the material, the correlation is not straightforward. This is because the spectral data are multivariate, with many reciprocal effects (Schwartz et al. 2011). Accordingly, the extraction of quantitative information on a given soil attribute using spectral information is not a simple task, especially if it is not a chromophore attribute. Malley et al. (2004) provided a summary to what soil attributes can be spectrally modelled in soils. In 2006, Rossel et al. extended the application and provided a long list of authors and soil attributes, most of which use chemometric approaches. The chemometric approach is an empirical (statically driven) method, and although no physical, chemical or other assumptions are made, the method has a strong spectroscopic basis, in which the selected bands in the model must have specific assignments. This method provides quantitative information about its chromophore that can be further used. It can be either an index, equation or a model that is extracted from the spectral information, usually combined with the reference information from traditional chemical analyses to "train" the system. Chemometrics also refers to proximate analysis of soil attributes related to the spectral analysis of soil. A sophisticated method of finding this relationship, also known as "data mining", has to be applied. As the final goal is to use the spectral model for practical remote sensing practices, it is crucial to extract the best model in a given population, rather than just finding a correlation.

Many methods for applying data-mining approach to soil spectral information have been used and developed, from multiple linear regression (MLR) analysis (of the spectra against the chemical/physical data) through principal component analysis regression (Chang et al. 2001), partial least squares regression (PLSR) (Zhao et al. 2015), artificial neural networks (ANN) (Carmon and Ben-Dor 2017) and random forest among others. The standard procedure for developing such models will be to divide the samples into calibration and validation sets. The model is then developed on the spectral and chemical data of the calibration group and is applied on the spectral data of the validation group to predict its chemical values. The quality of the model is determined by its prediction accuracy using various statistical parameters. When a prediction model with good quality is found, it can be used to predict the chemical values of new samples with just a spectral measurement, either from point or from imaging spectrometers.

As already mentioned, spectral data are affected by various components in the soil, some of which are connected to the chemical property in question, and some not. Thus, applying preprocessing algorithms on the spectra prior to developing the model can amplify relevant spectral features and is traditionally taking place. Manipulation of spectra using derivatives and transformations to log space enables the enhancement of weak spectral features as well as minimizes physical effects (Demetriades-Shah et al. 1990). As a given dataset can be executed using several manipulation stages in a process chain, it is impossible to check many preprocessing combinations manually. Ben-Dor and Banin (1995a) suggested developing a "whole-process" possibility chain in an automated environment to enable optimal data mining, such that the best preprocessing combination could be selected. This concept is termed all possibilities approach (APA), in which all possible combinations are evaluated. Moreover, they concluded that aside from good statistical parameters and a selected processing chain, a reliable model must have solid spectral assignments for the spectral region/channels selected by the analysis. This is done by finding the important spectral ranges used by the model and examining whether the selected wavelengths have a meaningful explanation based on the physical processes described earlier.

To cope with these challenges, Schwartz et al. (2011) developed a data-mining machine termed "PARACUDA®" which runs several preprocessing spectral data manipulations. Their concept was based on a smart and single selection of calibration and validation groups from the population in question, using a cubic Latin hypercube sampling algorithm for semi-randomized grouping (Minasny and McBratney 2006). Remarkable results were obtained using the PARACUDA machine, mainly due to its automated capability to parallel-check 120 preprocessing combinations. The continued development of the system resulted in the notion that the model quality is sensitive also to the grouping stage and not only to the preprocessing combination. Moreover, the system did not have a viable spectral assignment output, which could significantly amplify the models' robustness and improve our understanding of the spectral correlations to various soil properties. As the main goal was to develop an accurate (reliable) prediction model, based on finding the best preprocessing combination and spectral assignments, a new system was recently developed to fully exploit the APA idea of the PARACUDA engine (Carmon and Ben-Dor 2017).

As the quantitative approach of spectral data mining of soils developed, many users are getting into this field and the usage is growing consistently. In this direction, the new machine learning software (such as the PARACUDA-II®, Carmon and Ben-Dor 2017) or computing approach (such as random forest, Gholizadeh et al. 2015) fosters the use of soil spectroscopy for the quantitative domain. In this case, we examined how many papers have been published during the past 10 years using some keywords of soil spectroscopy (point and image) using Google. An exponential growth in both technologies was found, whereas the imaging is still lagging behind with respect to point spectroscopy. This is because the soil imaging spectroscopy started later than the point spectroscopy, and only recently has the image technology became more available to more users and the number of papers started to grow. The exponential pattern is not surprising and provides a promising future for the practical use of soil spectroscopy.

It should be pointed out that recently not only the optical domain (Vis–NIR–SWIR) is used for soils but also the thermal infrared part, also termed mid-infrared (mid-IR: 3–12 μm), which shows promising capabilities. From the first paper based on Janik et al. (1998) who raised the question "Can mid-IR diffuse reflectance analysis replace soil extractions?", it is clear today that this region is important, especially for detecting silicate-bearing minerals (Weksler et al. 2017). Chang et al. (2001) used a principal components regression method to determine the soil attributes from the thermal region in the laboratory. They had success with some attributes such as calcium, but not with micronutrients such as zinc or sodium. Chang and Islam (2000) constructed an ANN model based on the physical linkages among the space–time distribution of brightness temperature, soil moisture and the soil media properties. They showed that it is possible to infer soil texture from spectral reflectance properties, based on the current activities and knowledge about soil spectroscopy and analysis. Eisele et al. (2012, 2015) demonstrated the advantages of the 8–12 μm domain for the quantification of several soil properties such as soil texture and in particular for sand content that is hardly retrievable based on the optical domain. Similarly, Kopacková et al. (2017) were able to show that the mid-IR between 3 and 12 μm is capable of providing quantitative information of selected samples of organic soil using the PARACUDA-II algorithm and demonstrated the added value of this region to the optical region. Most studies until now in IS for soils in the mid-IR were based on laboratory spectroscopy due to the lack of availability of airborne data and adapted softwares. Nevertheless, nowadays airborne thermal IS is getting more and more attention based on the release of new technology. New airborne imaging thermal spectrometers are becoming available commercially (e.g., the TASI-600 from ITRES, recently the pushbroom AisaOWL and FTIR HyperCam from Specim and Telops), and new orbital initiatives including thermal bands are now started or planned (e.g., ECOSTRESS the ECOsystem Spaceborne Thermal Radiometer Experiment on Space Station, launched in June 2018, and the planned NASA-SBG) (Hook et al. 2017). Thus, the thermal region of the electromagnetic spectrum aimed to be adopted for soils in addition to the optical region in future.

## 2.2 Effect of Surface Roughness on Soil Reflectance

The roughness of soils, regarded here as irregularities of their surfaces resulting from the existence of soil particles, aggregates, rock fragments and micro-relief configuration, significantly affects soil spectral reflectance. Although this impact was noticed and examined many years ago (Bowers and Hanks 1965; Brennan and Bandeen 1970; Stoner and Baumgardner 1981; Cierniewski 1987; Cierniewski and Courault 1993), it is still underappreciated (Cierniewski et al. 2015).

The spectral reflectance of soil surfaces as with many other Earth objects is anisotropic. Irregularities in soil surfaces produce shadow areas, where solar beams in field conditions or beams from an artificial light source in laboratory conditions do not directly reach the surfaces. Wave energy leaving the areas is many orders of magnitude smaller than energy reflected from directly illuminated soil fragments. Cierniewski et al. (2010) showed spectra of a ploughed soil obtained by a hyperspectral camera. The overall reflectance level of the shaded soil fragments was clearly lower than that related to sunlit fragments, although the shape of the two categories of these spectra was quite similar to each other.

Cultivated bare soils with dominant diffuse features usually appear brightest from the direction which gives the lowest proportion of shaded fragments. Those soil surfaces usually display a clear backscattering character with a reflectance peak toward the Sun position ('hot spot' direction) and decreasing reflectance in the direction away from the peak (Brennan and Bandeen 1970; Kriebel 1976; Milton and Webb 1987). Desert surfaces show that soil reflectance can clearly have both a backscattering and a forward-scattering character (Deering et al. 1990). The surfaces display maximum reflectance in the extreme forward-scattering direction near horizon if they are relatively smooth with a strong specular behaviour. Shoshany (1993) found that different types of desert stony pavements and rocky surfaces under varied illumination conditions exhibited an anisotropic reflection with a clear backscattering component.

Non-Lambertian behaviour is presented in Fig. 2 for two soil surfaces, one uncultivated and smooth and another cultivated and moderately rough, both with non-directional spreading of their height irregularities. Their reflectance distributions normalized to the nadir viewing in all possible directions for the chosen wavelength of 850 nm under clear-sky conditions at various solar zenith ($\theta_s$) and azimuth ($\phi_s$) angles were predicted by a hemispherical-directional reflectance model (Cierniewski et al. 2004). The larger the soil surface irregularities and the higher the $\theta_s$, the higher the variation of the soil directional reflectance. The variation is the most visible along the solar principal plane. Croft et al. (2012) and Wang et al. (2012) also reported this non-Lambertian behaviour of soil surfaces under laboratory conditions. How a directional furrow micro-relief can additionally complicate the reflectance of soils is shown by the results of a laboratory measuring experiment simulating the reflectance behaviour of sandy soils with furrows treated by a harrow or a seeder (Cierniewski and Guliński 2010). They found that the spectrum level (between 400 and 2300 nm) for the mostly deeply furrowed surface viewed at the nadir and illuminated by sunbeams coming along the furrows was 5–10% higher than for the same surface but illuminated by sunbeams coming perpendicular to them. Under the same illumination and viewing conditions of the furrows, the level of the spectra for the surface with the deepest furrows was about 2% lower than for the three times shallower furrows.

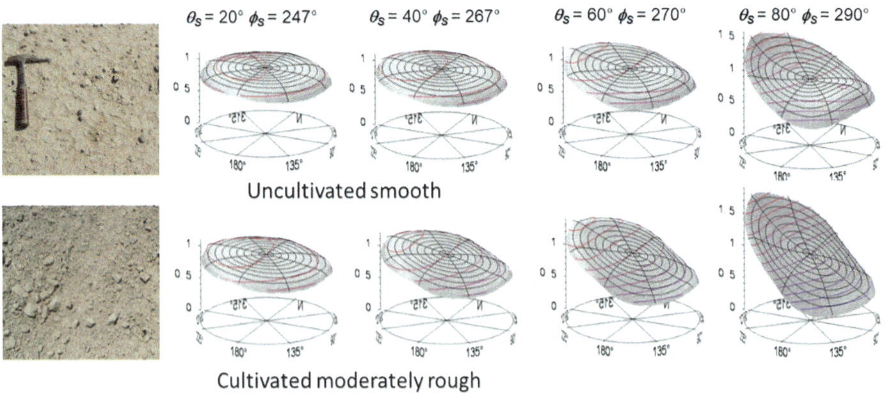


**Fig. 2** Normalized directional reflectance distributions of soils for chosen wavelength of 850 nm under clear-sky conditions at various solar zenith ($\theta_s$) and azimuth ($\phi_s$) angles predicted by a hemispherical-directional reflectance model (after Cierniewski et al. 2004)

Soil roughness also clearly affects albedo. Analysing broadband soil albedo variation during the day, it was found that soil roughness affected not only the overall level of that variation (Cierniewski et al. 2015), but also the intensity of the albedo, which increased from $\theta_s$ at the local noon to about 75°–80°. Rough, deeply ploughed soil surfaces showed almost no rise in albedo values for $\theta_s$ lower than 75°, while the same soil, but smoothed, exhibited a gradual albedo increase at these angles. Clearly, spectral reflectance of soils covered with a crust formed by the sequential wetting and drying of their surfaces is higher than that of soils without a crust (Cipra et al. 1971; Goldshlager et al. 2010). The reflectance of disturbed soil samples increases as soil particle size decreases (Piech and Walker 1974). The smaller aggregates are more spherical in shape, but the larger ones are irregular in shape with a higher number of inter-aggregate spaces and cracks where the incident light is trapped (Coulson and Reynolds 1971). Especially after tillage treatments, the impact of soil roughness on spectral reflectance can be very variable. Matthias et al. (2000) found that, when the relatively smooth surface of a fine sandy soil was ploughed, its reflectance decreased by about 25%. Potter et al. (1987) reported that, conversely, the reflectance of ploughed sandy soils increased by about 25% after rain and subsequent drying of their surface.

With the intention to as precisely as possible model processes associated with the flow of radiation between the Earth's surface and the atmosphere in longer periods of several days, a month, a season or a year, average diurnal albedo values ($\alpha_d$) appear to be more useful than instantaneous values. Cierniewski et al. (2013) considered the optimal time (To) to obtain spectral data of bare soils for approximating their $\alpha_d$ values using satellite technology. They supposed that raw satellite data for the Earth's surfaces obtained in that time do not need be corrected due to the Sun position expressed by $\theta_s$. It was anticipated that the correctness of soil $\alpha_d$ estimation could be increased by eliminating at least one of the factors having a significant impact on the approximation of soil's $\alpha_d$ (the effects of the atmosphere, the direction of the soil observation by satellite, which together with the direction of the Sun position determine the bidirectional reflectance of studied surfaces) and extrapolating the narrowband albedo to its broadband value, and the correctness of soil $\alpha_d$ estimation would be higher. Struggling to minimize errors of this approximation becomes especially important in the context of the statement by Sellers et al. (1995) that global climate change models required albedo values with errors of less than ±2%. Cierniewski et al. (2013) analysed how strongly the roughness of soils (smooth, moderately rough and very rough) and their location in the world affect To, taking into account their latitudinal position and assuming that they were observed by a satellite in Sun-synchronous orbit at chosen dates with errors ±2%. It was found that morning To was expected for very rough soil earliest and for the smooth soil latest. In the afternoon, this trend was reversed. The usefulness of the orbit during the analysed dates was expressed by its length, from which observation of the soils was available with the acceptable error ±2%. The longest parts of the orbits, estimated as larger than 90°, were predicted for the morning in mid-April, while the shortest length of them, reaching only about 20°, was expected for the afternoon in the beginning of the astronomical summer. An attempt was also made to compare the usefulness of satellite orbits crossing the equator at local solar time 7:30 and 10:30 such as for the NOAA-15 and the MODIS (Cierniewski 2012). The earlier orbit proved to be the much more useful for observing bare soils than the later one.

A few decades ago, variation of the soil surface height was only measured along a direction using a profile meter with needles or a chain set (Gilley and Kottwitz 1995). Many years later, Moreno et al. (2008) drew attention to the fact that the use of these simple tools can be successfully replaced or supplemented by analysing the shading of soil surface irregularities with their directional and non-directional spreading. Now, the soil surface height is automatically recorded using laser scanners along a single or multiple straight lines (Thomsen et al. 2015). This allows for analysis of soil irregularities in the two-dimensional or three-dimensional space, respectively. Currently, the soil surface roughness is most often investigated in a three-dimensional space using methods of close range digital photogrammetry (Heng et al. 2010; Rieke-Zapp and Nearing 2005). Recently, Gilliot et al. (2017) have presented the use of photographs of a studied soil surface taken from over a dozen directions by a hand-held digital camera that moved around the area of interest. These photographs, taken together with horizontal slats, allow for the creation of digital elevation models (DEMs) of the studied surfaces, which are the basis for calculating the roughness indices of soil surfaces.

The standard deviation computed of the surface height data collected along a direction or within DEM units is the most common index for describing the soil surface roughness (Ulaby et al. 1982; Marzahn et al. 2012). Boiffin (1986) proposed the use of the description of the turtle's index, which represents the ratio between the actual length of the soil surface profile and the projected horizontal length of this profile. Later, Taconet and Ciarletti (2007) modified the reference of the index to the two-dimensional space, defining it (T3D) as the ratio of the real surface area within its basic DEM unit to its flat horizontal area. Other indices used mainly to quantify the temporal evolution of the soil surface roughness due to rainfall events are based on a semivariogram analysis (Croft et al. 2013; Rosa et al. 2012; Vermang et al. 2013).

# 3 Applications of Imaging Spectroscopy for Soil Mapping

## 3.1 Common Soil Properties Mapping

### 3.1.1 Examples of Soil Properties that have been Predicted with Airborne Spectroscopy

The mapping of soil surface properties from imaging spectroscopy in the Vis–NIR–SWIR region has emerged in the early 2000s. Since then, it has been further demonstrated and extended in many environments and for many soil properties based on different sensors and different methods. Nowadays, this technique is commonly used over bare soils on cultivated areas where: (1) soil is regularly ploughed inducing topsoil homogenization (usually 20–30 cm), (2) large proportion of bare soil and no soil crust formation are exposed, which may be ensured with a flight window during seeding of summer crops (such as maize, sugar beet or potato) or winter crops (such as winter cereals), (3) top-surface water content is low.

One of the first examples of soil property mapping was obtained using the DAIS 7915 airborne sensor in Israel (Ben-Dor et al. 2002). These authors used a two-step approach. First, they acquired field spectra using an ASD spectrometer and analysed the samples from the corresponding surfaces by means of conventional wet chemistry techniques for soil properties such as OM, soil moisture content in field conditions, saturated soil moisture content, salinity and pH. They constructed a spectral model using MLR of the 38 wavelengths that showed both the highest correlation between reflectance and measured soil properties and were known for their physical interaction with the soil property concerned. Then, they applied these models using the reflectance of the airborne sensor. The authors demonstrated that the organic content, hygroscopic moisture and electric conductivity (EC) (all "surface" properties) can be predicted. Applying their models on a pixel-by-pixel basis revealed the spatial distribution for each property. Some years after, Selige et al. (2006) were also able to predict the organic carbon ($C_{org}$), total nitrogen ($N_{tot}$), sand and clay content for 12 bare cropland fields covering 7000 ha within a Hymap (128 wave bands from 420 to 2480 nm) flight line of 200 $km^2$. Although it is generally assumed that there is a strong correlation between C and N, the spectral models based on a multiple linear regression used distinctly different wavelengths for the two soil properties. This difference in spectral features is not in agreement with Rule 1.2, but fits within Rule 1.1 and 2 as referred to in chapter 2.1. Moreover, Selige et al. (2006) demonstrated that the maps of $N_{tot}$ showed patterns caused by historical differences in land management dating from the period before land consolidation that were not visible in the $C_{org}$ map.

Later, Stevens et al. (2010, 2012) predicted the SOC content in croplands of a N–S light strip (420 $km^2$) in Luxembourg, as this soil property obeys Rule 1.1 and Rule 2 as referred to in Sect. 2.1. Data from the hyperspectral airborne AHS 160 sensor were used with 20 bands between 430 and 1030 nm and 42 bands between 1994 and 2540 nm. The PLSR and PSR methods showed highest correlations between SOC content and reflectance in the visible ranges (600–750 nm), while the correlations in the SWIR were noisy for the PLSR and more consistent around 2100 nm for the PSR. The correlations around 700 nm and 2100 nm were also found by Ben-Dor et al. (2002) and confirm that the prediction models are based on spectral features.

Further, Gomez et al. (2012b) used the hyperspectral airborne Aisa-Dual sensor (260 hyperspectral bands from 450 to 2500 nm) to map eight soil properties in a 300 $km^2$ area in Northern Tunisia. They obtained good results for four properties (iron, cation exchange capacity (CEC), clay and sand) which followed Rules 1.1 or 1.2, and Rule 2 and incorrect estimations for four other properties ($CaCO_3$, pH, SOC and Silt) which did not follow at least two out of the three rules. Given the large proportion of bare soil in this semiarid cropland environment, they were able to show a complex regional soil pattern reflecting the variations in lithology.

### 3.1.2 Quality of the Predicted Soil Property Maps

In this section, we are looking at the figures of merit of the predicted soil properties maps in terms of error measures ($R^2$, RMSE, ...), the analysis of the spatial patterns in the predicted soil maps and the uncertainties analysis which could be associated with a quality analysis of predicted soil properties maps. Most papers on soil property mapping using imaging spectroscopy (cross)-validate the spectra of pixels registered by the airborne sensor against the conventional chemical analysis of soil samples collected in the same pixel. Table 1 shows examples of model performances including errors of prediction based on ground truth validation for the prediction of common soil properties maps using imaging spectroscopy. Different validation techniques are derived similar to chemometric approaches in the laboratory. Stevens et al. (2012), for example, examine different cross-validation techniques and their impact on prediction accuracy. They further compare cross-validation with real validation. This study shows that prediction accuracy of a soil property map derived from IS is also a question of the applied validation strategy.

However, in contrast to the laboratory where the mean error of prediction on each sample is the most important metric for the evaluation of the quality of the model, for soil maps there are more sources of uncertainty such as GPS position, atmospheric disturbance,

**Table 1** Examples of prediction performances obtained from hyperspectral imaging sensors for common soil properties (extracted from the literature)

| Property | References | Method | Number of samples | Data range | Imaging performances per method (Sensor name) |
|---|---|---|---|---|---|
| Iron | Bartholomeus et al. (2007) | Spectral index | 35 | 7–20% | $R^2=0.22$<br>(ROSIS) |
| Textural clay | Gomez et al. (2008b) | PLSR/spectral index | 52 | 6.5–45.2% | $R^2_{cv}=0.64/0.58$<br>$RMSE_{CV}=49.6/82$ g/kg<br>(HyMap) |
| N total | Selige et al. (2006) | PLSR/MLR | 72 | 0.07–0.37% | $R^2=0.92/0.87$<br>$RMSE_{CV}=0.03/0.02\%$ |
| $CaCO_3$ | Gomez et al. (2008b) | PLSR/spectral index | 52 | 2.6–47.3% | $R^2_{cv}=0.77/0.47$<br>$RMSE_{CV}=76.6/132$ g/kg<br>(HyMap) |
| Sand | Selige et al. (2006) | PLSR/MLR | 72 | 16–84% | $R^2=0.95/0.87$<br>$RMSE_{CV}=9.7/12.9\%$ |
| pH | Ben-Dor et al. (2002) | MLR | 62 | 7.5–8.2 | $R^2_{cv}=0.53$<br>(DAIS-7915) |
| Salinity | Dehaan and Taylor (2003) | Salinity index | Not specified | Not specified | Correspondence with image-derived indicators for halophytic vegetation<br>(Hymap) |
| SOC | Schwanghart and Jarmer (2011) | PLSR | 61 | 0.24–1.13% | $R^2=0.767$<br>$RMSE_v=0.129\%$<br>(Hymap) |
| | Stevens et al. (2012) | PLSR/PSR/SVMR | 400 | 7–46.5 g/kg | $R^2_{cv}=0.7–0.81$<br>$RMSE_{CV}=3.8–4.8$ g/kg<br>(AHS 160) |

variation in soil surface conditions (e.g., moisture residue or roughness) and within-pixel variation of the soil property (see Sect. 4.2). Thus, spatial patterns of the soil property and uncertainty of the soil property maps are in addition to the model metrics important criteria to evaluate the quality of produced maps.

Imaging spectroscopy produces high-resolution (2–8 m pixels) maps of soil surface properties covering large areas that were hitherto not available. The clay maps of, e.g., Gomez et al. (2012b) cover an area of ca. 300 km$^2$ in Northern Tunisia. Due to the low vegetation cover in these Mediterranean croplands, a large area could be mapped producing a clay map showing strong similarities to the geological map. Although extensive areas of bare cropland soil are rarer in temperate regions, Selige et al. (2006) and Stevens et al. (2010) also pointed to the spatial patterns that could be identified within or between fields. They related these patterns to historical differences in manure application reflected in the total Nitrogen ($N_{tot}$) maps or to rotations with temporary grassland reflected in the SOC maps that cannot be observed anymore after land consolidation. SOC patterns in fields covering the hill slope from the crest to the foot slope showed patterns related to erosion of C associated with the topsoil. Stevens et al. (2015) further demonstrated the importance of land management on SOC content. Their residual maximum likelihood model (REML) applied to the SOC map produced from the AHS160 flight over Luxembourg predicted that a field effect accounted for 48 ± 8% of the variance in SOC content in a cluster of fields (Fig. 3).

The performance of the estimations obtained from regression models (MLR, PLSR, PSR, SVM…) is usually assessed with figures of merit such as the standard error of calibration, the standard error of prediction or the ratio of performance deviation, and these figures of merit evaluate the global model performance, as they are calculated during the model building and validation stages. A first evaluation of uncertainty that affects each prediction obtained by a hyperspectral airborne sensor has been realized by Gomez et al. (2015a). An evaluation of different types of uncertainties (i.e. variance of predictions) has been done: (1) prediction variance due to the regression model, (2) prediction variance due to the spectra and (3) prediction variance due to the interaction between these two effects, prediction variance due to differences between spectral predictors and spectral calibration samples. This paper showed that these prediction uncertainty maps may be used to better

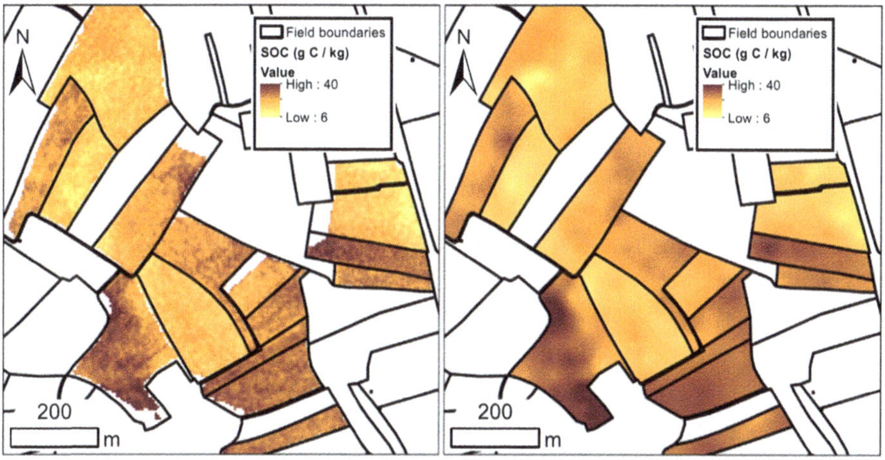


**Fig. 3** Predicted SOC maps (left) from the AHS 160 campaign in Luxembourg (Stevens et al. 2010) and (right) using a residual maximum likelihood model (Stevens et al. 2015)

characterize the quality of the soil properties mapping results, mask no-soil pixels and define the soil sampling and the calibration dataset.

### 3.2 Soil Moisture

Water is considered to be one of the most significant chromophores in the soil system (Idso et al. 1975; Stoner and Baumgardner 1981; Baumgardner et al. 1985; Hummel et al. 2001; Lobell and Asner 2002). Bowers and Hanks (1965) described the effect of soil moisture content on reflectance for the first time. The main effect is the decrease in reflectance with increasing soil moisture, and some spectral features are more affected than others. Dalal and Henry (1986) isolated the main differences in absorbance (log 1/reflectance) and found them to be related to the variation in the moisture contents used across the 1100–2500-nm SWIR spectral region. Bishop et al. (1994) show the features directly associated with the OH group in the water molecule (at 1400 and 1900 nm), and some are indirectly associated with the strong OH group in the thermal infrared region (around 2750–3000 nm) that affect the lattice OH in clay (at 2200 nm) and $CO_3$ in carbonates (at 2330 nm). Ben-Dor et al. (1999) have noted the diminishing of the 2200-nm absorption feature in Ca-montmorillonite mineral at various relative humidity conditions. The highly sensitive 1900-nm region, a water OH combination band, showed an excellent nonlinear fit to the increase in water content. Muller and Decamps (2000) determined that the impact of soil moisture on reflectance could be greater than the differences in reflectance due to the soil categories, as it affects the baseline height (albedo) as well as several spectral features across the entire spectral range, as can be seen in Fig. 4a.

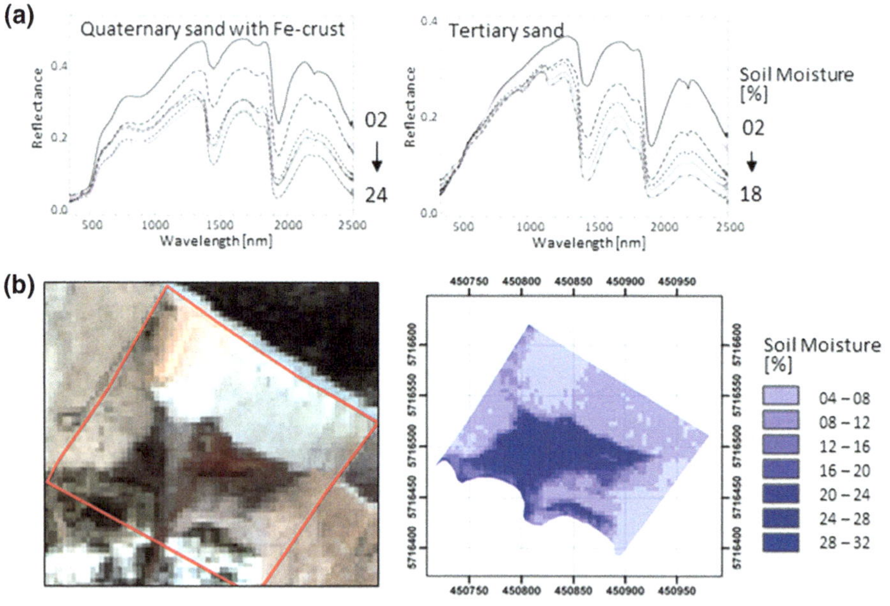


**Fig. 4** **a** Influence of soil moisture on soil spectra from quaternary and tertiary sand; **b** determination of gravimetric water content based on HyMap imagery: HyMap RGB image (left) and modelled gravimetric soil water content (in %) (right). Modified from Haubrock et al. (2008a, b)

Several studies focused on the modelling of soil spectral reflectance and soil water content. Bach and Mauser (1994) were able to simulate the reflectance change of the soil spectra from dry to moist through the use of various water film depths related to moisture content. They combined the model of Lekner and Dorf (1988) for internal reflectance with the absorption coefficients from Palmer and Williams (1974) into Lambert's law (or Beer's law) to simulate the moist reflectance $R$ from the dry reflectance Ro by the exponential of the absorption coefficient and an empirically determined active thickness. They applied this process for predicting the water contents within an AVIRIS image of a partially irrigated field and a dark organic soil field in the Freiburg test site in Germany from the image pixel spectra of dry and moist soils. The active thickness and water absorption coefficients predicted the amount of the soil's water content to a high $R^2$ of 0.88. Most recently, Bablet et al. (2018) have developed MARMIT, a multilayer radiative transfer model of soil reflectance to estimate soil water content that could be applied on imaging spectroscopy data in the laboratory, but not yet from remote sensing data.

Nowadays, robust spectral techniques to quantify and/or correct soil moisture content from surface reflectance exist that have been developed by several authors at the laboratory level and tested successfully on hyperspectral imagery (e.g., Bryant et al. 2003; Whiting et al. 2004a, b; Haubrock et al. 2008a, b; Fabre et al. 2015; Diek et al. 2019). The approaches range from spectral indices, exponential or Gaussian models, to geostatistical models. Whiting et al. (2004a) fitted an inverted Gaussian function centred on the assigned fundamental water absorption region at 2800 nm, beyond the limit of commonly used instruments, over the logarithmic soil spectra continuum found with convex hull boundary points. The area of the inverted function, soil moisture Gaussian model (SMGM), accurately estimated the water content with coefficients of determinations ($R^2$) of 0.94–0.98 when samples were separated according to landform position (Spain) and salinity (USA). Using AVIRIS hyperspectral images of these soil regions in an air-dried status, they improved the abundance estimates of clay and carbonate abundance by 10% of the regression mean by including the SMGM area as a parameter in the empirical determination (Whiting et al. 2005).

Haubrock et al. (2008a, b) developed a successful simple approach termed the Normalized Soil Moisture Index (NSMI) based on the edges of the water band centred at 1900 nm. It was developed in the laboratory and tested in the field for the best spectral prediction of soil moisture content taking into account the influence of different environmental factors, such as variable soil types, soil water profiles and the presence of soil crust and vegetation cover; and in remote sensing data based on hyperspectral imagery. The NSMI allowed the production of surface soil moisture maps, generated from HyMap airborne images (Fig. 4b), which were found to be highly correlated with the field moisture content measured at the time of the overflight (Haubrock et al. 2008b).

The NSMI and SMGM methods were successfully implemented in the soil mapping software tools HYSOMA and EnSoMAP (Chabrillat et al. 2011, 2016). Chabrillat et al. (2012) demonstrated that the two independent methods provide similar performances based on hyperspectral images for a same field, validated with ground data. Fabre et al. (2015) recognized that the water index soil (WISOIL) (Bryant et al. 2003) and the NSMI performed as good as the best soil moisture indices based on laboratory data. However, these indices are based on wavelengths located in or near the water absorption bands, which make them very sensitive to the quality of the atmospheric correction, and adapted indices were used by many authors (Fabre et al. 2015; Liu et al. 2003; Nocita et al. 2014). For that reason, Fabre et al. (2015), working on simulated spectral radiances at the sensor level, developed two additional soil moisture indices based on the wavelengths 2000–2200 nm that are less sensitive to atmospheric correction, but more sensitive to variations in soil texture. The two indices are the normalized index of the NIR and SWIR domain for soil moisture content estimation from linear regression (NINSON) and the normalized index of the NIR and SWIR domain for soil moisture content estimation from nonlinear regression (NINSOL). Diek et al. (2019) present a first attempt to reduce the effect of soil moisture (SM) variation on reflectance data of soils, independent of new laboratory measurements. SM was determined by a newly developed SM index, similar to the existing NINSOL index, and was successfully applied to both field and airborne spectroscopy data, with $R^2$ of 0.75 and 0.59, respectively.

Nowadays, surface water content in soils can be cautiously estimated based on spectral reflectance, providing topsoil information that is not available at this scale through other systems and that could complement in bare areas the global estimates of near-surface soil moisture provided by microwave satellites. However, completely correcting the effect of soil water content on the reflectance mineral and organic band depths for an improved retrieval of other soil covariate continues to elude us. Further work in reconstructing the spectra will combine the spectral relationship of water content and soil components based on the physical nature of the materials and photon absorptions.

### 3.3 Soil Crust

Soil crust refers to a thin layer on top of the soil surface that is exposed to solar radiation and hence to the remote sensing sensors (Agassi et al. 1981). The arrangement of the upper thin layer of the soils, which actually formed the crust, is controlled by biogenic/organic process as well as by mechanical forces. Depending on the processes dominating the crust formation, soil crusts are termed biological/organic or physical. Biological/organic crusts are formed by communities of microorganisms that live on the soil surface, whereas physical crusts are formed by physical impact such as that of raindrops.

Biological soil crusts, or biocrusts, are generated from live materials, such as complex communities of cyanobacteria, algae, lichen and mosses. Biocrusts modify the surface spectral response and can mask soil spectral properties (Weber et al. 2008). For recent reviews on the effect of biocrusts on soil spectral signal and the biocrust mapping methods based on optical remote sensing, we refer to Escribano et al. (2017) and Weber and Hill (2016). In particular, it was shown that many studies referred to biocrust spectral traits at the laboratory and field scale, and approaches based on hyperspectral indices or spectral mixing analyses were developed to spectrally identify biocrusts or for biocrust surface cover quantification. For example, Weber et al. (2008) and Chamizo et al. (2012) developed two biocrust mapping hyperspectral indices that were used to map cyanobacteria-dominated biocrust in Soebatsfontein (South Africa) and in southern Spain, respectively, based on hyperspectral imagery. With this, the subtle spectral differences between sparse vegetation, bare soil and biocrust could be identified. Nevertheless, important classification errors were observed in heterogeneous areas where each pixel is covered by a mixture of semiarid vegetation, bare soil and biocrusts (Alonso et al., 2014). Similarly, Hill et al. (1999) and Rodríguez-Caballero et al. (2014) used spectral unmixing methods to successfully quantify the amount of biocrust coverage within a pixel based on hyperspectral imagery in the Nitzana region (Israel), and in El Cautivo area (Spain). Some issues were observed in discriminating biocrusts in the areas dominated by vegetation, and an improved discrimination could be obtained by the prior definition of areas into different land units.

Karnieli et al. (1999) demonstrated that cyanobacteria provide unique spectral features on sand dune at the Negev area in Israel just as decomposed OM, whereas Ben-Dor et al. (1997) demonstrated that the same fresh OM may hold different spectral fingerprints based on the decomposition stage at the laboratory level. One of the important organic crusts that affect hydraulic conductivity of the thin surface layer is the OM with hydrophobic characteristics. This crust dramatically diminishes the infiltration rate of water to the soil profile and hence increases erosion and soil degradation and loss. Recently Ben-Dor et al. (2017) demonstrated a significant correlation between reflectance properties of the soil crust measured in the field by using the SoilPRO apparatus and hydrophobicity field values as measured by the water droplet penetration time (WDPT) method over orchards trees soils, by thus demonstrating a possible application based on future IS data.

The soil physical crust mostly refers to the rearrangement of soil particles as a result of the raindrop energy which disintegrates soil aggregates into a microstructure where fine particles are exposed at the top surface facing the Sun, and coarser particles are found in the soil profile. In general, the factors leading to the formation of the structural crust (raindrop energy driven) are a combination of the kinetic energy impact of raindrops and the level of stability of the soil aggregates (Agassi et al. 1981, 1985). The structural crust is generated within minutes and significantly reduces the soil infiltration rate (IR). The crust hydraulic conductivity (HC) is lowered by a few orders of magnitude compared to the underlying soil (McIntyre 1958; Benyamini and Unger 1984). As the structural crust affects physical properties of the soil and can be observed by significant colour changes on the soil's surface, spectral information, and especially on the surface, may be an excellent tool to monitor its status. To that end, spectral differences observed in the structural crust demonstrated that the upper microstructure is composed of fine texture with some significant increase of the clay fraction (Ben-Dor et al. 2004). Whenever the HC of the crust is lower than the rainfall intensity, ponding, runoff and soil erosion will follow crust (Agassi et al. 1981, 1985). These processes may further change the upper soil surface properties as seen by hyperspectral sensor pointing on erosion, deposition and sedimentation effect over the soil in question.

Monitoring the soil crust condition is essential for the proper management of soils, from both an agricultural and land degradation perspective, and much work toward this end was done using reflectance spectroscopy at the laboratory level. Several studies demonstrated that reflectance spectroscopy can provide a valuable method for assessing the condition of the soil crust and estimating the related problem (Demattê et al. 2004; Goldshleger et al. 2001; De Jong 1992; Eshel and Levey 2004; Ben-Dor et al. 2004). More studies toward this direction by Goldshleger et al. (2002, 2004) and Ben-Dor et al. (2003) demonstrated that the reflectance properties of soils can be an indicator for temporal processes that evolve on the soil surface due to runoff and erosion that were driven by the physical crust formation, based on laboratory experiments. An ambitious and very interesting study to upscale the physical soil crust spectral properties from the laboratory (rain simulator) to the field and to remote sensing was done by Ben-Dor et al. (2004). In this study, loess (agricultural) soils from Israel were examined for the soil crust at the laboratory which was then applied to the AisaEAGLE sensor to generate a soil infiltration map on a pixel-by-pixel basis (Fig. 5). Validation in the field confirmed the results and suggested that this idea could be extrapolated to other areas in order to prevent soil degradation processes by water erosion.

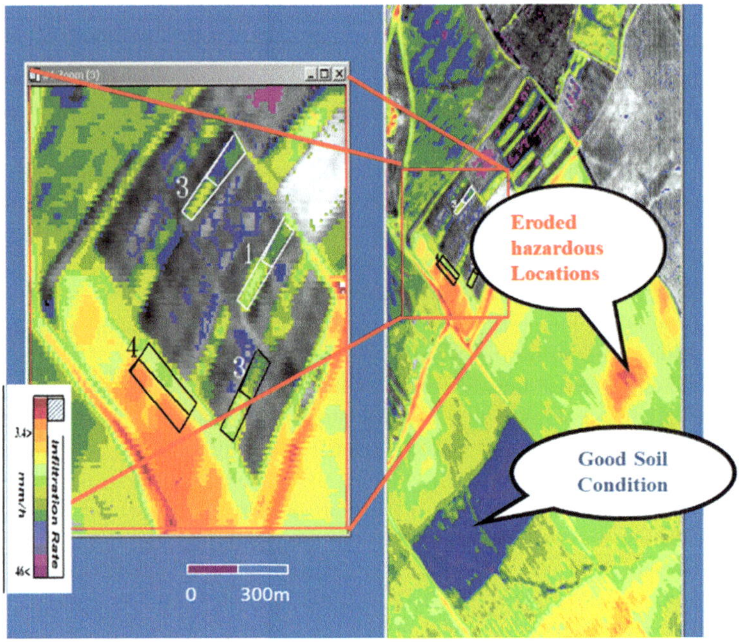


**Fig. 5** Mapping of soil infiltration on a pixel-by-pixel basis derived from AisaEAGLE hyperspectral imagery (after Ben-Dor et al. 2004)

## 3.4 Soil Erosion and Degradation

Soils are considered non-renewable resources which have a limited extent, are unequally distributed geographically, are affected by degradation due to inappropriate land management and use, but are essential to terrestrial life and human well-being (Lal 2015). Adverse climatic conditions and inappropriate human activities on land use can lead to a loss of soil quality and as a consequence to degradation of soil properties (Stolte et al. 2016). Soil degradation is a change in the soil health status resulting in a diminished capacity of the ecosystem to provide goods and services for its beneficiaries (FAO and ITPS 2015). Degraded soils do not provide services according to their original potential in an ecosystem. Soil erosion, salinization, desertification and pollution are some of the main processes affecting soil degradation, which occur over space and time. Soil erosion is a land degradation process which is often found in cultivated environments due to natural processes (e.g., climate events) and accelerated by human activities (e.g., extensive tillage). Tillage-induced soil erosion brings about the progressive removal of soil horizons and the corresponding accumulation of soil materials at the slope (Previtali 2014). Furthermore, soil erosion may reduce crop production potential, lower surface water quality and damage tile drainage systems (Toy et al. 2002). Erosion is associated with about 85% of land degradation in the world, causing up to a 17% reduction in crop productivity (Oldeman et al. 1990). In Europe, soil loss is estimated at 2.46 t $ha^{-1}$ $year^{-1}$ with a total loss of 970 Mt annually within agricultural, forests and semi-natural areas that are most affected by erosion (Panagos et al. 2015). To counteract land degradation, improvements are needed in methods of managing and monitoring of soil resources (Eswaran et al. 2001). Mapping the extent of degraded lands and monitoring the situation in erosion-threatened soils are important in order to achieve the sustainable development goals set out by the United Nations. This implies sustainable management practices, soil conservation, restoration of degraded soils and improved land management practices (Ussiri and Lal 2018).

Remote sensing technologies can provide the necessary information for carrying out the assessment and monitoring of soil degradation processes such as soil erosion (Shoshany et al. 2013). Within the Vis–NIR and SWIR spectral regions, inorganic and organic components such as clay minerals, soil organic matter (SOM), iron oxides or calcium carbonate ($CaCO_3$) interact with the electromagnetic radiation and produce characteristic absorption features in soil reflectance spectra that can be used to identify soil properties when soils are exposed at the surface and vegetation cover is low (Chabrillat et al. 2002; Nanni and Demattê 2006; Stevens et al. 2013). Hyperspectral Vis–NIR–SWIR imaging has proven to be a promising tool to characterize and map topsoil properties (Chabrillat et al. 2002; Stevens et al. 2013; Schmid et al. 2016). This is especially relevant within Mediterranean regions where the exposure of bare-soil surfaces is common due to sparse vegetation cover and management practices such as leaving fields fallow (Bartholomeus et al. 2007; Gómez et al. 2012). Nowadays, it is well established that the quality of hyperspectral data is very important for quantitative assessment of key soil properties (Ben-Dor et al. 2009), and the soil properties determined by IS can be related to soil degradation processes. At present, there are a number of hyperspectral airborne sensors available worldwide such as HyMap (www.hyvista.com), HySpex (www.hyspex.no) or Aisa (www.specim.fi) that are suitable for soil studies. Furthermore, data from satellite-borne hyperspectral sensors such as Hyperion and CHRIS-PROBA exist and could be implemented to determine common soil properties such as clay minerals, iron oxide and calcium carbonate content. These two sensors were operating since 2000 and 2001, respectively, but the Hyperion sensor was decommissioned at the beginning of 2017. Then, as summarized in Sect. 1, there are a series of upcoming hyperspectral satellite-borne sensors that will be ideal for the monitoring of surface soil properties and related soil degradation.

Studies on soil properties and conditions have been carried out using data from field spectroradiometer, hyperspectral and multispectral airborne and satellite-borne sensors to determine the spatial distribution of surface soil properties and soil erosion on a local to regional scale (Ben-Dor et al. 2006; Chabrillat 2006; Corbane et al. 2008; Hill and Schütt 2000; Schmid et al. 2008; Vrieling et al. 2008). This approach is of interest when studying soil erosion as given in the example from Schmid et al. (2016). Here, hyperspectral data supported by morphological and physicochemical ground data were able to identify, define and map soil properties that could be used as indicators to assess soil erosion and accumulation stages (SEAS) in a Mediterranean rainfed cultivated region (Camarena, near Madrid, Spain). These properties were characterized by different soil horizons that emerge at the surface as a consequence of the intensity of the erosion processes, or the result of accumulation conditions. Therefore, selected sites representing different soil properties for the SEAS were used to train the SVM classifier to obtain their spatial distribution (Fig. 6).

## 4 Challenges and Gaps

There is a known decrease of soil properties' prediction performances from laboratory to airborne imaging spectroscopy that has been demonstrated by several authors. This performance decrease is due to the combination of several effects such as differences in

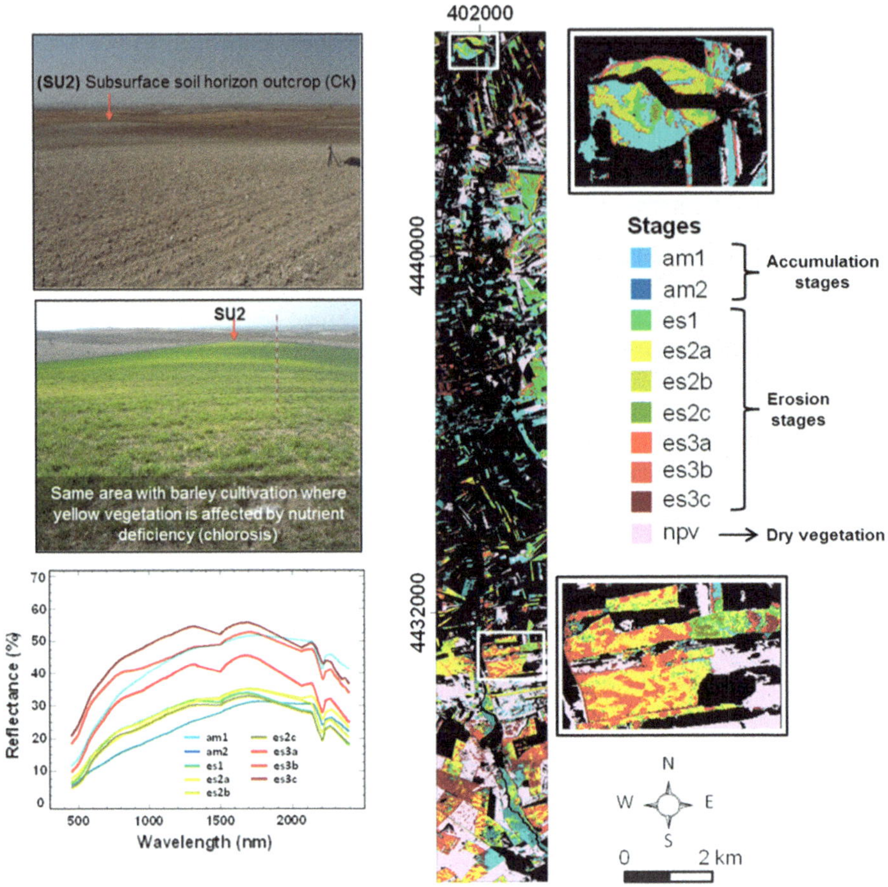


**Fig. 6** Mapping of soil erosion and degradation stages derived from AisaEAGLE and AisaHAWK hyperspectral imagery. Modified from Schmid et al. (2016)

the measurement conditions (sensors performances, different acquisition altitudes, optical angles and lighting conditions), target structure and composition differences (laboratory soil samples are dried, crushed and sieved while hyperspectral airborne targets are natural surfaces with heterogeneous surface temperatures and mixture in the field of view, moisture levels and roughness). The three main factors affecting the soil reflectance and so the model prediction accuracy have been assigned to (e.g., Stevens et al. 2008; Ben-Dor et al. 2009; Lagacherie et al. 2008): (1) the atmosphere; (2) a low signal-to-noise ratio of the imaging sensor data and (3) disruptive factors affecting the soil surface (partial vegetation cover, moisture, crust). Additionally, the availability of soil data, linked to the ad hoc need for calibration data, is a global challenge for the modelling of soil properties. Below, we present a summary of the main challenges and avenues of future research for global soil mapping and monitoring applications based on imaging spectroscopy data. They are grouped into three categories of challenges associated with different sensors, platforms, and reflectance data preprocessing, different surface conditions and different methodologies used for soil properties extraction.

### 4.1 Sensors, Platforms and Reflectance Data Preprocessing

Imaging spectroscopy data in the Vis–NIR–SWIR may be acquired by three types of platforms: unmanned aerial vehicles (UAV), satellite, and airborne which is the most common. Hyperspectral cameras that can be mounted on UAV are of recent manufacture and cover only the VNIR spectral domain. Although SWIR cameras for UAVs still need to be optimized in terms of size and weight, and the auxiliary devices needed during flight have to be miniaturized, first models are just becoming available (e.g., HySpex Mjolnir S-620 from Norsk Elektro Optikk, NEO). The UAV low altitude allows a high spatial (<50 cm) and temporal resolution which bears a very high potential for applications in precision agriculture at the field scale, with cost-effective and flexible deployment capability regarding meteorological conditions. The performance of soil properties estimation by hyperspectral UAV sensors has still to be evaluated. A first initiative of soil mapping by a UAV platform focused on a VNIR multispectral UAV sensor for organic carbon content estimation and gave encouraging results (Aldana-Jague et al. 2016).

Only one hyperspectral satellite sensor covering the full Vis–NIR–SWIR has been in orbit from 2000 to early 2017, the Hyperion sensor aboard EO-1 satellite, which provided a spatial resolution of 30 m, a spectral resolution of 10 nm, a medium–low signal-to-noise ratio and a swath of 7.5 km (Folkman et al. 2001). Few studies were carried out on soil properties mapping by Hyperion data. Gomez et al. (2008a) and Lu et al. (2013) mapped SOC content with moderate performances using Hyperion data. Lu et al. (2013) mapped total phosphorus and pH with moderate performances, whereas the cation exchange capacity prediction exhibited low efficiency. More hyperspectral Vis–NIR–SWIR satellite sensors are planned to be launched within the next years, as already noted earlier in this paper, with high radiometric quality (higher signal-to-noise ratio). Below, we review the impact of several issues related to sensors and data preprocessing on soil properties mapping.

#### 4.1.1 Impact of Spatial Resolution

Gomez et al. (2015b) studied the sensitivity of the clay content prediction with respect to the degradation of the spatial resolution on hyperspectral Vis–NIR–SWIR airborne data. Their results showed that the spatial resolution impact on clay content mapping may depend on both the spatial structure of the studied soil property and the size of the studied fields in cultivated areas. In their study area characterized by small fields (around 0.56 ha) and strong spatial structure of clay due to short-scale successions of the parent material, clay mapping performances using hyperspectral data were accurate with an $R^2$ of 0.74 at 5 m resolution and 0.66 at 30 m resolution (Fig. 7). Therefore, even over small-scale agriculture fields, the spatial resolution planned for future PRISMA, EnMAP, SHALOM and HypXIM (Lefèvre-Fonollosa et al. 2016) satellite sensors seems to be promising for clay content mapping.

#### 4.1.2 Impact of Spectral Resolution

Using laboratory spectroscopic Vis–NIR–SWIR data, Castaldi et al. (2016) obtained accurate clay prediction performances using spectra with bandwidths from 10 to 160 nm. In addition, Adeline et al. (2017) obtained accurate clay prediction performances using laboratory spectra with bandwidths of up to 200 nm. Adeline et al. (2017) observed that (1)

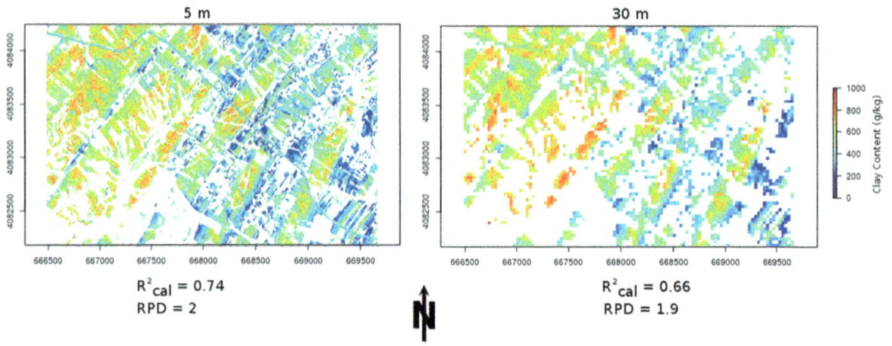


**Fig. 7** Sensitivity of clay content prediction with respect to degradation of the spatial resolution on hyperspectral Vis–NIR–SWIR airborne data. Extracted from Gomez et al. (2015b)

soil properties with large and pronounced spectral features (e.g., iron) are only slightly impacted by a decrease in spectral resolution and (2) soil properties with no spectral features (e.g., pH) may be strongly impacted by a decrease in spectral resolution, and this impact may be reduced if these soil properties rely on the beneficial effects of correlations with soil properties and with spectral features (e.g., correlation between pH and iron).

### 4.1.3 Impact of Atmospheric Attenuation

Rectification of atmospheric attenuation is critical in the IS domain, since if it is not properly done, it may influence the spectral signals that mistakenly can be assumed as part of the soil chromophores and increase noise in the signal. The issue of precise atmospheric correction has been studied by many researchers, and a combination of radiative transfer modelling and empirical methods is found to solve most of the past problems. In recent publications, the sensor radiometric instability problems as well as new and simple indicators to judge and correct systematic effects in airborne IS sensors have been developed by Brook and Ben-Dor (2014, 2015). Also, Gomez et al. (2015b) studied the sensitivity of the clay content prediction due to atmospheric effects on hyperspectral Vis–NIR–SWIR airborne data and simulated satellite data. They observed that when a correct compensation of atmosphere effects was performed using an inverse radiative transfer model, only slight differences were observed between clay content maps obtained using airborne imagery and simulated satellite imagery.

## 4.2 Soil Surface Conditions: Soil Moisture and Soil Roughness, Partial Vegetation Cover

Short-term variations in soil surface conditions due to meteorological conditions or land management practices such as ploughing and greening of the crop fields to prevent soil erosion have all an effect on soil reflectance. Soil reflectance decreases with increasing soil moisture and with increasing soil roughness, and nonlinear effects on soil spectral features are also observed. The increase in partial vegetation cover changes completely the shape of soil spectral reflectance. In turn, variations such as in soil moisture, soil surface roughness and vegetation cover have a strong impact on the quality and accuracy of the prediction of the soil properties. Here, we describe the latest developments in the assessment of the impact of these effects, and on the development of procedures to take into account or correct for these effects, in the frame of soil mapping and monitoring applications.

#### 4.2.1 Soil Moisture

The impact of soil moisture on the surface reflectance and examples of the successful quantification of topsoil gravimetric water content based on imaging spectroscopy data are presented in detail in Sect. 3.2. Mainly, three types of approaches based on empirical models are commonly used for SM estimation: spectral indices (Haubrock et al. 2008a, b; Fabre et al. 2015), exponential or Gaussian spectral models (Bryant et al. 2003; Whiting et al. 2004a; Liu et al. 2015) and geostatistical models (Fabre et al. 2015). The use of these methods for the inverse retrieval of the impact of soil moisture on the surface reflectance is not obvious as they need a priori knowledge or field data for calibration or need to be tested for wider applications in the case of the spectral indices. Several authors (Bogrekci and Lee 2006; Minasny et al. 2011; Castaldi et al. 2015; Ge et al. 2014; Liu et al. 2015) focused on the development of accurate procedures to correct or reduce the effect of soil moisture on reflectance data. Nevertheless, these studies are not commonly applicable on airborne IS. They are complex and focus on only one single soil property, or they need additional laboratory analyses that make them not yet adequate for remote sensing applications. Very recent further developments, however, use physical modelling (Bablet et al. 2017) or an adapted soil spectral index (Diek et al. 2017a) to successfully derive bare-soil spectra corrected for soil moisture and soil roughness effects. Furthermore, Diek et al. (2017a) were able to apply it on APEX imagery and validate the values against laboratory dry-soil spectra, showing the potential for further application in soil mapping based on airborne IS data.

#### 4.2.2 Soil Roughness

Although there is a full literature on the categories of soil roughness, on the effect of variable soil roughness on reflectance data and on the BRDF simulation of rough soil (shown in Sect. 2.2), the applicability of these studies for the correction of soil roughness effects on airborne imaging spectroscopy data is not yet resolved. Several studies focus on the decrease of the effect of soil roughness on reflectance data (Wu et al. 2009; Croft et al. 2014; Denis et al. 2014; Ji et al. 2015). But they are not applicable to remote sensing spectroscopy applications due to their complexity, or limitations in terms of not including all categories or all aspects of soil roughness effects. Nevertheless, one can cite the recent efforts of Bablet et al. (2017) and Diek et al. (2017a) for the correction of the combined effects of soil moisture and soil surface roughness on soil reflectance and the potential for application on airborne IS data (Diek et al. 2017a).

Current studies on the impact of soil crust property on soil reflectance demonstrated that the upper soil surface and the undisturbed surface properties are actually seen by the IS sensor. The soil proxy analysis that is being developed using soil spectral libraries is problematic to be directly implemented at the airborne scale to the hyperspectral sensors, and therefore, the soil crust remains one of the properties that needs to be spectrally modelled in order to combine laboratory and field to the airborne domains. This is mainly because the soil spectral library is sampled within 0–5 cm and may disturb the soil upper surface. Accordingly, the reflectance seen by the IS sensor is not the reflectance that is used to generate the proxy model and thus the spectral library model cannot be directly used for the IS data. A precaution then has to be taken where soil crust may evolve in the soil measured by IS sensors. Nonetheless, as a generic model can be obtained from different soils, developing a soil physical crust spectral library can be an added value to the current soil spectral libraries and should be further studied, developed and used.

### 4.2.3 Partial Vegetation Cover

Soils are not bare in most cases while vegetation, organic and inorganic debris are partially or completely covering the soil in question. According to Bartholomeus et al. (2007), vegetation cover of more than 20% prevents accurate soil properties estimation. In situations with partial soil exposure variations, Rodger and Cudahy (2009) showed that the depth of the clay absorption feature can be attributed to varying clay contents as well as differing proportions of soil exposure. Bartholomeus et al. (2011) show that with only a few per cent fractional maize cover, the accuracy of a PLSR-based SOC prediction model drops dramatically. As a result, most studies on soil properties mapping focus on bare or semi-bare areas for soil properties prediction, associated with natural bare areas or with crop areas just after harvesting or just before or after sowing. Then, two-step approaches are mostly used for operational soil properties mapping based on imaging spectroscopy, first removing the pixels with high vegetation cover and then applying soil algorithms in bare and semi-bare areas. For example, Chabrillat et al. (2011, 2016) in the HYSOMA and EnSoMAP soil mapping tools derive green and dry vegetation cover based on vegetation indices and develop a soil dominant pixel image on which the soil mapping tools will be applied. The use of multi-temporal data composites from airborne imaging spectroscopy (e.g., Gerighausen et al. 2012; Diek et al. 2016) is advocated to develop larger bare-soil maps. Experience in this domain from multispectral satellites (e.g., Blasch et al. 2015; Rogge et al. 2017; Diek et al. 2017b; Demattê et al. 2018) might help to develop in future such strategies based on hyperspectral satellites.

Simultaneously, intensive research is performed on the reduction of the impact of vegetation cover in mixed soil–vegetation spectra. Many studies look at the removal of vegetation component in the soil reflectance spectrum by spectral modelization. Bartholomeus et al. (2011) used the residual spectral unmixing (RSU) to filter out the influence of maize and could do SOC mapping with accuracies comparable to studies performed on bare-soil surfaces. Then, the "double-extraction" approach was introduced by Ouerghemmi et al. (2011), which consists of a first extraction of an estimated soil spectrum from mixed Vis–NIR–SWIR spectra using a semi-blind source separation method and a second extraction of the soil property content from the estimated soil spectrum, using the PLSR method calibrated over bare-soil pixels. First results, obtained by Ouerghemmi et al. (2016) were encouraging, as this "double-extraction" approach using a nonnegative matrix factorization method for the second extraction and applied to HyMap Vis–NIR–SWIR imaging data, allowed to map clay content over more than 63% of study area (59.5% of semi-vegetated area and 3.5% of bare soil). The nonnegative matrix factorization method was also tested by Li et al. (2017). They demonstrated that this method is feasible to discriminate the vegetation signals from the soil backgrounds in varying densities ranging from 14 to 94.6%. Although implemented to retrieve a pure vegetation spectrum from every pixel, the other way can be done as well, i.e. retrieve soil spectra out of dense vegetation coverage. Similarly, Kuester et al. (2017) use the HySimCaR (hyperspectral simulation of canopy reflectance) model (Kuester et al. 2014) to generate virtual crop landscapes with variable type of cover (early green phenological stages and dry stubbles), fractional cover and different soil backgrounds with varying SOC and textural composition and to assess the impact of vegetation cover on soil spectral features. They show that the impact of vegetation cover is variable depending on the soil background, and furthermore, correction procedures are proposed that need to be further tested and validated in wider context.

### 4.3 Methodologies for Soil Properties Extraction

Two intrinsic limitations in the use of IS for soil mapping are that (1) the sensor sees only the upper surface (about 50 mm) and thus cannot do a pedogenic mapping where the entire soil profile is needed in order to classify the soil order and (2) the high costs and complexity of commonly used airborne campaigns and associated data analyses and processing hinder an extensive use of the technology. It requires expensive sensors, flight hours, professional manpower, expert knowledge and a sophisticated infrastructure that cannot be regularly used. This requires that the problems that IS should solve are of high importance to the end users and that IS can provide a better view and economic benefits than the traditional methods such as wet chemistry. As such, IS can pick important surface properties that evolve on the soil surface such as OM, surface moisture, dust accumulation, soil deposition or erosion, physical or biogenic crust sand compaction.

With the upcoming availability of the next generation of spaceborne imaging spectroscopy sensors, an important issue is linked with the development of accessible methods for the generation and delivery of soil image products to the geoscience community. Established techniques for soil property determination from airborne imaging spectroscopy are mainly either based on multivariate statistical methods such as PLSR and SVM or based on the direct analyses of the spectral reflectance signal such as spectral feature analyses (measurements of spectral features, slopes, Gaussian modelling). Both types of methodologies present different conditions of use and a priori expected performances which make them use on a per case basis (e.g., see analyses of Chabrillat et al. 2014; Escribano et al. 2017), and in all case both need ground data for quantitative soil properties mapping. Although PLSR is by far the most used method in quantitative soil spectroscopy, direct reflectance-analyses methods present the advantage that they are more generic and can be implemented on an operational basis in software interfaces, and repeatability and robustness of the soil prediction models can be better provided. Furthermore, such methods allow fully automatic derivation of semi-quantitative soil maps that do not require the need for ground data and do not require expert knowledge. For example, methods based on the direct analyses of spectral reflectance (absorption band analyses, slopes, spectral indices, Gaussian modelling) were implemented in the HYSOMA and EnSoMAP software interfaces for soil mapping (e.g., Chabrillat et al. 2011, 2016). HYSOMA (Fig. 8), developed originally as demonstrator of end-to-end processing chain for the delivery of soil maps to the users and provided since 2012 on the Web for free as stand-alone interface, stands nowadays with worldwide > 130 users and demonstrated the need for such tools. Also, further implemented as EnSoMAP in the EnMAP-box tool (Van der Linden et al. 2015), it shows the potential of such soil products to be commonly provided for future soil users of upcoming IS spaceborne imagery.

An important methodological issue in quantitative soil properties mapping based on IS is linked with the need for ad hoc ground truth data for model calibration and validation, and standards and protocols for ground data acquisition. Since the pixel size cannot really represent point measurements, at least 3 × 3 pixels have to be averaged for both spectral ground truth measurements and chemical analysis. Different strategies for soil sampling at the scale of the IS sensor are developed that are mostly similar in their concept but also

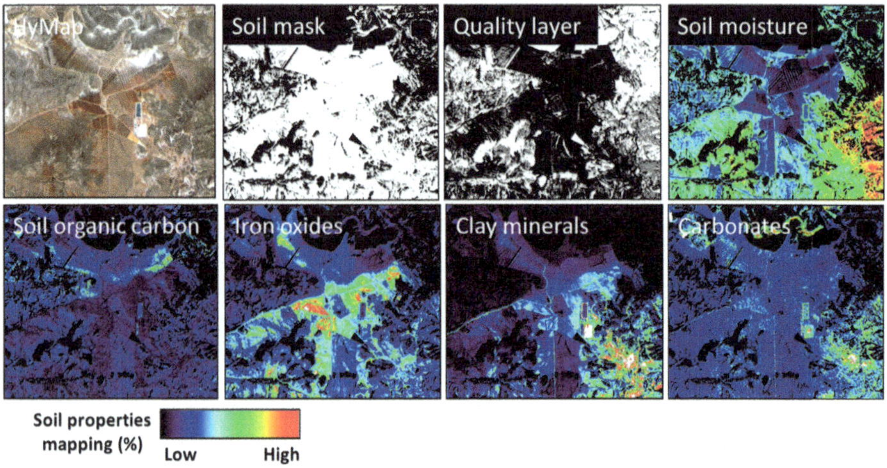


**Fig. 8** Operational semi-quantitative soil properties mapping derived from airborne HyMap hyperspectral imagery. Map outputs were created using the HYSOMA software (Chabrillat et al. 2011, 2016)

with differences such as in Stevens et al. (2013), Schmid et al. (2016), or Castaldi et al. (2019). Overall, in most reviewed studies that successfully applied the quantitative chemometric approach to IS data, it was found that these studies lean heavily mostly on homogeneous areas (in terms of parent material and soil-forming processes "soilscapes"). Ground truth measurement and sampling to represent pixel in the imagery would need to sample area in a standard way that will cope for all the issues previously mentioned such as representativity of the sampled area, spatial resolution, bare and dry area (not affecting the soil surface and controlling the BRDF effects). Apparently, such an approach is not yet available, whereas Ben-Dor et al. (2017) developed an apparatus that is capable of maintaining a very stable and standard spectral measurement of soil in the field without relying on the natural illumination and atmosphere conditions while overcoming the above-mentioned effects.

A novel strategy to overcome the need for the acquisition of large sets of ground measurements is to consider the use of large soil databases and/or legacy soil data. Although most of the published research nowadays concerns the calibration of prediction models using local soil datasets, recently this strategy has attracted the attention of many scientists due to the higher potential for large-scale soil mapping and for reduction of the soil sample collection cost and time. Gomez et al. (2016) showed that using legacy soil data collected over and/or around a large area in the Mediterranean region, they could calibrate a clay spectral index and derive topsoil clay content maps based on Aisa-Dual IS data. Furthermore, a possible way to solve the issues related to the collection of a representative ad hoc local dataset is the use of large soil spectral libraries. Castaldi et al. (2018a, b), exploiting the potential of the EU-wide LUCAS soil database described in, e.g., Toth et al. (2013), proposed a new approach based on uniform analysis protocol and PLSR models for SOC mapping in croplands in Europe. They demonstrated the capability to be applied for laboratory prediction of SOC but also for quantitative SOC mapping based on APEX hyperspectral imagery.

The challenges posed by the use of multivariate statistical models for soil mapping raise questions as to what extent can the quantitative models provide reliable and useful information, transferability of the models, dependency on the availability of soil data, or whether is it always possible to adopt the chemometric strategy for real IS remote sensing data regarding the difference between airborne and laboratory spectroscopy. In general, it has been shown that this technique can be used if the soil samples preserved the exact surface properties seen by the sensor but then, only some soil properties can be modelled (e.g., crust dust, OM and surface moisture). If the soils are ploughed and the soil surface is broken to represent the 0–5 cm that is usually measured in soil spectral libraries, it is more likely that the laboratory reflectance measurements will be comparable to the soil measurements from the air and space. Furthermore, in the frame of the upcoming availability of high-quality spaceborne IS data at global scale covering the whole Earth, an additional challenge linked to the transfer of the soil spectral models from regional to global scale is with respect to the different global soil regions and resulting implications for the empirical and statistical modelling. In that regard, extensive efforts are being devoted to the development of global soil databases and global coherent soil spectral libraries (SSLs) that will allow to develop robust calibrations for multivariate prediction of soil properties applicable at the regional and the global scale. For example, SSLs have been developed recently, especially in the large-scale domain at different levels including national level (e.g., Brodsky et al., 2011; Knadel et al. 2012; Tsakiridis et al. 2017), European or continental level (Toth et al. 2013; Stevens et al. 2013; Romero et al. 2018) and global level (Viscarra Rossel et al. 2016). Then, associated with the upcoming availability of the next generation of hyperspectral Vis–NIR–SWIR satellite data for the entire globe, and the upcoming availability of improved tools for the extraction of soil properties, it shall pave the way for inexpensive methods for the delivery of high spatial-resolution, fully quantitative, soil properties maps to soil users.

## 5 Conclusion: Observational Requirements and Future Directions in Imaging Spectroscopy

### 5.1 Summary of Achievements, Challenges and Prospect for the Use of Future Spaceborne Imaging Spectroscopy for Soil Applications

Imaging spectroscopy of soils has proven the capability to enable the quantitative evaluation of many soil properties such as SOC, clay, iron, carbonates, sand and pH, due to the presence of soil chromophores in the reflectance spectrum. Common properties modelling approaches use multivariate statistics based on soil data for calibration/validation or direct spectral features for the extraction of the soil properties. Then, good soil properties predictions can be achieved in local areas for exposed soils in appropriate surface conditions, as demonstrated in many papers from local-scale (airborne) to regional-scale (spaceborne) applications. With the upcoming availability of the next generation of orbiting hyperspectral sensors and routinely delivered high spectral resolution images for the entire globe, a new opportunity toward the operational quantitative monitoring of soil surfaces at large scales arises. These satellites would provide at the regional to the global scale (spatial resolution from 10 to 30 m) routinely soil maps at different times. Besides the proven capabilities of IS air- and spaceborne technology for quantitative soil properties mapping, other specific fields in the soil science discipline can learn from this technology, such as soil mineralogy, soil chemistry, soil fertility, soil physics, soil microbiology, soil pollutions, fertilizers, management, weathering and conservation.

Nonetheless, advances are still necessary to fully develop imaging spectroscopy soil products that can support, in a credible manner, global digital mapping and monitoring of soils as a support of societal needs for soil information. Challenges and gaps are identified due to low signal-to-noise ratio, geometric and optical distortions, atmospheric attenuation, vegetation cover, soil moisture and roughness effects, and sensing only of the first ~ 100–200 μm (Ciani et al. 2005) of the soil body. Further identified challenges are the transfer of modelling approaches from the local to global scale and the availability of global soil data including harmonized soil spectral libraries, Attempts to overcome these problems are areas of intense research nowadays.

Precision agriculture and food security issues are two good examples where both point and imaging spectroscopy can be exploited. The development of the IS technology leans toward better quality sensors and miniaturization, but also on new development of platforms such as UAVs, balloons and micro-satellites. Together, this direction will enable more users to use the IS technology in general and for soil science in particular and will promise that this technology has a bright future for soil scientists, farmers and stakeholders.

## 5.2 User and Observational Requirements

Several recent studies looked at the potential of upcoming IS satellite missions for soil mapping, based on satellite-simulated data (Chabrillat et al. 2014; Castaldi et al. 2015, 2016; Gomez et al. 2015b; Steinberg et al. 2016). It was shown, e.g., that using simulated EnMAP data at 30 m resolution applying semi-operational methods, it is possible to predict soil properties such as SOC, clay and iron oxide content, with slightly reduced accuracy compared to airborne hyperspectral imagery at spatial scales < 5 m. The spatial distribution of the soil properties was in general coherent between the simulated EnMAP and the airborne mapping, although the variability in spatial, spectral and radiometric resolutions. Castaldi et al. (2016) also clearly demonstrated the improvement in accuracy for the estimation of soil variables over bare soils using forthcoming hyperspectral imagers, as compared to current-generation multispectral sensors such as ALI, Landsat 8, Sentinel-2, covering spatial resolutions from 10 to 30 m, but providing much lower spectral coverage and spectral resolution.

In this regard, we can define mission scenarios for two general user and observational requirements associated with two different spatial scales of user needs, linked to food security, resource management and sustainable agriculture (Table 2), such as (1) precision farming and (2) global mapping and monitoring of our soils status. These scenarios represent current technology trade-offs associated with IS technologies that could be used as mission requirements.

The first mission would provide the link between laboratory- and field-scale soil spectroscopy and will allow to support precision farming activities with the production of accurate soil properties maps at field scale for farmers and regional administrations to support, e.g., EU subsidy policies on fertilizer use, and improvement in SOC storage. This first user need is currently performed by laboratory analyses, field sensors and UAV and airborne sensors when possible. The given temporal requirement of 1 week would be useful to monitor short-term surface changes and is needed to catch the short bare-soil period between two cropping seasons. The second mission would focus on the production of large-scale soil maps for regional and national administration and impact at decision-making level of regional-scale, national and EU policies.

**Table 2** User and observational requirements for an imaging spectroscopy mission focused on soil properties mapping and food security

| User objective | Spatial requirements | Temporal requirements | Spectral requirements | Imaging spectroscopy capabilities |
|---|---|---|---|---|
| Precision farming (field scale) | GSD < 5 m | ~ 1 week | Vis–NIR–SWIR<br>High spectral resolution < 10 nm | Target mission/regional mapping |
| Large-scale soil mapping and monitoring | GSD 20–30 m | ~ 2 weeks | Vis–NIR–SWIR<br>High spectral resolution ~ 10 nm | Global mapping mission |

Soils are relevant to a long list of national, European and international policies, including the EU Soil Thematic Strategy, Common Agricultural Policy, Food Safety and Food security (FAO), UN Convention to Combat Desertification, and SOC is the common denominator. Overall, there is a need for more soil mapping at local level (target mission) and regional level (mapping mission). It has been widely recognized that a global effort for a better link between producers and users of soil and land information is needed that takes into account the ongoing developments in new and emerging technologies (such as remote sensing, mobile soil testing and digital soil mapping), availability of digital data (Web-based, app-based applications), ability of mid-users to utilize new soil information in policy- and decision- making relevant information, needs of the different stakeholders (farmer, regional administrator, country ministry) and impact at decision-making level (local and regional scale, national and EU policies) (Chabrillat et al. 2015).

New soil information to support societal needs can be brought by soil spectroscopy first at the laboratory and field scale, as it can be used in a preliminary way to analyse soil samples quickly and cheaply, and it combines well with digital soil mapping, secondly at the remote sensing scale from air- and spaceborne sensors for global soil mapping and monitoring. For this, the IS technology is available currently from hand-held, airborne and few spectrally limited spaceborne instruments, and soon potentially with high-quality data from global satellite missions at high spatial resolution (< 30 m), which will no doubt support regional-scale soil mapping and soil monitoring combined with digital soil mapping approaches and soil datasets over the world. Nevertheless, further efforts are needed from the imaging spectroscopy community toward more methodological developments to fulfil the entire potential of soil spectroscopy from airborne and orbital vehicles.

**Acknowledgements** This paper is an outcome of a Workshop on requirements, capabilities and directions in spaceborne imaging spectroscopy held at the International Space Science Institute (ISSI) in Bern, Switzerland, in November 2016. The support of ISSI is gratefully acknowledged. The EnMAP science preparation program and EnMAP coordination team are gratefully acknowledged without which the ISSI Workshop would not have taken place.

## Compliance with Ethical Standards

**Conflict of interest** The authors declare that they have no conflict of interest.

## References

Adeline K, Gomez C, Gorretta N, Roger JM (2017) Predictive ability of soil properties to spectral degradation from laboratory Vis–NIR spectroscopy data. Geoderma 288:143–153

Agassi M, Shainberg I, Morin J (1981) Effect of electrolyte concentration and soil sodicity on infiltration rate and crust formation. Soil Sci Soc Am J 45(5):848–851
Agassi M, Morin J, Shainberg I (1985) Effect of raindrop impact energy and water salinity on infiltration rates of sodic soils. Soil Sci Soc Am J 49:186–190
Aldana-Jague E, Heckrath G, Macdonald A, Van Wesemael B, Van Oost K (2016) UAV-based soil carbon mapping using VIS–NIR (480–1000 nm) multi-spectral imaging: potential and limitations. Geoderma 275(1):55–66
Alonso M, Rodríguez-Caballero E, Chamizo S, Escribano P, Cantón Y (2014) Evaluación de los diferentes índices para cartografiar biocostras a partir de información espectral. Rev Esp Teledetec 42:63–82
Arrouays D, Lagacherie P, Hartemink AE (2017) Digital soil mapping across the globe. Geoderma Reg 9:1–4
Bablet A, Jacquemoud S, Viallefond F, Fabre S, Briottet X (2017) Modeling bare soil reflectance in the solar domain as a function of water content and surface roughness. In: Abstract 10th EARSeL SIG Imaging Spectroscopy Workshop, 19–21 April 2017, Zurich, Switzerland
Bablet A, Vu PVH, Jacquemoud S, Viallefont-Robinet F, Fabre S, Briottet X, Sadeghid M, Whiting ML, Baret F, Tian J (2018) MARMIT: a multilayer radiative transfer model of soil reflectance to estimate surface soil moisture content in the solar domain (400–2500 nm). Remote Sens Environ 217:1–17
Bach H, Mauser W (1994) Modelling and model verification of the spectral reflectance of soils under varying moisture conditions. In: Proceedings IGARSS 94. 'Surface and atmospheric remote sensing: technologies, data analysis and interpretation', Pasadena, California, 8–12 August 1994, IEEE international. IEEE, Piscataway, NJ, pp 2354–2356
Bartholomeus H, Epema G, Schaepman M (2007) Determining iron content in Mediterranean soils in partly vegetated regions, using spectral reflectance and imaging spectroscopy. Int J Appl Earth Obs Geoinf 9:194–203
Bartholomeus H, Kooistraa L, Stevens A, van Leeuwen M, van Wesemael B, Ben-Dor E, Tychon B (2011) Soil organic carbon mapping of partially vegetated agricultural fields with imaging spectroscopy. Int J Appl Earth Obs Geoinf 13:81–88
Baumgardner MF, Silva LF, Biehl LL, Stoner ER (1985) Reflectance properties of soils. Adv Agron 38:1–44
Ben-Dor E, Banin A (1995a) Near infrared analysis (NIRA) as a method to simultaneously evaluate spectral featureless constituents in soils. Soil Sci 159(4):259–270
Ben-Dor E, Banin A (1995b) Near infrared analysis (NIRA) as a rapid method to simultaneously evaluate several soil properties. Soil Sci Soc Am J 59:364–372
Ben-Dor E, Inbar Y, Chen Y (1997) The reflectance spectra of organic matter in the visible near-infrared and short wave infrared region (400–2500) during a controlled decomposition process. Remote Sens Environ 61:1–15
Ben-Dor E, Irons JR, Epema GF (1999) Remote sensing of the earth sciences: manual of remote sensing. Soil Reflectance 3:111
Ben-Dor E, Patkin K, Banin A, Karnieli A (2002) Mapping of several soil properties using DAIS-7915 hyperspectral scanner data: a case study over clayey soils in Israel. Int J Remote Sens 23:1043–1062
Ben-Dor E, Goldlshleger N, Benyamini Y, Agassi M, Blumberg DG (2003) The spectral reflectance properties of soil structural crusts in the 1.2 to 2.5 µm spectral region. Soil Sci Soc Am J 67(1):289–299
Ben-Dor E, Goldshalager N, Braun O, Kindel B, Goetz AFH, Bonfil D, Agassi M, Margalit N, Binayminy Y, Karnieli A (2004) Monitoring of infiltration rate in semiarid soils using airborne hyperspectral technology. Int J Remote Sens 25:1–18
Ben-Dor E, Levin TN, Singer A, Karnieli A, Braun O, Kidron GJ (2006) Quantitative mapping of the soil rubification process on sand dunes using an airborne hyperspectral sensor. Geoderma 131:1–21
Ben-Dor E, Taylor GR, Hill J, Demattê JAM, Whiting ML, Chabrillat S, Sommer S (2008) Imaging spectrometry for soil applications. Adv Agron J 97:321–392
Ben-Dor E, Chabrillat S, Demattê JAM, Taylor GR, Hill J, Whiting ML, Sommer S (2009) Using imaging spectroscopy to study soil properties. Remote Sens Environ 113:S38–S55
Ben-Dor E, Kafri A, Varacalli G (2014) SHALOM: an Italian–Israeli hyperspectral orbital mission—update. In: International geoscience and remote sensing symposium, Quebec, Canada, 13–18 July 2014
Ben-Dor E, Granot A, Notesco G (2017) A simple apparatus to measure soil spectral information in the field under stable conditions. Geoderma 306:73–80
Ben-Dor E, Chabrillat S, Demattê JAM (2018) Characterization of soil properties using reflectance spectroscopy. In: Thenkabail PS, Lyon JG, Huete A (eds) Hyperspectral remote sensing of vegetation, 2nd edn. Four volume set—volume I: fundamentals, sensor systems, spectral libraries, and data mining for vegetation. CRC Press, Boca Raton, pp 187–247. ISBN 978-1-138-05854-5
Benyamini Y, Unger PW (1984) Crust development under simulated rainfall on four soils. In: Agronomy abstracts. ASA, Madison, WI, pp 243–244

Bishop JL, Pieters CM, Edwards JO (1994) Infrared spectroscopic analyses on the nature of water in montmorillonite. Clays Clay Miner 42:702–716
Blasch G, Spengler D, Hohmann C, Neumann C, Itzerott I, Kaufmann H (2015) Multitemporal soil pattern analysis with multispectral remote sensing data at the field-scale. Comput Electron Agric 113:1–13
Bogrekci I, Lee WS (2006) Effects of soil moisture content on absorbance spectra of sandy soils in sensing phosphorus concentrations using UV–VIS–NIR spectroscopy. Trans ASABE 49:1175–1180. https://doi.org/10.13031/2013.21717
Boiffin J (1986) Stages and time-depedency of soil crusting in situ. In: Callebaut F, Gabriels D, De Boodt M (eds) Assessment of soil surface sealing and crusting. Flanders Research Center for Soil Erosion and Soil Conservation, Ghent, pp 91–98
Bowers S, Hanks RJ (1965) Reflectance of radiant energy from soils. Soil Sci 100:130–138
Brennan B, Bandeen WR (1970) Anisotropic reflectance characteristics of natural Earth surfaces. Appl Opt 9:405–412
Briottet X et al (2017) European Hyperspectral Explorer: HYPEX-2—a new space mission for vegetation biodiversity, bare continental surfaces, coastal zones and urban area ecosystems. In: Abstract 10th EARSeL SIG imaging spectroscopy workshop, 19–21 April 2017, Zurich, Switzerland
Brodsky L, Klement A, Penizek V, Kodesova R, Boruvka L (2011) Building soil spectral library of the Czech soils for quantitative digital soil mapping. Soil Water Res 6:165–172
Brook A, Ben-Dor E (2014) Practical example of the supervised vicarious calibration (SVC) method: VALCALHYP airborne hyper spectra campaigns under EUFAR. In: EaRSeL eProceedings, no 2, pp 38–48
Brook A, Ben-Dor E (2015) Supervised vicarious calibration (SVC) of multi-source hyperspectral remote-sensing data. Remote Sens 7:6196–6223
Bryant R, Thoma D, Moran S, Holifield C, Goodrich D, Keefer T, Paige G, Williams D, Skirvin S (2003) Evaluation of hyperspectral, infrared temperature and radar measurements for monitoring surface soil moisture. In: Proceedings of the first interagency conference on research in the watersheds, Benson (USA), pp 528–533
Carmon N, Ben-Dor E (2017) An advanced analytical approach for spectral-based modelling of soil properties. Int J Emerg Technol Adv Eng 7:90–97
Castaldi F, Palombo A, Pascucci S, Pignatti S, Santini F, Casa R (2015) Reducing the influence of soil moisture on the estimation of clay from hyperspectral data: a case study using simulated PRISMA data. Remote Sens 7:15561–15582. https://doi.org/10.3390/rs71115561
Castaldi F, Palombo A, Santini F, Pascucci S, Pignatti S, Casa R (2016) Evaluation of the potential of the current and forthcoming multispectral and hyperspectral imagers to estimate soil texture and organic carbon. Remote Sens Environ 179:54–65
Castaldi F, Chabrillat S, Jones A, Vreys K, Bomans B, van Wesemael B (2018a) Soil organic carbon estimation in croplands by hyperspectral remote APEX Data using the LUCAS topsoil database. Remote Sens 10:153. https://doi.org/10.3390/rs10020153
Castaldi F, Chabrillat S, Chartin C, Genot V, Jones AR, van Wesemael B (2018b) Using LUCAS topsoil database to estimate soil organic carbon content in croplands sampled in Belgium and Luxembourg. Eur J Soil Sci 69(4):592–603
Castaldi F, Chabrillat S, van Wesemael B (2019) Sampling strategies for soil property mapping using multispectral Sentinel-2 and hyperspectral EnMAP satellite data. Remote Sens (in press)
Chabrillat S (2006) Land degradation indicators: spectral indices. Ann Arid Zone Spec Issue Land Special Assess 45(3–4):331–354
Chabrillat S, Goetz AFH, Krosley L, Olsen HW (2002) Use of hyperspectral images in the identification and mapping of expansive clay soils and the role of spatial resolution. Remote Sens Environ 82:431–445
Chabrillat S, Eisele A, Guillaso S, Rogaß C, Ben-Dor E, Kaufmann H (2011) HYSOMA: an easy-to-use software interface for soil mapping applications of hyperspectral imagery. In: Proceedings 7th EARSeL SIG imaging spectroscopy workshop, Edinburgh, Scotland
Chabrillat S, Whiting ML, Guillaso S, Eisele A, Haubrock SN, Kaufmann H (2012) Quantitative mapping of surface soil moisture with hyperspectral imagery using the HYSOMA interface. In: Abstract 2012 IEEE international geoscience and remote sensing symposium, IGARSS 2012, Muenchen, Germany
Chabrillat S, Foerster S, Steinberg A, Segl K (2014) Prediction of common surface soil properties using airborne and simulated EnMAP hyperspectral images: impact of soil algorithm and sensor characteristic. In: Proceedings 2014 IEEE international geoscience and remote sensing symposium, IGARSS 2014, Québec city, QC, Canada, pp 2914–2917
Chabrillat S, Hanegraaf M, Sommer R, van den Bosch R, Montanarella L, van Wesemael B, Schwilch G, Harthoorn J, Skalsky R, Braslow J, Mills J, Verzandvoort S, Obersteiner M (2015) Soil and

land information: how to support decision making? Rapporteurs report, Global Soil Week 2015, Germany, 20–23 April 2015
Chabrillat S, Guillaso S, Rabe A, Foerster S, Guanter L (2016) From HYSOMA to ENSOMAP—a new open source tool for quantitative soil properties mapping based on hyperspectral imagery from airborne to spaceborne applications. General Assembly European Geosciences Union, Vienna, Austria, 2016, geophysical research abstracts, vol 18, EGU2016-14697
Chamizo S, Stevens A, Cantón Y, Miralles I, Domingo F, Van Wesemael B (2012) Discriminating soil crust type, development stage and degree of disturbance in semiarid environments from their spectral characteristics. Eur J Soil Sci 63:42–53
Chang DH, Islam S (2000) Estimation of soil physical properties using remote sensing and artificial neural network. Remote Sens Environ 74(3):534–544
Chang CW, Laird DA, Mausbach MJ, Hurburgh CR (2001) Near-infrared reflectance spectroscopy–principal components regression analyses of soil properties. Soil Sci Soc Am J 65:480–490
Ciani A, Goss KU, Schwarzenbach RP (2005) Light penetration in soil and particulate minerals. Eur J Soil Sci 56(5):561–574
Cierniewski J (1987) A model for soil surface roughness influence on the spectral response of bare soils in the visible and near-infrared range. Remote Sens Environ 23:97–115
Cierniewski J (2012) Satellite observation of bare soils for their average diurnal albedo approximation. In: Ist international conference on sensor networks, 24–26 February, 2012, Rome, Italy, pp 1–6
Cierniewski J, Courault D (1993) Bidirectional reflectance of bare soil surface in the visible and near-infrared range. Remote Sens Rev 7:321–339
Cierniewski J, Guliński M (2010) Furrow microrelief influence on the directional hyperspectral reflectance of soil at various illumination and observation conditions. IEEE Trans Geosci Remote Sens 48(11):4143–4148
Cierniewski J, Gdala T, Karnieli A (2004) A hemispherical–directional reflectance model as a tool for understanding image distinctions between cultivated and uncultivated bare surfaces. Remote Sens Environ 90:505–552
Cierniewski J, Karnieli A, Hermann I, Królewicz S, Kuśnierek K (2010) Soil surface illumination at micro-relief scale and soil BRDF data collected by a hyperspectral camera. Int J Remote Sens 31:2151–2157
Cierniewski J, Karnieli A, Kuśnierek K, Herrmann I (2013) Approximating the average daily surface albedo with respect to soil roughness and latitude. Int J Remote Sens 34(9–10):3416–3424
Cierniewski J, Kaźmierowski C, Królewicz S (2015) Evaluation of the effects of surface roughness on the relationship between soil BRF data and broadband albedo. IEEE J Sel Top Appl Earth Obs Remote Sens 8(4):1528–1533
Cipra JE, Baumgardner MF, Stoner ER, MacDonald RB (1971) Measuring radiance characteristics of soil with a field spectroradiometer. Soil Sci Soc Am J 35:1014–1017
Corbane C, Raclot D, Jacob F, Albergel J, Andrieux P (2008) Remote sensing of soil surface characteristics from a multiscale classification approach. CATENA 75:308–318
Coulson KL, Reynolds DW (1971) The spectral reflectance of natural surfaces. J Appl Meteorol 10:1285–1295
Croft H, Anderson K, Kuhn NJ (2012) Reflectance anisotropy for measuring soil surface roughness of multiple soil types. CATENA 93:87–96
Croft H, Anderson K, Brazier ER, Kuhn NJ (2013) Modeling fine-scale soil surface structure using geostatistics. Water Resour Res 49:1858–1870
Croft H, Anderson K, Kuhn NJ (2014) Evaluating the influence of surface soil moisture and soil surface roughness on optical directional reflectance factors. Eur J Soil Sci 65:605–612. https://doi.org/10.1111/ejss.12142
Dalal RC, Henry RJ (1986) Simultaneous determination of moisture, organic carbon and total nitrogen by near infrared reflectance spectroscopy. Soil Sci Soc Am J 50:120–123
De Jong SM (1992) The analysis of spectroscopical data to map soil and types and soil crust of Mediterranean eroded soil. Soil Technol 5:199–211
Deering DW, Eck TF, Otterman J (1990) Bidirectional reflectance of selected desert surfaces and their three-parameter soil characterization. Agric For Meteorol 52:71–90
Dehaan R, Taylor GR (2003) Image-derived spectral endmembers as indicators of salinization. Int J Remote Sens 24:775–794
Demattê JA, Campos RC, Alves MC, Fiorio PR, Nanni MR (2004) Visible–NIR reflectance: a new approach on soil evaluation. Geoderma 121(1):95–112
Demattê JAM, Morgan CLS, Chabrillat S, Rizzo R, Franceschini MHD, FdaS Terra, Vasques GM, Wetterlind J (2015) Spectral sensing from ground to space in soil science: state of the art, applications,

potential and perspectives. In: Thenkabail PS (ed) Remote sensing handbook—three volume set: land resources monitoring, modeling, and mapping with remote sensing. CRC Press, Boca Raton, pp 661–732. ISBN-10: 1482217953

Demattê JAM, Fongaro CT, Rizzo R, Safanelli JL (2018) Geospatial soil sensing system (GEOS3): a powerful data mining procedure to retrieve soil spectral reflectance from satellite images. Remote Sens Environ 212:161–175

Demetriades-Shah TH, Steven MD, Clark JA (1990) High resolution derivative spectra in remote sensing. Remote Sens Environ 33:55–64

Denis A, Stevens A, van Wesemael B, Udelhoven T, Tychon B (2014) Soil organic carbon assessment by field and airborne spectrometry in bare croplands: accounting for soil surface roughness. Geoderma 226–227:94–102. https://doi.org/10.1016/j.geoderma.2014.02.015

Diek S, Schaepman ME, de Jong R (2016) Creating multi-temporal composites of airborne imaging spectroscopy data in support of digital soil mapping. Remote Sens 8:906–934. https://doi.org/10.3390/rs8110906

Diek S, Schaepmann M, de Jong R (2017a) Correcting airborne imaging spectroscopy data for soil moisture and soil surface roughness effects in support of digital soil mapping. In: Abstract 10th EARSeL SIG imaging spectroscopy workshop, 19–21 April 2017, Zurich, Switzerland

Diek S, Fornallaz F, Schaepman ME, de Jong R (2017b) Barest pixel composite for agricultural area using time series of Landsat data. Remote Sens 9(12):1245. https://doi.org/10.3390/rs9121245

Diek S, Chabrillat S, Nocita M, Schaepman ME, de Jong R (2019) Minimizing soil moisture variations in multi-temporal airborne imaging spectrometer data for digital soil mapping. Geoderma 337:607–621

Eisele A, Lau IC, Hewson R, Carter D, Wheaton B, Ong C, Cudahy TJ, Chabrillat S, Kaufmann H (2012) Applicability of the thermal infrared spectral region for the prediction of soil properties across semi-arid agricultural landscapes. Remote Sens 4(11):3265–3286. https://doi.org/10.3390/rs4113265

Eisele A, Chabrillat S, Hecker C, Hewson R, Lau IC, Rogass C, Segl K, Cudahy TJ, Udelhoven T, Hostert P, Kaufmann H (2015) Advantages using the longwave infrared (LWIR) to detect and quantify semi-arid soil properties. Remote Sens Environ 163:296–311. https://doi.org/10.1016/j.rse.2015.04.001

Escribano P, Schmid T, Chabrillat S, Rodríguez-Caballero E, García M (2017) Optical remote sensing for soil mapping and monitoring. In: Pereira P, Brevik E, Muñoz-Rojas M, Miller B (eds) Soil mapping and process modelling for sustainable land use management. Elsevier, Amsterdam, pp 87–125. ISBN 978-0-12-805200-6

Eshel G, Levey G (2004) Spectral reflectance properties of crusted soils under solar illumination. Soil Sci Soc Am J 66:1982–1991

Eswaran H, Lal R, Reich PF (2001) Land degradation: an overview. In: Bridges EM, Penning de Vries FWT, Oldeman LR, Sombatpanit S, Scherr SJ (eds) Response to land degradation. Science Publishers, Inc., Enfield, NH, pp 20–35

European Commission (EC) (2006) Soil protection: the long story behind the strategy. Office for Official Publications of the European Communities, Luxembourg

Fabre S, Briottet X, Lesaignoux A (2015) Estimation of soil moisture content from the spectral reflectance of bare soils in the 0.4–2.5 μm domain. Sensors 15:3262–3281

FAO and ITPS (2015) Status of the world's soil resources (SWSR): main report. Food and Agriculture Organization of the United Nations, Rome

FAO/IIASA/ISRIC/ISS-CAS/JRC (2012) Harmonized World Soil Database (version 1.2). FAO, Rome

Folkman M, Pearlman J, Liao L, Jarecke P (2001) EO1/Hyperion hyperspectral imager design, development, characterization and prediction. In: Smith WL, Yasuoka Y (eds) Hyperspectral remote sensing of the land and atmosphere. SPIE proceeding, vol 4151, pp 40–51

Ge Y, Morgan CLS, Ackerson JP (2014) VisNIR spectra of dried ground soils predict properties of soils scanned moist and intact. Geoderma 221–222:61–69. https://doi.org/10.1016/j.geoderma.2014.01.011

Gerighausen H, Menz G, Kaufmann H (2012) Spatially explicit estimation of clay and organic carbon content in agricultural soils using multi-annual imaging spectroscopy data. Appl Environ Soil Sci. https://doi.org/10.1155/2012/868090

Gholizadeh A, Borůvka L, Saberioon MM, Kozák J, Vašát R, Němeček K (2015) Comparing different data preprocessing methods for monitoring soil heavy metals based on soil spectral features. Soil Water Res 10(4):218–227

Gilley JE, Kottwitz ER (1995) Random roughness assessment by the pin and chain method. Appl Eng Agric 12(1):39–43

Gilliot JM, Vaudour E, Michelin J (2017) Soil surface roughness measurement: a new fully automatic photogrammetric approach applied to agricultural bare fields. Comput Electron Agric 134:63–78

Goetz AFH, Vane G, Solomon JE, Rock BN (1985) Imaging spectrometry for Earth remote sensing. Science 228:1147–1153

Goldshleger N, Ben-Dor E, Benyamini Y, Blumberg D, Agassi M (2001) The spectral reflectance of soil's structural crust in the SWIR region 1.2–2.5 μm. Terra-Nova 13(1):12–17
Goldshleger N, Ben-Dor E, Benyamini Y, Blumberg DG, Agassi M (2002) Soil crusting and infiltration process as monitored by soil reflectance spectroscopy in the SWIR Region. Int J Remote Sens 23(19):3909–3920
Goldshleger NE, Ben-Dor E, Benyamini Y, Agassi M (2004) Soil reflectance as a tool for assessing physical crust arrangement of four typical soils in Israel. Soil Sci 169:677–687
Goldshlager N, Ben-Dor E, Chudnovsky A, Agassi M (2010) Soil reflectance as a generic tool for assessing infiltration rate induced by structural crust for heterogeneous soils. Eur J Soil Sci 60:1038–1951
Gomez C, ViscarraRossel RA, McBratney AB (2008a) Soil organic carbon prediction by hyperspectral remote sensing and field VIS–NIR spectroscopy: an Australian case study. Geoderma 146:403–411
Gomez C, Lagacherie P, Coulouma G (2008b) Continuum removal versus PLSR method for clay and calcium carbonate content estimation from laboratory and airborne hyperspectral measurements. Geoderma 148:141–148
Gomez C, Lagacherie P, Coulouma G (2012a) Regional predictions of eight common soil properties and their spatial structures from hyperspectral Vis–NIR data. Geoderma 189–190:176–185
Gomez C, Lagacherie P, Bacha S (2012b) Using Vis–NIR hyperspectral data to map topsoil properties over bare soils in the Cap Bon region, Tunisia. In: Digital soil assessments and beyond—proceedings of the fifth global workshop on digital soil mapping, pp 387–392
Gomez C, Drost APA, Roger JM (2015a) Analysis of the uncertainties affecting predictions of clay contents from VNIR/SWIR hyperspectral data. Remote Sens Environ 156:58–70
Gomez C, Oltra Carrio R, Lagacherie P, Bacha S, Briottet X (2015b) Sensitivity of soil property prediction obtained from hyperspectral Vis–NIR imagery to atmospheric effects and degradation in image spatial resolutions. Remote Sens Environ 164:1–15
Gomez C, Gholizadeh A, Borůvka L, Lagacherie P (2016) Using legacy data for predicting soil surface clay content from VNIR/SWIR hyperspectral airborne images. Geoderma 276:84–92
Green R (2018) Global VSWIR imaging spectroscopy and the 2017 Decadal Survey. In: Proceedings of IGARSS 2018, IEEE international geoscience and remote sensing symposium, Valencia, Spain, 22–27 July 2018
Grunwald S, Thompson JA, Boettinger JL (2011) Digital soil mapping and modeling at continental scales: finding solutions for global issues. Soil Sci Soc Am J 75:1201
Guanter L, Kaufmann H, Segl K, Förster S, Rogass C, Chabrillat S, Küster T, Hollstein A et al (2015) The EnMAP spaceborne imaging spectroscopy mission for Earth observation. Remote Sens 7(7):8830–8857
Hapke BW (1981) Bidirectional reflectance spectroscopy I. Theory. J Geophys Res 86:3039–3054
Hartemink AE (2008) Soils are back on the global agenda. Soil Use Manag 24(4):327–330
Hartemink AE, McBratney AB (2008) A soil science renaissance. Geoderma 148:123–129
Haubrock S-N, Chabrillat S, Lemmnitz C, Kaufmann H (2008a) Surface soil moisture quantification models from reflectance data under field conditions. Int J Remote Sens 29(1):3–29
Haubrock S-N, Chabrillat S, Kuhnert M, Hostert P, Kaufmann H (2008b) Surface soil moisture quantification and validation based on hyperspectral data and field measurements. J Appl Remote Sens 2:023552
Heng BCP, Chandler JH, Armstrong A (2010) Applying close range digital photogrammetry in soil erosion studies. Photogramm Rec 25(131):240–265
Hill J, Schütt B (2000) Mapping complex patterns of erosion and stability in dry Mediterranean ecosystems. Remote Sens Environ 74:557–569
Hill J, Udelhoven T, Schütt B, Yair A (1999) Differentiating biological soil crusts in a sandy arid eco-system based on multi- and hyperspectral remote sensing data. In: Schaepmann M, Schläpfer D, Itten K (Eds) 1st EARSEL workshop on imaging spectroscopy. Proceedings of the EARSEL workshop, Zürich, 6–8 October 1998. EARSEL Secretariat, Paris, pp 427–436
Hook SJ, Hulley G, Cawse-Nicholson K (2017) HyTES, ECOSTRESS and HyspIRI—imaging spectroscopy and broad band imaging in the thermal infrared. In: Abstract 10th EARSeL SIG imaging spectroscopy workshop, 19–21 April 2017, Zurich, Switzerland
Hummel JW, Sudduth KA, Hollinger SE (2001) Soil moisture and organic matter prediction of surface and subsurface soils using an NIR soil sensor. Comput Electron Agric 32:149–165
Idso SB, Jackson RD, Reginato RJ, Kimball BA, Nakayama FS (1975) The dependence of bare soil albedo on soil water content. J Appl Meteorol 14:109–113
Jackson RD, Moran S, Slater PN, Biggar SF (1987) Field calibration of reflectance panels. Remote Sens Environ 22:145–158

Janik LJ, Merry RH, Skjemstad JO (1998) Can mid infrared diffuse reflectance analysis replace soil extractions? Anim Prod Sci 38(7):681–696
Ji W, Viscarra Rossel RA, Shi Z (2015) Accounting for the effects of water and the environment on proximally sensed vis–NIR soil spectra and their calibrations. Eur J Soil Sci 66:555–565. https://doi.org/10.1111/ejss.12239
JRC, Jones A, Panagos P et al (2012) The state of soil in Europe: a contribution of the JRC to the European environments agency's environment state and outlook report—SOER 2010, JRC68418
Karnieli A, Kidron G, Ghassler C, Ben-Dor E (1999) Spectral characteristics of cyanobacteria soil crust in the visible near infrared and short wave infrared (400–2,500 nm) in semiarid environment. Int J Remote Sens 69:67–77
Knadel M, Deng F, Thomsen A, Greve HM (2012) Development of a Danish national Vis–NIR soil spectral library for soil organic carbon determination. In: Digital soil assessments and beyond: proceedings of the 5th global workshop on digital soil mapping 2012, Sydney, Australia, vol 43. CRC Press, Boca Raton
Kopačková V, Ben-Dor E, Carmon N, Notesco G (2017) Modelling diverse soil attributes with visible to longwave infrared spectroscopy using PLSR employed by an automatic modelling engine. Remote Sens 9(2):134
Kriebel KT (1976) On the variability of the reflected radiation field due to differing distributions of the irradiation. Remote Sens Environ 4:257–264
Kuester T, Spengler D, Barczi J-F, Segl K, Hostert P, Kaufmann H (2014) Simulation of multitemporal and hyperspectral vegetation canopy bidirectional reflectance using detailed virtual 3-D canopy models. IEEE Trans Geosci Remote Sens 52(4):2096–2108
Kuester T, Foerster S, Chabrillat S, Spengler D, Guanter L (2017) Assessing the influence of variable fractional vegetation cover on soil spectral features using simulated canopy reflectance modelling. In: Abstract 10th EARSeL SIG Imaging spectroscopy workshop, 19–21 April 2017, Zurich, Switzerland
Lagacherie P, Baret F, Feret JB, Netto JM, Robbez-Masson JM (2008) Estimation of soil clay and calcium carbonate using laboratory, field and airborne hyperspectral measurements. Remote Sens Environ 112:825–835
Lal R (2015) Restoring soil quality to mitigate soil degradation. Sustainability 7(5):5875–5895
Lee CM, Cable ML, Hook SJ, Green RO, Ustin SL, Mandl DJ, Middleton EM (2015) An introduction to the NASA Hyperspectral InfraRed Imager (HyspIRI) mission and preparatory activities. Remote Sens Environ 167:6–19
Lefèvre-Fonollosa MJ, Bajouk T, Briottet X, Carrère V, Delacourt, C, Feret JB, Gastellu-Etchegorry JP, Gomez C, Jacquemoud S, Le Dantec N, Marion R, Petit T, Weber C (2016) Preparing the future: the HYPXIM mission. In: 36th EARSeL symposium 2016, Bonn, Germany, 20–24 June 2016
Lekner J, Dorf MC (1988) Why some things are darker when wet. Appl Opt 27:1278–1280
Li Y, Liu Y, Wu S, Wang C, Xu A, Pan X (2017) Hyperspectral estimation of wheat biomass after alleviating of soil effects on spectra by non-negative matrix factorization. Eur J Agron 84:58–66
Liang S, Townshend JR (1996) A modified Hapke model for soil bidirectional reflectance. Remote Sens Environ 55(1):1–10
Liu W, Baret F, Gu XF, Zhang B, Tong Q, Zheng L (2003) Evaluation of methods for soil surface moisture estimation from reflectance data. Int J Rem Sens 24(10):2069–2083
Liu Y, Pan X, Wang C, Li Y, Shi R (2015) Predicting soil salinity with VisNIR spectra after removing the effects of soil moisture using external parameter orthogonalization. PLoS One. https://doi.org/10.1371/journal.pone.0140688
Lobell DB, Asner GP (2002) Moisture effects on soil reflectance. Soil Sci Soc Am J 66:722–727
Loizzo R, Guarini R, Longo F, Scopa T, Formaro R, Facchinetti C, Varavalli G (2018), PRISMA: the Italian hyperspectral mission. In: Proceedings of IGARSS 2018, IEEE international geoscience and remote sensing symposium, Valencia, Spain, 22–27 July 2018
Lu P, Wang L, Niu Z, Li L, Zhang W (2013) Prediction of soil properties using laboratory VIS–NIR spectroscopy and Hyperion imagery. J Geochem Explor 132:26–33
Malley DF, Martin P, Ben-Dor E (2004) Application in analysis of soils. Chapter 26. In: Craig R, Windham R, Workman J (eds) Near infrared spectroscopy in agriculture, vol 44. A three Societies Monograph. ASA, SSSA, CSSA, Madison, pp 729–784
Marzahn P, Rieke-Zapp D, Ludwig L (2012) Assessment of soil surface roughness statistics for microwave remote sensing applications using a simple photogrammetric acquisition system. ISPRS J Photogramm Remote Sens 72:80–89
Matsunaga T, Iwasaki A, Tsuchida S, Iwao K, Tanii, J, Kashimura O, Nakamura R, Yamamoto H, Kato S, Obata K, Mouri K, Tachikawa, T (2018) HISUI status toward FY2019 launch. In: Proceedings

of IGARSS 2018, IEEE international geoscience and remote sensing symposium, Valencia, Spain, 22–27 July 2018
Matthias AD, Fimbres A, Sano EE, Post DF, Accily L, Batchily AK, Ferreira LG (2000) Surface roughness effects on soil albedo. SSSA J 63(3):1035–1041
McIntyre DS (1958) Permeability measurements of soil crusts formed by raindrop impact. Soil Sci 85(4):185–189
Meine C, Knight RL (eds) (1999) The essential Aldo Leopold: quotations and commentaries. University of Wisconsin Press, Madison
Milton EJ, Webb JP (1987) Ground radiometry and airborne multispectral survey of bare soils. Int J Remote Sens 18:3–14
Minasny B, McBratney AB (2006) A conditioned Latin hypercube method for sampling in the presence of ancillary information. Comput Geosci 32(9):1378–1388
Minasny B, McBratney AB, Bellon-Maurel V, Roger J-M, Gobrecht A, Ferrand L, Joalland S, Minasny B, Bellon-Maurel V, Gobrecht A, Roger J-M, Ferrand L, Joalland S (2011) Removing the effect of soil moisture from NIR diffuse reflectance spectra for the prediction of soil organic carbon. Geoderma 167–168:118–124. https://doi.org/10.1016/j.geoderma.2011.09.008
Moreno RG, Requejo AS, Alonso AMT, Barrington S, Alvarez MCD (2008) Shadow analysis: a method for measuring soil surface roughness. Geoderma 146:201–208
Muller E, Decamps H (2000) Modeling soil moisture-reflectance. Remote Sens Environ 76:173–180
Nanni MR, Demattê JAM (2006) Spectral reflectance methodology in comparison to traditional soil analysis. Soil Sci Soc Am J 70:393–407
Nocita M, Stevens A, Toth G, Panagos P, van Wesemael B, Montanarella L (2014) Prediction of soil organic carbon content by diffuse reflectance spectroscopy using a local partial least square regression approach. Soil Biol Biochem 68:337–347
Oldeman L, Hakkeling R, Sombroek W (1990) World map of the status of soil degradation, an explanatory note. International Soil Reference and Information Center, Wageningen
Ouerghemmi W, Gomez C, Naceur S, Lagacherie P (2011) Applying blind source separation on hyperspectral data for clay content estimation over partially vegetated surfaces. Geoderma 163(3–4):227–237
Ouerghemmi W, Gomez C, Nacer S, Lagacherie P (2016) Semi-blind source separation for estimation of clay content over semi-vegetated areas, from VNIR/SWIR hyperspectral airborne data. Remote Sens Environ 181:251–263
Palmer JM (1982) Field standards of reflectance. Photogramm Eng Remote Sens 48:1623–1625
Palmer KF, Williams D (1974) Optical properties of water in the near infrared. J Opt Soc Am 64:1107–1110
Panagos P, Borrelli P, Poesen J, Ballabio C, Lugato E, Meusburger K, Montanarella L, Alewell C (2015) The new assessment of soil loss by water erosion in Europe. Environ Sci Policy 54:438–447
Piech KR, Walker JE (1974) Interpretation of soils. Photogramm Eng 40:87–94
Potter KN, Horton R, Cruse RM (1987) Soil surface roughness effects on radiation reflectance and soil heat flux. Soil Sci Soc Am J 51:855–860
Previtali F (2014) Pedoenvironments of the Mediterranean countries: resources and threats. In: Kapur S, Ersahin S (eds) Soil Security for ecosystem management, chap 4. Springer, New York, NY, pp 61–82
Rast M, Nieke J, Ananasso C, Bach H, Ben Dor E, Chabrillat S, Colombo R, Feret JB, Giardino C, Green RO, Guanter L, Marsh S, Ong C, Rum G, Schaepman M, Schlerf M, Skidmore AK, Strobl P, Gascon F, Adams J, Isola C, Del Bello U (2019) The copernicus hyperspectral imaging mission for the environment (CHIME). In: Abstract 2019 ESA living planet symposium, Milan, Italy, 13–17 May 2019
Rieke-Zapp DH, Nearing MA (2005) Digital close range photogrammetry for measurement of soil erosion. Photogramm Rec 20(109):69–87
Rodger A, Cudahy T (2009) Vegetation corrected continuum depths at 2.20 μm: an approach for hyperspectral sensors. Remote Sens Environ 113:2243–2257
Rodríguez-Caballero E, Escribano P, Cantón Y (2014) Advanced image processing methods as a tool to map and quantify different types of biological soil crust. ISPRS J Photogramm Remote Sens 90:59–67
Rogge D, Bauer A, Zeidler J, Mueller A, Esch T, Heiden U (2017) Building an exposed soil composite processor (SCMaP) for mapping spatial and temporal characteristics of soils with Landsat imagery (1984–2014). Remote Sens Environ 205:1–17
Romero DJ, Ben-Dor E, Demattê JAM, Souza AB, Vicente LE, Tavares TR, Martello M, Strabeli TF, da Silva Barros PP, Fiorio PR, Gallo BC, Sato MV, Eitelweind MT (2018) Internal soil standard method for the Brazilian soil spectral library: performance and proximate analysis. Geoderma 312:95–103
Rosa JD, Cooper M, Darboux F, Medeiros JC (2012) Soil roughness evolution in different tillage systems under simulated rainfall using a semivariogram-based index. Soil Tillage Res 124:226–232

Rossel RV, Walvoort DJJ, McBratney AB, Janik LJ, Skjemstad JO (2006) Visible, near infrared, mid infrared or combined diffuse reflectance spectroscopy for simultaneous assessment of various soil properties. Geoderma 131(1):59–75
Schmid T, Koch M, Gumuzzio J (2008) Application of hyperspectral imagery to map soil salinity. In: Metternicht G, Zinck A (eds) Remote sensing of soil salinization: impact and land management. CRC Press, Boca Raton, pp 113–139 **Chapter 7**
Schmid T, Rodríguez-Rastrero M, Escribano P, Palacios-Orueta A, Ben-Dor E, Plaza A, Milewski R, Huesca M, Bracken A, Cicuéndez V, Pelayo M, Chabrillat S (2016) Characterization of soil erosion indicators using hyperspectral data from a Mediterranean rainfed cultivated region. IEEE J Sel Top Appl Earth Obs Remote Sens 9(2):845–860
Schwanghart W, Jarmer T (2011) Linking spatial patterns of soil organic carbon to topography: a case study from south-eastern Spain. Geomorphology 126:252–263
Schwartz G, Eshel G, Ben-Dor E (2011) Reflectance spectroscopy as a tool for monitoring contaminated soils. In: Pascucci S (ed) Soil contamination. InTech, Manhattan, pp 67–90
Selige T, Böhner J, Schmidhalter U (2006) High resolution topsoil mapping using hyperspectral image and field data in multivariate regression modeling procedures. Geoderma 136:235–244
Sellers PJ et al (1995) Remote sensing of the land-surface for studies of global change: models-algorithms-experiments. Remote Sens Environ 51:3–26
Shoshany M (1993) Roughness-reflectance relationships of bare desert terrain: an empirical study. Remote Sens Environ 45:15–27
Shoshany M, Goldshleger N, Chudnovsky A (2013) Monitoring of agricultural soil degradation by remote-sensing methods: a review. Int J Remote Sens 34(17):6152–6181
Steinberg A, Chabrillat S, Stevens A, Segl K, Foerster S (2016) Prediction of common surface soil properties based on Vis–NIR airborne and simulated EnMAP imaging spectroscopy data: prediction accuracy and influence of spatial resolution. Remote Sens 8(7):613
Stenberg B, Viscarra Rossel RA, Mouazen AM, Wetterlind J (2015) Visible and near infrared spectroscopy in soil science. In: Sparks DL (ed) Advances in agronomy, vol 107. Academic Press, Burlington, pp 163–215. https://doi.org/10.1016/s0065-2113(10)07005-7
Stevens A, van Wesemael B, Bartholomeus H, Rosillon D, Tychon B, Ben-Dor E (2008) Laboratory, field and airborne spectroscopy for monitoring organic carbon content in agricultural soils. Geoderma 144:395–404
Stevens A, Udelhoven T, Denis A, Tychon B, Lioy R, Hoffmann L, van Wesemael B (2010) Measuring soil organic carbon in croplands at regional scale using airborne imaging spectroscopy. Geoderma 158:32–45
Stevens A, Miralles I, Van Wesemael B (2012) Soil organic carbon predictions by airborne imaging spectroscopy: comparing cross-validation and validation. Soil Sci Soc Am J 76:2174–2183
Stevens A, Nocita M, Tóth G, Montanarella L, Van Wesemael B (2013) Prediction of soil organic carbon at the European scale by visible and near infrared reflectance spectroscopy. PLoS ONE 8(6):1–13
Stevens F, Bogaert P, van Wesemael B (2015) Detecting and quantifying field-related spatial variation of soil organic carbon using mixed-effect models and airborne imagery. Geoderma 259–260:93–103
Stolte J, Tesfai M, Øygarden L, Kværnø S, Keizer J, Verheijen F, Panagos P, Ballabio C, Hessel R (2016) Soil threats in Europe; EUR 27607
Stoner ER, Baumgardner MF (1981) Characteristic variations in reflectance of surface soils. Soil Sci Soc Am J 45:1161–1165
Taconet OV, Ciarletti V (2007) Estimating soil roughness indices on a ridge-and-furrow surface using stereo photogrammetry. Soil Tillage Res 93:64–76
Thomsen LM, Baartman JEM, Barneveld RJ, Starkloff T, Stolte J (2015) Soil surface roughness: comparing old and new measuring methods and application in a soil erosion model. Soil 1:399–410
Toth G, Jones A, Montanarella L (2013) The LUCAS topsoil database and derived information on the regional variability of cropland topsoil properties in the European Union. Environ Monit Assess 185:7409–7425
Toy TJ, Foster GR, Renard KG (2002) Soil erosion: processes, prediction, measurement, and control. Wiley, New York
Tsakiridis NL, Tziolas N, Dimitrakos A, Galanis G, Ntonou E, Tsirika A, Terzopoulou E, Kalopesa E, Zalidis GC (2017) Predicting soil properties for sustainable agriculture using Vis–NIR spectroscopy: a case study in northern Greece. In: Proceedings of SPIE 10444, fifth international conference on remote sensing and geoinformation of the environment (RSCy2017). https://doi.org/10.1117/12.2277905
Ulaby FT, Moore RK, Fung AK (1982) Microwave remote sensing active and passive. Addison-Wesley, Reading

UNEP (2012) One planet, how many people? A review of Earth's carrying capacity, A discussion paper for the year of RIO+20. UNEP Global Environmental Alert Service
Ussiri DAN, Lal R (2018) The role of soil management and restoration in advancing sustainable development goals. In: Lal R, Horn R, Kosaki T (eds) Soil and sutainable development goals, vol 5. Catena, Stuttgart, pp 61–81
Van der Linden S, Rabe A, Held M, Jakimow B, Leitão PJ, Okujeni A, Schwieder M, Suess S, Hostert P (2015) The EnMAP-Box—a toolbox and application programming interface for EnMAP data processing. Remote Sens 7:11249–11266
Vermang J, Norton LD, Baetens JM, Huang C, Cornelis WM, Gabriels D (2013) Quantification of soil surface roughness evolution under simulated rainfall. Trans Am Soc Agric Biol Eng 56(2):505–514
Viscarra Rossel R, Behrens T, Ben-Dor E, Brown D, Demattê J, Shepherd K, Shi Z, Stenberg B, Stevens A, Adamchuk V (2016) A global spectral library to characterize the world's soil. Earth Sci Rev 155:198–230
Vrieling A, De Jong SM, Sterk G, Rodrigues SC (2008) Timing of erosion and satellite data: a multi-resolution approach to soil erosion risk mapping. Int J Appl Earth Obs Geoinf 10(3):267–281
Wang Z, Coburn CA, Ren X, Teillet PM (2012) Effect of soil surface roughness and scene components on soil surface bidirectional reflectance factor. Can J Soil Sci 92:297–313
Weber B, Hill J (2016) Remote sensing of biological soil crusts at different scales. In: Weber B et al (eds) Biological soil crusts: an organizing principle in drylands. Springer, Cham, pp 215–234
Weber B, Olehowski C, Knerr T, Hill J, Deutschewitz K, Wessels DCJ et al (2008) A new approach for mapping of biological soil crusts in semidesert areas with hyperspectral imagery. Remote Sens Environ 112:2187–2201
Weksler S, Notespo G, Ben-Dor E (2017) an automated procedure for reducing atmospheric features and emphasizing surface emissivity in hyperspectral longwave infrared (LWIR) images. Int J Remote Sens. https://doi.org/10.1080/01431161.2017.1325535
Whiting ML, Li L, Ustin SL (2004a) Predicting water content using Gaussian model on soil spectra. Remote Sens Environ 89:535–552
Whiting ML, Li L, Ustin SL (2004b) Correcting mineral abundance estimates for soil moisture. In: Green RO (ed) 13th annual JPL airborne Earth science workshop, Pasadena, California, March 30–April 3, 2004. JPL Publication 05-3-1
Whiting ML, Palacios-Orueta A, Li L, Ustin SL (2005) Light absorption model for water content to improve soil mineral estimates in hyperspectral imagery. In: Pecora 16, global priorities in land remote sensing, Sioux Falls, South Dakota, 23–27 October 2005. American Society of Photogrammetry and Remote Sensing
Wu CY, Jacobson AR, Laba M, Baveye PC (2009) Accounting for surface roughness effects in the near-infrared reflectance sensing of soils. Geoderma 152:171–180. https://doi.org/10.1016/j.geoderma.2009.06.002
Zhao X, Xue J-F, Zhang X-Q, Kong F-L, Chen F, Lal R et al (2015) Stratification and storage of soil organic carbon and nitrogen as affected by tillage practices in the North China Plain. PLoS ONE 10(6):e0128873



## Affiliations

**S. Chabrillat[1] · E. Ben-Dor[2] · J. Cierniewski[3] · C. Gomez[4] · T. Schmid[5] · B. van Wesemael[6]**

[1] Section Remote Sensing, Helmholtz Center Potsdam GFZ German Research Center for Geosciences, Telegrafenberg, 14473 Potsdam, Germany

[2] Department of Geography and Human Environment, Tel Aviv University, 69978 Tel Aviv, Israel

[3] Department of Soil and Remote Sensing Soil Science, Adam Mickiewicz University, Poznan, Poland

[4] UMR LISAH (INRA-IRD-SupAgro), IRD, 34060 Montpellier, France

[5] Department of Environment, Centro de Investigaciones Energéticas, Medioambientes y Tecnológicas (CIEMAT), 28040 Madrid, Spain

[6] Georges Lemaître Centre for Earth and Climate Research, Earth and Life Institute, Université Catholique de Louvain, 1348 Louvain-la-Neuve, Belgium

Surv Geophys (2019) 40:401–429
https://doi.org/10.1007/s10712-018-9476-0

# Imaging Spectrometry of Inland and Coastal Waters: State of the Art, Achievements and Perspectives

**C. Giardino[1] · V. E. Brando[2] · P. Gege[3] · N. Pinnel[4] · E. Hochberg[5] · E. Knaeps[6] · I. Reusen[6] · R. Doerffer[7] · M. Bresciani[1] · F. Braga[8] · S. Foerster[9] · N. Champollion[10,11] · A. Dekker[12]**

Received: 6 October 2017 / Accepted: 8 May 2018 / Published online: 5 June 2018


**Abstract** Imaging spectrometry of non-oceanic aquatic ecosystems has been in development since the late 1980s when the first airborne hyperspectral sensors were deployed over lakes. Most water quality management applications were, however, developed using multispectral mid-spatial resolution satellites or coarse spatial resolution ocean colour satellites till now. This situation is about to change with a suite of upcoming imaging spectrometers being deployed from experimental satellites or from the International Space Station. We review the science of developing applications for inland and coastal aquatic ecosystems

✉ C. Giardino
giardino.c@irea.cnr.it

[1] CNR-IREA, National Research Council of Italy, Institute for Electromagnetic Sensing of the Environment, 20133 Milan, Italy

[2] CNR-ISAC, National Research Council of Italy, Institute of Atmospheric Sciences and Climate, 00133 Rome, Italy

[3] Remote Sensing Technology Institute, German Aerospace Center (DLR), Münchener Straße 20, 82234 Weßling, Germany

[4] German Aerospace Center (DLR), German Remote Sensing Data Center (DFD), Münchener Straße 20, 82234 Weßling, Germany

[5] Bermuda Institute of Ocean Sciences, 17 Biological Station, St. George's GE01, Bermuda

[6] Vlaamse Instelling voor Technologisch Onderzoek (VITO), Boeretang 200, 2400 Mol, Belgium

[7] Helmholtz-Zentrum Geesthacht, Zentrum für Material- und Küstenforschung GmbH, Max-Planck-Straße 1 I, 21502 Geesthacht, Germany

[8] CNR-ISMAR, National Research Council of Italy, Institute of Marine Science, 20133 Venice, Italy

[9] Section 1.4 Remote Sensing, Helmholtz Center Potsdam GFZ German Research Center for Geosciences, Telegrafenberg, 14473 Potsdam, Germany

[10] International Space Science Institute (ISSI), Hallerstrasse 6, 3012 Bern, Switzerland

[11] Climate Lab, University of Bremen, Celsiusstrasse 2 FVG-M2040, 28359 Bremen, Germany

[12] Fenner School of Environment and Society, The Australian National University, Building 141, Linnaeus Way, Canberra, ACT 2601, Australia

that often are a mixture of optically shallow and optically deep waters, with gradients of clear to turbid and oligotrophic to hypertrophic productive waters and with varying bottom visibility with and without macrophytes, macro-algae, benthic micro-algae or corals. As the spaceborne, airborne and in situ optical sensors become increasingly available and appropriate for aquatic ecosystem detection, monitoring and assessment, the science-based applications will need to be further developed to an operational level. The Earth Observation-derived information products will range from more accurate estimates of turbidity and transparency measures, chlorophyll, suspended matter and coloured dissolved organic matter concentration, to more sophisticated products such as particle size distributions, phytoplankton functional types or distinguishing sources of suspended and coloured dissolved matter, estimating water depth and mapping types of heterogeneous substrates. We provide an overview of past science, current state of the art and future directions so that early career scientists as well as aquatic ecosystem managers and associated industry groups may be prepared for the imminent deluge of imaging spectrometry data.

**Keywords** Inland and coastal water · Imaging spectrometry · Biophysical parameters · Spectral, spatial and temporal resolutions · Satellite observations

## 1 Introduction

Inland and coastal water ecosystems are environmentally important, provide multiple ecosystem services and are vital for human consumption, irrigation, sanitation, transportation, recreation and industry. In the past decades, these ecosystems have experienced high stress from various human impacts as well as climate change (Hartmann et al. 2013). Among them, the increasing eutrophication and pollution of many of these environments are major environmental threats.

There is an increasing need for regular monitoring of inland and coastal waters to support national and international directives and conventions such as the European Water Framework Directive, the US Clean Water Act & Safe Drinking Water Act and the Australian Reef Water Quality Protection Plan, requiring biological, hydro-morphological and physico-chemical parameters of water bodies to be monitored on a regular basis. Information on the state and development of water bodies is also a prerequisite to meet the United Nations Sustainable Development Goal No. 6 to "ensure availability and sustainable management of water and sanitation for all" by 2020.

Inland and coastal waters are natural and man-made systems, with differing characteristics that show very large dynamics in time and space. For example, freshwater ecosystems include lakes, ponds, streams and rivers, as well as wetlands, comprised of marshes, mangroves, floodplains and swamps. The overall number of lakes larger than 0.002 $km^2$ is estimated to be more than 117 million making up 3.7% of the Earth's non-glaciated land area (Verpoorter et al. 2014). However, out of these only a small proportion of water bodies is currently included in a regular and consistent monitoring programme biased towards larger water bodies, whilst the majority of small lakes and water bodies is never monitored.

Earth Observation (EO) may be used for acquiring timely, frequent synoptic information, from local to global scales, of inland and coastal waters. EO based measurements of physical and biochemical parameters in these waters mainly rely on the interpretation of the spectral reflectance, which is used to retrieve water components, water depth and bottom properties. EO data have been successfully applied for mapping inland and coastal waters for about 50 years (e.g., Strong 1974; Wang and Shi 2008; Stumpf et al. 2012; Bresciani et al. 2014; Matthews and Odermatt 2015). Landsat data have been continuously used to have long-term lake water quality information on synoptic scale (Olmanson et al. 2008), while for large water bodies many applications used coarse spatial resolution data gathered from of ocean colour satellite sensors because they provide the most suited spectral, radiometric and temporal resolutions for aquatic resolution. In particular, MERIS ocean colour sensor, with its spatial resolution of about 300 m and dedicated spectral bands for detecting chlorophyll-a (Chl-a) in turbid waters, has been successfully used for studying lakes and coastal waters from 2002 to 2012 (Odermatt et al. 2012). These studies focused on the retrieval of water reflectance, physical parameters (e.g., turbidity, diffuse attenuation coefficient, water clarity), suspended and dissolved water quality components such as Chl-a concentration (commonly used as a proxy of phytoplankton biomass), coloured dissolved organic matter (CDOM) and total suspended matter (TSM). More recently, the synoptic, fine-scale and high-frequency retrieval of these parameters has become possible because of the latest generation of medium to high spatial resolution multispectral sensors onboard Landsat-8 and Sentinel-2 satellites. These land imaging sensors are not specifically designed for observing aquatic ecosystems but they are suitable for detailed water quality analysis (Pahlevan et al. 2014; Toming et al. 2016) due to their: (1) radiometric sensitivity (> 12-bit quantisation) (Hedley et al. 2012; Dörnhöfer and Oppelt 2016); (2) spatial resolution of 10–30 m; (3) revisit times of 16 and 5 days, respectively; and (4) improved spectral band setting in the visible/near-infrared wavelengths range. Successful examples of the use of these new sensors are provided by Hedley et al. (2016) for mapping corals, Kutser et al. (2016) and Giardino et al. (2014) for boreal and subalpine lakes, Dörnhöfer et al. (2016) for oligotrophic lakes, Vanhellemont and Ruddick (2014) for coastal turbid waters, Lobo et al. (2015) and Giardino et al. (2017) for rivers and Brando et al. (2015) for river plumes.

Nevertheless, according to Mouw et al. (2015), the current satellite radiometers are designed for observing the global ocean or land surface and thus not specifically suited for observing coastal and inland waters, so that deriving coastal and inland aquatic applications from these existing sensors remains challenging. Tyler et al. (2016) reviewed the developments and technological advances in EO for the assessment and monitoring of inland, transitional, coastal and shelf-sea waters. They summarised the opportunities that the next generation of satellites offer for water quality monitoring and concluded that the operational use of satellites for the assessment and monitoring of freshwater and coastal ecosystems is still evolving. Muller-Karger et al. (2018) recently outlined specifications for satellite requirements that offer the potential for rapid, frequently repeated, and consistent high-quality observations to characterise changes of essential biodiversity variables of coastal ecosystems.

Most of the challenges in inland and coastal water observations are due to their optical complexity. These aquatic ecosystems can be a mixture of optically shallow and optically deep waters, with gradients of clear to turbid and oligotrophic to hypertrophic productive waters and varying bottom visibility with and without macrophytes. For example, a lake receives and recycles organic and inorganic substances from within the lake, from its watershed and beyond (atmospheric deposition). Hence, a large range in optical absorption and backscattering resulting in high optical variability can be found among and within lakes, estuaries and coastal waters. This poses a challenge for bio-optical algorithms applied to optical remote sensing for water quality monitoring applications. Another challenge is performing atmospheric corrections over such variable aquatic ecosystems as their complexity requires different approaches than those for land and ocean applications.

The effect of the atmosphere on the signal received by EO satellite sensors is significant and usually represents 80% or more of the total at-sensor radiance, especially in the shorter (blue) wavelengths. Accurate modelling of atmospheric absorption and scattering effects, in addition to the specular water surface reflection effects, is required to derive the water leaving reflectance from hyperspectral image data of water (Gao et al. 2009). Moses et al. (2017) indicated further issues that make the estimation of water leaving reflectance at the surface challenging: (1) the proximity to terrestrial sources of atmospheric pollution which results in an optically heterogeneous atmosphere that is difficult to model; (2) the adjacency effects from neighbouring land pixels around the water body; and (3) the non-negligible reflectance of water in the near-infrared region due to high sediment concentrations in inland and near-coastal waters preventing the use of atmospheric correction schemes adopted for oceans.

Despite the complexity of using imaging spectrometry for inland and coastal water ecosystems monitoring, Hestir et al. (2015) showed how bio-optical algorithms, specifically developed for assessing water quality in optically complex water systems, perform best if they can make use of hyperspectral data. Hyperspectral remote sensing, or imaging spectrometry, provides measurements across numerous discrete narrow bands, forming a contiguous spectrum that enables detection and identification of key biophysical properties of water column and bottom. For over three decades, the acquisition of the contiguous spectrum for each image pixel has been illustrated in the form of hypercubes that are commonly used to illustrate the concept of hyperspectral imaging. In the case of inland and coastal waters applications, hypercubes (Fig. 1) show an amount hyperspectral bands bringing data from the water systems that compared to land targets are mostly limited to the first part of the spectrum. Between 400 and 800 nm, Lee et al. (2007) suggested a minimum of 17 spectral bands to observe subtle changes in the remote sensing reflectance of water, given the high variability in the influence of turbidity, bottom reflectance, and complex atmospheric conditions. Furthermore, when also considering interdisciplinary studies to identify shorelines, floating substances and benthic or emergent communities the required number and range of bands increases substantially. Overall, work is still required on the development and validation of remote sensing algorithms for these optically complex waters (Tyler et al. 2016) for which the large ranges of concentrations of water components (e.g., in Eleveld et al. (2017), Chl-a varies from 0.1 to 940 mg $m^{-3}$, the TSM from 0.1 to 290 g $m^{-3}$, and the absorption CDOM at 440 nm from 0.04 to 10 $m^{-1}$), also affect algorithm performances.

**Fig. 1** Imaging spectrometer data hypercubes, where colours in the third dimension represent the reflectance of the edge pixels. On the left the fine-scale details in shallow waters (Sirmione Penisula, Lake Garda) captured from airborne AISA Eagle data; on the right the optical complexity from transitional to coastal and pelagic waters seen from the HICO (Venice lagoon and North Adriatic Sea) onboard the International Space Station

This study aims to present the fundamentals, state of the art, benefits and limitations as well as future perspectives of imaging spectrometry of inland and coastal water monitoring building forth on the review on aquatic ecosystem from both airborne imaging spectrometry by, e.g., Dekker et al. (2001) and satellite technologies (e.g., Mouw et al. 2015; Bracher et al. 2017). The paper is divided into three sections: Sect. 2 presents the principles of imaging spectrometry and the underpinning bio-optical models that enables simulation and inversion of the concentration of water constituents, water physical properties (turbidity, transparency and vertical attenuation of light), and in the case of optically shallow waters the water depth and bottom type and cover.

Section 3 summarises the main achievements of imaging spectrometry for the study of optically deep and optically shallow waters. Section 4 considers current imaging spectrometry capabilities and future directions based on user requirements, state-of-the-art science as well as remaining challenges.

# 2 Principles of Imaging Spectrometry of Inland and Coastal Waters

## 2.1 Bio-optical Models

Natural waters are composed of pure water and various molecules and particles differing in size, shape and chemical composition. Besides inorganic components like salts, quartz sand, clay minerals, metal oxides, and gas bubbles of various sizes, a large variety of living organisms and dead organic material is observed in most open waters. Imaging spectrometry utilises the fact that some constituents affect the spectral composition of the reflected light sufficiently for quantitative determination of these components. If the reflected light is furthermore influenced by the bottom, in so-called *optically shallow* waters, imaging spectrometry can also be used to determine water depth and classify benthic substrates in the case of clear water.

The equations describing the relationship between the water constituents and the optical properties of a water body are called *bio-optical models* to acknowledge the omnipresence of biological components (Morel 2001). Such models connect measured quantities such as reflectance or attenuation with concentrations of different types of material present and their optical properties such as the specific absorption or backscattering coefficient. Optical properties depending only on the material are called *inherent optical properties* (IOPs), while illumination-dependent optical properties are termed *apparent optical properties* (AOPs) (Preisendorfer 1961). IOPs are additive, but not AOPs, except under certain simplifying conditions. Concentration-normalised IOPs are called *specific inherent optical properties* (SIOPs) and labelled by a star symbol.

### 2.1.1 Inherent Optical Properties

Bio-optical models group the huge variety of water constituents usually into three classes: coloured dissolved organic matter (CDOM), phytoplankton (phy), and non-algal particles (NAP). Their contributions to the absorption coefficient, $a(\lambda)$, and the backscattering coefficient, $b_b(\lambda)$, of the water body can be expressed as follows:

$$a(\lambda) = a_w(\lambda) + C_{CDOM} \times a^*_{CDOM}(\lambda) + C_{phy} \times a^*_{phy}(\lambda) + C_{NAP} \times a^*_{NAP}(\lambda) \tag{2.1}$$

$$b_b(\lambda) = b_{b,w}(\lambda) + C_{phy} \times b^*_{b,phy}(\lambda) + C_{NAP} \times b^*_{b,NAP}(\lambda) \tag{2.2}$$

where $a_w(\lambda)$ and $b_{b,w}(\lambda)$ are the absorption and backscattering coefficients of pure water, respectively. $C_i$ is the concentration of class $i$, $a_i^*(\lambda)$ its specific absorption coefficient, and $b_{b,i}^*(\lambda)$ its specific backscattering coefficient. CDOM backscattering is negligible, thus it is omitted in Eq. (2.2). The sum of all particulate components is called *total suspended matter* (TSM).

Examples of absorption and backscattering coefficients are shown in Fig. 2. They represent water types with relatively low concentrations of CDOM, phytoplankton and NAP: $C_{CDOM}=1$ g m$^{-3}$, $C_{phy}=5$ mg m$^{-3}$, $C_{NAP}=1$ g m$^{-3}$. The SIOPs were chosen as follows: $a^*_{CDOM}(\lambda)$ from Eq. (2.3) with $a^*_{CDOM}(380)=0.44$ m$^2$ g$^{-1}$ and $S=0.014$ nm$^{-1}$, phy are green algae, $a^*_{NAP}(\lambda)$ from Eq. (2.4) with $S_{NAP}=0.011$ nm$^{-1}$, $a^*_{NAP}(440)=0.027$ m$^2$ g$^{-1}$, $b^*_{b,NAP}(555)=0.011$ m$^2$ g$^{-1}$, $n=0.75$.

*2.1.1.1 Colored Dissolved Organic Matter (CDOM)* Light absorption in the blue to yellow wavelengths of inland and coastal waters is frequently dominated by CDOM. It is the coloured dissolved fraction of the water constituents passing a filter with a pore size of 0.2 µm. Historical names are, besides others, gelbstoff (Kalle 1938) and yellow matter. Most of its compounds are produced during the decay of plant matter and consist of various humic and fulvic acids. Their absorption is mainly caused by aromatic groups with various degrees and types of substitution, including monosubstituted and polysubstituted phenols and various aromatic acids (Korshin et al. 1997). The absorption maxima of these so-called chromophores lie in the UV, for example at 180, 203 and 253 nm for benzene (Korshin et al. 1997), which is a common aromatic ring structure in humic matter. The longwave tail of their peaks is commonly approximated by an exponential function for wavelengths greater than 300 nm:

$$a^*_{CDOM}(\lambda) = a^*_{CDOM} \times \exp\left\{-S \times \left(\lambda - \lambda_0\right)\right\} \tag{2.3}$$

The specific absorption coefficient of CDOM, $a^*_{CDOM}$, depends on its composition and is highly variable within a reported range from 0.27 to 2.16 m$^2$ g$^{-1}$ at $\lambda_0=380$ nm (Sipelgas et al. 2003). The mass-specific concentration of CDOM is rarely measured together with CDOM absorption. Because it is linked to remote sensing data only indirectly, the absorption coefficient at a reference wavelength $\lambda_0$ is frequently used to express CDOM concentration, which is a common observable of field measurements and directly linked to the optical properties of water. For $\lambda_0=440$ nm, CDOM absorption ranges from ~0.004 to 4 m$^{-1}$ in coastal areas and from ~0.06 to 19 m$^{-1}$ in inland waters (Kirk 2011). The slope $S$ depends on CDOM composition. Values ranging from 0.004 to 0.053 nm$^{-1}$ were observed

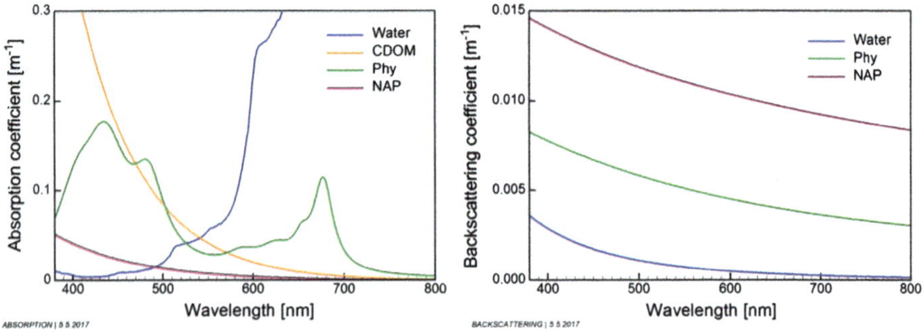


**Fig. 2** Examples of absorption (left) and backscattering (right) coefficients of water, CDOM, phytoplankton (Phy) and non-algal particles (NAP)

(Aas 2000), but in most cases $S$ is in the range between 0.010 nm$^{-1}$ for humic acid dominated waters and 0.020 nm$^{-1}$ when fulvic acids prevail (Bricaud et al. 1981; Zepp and Schlotzhauer 1981; Carder et al. 1989; Dekker and Peters 1993; Schwarz et al. 2002; Babin et al. 2003b; Laanen 2007; Binding et al. 2008; Kirk 2011).

Since Eq. (2.3) is an approximation to a function with a maximum in the UV, the slope $S$ decreases with increasing $\lambda_0$. Gege (2000) has derived an equation for the dependence of $S$ on $\lambda_0$, and he studied the accuracy of the approximation (2.3) for Lake Constance. The error exceeded 10% for $|\lambda - \lambda_0| > 60$ nm. As CDOM dominates absorption in many coastal and inland waters, an accurate spectral model of $a^*_{\mathrm{CDOM}}(\lambda)$ is important. Scattering of CDOM is small and commonly ignored, yet fluorescence may be significant and should be accounted for at high concentrations (Pozdnyakov et al. 2002). CDOM fluorescence is, however, not well studied.

*2.1.1.2 Phytoplankton* Phytoplankton is one of the most important water constituents for many aquatic disciplines. Its "fingerprint molecule" Chl-a is frequently used as a measure of concentration since it is an inevitable molecule in the photosynthesis process and present in all species. Natural waters can contain dozens or hundreds of species, most of them with similar optical properties. Perhaps the first optical classification for integration into a bio-optical model was made for Lake Constance (Gege 1998). Four spectral classes could be distinguished (Cryptophyta, Diatoms, Chlorophyte, Dinoflagellates) and their concentrations determined if these were above certain thresholds. A focus of remote sensing applications is currently on phytoplankton functional types (PFTs) (Sathyendranath et al. 2014; Mouw et al. 2017), each PFT representing a group of species aggregated according to distinct functional characteristics, in order to increase our understanding of the role of phytoplankton in the global carbon cycle (Nair et al. 2008).

Except for blooms and certain species, phytoplankton backscattering is often small compared to NAP and has a similar spectral signature (see Fig. 2), thus phytoplankton and NAP backscattering are frequently combined neglecting spectral differences. A detailed description of bio-optical properties of phytoplankton is available in Matthews (2017).

*2.1.1.3 Non-algal Particles (NAP)* All natural waters contain high numbers of mineral particles derived from the land or from bottom sediments, but also organic components like bacteria, dead cells and fragments of cells (Kirk 2011). As the particles are of various origins, the spectral properties of NAP can change significantly. Nevertheless, useful approximations have been found for both absorption and scattering.

NAP absorption frequently follows an exponential equation similar to CDOM absorption,

$$a^*_{\mathrm{NAP}}(\lambda) = a^*_{\mathrm{NAP}}(\lambda_0) \times \exp\{-S_{\mathrm{NAP}} \times (\lambda - \lambda_0)\} \tag{2.4}$$

with a spectral slope typically less than for CDOM. The spectrum sometimes shows shoulders due to the breakdown products of photosynthetic pigments (Kirk 2011). Studies in coastal waters report $S_{\mathrm{NAP}}$ values of $0.0123 \pm 0.0013$ nm$^{-1}$ (Babin et al. 2003b) and 0.011 nm$^{-1}$ (D'Sa et al. 2006), and $a^*_{\mathrm{NAP}}(\lambda_0)$ in the order of 0.027 m$^2$ g$^{-1}$ at $\lambda_0 = 440$ nm (Babin et al. 2003b).

The particle size distribution is roughly hyperbolic in many waters (Bader, 1970), i.e. the number of particles with a diameter greater than $D$ is proportional to $D^{-\gamma}$, where $\gamma$ varies widely from 0.7 to 6 (Jerlov 1976). Although small particles are numerous, their low scattering efficiency makes scattering in natural waters frequently dominated by particles with $D > 2$ μm (Jerlov 1976). Scattering and backscattering follow approximately the law of Angström:

$$b^*_{b,NAP}(\lambda) = b^*_{b,NAP}(\lambda_s) \times \left(\frac{\lambda}{\lambda_s}\right)^{-n} \quad (2.5)$$

The Angström exponent $n$ is typically between 0 and 1 in the open ocean and in clear coastal waters (Morel and Maritorena 2001; Blondeau-Patissier et al. 2009), and around zero in shallow and inland waters (Babin et al. 2003a; Chami et al. 2005). The specific backscattering coefficient $b^*_{\mathrm{b,NAP}}(\lambda_s)$ varies by an order of magnitude or more (Chami et al. 2005; Antoine et al. 2011) and is in the order of 0.01 $m^2\ g^{-1}$.

### 2.1.2 AOPs for Optically Deep Waters

Regions of coastal and inland waters that are not noticeably affected by light reflected at the sea floor are called *optically deep.* The reflected light depends on the optical properties of the water body, but also on the illumination conditions, on the field of view of the sensor and on sensor orientation. That means reflectance is an AOP. To account for different illumination and viewing geometries, various definitions of reflectance exist (Schaepman-Strub et al. 2006). Most useful for remote sensing is *radiance reflectance,*

$$r_{\mathrm{rs}}(\lambda) = \frac{L_{\mathrm{u}}(\lambda)}{E_{\mathrm{d}}(\lambda)}\left[sr^{-1}\right] \quad (2.6)$$

since sensors on satellite and aircraft measure upwelling radiances, $L_{\mathrm{u}}$, that are induced by the downwelling irradiance, $E_{\mathrm{d}}$, which is the hemispherical integral of the downwelling radiance.

Ignoring non-solar sources such as bioluminescence, sun-induced fluorescence signals from Chl-a, phycobiliproteins, phaeopigments, CDOM as well as Raman scattering, the upwelling radiance consists of two terms:

$$L_{\mathrm{u}}(\lambda) = L_{\mathrm{w}}(\lambda) + L_{\mathrm{r}}(\lambda) \quad (2.7)$$

The first, the *water leaving radiance* $L_{\mathrm{w}}$, carries information about the water body, while the second, the *reflected sky radiance* $L_{\mathrm{r}}$, is radiation from specular reflection of light from the sun, the sky and clouds at the water surface. Only $L_{\mathrm{w}}$ is relevant for bio-optical models of the water body, hence the ratio

$$R_{\mathrm{rs}}(\lambda) = \frac{L_{\mathrm{w}}(\lambda)}{E_{\mathrm{d}}(\lambda, 0+)}\left[sr^{-1}\right] \quad (2.8)$$

is an important quantity for remote sensing of water bodies. It is called *remote sensing reflectance.* The symbol $0^+$ indicates a measurement just above the water surface. The quantity $L_w$ cannot be measured directly but has to be derived from a measurement of $L_{\mathrm{u}}$ by subtracting $L_{\mathrm{r}}$. Then, $R_{\mathrm{rs}}$ is related to subsurface $r_{\mathrm{rs}}$ as follows (Mobley 1994; Lee et al. 1998):

$$R_{rs}(\lambda) = \frac{\zeta \times r_{rs}(\lambda)}{1 - \Gamma \times r_{rs}(\lambda)} \quad (2.9)$$

where $\zeta \approx 0.52$ is the water-to-air radiance divergence factor, and the denominator with $\Gamma \approx 1.6$ accounts for the effects of internal reflection from water to air. Monte Carlo simulations and analytical models have demonstrated the usefulness of the function

$$u(\lambda) = \frac{b_b(\lambda)}{a(\lambda) + b_b(\lambda)} \tag{2.10}$$

for modelling reflectance of water (Gordon et al. 1975). It allows expressing reflectance as the product of an AOP, $f_{rs}(\lambda)$, and an IOP, $u(\lambda)$:

$$r_{rs}^{deep}(\lambda) = f_{rs}(\lambda) \times u(\lambda). \tag{2.11}$$

The superscript "deep" indicates a measurement of optically deep water. A number of analytical models of the $f_{rs}$ factor have been derived (overview in Sokoletsky and Shen 2014). Here, the model of Albert is used (Albert and Mobley 2003; Albert 2004):

$$\begin{aligned} f_{rs}(\lambda) &= 0.0512 \times \left(1 + 4.6659 \times u(\lambda) - 7.8387 \times u(\lambda)^2 + 5.4571 \times u(\lambda)^3\right) \\ &\times \left(1 + \frac{0.1098}{\cos\theta'_{sun}}\right) \times \left(1 + \frac{0.4021}{\cos\theta'_{v}}\right) \times \left(1 - 0.0044 v_w\right). \end{aligned} \tag{2.12}$$

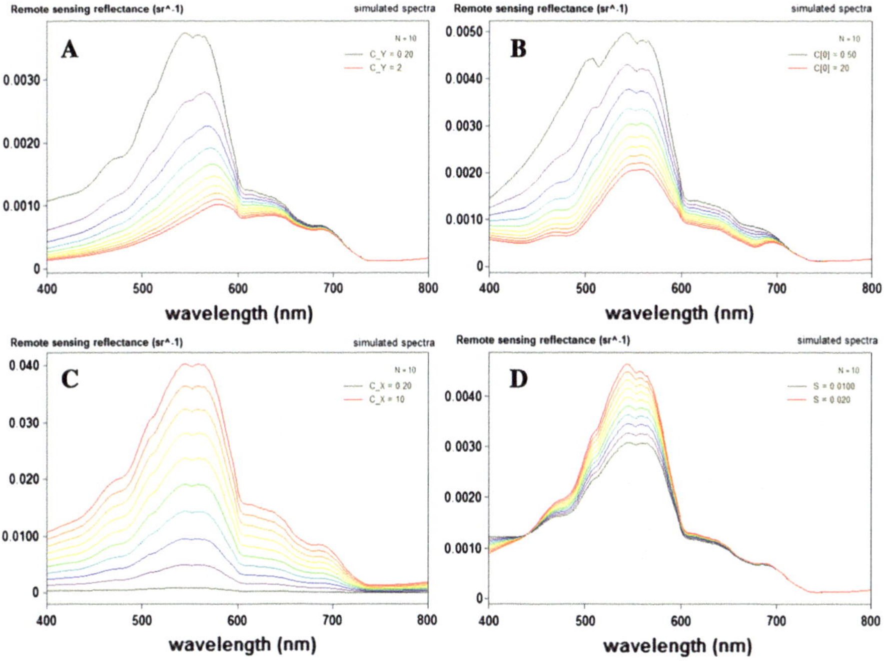


**Fig. 3** Variability of remote sensing reflectance of optically deep water. The IOPs of Fig. 2 are used for each simulation. Altered quantities: **a** CDOM absorption at 440 nm, range 0.2–2 m$^{-1}$; **b** phytoplankton concentration, range 0.5–20 mg m$^{-3}$; **c** non-algal particle concentration, range 0.2–10 g m$^{-3}$; **d** spectral slope of CDOM absorption, range 0.01–0.02 nm$^{-1}$. Screenshots of WASI (Gege 2017b). The spacing of the ten lines is equidistant on a linear scale for each plot. Hence the spacing is 0.2 m$^{-1}$ for Fig. 3a, 0.5 mg m$^{-3}$ for Fig. 3b, 0.2 g m$^{-3}$ for Fig. 3c, and 0.001 nm$^{-1}$ for Fig. 3d

Here, $\theta'_{\mathrm{sun}}$ is the sun zenith angle in water, $\theta'_{\mathrm{v}}$ the viewing zenith angle in water, and $v_{\mathrm{w}}$ the wind speed in units of [m s$^{-1}$].

These equations allow simulation of $R_{\mathrm{rs}}(\lambda)$ measurements. The variability of $R_{\mathrm{rs}}(\lambda)$ in optically deep waters is illustrated in Fig. 3. The simulations are based on the IOPs shown in Fig. 2 and of concentrations shown in Fig. 3.

### 2.1.3 AOPs for Optically Shallow Waters

When the bottom reflectance signal affects the water leaving radiance, the water is called *optically shallow*. Analytic models of remote sensing reflectance of such waters have been developed by Lee et al. (1998, 1999) and Albert (Albert and Mobley 2003; Albert 2004). They used the same approach,

$$\begin{aligned} r_{\mathrm{rs}}(\lambda) = r_{\mathrm{rs}}^{\mathrm{deep}}(\lambda) \times \left[1 - A_{\mathrm{rs},1} \times \exp\left\{-\left(K_{\mathrm{d}}(\lambda) + k_{\mathrm{uW}}(\lambda)\right) \times z_{\mathrm{B}}\right\}\right] \\ + A_{\mathrm{rs},2} \times R_{\mathrm{rs}}^{\mathrm{b}}(\lambda) \times \exp\left\{-\left(K_{\mathrm{d}}(\lambda) + k_{\mathrm{uB}}(\lambda)\right) \times z_{\mathrm{B}}\right\} \end{aligned} \tag{2.13}$$

but different expressions for the model components (see comparisons in Gege 2017a). The first term on the right is the contribution of a water layer of thickness $z_{\mathrm{B}}$, the second term of the bottom. The reflected light has passed the water column twice. The corresponding extinction is described by the attenuation coefficients $K_{\mathrm{d}}$ for downwelling irradiance, $k_{\mathrm{uW}}$ for upwelling radiance originating from the water layer, and $k_{\mathrm{uB}}$ for upwelling radiance from the bottom. $A_{\mathrm{rs},1}$ and $A_{\mathrm{rs},2}$ are empirical constants close to 1.

The reflectance properties of the water body bottom (or benthos) are specified by the irradiance reflectance of the bottom, $R^{\mathrm{b}}(\lambda)$. Since the benthos composition and cover type can be heterogeneous on small scales due to mixtures of different benthic substrate type and covers (Fig. 4), $R^{\mathrm{b}}(\lambda)$ is expressed as a weighted sum of the contributions of the different substrates:

$$R^{\mathrm{b}}(\lambda) = \sum_{n=1}^{N} f_n \times A_n(\lambda). \tag{2.14}$$

**Fig. 4** Heterogeneity of bottom substrates: **a** seagrass in Baltic Sea; **b** a mussel bed in Baltic Sea; **c** a ramified coral of Lampi Island (Myanmar); **d** seagrass and green algae in South Australia; and submerged macrophytes in Lake Constance (Europe): **e** *Potamogeton* sp. and **f** *Chara tomentosa*

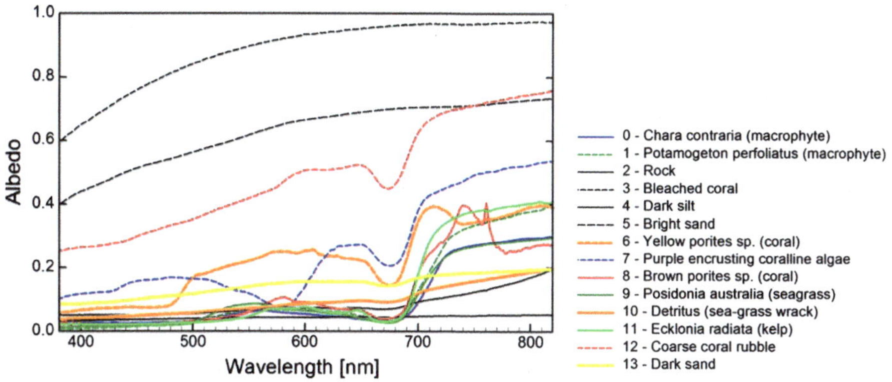


**Fig. 5** Spectral variability of bottom substrates (source CEOS 2017)

Irradiance reflectance of a surface is called *albedo*, thus the symbol $A_n(\lambda)$ is used to specify the albedo of substrate number $n$. The weights $f_n$ are the areal fractions of the substrates within an image pixel ($\Sigma f_n = 1$). Figure 5 illustrates the diversity of some bottom substrate albedos.

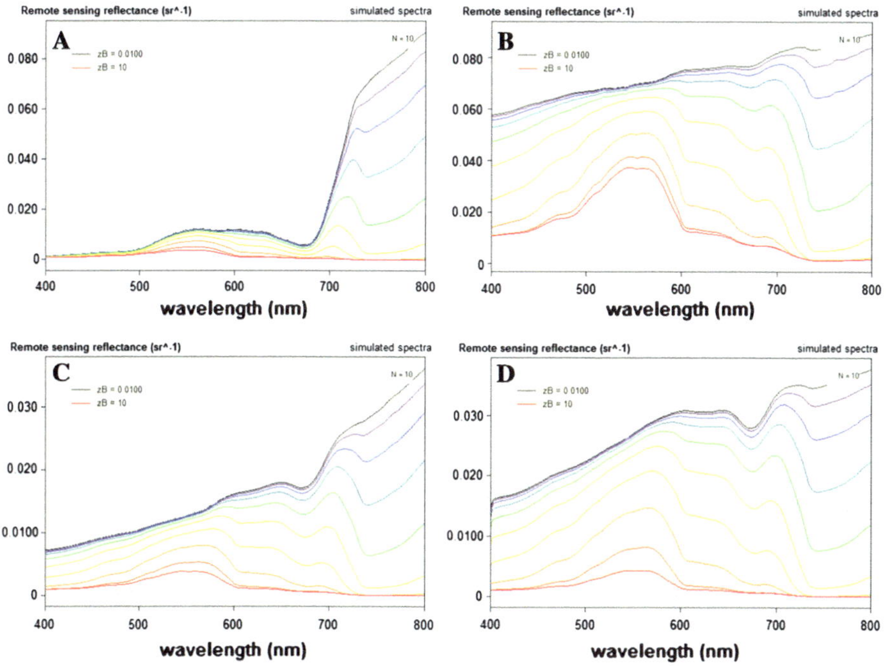


**Fig. 6** Dependence of remote sensing reflectance of optically shallow water on water depth and benthos type. Water depth is (from black to red) 0.01, 0.02, 0.05, 0.1, 0.2, 0.5, 1, 2, 5, 10 m. Bottom types are macrophytes (**a**), dark silt (**b**), detrital seagrass wrack (**c**), and dark sand (**d**). Screenshots of WASI (Gege 2017b)

The strong dependence of reflectance of optically shallow waters on benthos type and water depth is illustrated in Fig. 6. The albedo spectra 1, 4, 10 and 13 from Fig. 5 were used for the simulations; the water layer was modelled using the IOPs of Fig. 2.

The simulations were made using the software WASI (Gege 2017b) and the Albert model (Albert and Mobley 2003; Albert 2004), which accounts for illumination and viewing geometry and has been developed for a large range of environmental parameters, including most of the water constituent concentrations observed in coastal and inland waters.

## 2.2 Retrieval Algorithms

Various algorithms exist to derive water quality parameters in both optically deep and optically shallow waters based on the relation between remote sensing reflectance, concentration and IOPs of water components, and bottom albedo. The approaches that are used to invert these relations, typically distinguish between empirical and analytical algorithms (Odermatt et al. 2012).

In many applications (e.g., TSM, Turbidity, Chl-a) the concentrations of water components are often retrieved by empirical or semi-empirical algorithms. They make use of single band, band ratio or band difference positioned at wavelengths where remote sensing reflectance is sensitive to variation of water components. The single band algorithms, such as for instance Nechad et al. (2010) for TSM and turbidity, have a sound theoretical background and perform well in a variety of coastal and inland waters; for waters with large variety in concentrations an additional switching has been proposed between several bands at longer wavelengths (Shen et al. 2010; Dogliotti et al. 2015). However, single band algorithms are generally sensitive to the variability in reflectance due to influences of other water constituents, the natural variability in the IOPs and to errors in the atmospheric correction (e.g., Sterckx et al. 2007). To overcome this issue, as well as to account for changes in the concentrations of water constituents and in the spectral shape of their IOPs, several band ratio and band difference models have been proposed for different variables. Gitelson et al. (2007) used band ratios of the value of signal at the secondary Chl-a absorption maximum (at around 675 nm) to that at adjacent spectral bands not affected by phytoplankton absorption. These algorithms are typically applicable to Chl-a concentrations above ~10 mg $m^{-3}$. As the backscattering coefficient has a relatively flat spectral signature, Doxaran et al. (2003) used red-to-green and near-infrared-(NIR)-to-red band ratios for estimating TSM and turbidity. In this case, band ratio algorithms demonstrate reduced effects of variable sediment types and become less sensitive to illumination conditions. Restricted to certain water types in which CDOM dominates absorption in the blue (i.e. high CDOM and low Chl-a) and in which the NIR is not affected much by TSM, CDOM has been estimated with band ratios of blue and NIR bands (Kutser et al. 2005).

Other algorithm approaches are based on the spectral inversion of analytical bio-optical models, such as the one presented in the sections above, and of numerical radiative transfer models (e.g., Hydrolight, Mobley 1994). The models are inverted through neural networks (e.g., Doerffer and Schiller 2007; Schroeder et al. 2007), use of Look Up Tables (Salem et al. 2017), matrix inversion (e.g., Brando and Dekker 2003), quasi-analytical approaches (e.g., Lee et al. 2002) or curve fitting techniques (e.g., Keller 2001) using the different spectral bands available. These approaches provide a robust mechanism for inverting reflectance into IOPs and water quality parameters through a combination of radiative transfer theory and a level of empiricism due to the need for "a priori" SIOP parametrisation (Werdell et al. 2013). These approaches are also widely used in shallow waters applications. In this field, most of the state-of-the-art approaches applied to hyperspectral imagery are based on the inversion-optimisation (i.e. curve fitting) of Eq. 2.13 to retrieve simultaneously depth, water quality parameters and $R^b(\lambda)$ (see Dekker et al. 2011; Hedley et al. 2016 for comprehensive reviews).

Based on robust physical knowledge, these spectral inversion approaches are more widely applicable than the empirical counterparts, particularly when they are coupled with methods allowing for spatial and temporal variations in water quality parameters typical of optically complex waters (e.g., Moore et al. 2009; Brando et al. 2012; Verpoorter et al. 2012; Eleveld et al. 2017). In particular, these approaches have high potential to work well for hyperspectral remote sensing observations as each single spectrum can be decomposed to estimate IOPs, water quality parameters, bottom types and depth simultaneously.

## 2.3 Uncertainties

An essential question asked by users of remote sensing data is how reliable and accurate are the maps of the distribution of water constituents, water depth and benthic composition, as they are derived from radiance spectra sampled at top of atmosphere or at aircraft altitudes. To answer this question, and by focusing on optically deep waters only, one has to consider a number of issues in optical remote sensing of water optical properties and in-water constituents, which cause and determine the uncertainties:

(1) The radiance spectra, from which concentrations of in-water constituents have been derived are determined by a large number of atmospheric variables, the optical processes at the air/sea interface and by the optical properties of the water. The number of variables that need to be retrieved often exceeds the number of spectral bands available in multispectral Earth observing sensors, thus leading to an underdetermined system. However, in the case of hyperspectral sensors, the number of spectral bands available more than the number of variables to be retrieved. As the optical properties at multiple wavelengths may not be independent of each other, a situation may arise where the inversion system is underdetermined even for hyperspectral observations (depending on the variables to be retrieved). It has to be realised that each of the water components (cf. Section Bio-optical models above) consists of a family of substances with different and varying optical properties, which implies a possible large uncertainty when these components are decomposed from the radiance spectrum observed by the sensor spectrum (but this will depend on whether the representativity of the model complexity and its parametrisation are adequate). Thus, there will always be a level of uncertainty and end-users want to know this level of uncertainty in the final information products to determine if they are fit for purpose.

(2) Water bodies belong to the darkest surfaces of the Earth. In such waters, the fraction of reflected light is very small, being of the order of 1% of the incident flux. Thus, we have to deal with a low signal-to-noise-ratio (SNR), which requires more accurate correction for the atmospheric and air-water interface contributions (Moses et al. 2017). This signal/noise ratio is wavelength dependent and changes with the concentration of the water constituents, the sky and sunlight reflected at the water surface, the composition of the atmosphere and the sensitivity of the EO sensors.

(3) The impact of each component of the water constituents on the water leaving radiance spectra depends on its optical properties (specific absorption and scattering coefficients and fluorescence) and its concentration. If one component, e.g., CDOM, dominates the other components it may fully mask the impact of the other components on the reflectance spectrum (masking effect), so that it may become impossible to detect the presence or determine the concentration of the other components (e.g., phytoplankton Chl-a) with the required accuracy.

Other major issues are the nonlinear relationship between the concentration of a water constituent and reflectance. High concentrations of highly-scattering substances might cause saturation, i.e. the reflectance does not increase significantly with increasing concentration. Also, the vertical distribution of the depth of the retrieved signal can lead to significant uncertainties. For multiband and hyperspectral algorithms, one has to consider that each band has a different penetration signal depth, i.e. the depth from which 90% of the reflected light comes (Gordon and McCluney 1975). This is, among other things, due to the absorption by pure water with its strong increase in the red and near-infrared spectral range. The consequence is that in case of a heterogeneous vertical distribution of the water constituents the spectral bands of a multiband algorithm get information from different layers and would therefore not be consistent with each other.

There are different ways to tackle the issue of uncertainties. The minimum and most important step is to inform the user of EO products about doubtful pixels and possible uncertainties. This requires co-algorithms, which detect cases/pixels, which are contaminated, e.g., by clouds or sun glint or which are out of scope of the algorithm for the atmospheric correction or for the retrieval of water constituents, water depth or benthic composition. These pixels have to be flagged by the co-algorithms as doubtful so that the user can exclude them.

A further step is to estimate the uncertainties. This has to be performed separately for different areas or seasons or even on a pixel by pixel basis because different optical types of waters and different atmospheric conditions cause different problems and thus modify uncertainties in the retrieval process.

If uncertainty numbers are available from comparisons with water samples from different optical water types, an optical classification of the pixels based on their reflectance spectra can be used to determine the uncertainty for each class (Belo Couto et al. 2016; Jackson et al. 2017; Sathyendranath et al. 2017).

In complex waters, where concurrent match-up samples and spectra are rare, the determination of uncertainties might also be based on simulated data. The forward model to simulate the reflectance spectra from the IOPs of water and its constituents obviously must be in agreement with the model of the inversion algorithm and should meet the natural conditions as close as possible. This method has been established for data of MERIS and OLCI of optically complex coastal waters (e.g., Doerffer and Schiller 1997, rev 2016), and for Hyperion by Brando and Dekker (2003) and Giardino et al. (2007).

# 3 Applications and User Requirements

## 3.1 Water Quality Monitoring and Assessment of Biophysical Parameters

Remote sensing instruments such as spaceborne optical spectrometers provide relevant information to monitor water quality as well as to estimate biophysical parameters. Interested users are both institutional and private. Institutional users include environment agencies, water management authorities and port authorities. Private users include agriculture, forestry, aquaculture, dredging companies and drinking water companies. Fishery and recreation industries are also interested, from the water quality and clarity as well as benthic habitat point of view.

From 1991 (Dekker et al. 1991) till the launch of MODIS in 1999 developments in algorithms for determination of water quality and biophysical properties (reviewed in Dekker et al. 2001) focused on airborne imaging spectrometry, while Landsat and SPOT have been continuously providing data for broader synoptic observation scales (e.g., Lindell et al. 1999; Olmanson et al. 2015) for more than three decades. More recently, the use of medium resolution sensor and of the ocean-coastal-focused MODIS and MERIS (with its successor OLCI) further supported the developments in water quality and biophysical parameter algorithms, while the multispectral Sentinel-2 along with Landsat-8, are offering advanced opportunities for synoptic, fine-scale and high-frequency monitoring applications in lakes (Pahlevan et al. 2014; Toming et al. 2016). Finally, finer scale studies of aquatic ecosystems have been based on higher spatial resolution satellite sensors (e.g., Rapide-Eye, WorldWiew-3) that are attractive for use in spatially heterogeneous areas for applications such as the mapping of benthic habitats (e.g., Paringit and Nadaoka 2012).

In such a context, hyperspectral imaging enables increases in the estimation accuracy of variables currently observed by multispectral sensors (e.g., Chl-a, TSM), as well as facilitates detection of new variables of interest (identification and quantification of particulate and dissolved matter: type and size of suspended particles, types of pigments, organic matter composition, cyanobacteria, inorganic pollutants, etc.) for multiple user-driven applications. For example, some dissolved components are markers of soil erosion while in tropical areas (e.g., Khan et al. 2018); inorganic particulate matter concentration can be used as an indicator of regional (deforestation) and global (climate) changes (e.g., Lal 2003); particulate matter and CDOM may be used as tracers of nutrients or pollutants (e.g., Ylöstalo et al. 2016); CDOM might play a significant role in the global carbon cycle (e.g., Tranvik et al. 2009). Hyperspectral measurements are particularly relevant to assess phytoplankton pigments that provide information for assessing the trophic status, evaluating the ecosystem's functioning and for distinguishing (potentially toxic) harmful algal blooms. For example, Dierssen et al. (2015a) used HICO data to capture a red tide ciliate bloom in coastal waters.

Related to the last case, the following example shows the mapping of phytoplankton types in inland waters. The approach presented in this study makes use of the spectral decomposition of phytoplankton absorption spectra as part of the bio-optical modelling of reflectance. This method allows to obtain information on the major phytoplankton functional types (PFT) present also under non-bloom conditions. The proposed methods, along with the difference in spectral absorption and scattering associated with phytoplankton types also considers the impact of their abundance as changes in phytoplankton quantity (e.g., Chl-a concentration) produce a shift in water reflectance peaks and dips, which only hyperspectral sensors are able to capture. According to different work (Nair et al. 2008;

Bracher et al. 2009; Sadeghi et al. 2012; Sathyendranath et al. 2014; Mouw et al. 2017) such methods have great potential for hyperspectral remote sensing. Airborne Prism Experiment (APEX) imagery (98 channels used from 426 to 905 nm) acquired on 27 September 2014 and corrected for atmosphere and air/water interface effects (Sterckx et al. 2011) was used for mapping PFT in Mantua Lakes in northern Italy. The airborne campaign was developed with coincident field measurements for validation. The Mantua Lakes are small, highly productive and hypertrophic. The lakes show covariant water constituents and transition from fluvial to lentic waters presenting a challenging case for both algorithm development and testing. More than 40 days of in situ measurements were performed in Mantua Lakes over 10 years to achieve a comprehensive dataset of concentrations of water quality parameters, of IOPs and AOPs (Bresciani et al. 2013; Manzo et al. 2015), leading to the parameterisation of a bio-optical model similar to the one described in Sect. 2 or to previously published case-2 models (e.g., Pierson and Strömbeck 2001; Albert and Mobley 2003). In particular, to model the absorption of phytoplankton, the decomposition approach described in Sect. 2 was adopted according with the four following phytoplankton groups: diatoms, chlorophyte, chrysophyta, cryptophyta and cyanophytes (*Planktothrix* sp. a cyanobacteria species dominated by phycoerythrin pigments, and *Cylindrospermopsis raciborskii* and *Anabaena* sp. a cyanobacteria species dominated by phycocyanin).

The bio-optical model was applied to the water reflectance from APEX imagery and all its contiguous bands were selected to estimate Chl-a concentrations and all of each PFT. A collection of input APEX spectra is presented in Fig. 7, illustrating the appearance and shift of spectral peaks and dips. This behaviour is not only due to the specific absorption of phytoplankton pigments but also to the abundance of other water components (TSM and CDOM). In particular, in the spectral region between 670 and 720 nm the relative Rrs maximum occurs in wavelength positions shifting from 693 nm to 709 nm, while the position of the reflectance minimum also 667 nm and 678 nm. These well-known features (Schalles

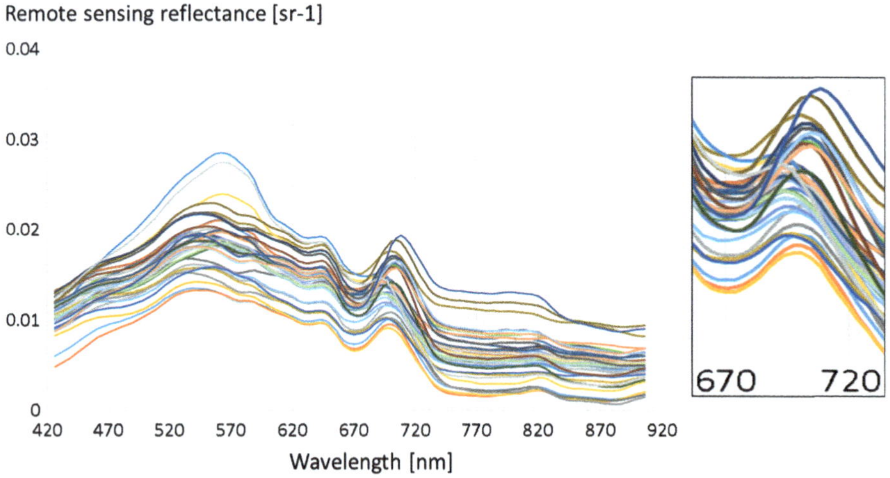


**Fig. 7** APEX spectra (corrected for atmosphere and air/water interface effects) from Mantua Lakes, Italy, flown on the 27 September 2014. In the near-infrared region the wavelength position of the relative Rrs maximum shifts from 693 to 709 nm depending on Chl-a concentration (broadening the 676 in vivo absorption feature) and the biomass of phytoplankton that causes increased scattering, all superimposed on a minimum in combined CDOM, NAP and pure water absorption). On the hand, the position of the reflectance minimum also shifts between 667 and 678 nm due to the Sun-induced Chl-a fluorescence signal

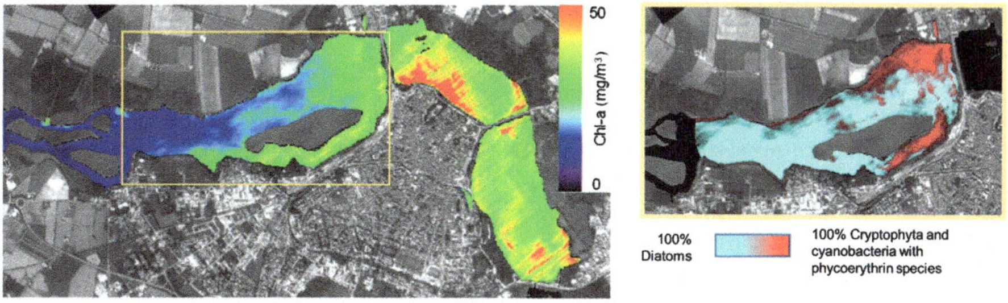


**Fig. 8** APEX data from Mantua Lakes, Italy, flown on the 27 September 2014. The map on the left shows the Chl-a concentrations obtained from the inversion of the bio-optical model. The same inversion provided, for the largest lake (yellow box), the distribution of phytoplankton types

and Yacobi 2000; Dall'Olmo and Gitelson 2005) are explained by an increase in Chl-a concentration and the biomass of phytoplankton and by the Sun-induced Chl-a fluorescence signal, that shifts the reflectance minimum towards shorter wavelengths (Dall'Olmo and Gitelson 2005). Therefore, Chl-a concentration is better estimated if contiguous bands are present allowing the switching between the different wavelengths (to capture the relative maximum/minimum of Rrs) or wavelength combinations are feasible (Bresciani et al. 2013). As found Dekker et al. (1991) and Dekker (1993) some spectra also show features due to phycoerythrin and phycocyanin pigments absorption around 590 nm and 620 nm, respectively, which can potentially be used to distinguish phytoplankton types (Hunter et al. 2016).

Figure 8 shows the output of the bio-optical model inversion. The Chl-a map (Fig. 8, on left) shows a gradient of increasing concentrations from 10 mg $m^{-3}$ to 50 mg $m^{-3}$ moving from west to east, thus from river inflow towards the lakes. The accuracy of this map, after the comparisons with water samples taken at nine stations across the three lakes, is high (average in situ: 24.9 mg $m^{-3}$; average from APEX: 25.2 mg $m^{-3}$, $r^2$ of 0.94). For the largest lake, a map of PFT was also produced by combining the algal pigments products obtained for the principal species measured in summer 2014. The PFT map (Fig. 8, on right) shows a west-to-east transition of phytoplankton types, so that diatoms are decreasing for a corresponding increasing of chrysophyta (*Chrysochromulina* sp.), cryptophyta and *Planktothrix* sp. a cyanobacteria species sometimes dominated by phycoerythrin (PE) and sometimes by phycocyanin (Dekker 1993). Such a behaviour was also observed in field visits as in the western station the diatoms count was 15,750 cells/mL, (so representing 66% of the total phytoplankton population), while in the eastern part of the lake, cryptophyta and *Planktothrix* sp. count was 18,500 cells/mL (so representing 68% of the total phytoplankton population while diatoms accounted to 25%).

## 3.2 Mapping of Shallow Water Habitats

Remote sensing offers a practical means of regularly mapping and monitoring optically shallow waters (those where the bottom is visible from the water surface and measurably influences the remote sensing reflectance, (Dekker et al. 2011)), which include inland waters, estuarine, tropical coral reef and temperate coastal ecosystems. In particular, information in the form of maps of bottom depth and benthic community are needed for science, resource management and defence operations.

EO has been used since the early 1990 s for mapping of bottom types and benthic communities (e.g., mud, sand-mud mixture, coral sands, coral reefs, seagrass, macrophytes): a thorough review of state of the art and future perspectives of remote sensing of aquatic macrophytes is given by Malthus (2017), while a recent review of remote sensing approaches of coral reefs mapping is provided by Hedley et al. (2016). With a spatial resolution of 10–30 m (up to 2–5 m from WorldView-like sensors) multispectral sensors are ideal for most of the application scales but they are limited to identify species with distinctively different spectral characteristics (e.g., Dekker et al. 2005; Villa et al. 2014). Imaging spectrometry provides the ability to differentiate submerged vegetation and coral reef taxa (Hedley et al. 2012; Botha et al. 2013) thus enabling several applications including fine tracking of biodiversity (Muller-Karger et al. 2018) as well as identification of invasive and resident species (Hestir et al. 2008). Then, Lee et al. (2013), also demonstrated that if WorldView band setting might be used for assessing depths lower than 5 m, then hyperspectral setting is needed for bottom depth mapping at improved accuracy, particularly for waters deeper than 10 m.

In coral reefs, benthic functional types (BFTs, equivalent to PFTs) can be discriminated, based on the presence/absence of diagnostic pigments in the types (Hochberg and Atkinson 2000, 2003; Hochberg et al. 2003b). Corals are host to photosynthetic dinoflagellates (zooxanthellae) which contain peridinin (Jeffrey and Haxo 1968; Prezelin 1987). Other reef algae do not contain peridinin but do have their own identifying pigments (e.g., Table 8.2 in Kirk 2011). Differential spectral absorption by those pigments results in fairly distinct signatures for the fundamental BFTs of coral and algae (Fig. 5). Sand is easily identified by its bright signal, even with fractional coverage by of benthic micro-algae (Roelfsema et al. 2002) present on the surface (Fig. 5).

Coral species cannot be spectrally distinguished (Hochberg et al. 2004) from each other, although Botha et al. (2013) have demonstrated that discriminating different coral colours at higher depth is possible when switching from multispectral to hyperspectral data. In addition to the photosynthetic pigments of the zooxanthellae, many corals possess host pigments, which influence the coral's spectral reflectance (Mazel 1990, 1995, 1996; Salih et al. 2000; Dove et al. 2001; Mazel and Fuchs 2003; Mazel et al. 2003). However, those spectral characteristics are highly variable, even within a single species (Hochberg et al., 2004). As Veron (2000) notes, coral colour is not predictive of taxonomy.

The spectral discrimination between coral and algae need not necessarily rely on fully hyperspectral data (Hochberg and Atkinson 2003). However, greater spectral information does improve the discrimination, while also improving water column correction (Lee and Carder, 2002). At least one near-infrared channel is required for correction of water surface sun and sky glint (Hochberg et al. 2003a; Hedley et al. 2005), and the availability of several channels across the shortwave infrared affords the best possible atmospheric correction (Gao et al., 2000, 2002, 2007).

Along with the challenges of surface effects and water column depth and turbidity, shallow water mapping using remote sensing platforms also includes issues associated with spatial, spectral, and radiometric resolutions. In defining the sensors requirements for inland and coastal water monitoring the needs for substrate mapping are often overlooked (Mouw et al. 2015), although both the CEOS report (2017) and the Muller-Karger et al. (2018) approach (namely H4 imaging), do explicitly deal with these requirements. As an example of trade-off in sensor and mission requirements, NASA's CORAL airborne Earth Venture mission rationale for 8 m pixels was largely logistical. It was reasoned that 10 m

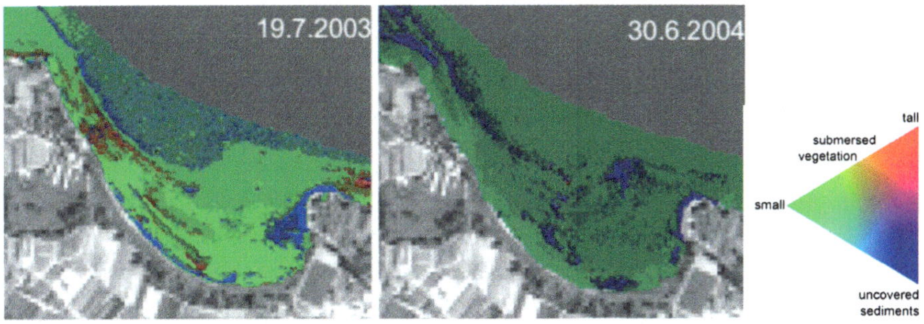


**Fig. 9** HyMap data from a Lake Constance (Richenau), flown on 19 July 2003 (left) and 30 June 2004 (right). Classification contains main bottom cover classes of sediment (blue), submersed-small vegetation (green) and emerged-tall vegetation (red). Grey areas in the image are masked as optically deep water areas

or better was required to resolve spatial variability of benthic reef communities. This was based on the science team's collective 9+ decades of experience studying coral reef ecology. At the same time, a survey airplane can only fly above or below commercial air traffic altitudes, >12,000 m or <9000 m, respectively. Flying at >12,000 m would result in pixels larger than 10 m. Thus, it was decided to fly at 8500 m, which gave pixels of roughly 8 m.

The following example shows a macrophyte mapping exercise using airborne hyperspectral remote sensing data from HyMAP acquired in July 2003 and June 2004 (Fig. 9). The data were corrected for atmospheric, air-water interface and water body effects using the physical based Modular Inversion Program (MIP) (Heege et al. 2003; Pinnel 2007). The various macrophyte taxa were classified to bottom cover classes by linear spectral unmixing. The result contains main bottom cover classes of submerged growing macrophytes (Characeae) in green, emerged macrophytes (here: mainly *Potamogeton perfoliatus* and *P. pectinatus*) in red and bottom sediments in blue (see colour triangle in Fig. 9). Mixed picture elements contain more than a single class, e.g., Characeae and bottom sediment. Differentiation of submerged and emerged vegetation could be achieved by subtracting the absolute water depth from the calculated water column on top of the plants. The bottom coverage for the two years could be mapped down to a water depth of 4.5 m, the maximum depth to which plausible reflectance spectra have been derived after water depth correction. The observed changes on benthic communities from July 2001 to June 2004 were mostly detected as an increase in small submersed vegetation further offshore and a reduction of tall vegetation close to shore.

**Table 1** Confusion matrix for the classification results of 2003 data

| | | HyMap Classification | | | |
|---|---|---|---|---|---|
| | Vegetation classes | small | tall | uncovered sediments | User's accuracy |
| Ground-Truth | small | 56 | 8 | 0 | 87.5% |
| | tall | 48 | 88 | 1 | 64.2% |
| | uncovered sediments | 2 | 0 | 13 | 86.7% |
| | Producer´s accuracy | 52.8% | 91.7% | 92.9% | 72.7% |

The classification results for 2003 were then validated with extensive ground truth measurements during the flight campaigns and traditional aerial photographic interpretation. The tall- and short-growing macrophytes and sediments were compared to the ground truth data as primary validation step. A plausibility control based on 216 ground truth measurements was statistically analysed. The results (Table 1) indicated a high correspondence (87.5%) between short-growing vegetation and sediment (86.7%); however, tall-growing macrophytes were less correctly classified (64.2%), as many of them were misidentified as short-growing macrophytes that had not grown up to the water surface, but were covered with a water column of at least 2 m. An overall accuracy of 73% was found. However, a spatial resolution of 4 m pixel might still be a bottleneck for macrophytes species recognition especially in smaller lakes where size and inhomogeneity of the patches require higher spatial information.

## 4 Observational Requirements and Future Directions in Imaging Spectrometry

Developing applications for inland and coastal surface waters, for the water column and for benthos from satellite remote sensing requires high spatial, spectral and temporal resolution and the highest possible radiometric resolution (CEOS 2017; Muller-Karger et al. 2018).

Imaging spectrometry has clear advantages over multispectral sensors for aquatic ecosystem monitoring. For example: a sensor with multiple contiguous spectral bands from blue to near-infrared wavelengths and for very turbid waters also to the short-wave infrared allows switching between the different wavelengths or wavelength combinations where reflectance features are present; a similar sensor is necessary to discriminate phytoplankton types and to assess the concentration of associated pigments, as suggested by Wolanin et al. (2016). Overall, the availability of a contiguous spectrum makes algorithms more effective in a wide variety of waters with varying water column depths and bottom reflectance and lead to more successful inversion of a larger number of properties (Hestir et al. 2015). This combination of factors points to the great utility -effectively a requirement- of hyperspectral data for inland and coastal waters retrievals.

Apart from spectral resolution, spatial resolution is also a key factor to be considered. Medium to high spatial resolution is required for smaller to larger areas (rivers, lakes, lagoons, estuaries, bays) which generally show patchy and inhomogeneous features (Ozesmi and Bauer 2002). In this study, an 8 m pixel was suggested as trade-off in sensor and mission requirements for regional scale mapping of corals, while a 4 m pixel would still be a bottleneck for macrophytes species recognition in a small lake.

The temporal resolution depends on the application but it can be critical for monitoring water quality and algal blooms so that data from geostationary platforms might be used to develop improved applications (Huang et al. 2015), while the time of passage can be critical for tides. Finally, a high SNR is also necessary because of the low reflectance by aquatic ecosystems.

In the case of shallow waters, hyperspectral satellite data enables one to identify species, to determine submerged habitat composition and to characterise environmental processes (e.g., macrophyte phenology, coral bleaching). Combined with a high spatial resolution (such as the 8 m pixel used in regional coral applications described in Sect. 3.2), hyperspectral data might provide species identification in areas characterised by high biodiversity. The temporal resolution is probably not critical for applications focused on benthic mapping and inventory while it is critical for tracking processes such as eutrophication and coral bleaching. The acquisition timing is, however, key for intertidal areas.

Further developments in science and applications in imaging spectrometry might be grouped by the platforms from which spectrometers are deployed driving the use of imaging spectrometry in specific applications.

- *Field and Near-Surface Imaging Spectrometry* Hyperspectral data gathered in the field have been used previously in support of remotely sensed observations (Goetz 2009). In inland and coastal waters they are in particular collected for understanding the link between AOP and quantities defined in the bio-optical models (such IOP), as well as to assist imagery interpretations in terms of reference data or algorithms calibrations (Eleveld et al. 2017). In particular, autonomous ship-based systems provide suitable platforms to collect spatially and temporally diverse continuous field reference data for addressing a minimum required number of spectral matchups for regional validation. A set of hyperspectral radiometer sensors that measure water surface, sky radiance and total irradiance have been successfully mounted to gather reference remote sensing reflectance for validating ocean colour sensors (e.g., Brando et al. 2016; Simis and Olsson 2013). In many cases, the ship-based systems include probes for gathering IOPs and water quality measures (e.g., fluorimeters, turbidimeters and thermometers) relevant for satellite products validation and to and improve the quality assessment of satellite retrievals (Brewin et al. 2016; Liu et al. 2018). Overall, the development in this field is progressing for both EO data processing and for other benefits: allowing the water management authority to apply the same algorithms as used for EO imagery in combination with their in situ sampling programs, continuous measurements (under daylight conditions) and measuring optical properties during time: Bresciani et al. (2013) showcased how continuous autonomous hyperspectral measurements obtained at a mooring are relevant to obtain intra- and inter-daily data in very productive waters for better understanding diurnal processes (e.g., phytoplankton processes such us light adaption, turbidity variations due to tides).
- *Airborne Imaging Spectrometry* High-resolution airborne hyperspectral sensors (e.g., AVIRIS, AISA, HyMAP, PRISM, APEX) represent enhanced mapping tools, but are limited in spatial coverage and revisit times. These airborne systems provide optimal resolutions for developing inland and water quality applications at fine scale (Dekker et al. 1991; Hoogenboom et al. 1998). In particular, airborne imaging spectrometry has been used to make medium-scale inventories of macrophytes, seagrasses and corals (Heege et al. 2003; Kutser et al. 2006; Hunter et al. 2010). The airborne technology has the merit of providing preliminary tests to design satellite-based systems. For example, Koponen et al. (2001) and Giardino et al. (2005) used airborne AISA and MIVIS images, respectively, for simulating MERIS data on lakes. More recently APEX data have been used to define and validate algorithms for mapping primary producers in eutrophic lake waters (Bolpagni et al. 2014), or to assess atmospheric correction procedures over a small eutrophic lake from the FENIX sensor (Markelin et al. 2016).
- *Satellites Imaging Spectrometry* Modern spaceborne hyperspectral sensors (e.g., HYPERION, CHRIS-PROBA and HICO from the ISS) show increasing capabilities in inland and coastal water applications (Brando and Dekker 2003; Giardino et al. 2007; Santini et al. 2010; Braga et al. 2013; Dierssen et al. 2015b), although their SNR might be limited in making accurate measurements of moderate changes in water quality parameters (e.g., CDOM with Hyperion in clear lakes (Giardino et al. 2007)). In

this context, a series of future satellite hyperspectral sensors are progressing showing similar (30 m spatial resolution of the satellite systems EnMAP and PRISMA, and DESIS and HISUI onboard of ISS), and additional (the thermal channels of HyspIRI only) characteristics. These missions will extend the knowledge that so far has been primarily driven by Hyperion with increased spectral and radiometric resolution. They will also allow to make progress in inland and coastal waters applications although, being designed for a variety of applications, might still have some limitations in measuring the remote sensing reflectance (e.g., low) over water bodies. The latter justifies the need for a dedicated atmospheric correction including adjacency effect correction (Sterckx et al. 2011; Moses et al. 2017). More specifically developed for ocean to coastal (optically deep) water applications is PACE (Pre-Aerosol, Cloud, ocean Ecosystem mission), which has been designed to take into account all requirements needed to assess biophysical parameters from space, which range from the atmospheric corrections, to phyto- and CDOM-pigment separation (5 nm spectral resolution between 380 and 800 nm, plus a spectral subsampling of ~1 to 2 nm resolution from 655 to 710 nm for refined characterisation of the Chl-a sun-induced fluorescence spectrum). Although designed to study fluorescence of land vegetation, the Earth Explorer ESA Mission FLEX (Fluorescence Explorer) might also have a key role for inland and water quality applications once launched. FLEX is on the same orbit as OLCI on Sentinel-3 and, albeit with a narrower swath, will measure in a narrow and contiguous spectral range of 500–780 nm, which is particular relevant for assessing the accessory pigments of cyanobacteria (phycocyanin and phycoerythrin) as well as Chl fluorescence.

Final recommendations for future directions in imaging spectrometry should also exploit using synergistically different sensor types (hyper- versus multispectral, high spatial versus high temporal resolution) as accurately described in Guanter et al. (in press).

**Acknowledgements** This paper is an outcome of a workshop on requirements, capabilities and directions in spaceborne imaging spectroscopy held at the International Space Science Institute (ISSI) in Bern, Switzerland, in November 2016. APEX data were provided by EUFAR TA. The activities in Lake Mantua were part of and co-funded by INFORM (Grant Agreement No. 606865) project funded under the European Commission's Seventh Framework Programme (FP7/2007–2013). We are very grateful to Shubha Sathyendranath and to the other anonymous reviewer for their detailed comments that significantly improved the manuscript.



## References

Aas E (2000) Spectral slope of yellow substance: problems caused by small particles. In: Proceedings of ocean optics XV conference, October 16–20, 2000, Monaco

Albert A (2004) Inversion technique for optical remote sensing in shallow water. Ph.D. dissertation, Universität Hamburg, Hamburg, Germany, pp 188

Albert A, Mobley CD (2003) An analytical model for subsurface irradiance and remote sensing reflectance in deep and shallow case-2 waters. Opt Express 11:2873–2890

Antoine D, Siegel DA, Kostadinov T, Maritorena S, Nelson NB, Gentili B, Vellucci V, Guillocheau N (2011) Variability in optical particle backscattering in three contrasting bio-optical oceanic regimes. Limnol Oceanogr 56(3):955–973
Babin M, Morel A, Fournier-Sicre V, Fell F, Stramski D (2003a) Light scattering properties of marine particles in coastal and open ocean waters as related to the particle mass concentration. Limnol Oceanogr 48:843–859
Babin M, Stramski D, Ferrari GM, Claustre H, Bricaud A, Obolensky G et al (2003b) Variations in the light absorption coefficients of phytoplankton, nonalgal particles, and dissolved organic matter in coastal waters around Europe. J Geophys Res 108:C7
Bader H (1970) The hyperbolic distribution of particle sizes. J Geophys Res 75:2822–2830
Belo Couto A, Brotas Mélin F, Groom S, Sathyendranath S (2016) Inter-comparison of OC-CCI chlorophyll-a estimates with precursor data sets. Int J Remote Sens 37(18):4337–4355
Binding CE, Jerome JH, Bukata RP, Booty WG (2008) Spectral absorption properties of dissolved and particulate matter in Lake Erie. Remote Sens Environ 112:1702–1711
Blondeau-Patissier D, Brando VE, Oubelkheir K, Dekker AG, Clementson LA, Daniel P (2009) Bio-optical variability of the absorption and scattering properties of the Queensland inshore and reef waters, Australia. J Geophys Res 114:C05003
Bolpagni R, Bresciani M, Laini A, Pinardi M, Matta E, Ampe EM, Giardino C, Viaroli P, Bartoli M (2014) Remote sensing of phytoplankton-macrophyte coexistence in shallow hypereutrophic fluvial lakes. Hydrobiologia 737(1):67–76
Botha EJ, Brando VE, Anstee JM, Dekker AG, Sagar S (2013) Increased spectral resolution enhances coral detection under varying water conditions. Remote Sens Environ 44:145–163
Bracher A, Vountas M, Dinter T, Burrows JP, Röttgers R, Peeken I (2009) Quantitative observation of cyanobacteria and diatoms from space using PhytoDOAS on SCIAMACHY data. Biogeosciences 6:751–764
Bracher A, Bouman HA, Bricaud A, Brewin RWJ, Brotas V, Ciotti AM, Clementson L, Devred E, Di Cicco AM, Dutkiewicz S, Hardman-Mountford NJ, Hickman AE, Hieronymi M, Hirata T, Losa SN, Mouw CB, Organelli E, Raitsos DE, Uitz J, Vogt M, Wolanin A (2017) Obtaining phytoplankton diversity from ocean color: a scientific roadmap for future development. Frontiers Mar Sci 4:00055
Braga F, Giardino C, Bassani C, Matta E, Candiani G, Strömbeck N, Adamo M, Bresciani M (2013) Assessing water quality in the northern Adriatic Sea from HICO™ data. Remote Sens Lett 4(10):1028–1037
Brando VE, Dekker AG (2003) Satellite hyperspectral remote sensing for estimating estuarine and coastal water quality. IEEE Trans Geosci Remote 41(6):1378–1387
Brando VE, Dekker AG, Park YJ, Schroeder T (2012) Adaptive semianalytical inversion of ocean color radiometry in optically complex waters. Appl Opt 51(15):2808–2833
Brando VE, Braga F, Zaggia L, Giardino C, Bresciani M, Matta E, Bellafiore D, Ferrarin C, Maicu F, Benetazzo A, Bonaldo D (2015) High-resolution satellite turbidity and sea surface temperature observations of river plume interactions during a significant flood event. Ocean Sci 11(6):909
Brando VE, Lovell JL, King EA, Boadle D, Scott R, Schroeder T (2016) The potential of autonomous ship-borne hyperspectral radiometers for the validation of ocean color radiometry data. Remote Sens 8(2):150
Bresciani M, Rossini M, Morabito G, Matta E, Pinardi M, Cogliati S, Julitta T, Colombo R, Braga F, Giardino C (2013) Analysis of within-and between-day chlorophyll-a dynamics in Mantua Superior Lake, with a continuous spectroradiometric measurement. Mar Freshw Res 64(4):303–316
Bresciani M, Adamo M, De Carolis G, Matta E, Pasquariello G, Vaičiūtė D, Giardino C (2014) Monitoring blooms and surface accumulation of cyanobacteria in the Curonian Lagoon by combining MERIS and ASAR data. Remote Sens Environ 146:124–135
Brewin RJ, Dall'Olmo G, Pardo S, van Dongen-Vogels V, Boss ES (2016) Underway spectrophotometry along the Atlantic Meridional Transect reveals high performance in satellite chlorophyll retrievals. Remote Sens Environ 183:82–97
Bricaud A, Morel A, Prieur L (1981) Absorption by dissolved organic matter of the sea (yellow substance) in the UV and visible domains. Limnol Oceanogr 26:43–53
Carder KL, Harvey GR, Ortner PB (1989) Marine humic and fulvic acids: their effects on remote sensing of ocean chlorophyll. Limnol Oceanogr 34:68–81
CEOS (2017) Feasibility study for an aquatic ecosystem earth observing sensor. In: Dekker AG (eds) CEOS report 2017. CSIRO, Canberra
Chami M, Shybanov EB, Churilova TY, Khomenko GA, Lee ME-G, Martynov OV, Berseneva GA, Korotaev GK (2005) Optical properties of the particles in the Crimea coastal waters (Black Sea). J Geophys Res 110:C11020

D'Sa EJ, Miller RL, Del Castillo C (2006) Bio-optical properties and ocean color algorithms for coastal waters influenced by the Mississippi River during a cold front. Appl Opt 45:7410–7428
Dall'Olmo G, Gitelson AA (2005) Effect of bio-optical parameter variability on the remote estimation of chlorophyll-a concentration in turbid productive waters: experimental results. Appl Opt 44(3):412–422
Dekker AG (1993) Detection of optical water quality parameters for eutrophic waters by high resolution remote sensing, Vrije Universiteit, Amsterdam, pp 1–240. dare.ubvu.vu.nl
Dekker AG, Peters SWM (1993) The use of the Thematic Mapper for the analysis of eutrophic lakes: a case study in The Netherlands. Int J Remote Sens 14:799–822
Dekker AG, Malthus TJ, Seyhan E (1991) Quantitative modeling of inland water quality for high-resolution MSS systems. IEEE Trans Geosci Remote 29(1):89–95
Dekker AG, Brando VE, Anstee JM, Pinnel N, Kutser T, Hoogenboom HJ, Pasterkamp R, Peters SWM, Vos RJ, Olbert C, Malthus TJ (2001) Imaging spectrometry of water, Ch. 11 in: imaging spectrometry: basic principles and prospective applications: remote sensing and digital image processing, v. IV. Kluwer Academic Publishers, Dordrecht, pp 307–359
Dekker AG, Brando VE, Anstee JM (2005) Retrospective seagrass change detection in a shallow coastal tidal Australian lake. Rem Sens Environ 97:415–433
Dekker AG, Phinn SR, Anstee J, Bissett P, Brando VE, Casey B, Fearns P, Hedley J, Klonowski W, Lee ZP, Lynch M (2011) Intercomparison of shallow water bathymetry, hydro-optics, and benthos mapping techniques in Australian and Caribbean coastal environments. Limnol Oceanogr-Meth 9(9):396–425
Dierssen H, McManus GB, Chlus A, Qiu D, Gao BC, Lin S (2015a) Space station image captures a red tide ciliate bloom at high spectral and spatial resolution. Proc Natl Acad Sci 112(48):14783–14787
Dierssen HM, Chlus A, Russell B (2015b) Hyperspectral discrimination of floating mats of seagrass wrack and the macroalgae Sargassum in coastal waters of Greater Florida Bay using airborne remote sensing. Remote Sens Environ 167:247–258
Doerffer R, Schiller H (1997) Pigment index, sediment and gelbstoff retrieval from directional water leaving radiance reflectances using inverse modelling technique, algorithm theoretical basis document (ATBD) ESA Doc. No. PO-TN-MEL-GS-0005
Doerffer R, Schiller H (2007) The MERIS case 2 water algorithm. Int J Remote Sens 28(3–4):517–535
Dogliotti AI, Ruddick KG, Nechad B, Doxaran D, Knaeps E (2015) A single algorithm to retrieve turbidity from remotely-sensed data in all coastal and estuarine waters. Remote Sens Environ 156:157–168
Dörnhöfer K, Oppelt N (2016) Remote sensing for lake research and monitoring—recent advances. Ecol Indic 64:105–122
Dörnhöfer K, Göritz A, Gege P, Pflug B, Oppelt N (2016) Water constituents and water depth retrieval from Sentinel-2a—a first evaluation in an oligotrophic lake. Remote Sens 8(11):941
Dove SG, Hoegh-Guldberg O, Ranganathan S (2001) Major colour patterns of reef-building corals are due to a family of GFP-like proteins. Coral Reefs 19(3):197–204
Doxaran D, Froidefond JM, Castaing P (2003) Remote-sensing reflectance of turbid sediment dominated waters. Reduction of sediment type variations and changing illumination conditions effects by use of reflectance ratios. Appl Opt 42:2623–2634
Eleveld MA, Ruescas AB, Hommersom A, Moore TS, Peters SW, Brockmann C (2017) An optical classification tool for global lake waters. Remote Sens 9(5):420
Gao BC, Montes MJ, Ahmad Z, Davis CO (2000) Atmospheric correction algorithm for hyperspectral remote sensing of ocean color from space. Appl Opt 39(6):887–896
Gao BC, Yang P, Han W, Li RR, Wiscombe WJ (2002) An algorithm using visible and 1.38-μm channels to retrieve cirrus cloud reflectances from aircraft and satellite data. IEEE Trans Geosci Remote 40(8):1659–1668
Gao BC, Montes MJ, Li RR, Dierssen HM, Davis CO (2007) An atmospheric correction algorithm for remote sensing of bright coastal waters using MODIS land and ocean channels in the solar spectral region. IEEE Trans Geosci Remote 45(6):1835–1843
Gao BC, Montes MJ, Davis CO, Goetz AFH (2009) Atmospheric correction algorithms for hyperspectral remote sensing data of land and ocean. Remote Sens Environ 113:17–24. https://doi.org/10.1016/j.rse.2007.12.015
Gege P (1998) Characterization of the phytoplankton Lake Constance for classification by remote sensing. In: Bäuerle E, Gaedke U (eds) Lake constance, characterization of an ecosystem in transition, Arch Hydrobiol Spec. Issues Advanc Limnol, vol 53, pp 179–193
Gege P (2000) Gaussian model for yellow substance absorption spectra. In: Proceedings of ocean optics XV conference, October 16–20, 2000, Monaco

Gege P (2017a) Radiative transfer theory for inland waters. In: Mishra DR, Ogashawara I (eds) Bio-optical modelling and remote sensing of inland waters. Elsevier, Amsterdam, pp 27–69. https://doi.org/10.1016/B978-0-12-804644-9.00002-1

Gege P (2017b) WASI (water colour simulator). http://www.ioccg.org/data/software.html. Accessed 04 Aug 2017

Giardino C, Candiani G, Zilioli E (2005) Detecting chlorophyll-a in Lake Garda using TOA MERIS radiances. Photogramm Eng Rem Sens 71(9):1045–1051

Giardino C, Brando VE, Dekker AG, Strömbeck N, Candiani G (2007) Assessment of water quality in Lake Garda (Italy) using Hyperion. Remote Sens Environ 109(2):183–195

Giardino C, Bresciani M, Cazzaniga I, Schenk K, Rieger P, Braga F, Matta E, Brando VE (2014) Evaluation of multi-resolution satellite sensors for assessing water quality and bottom depth of Lake Garda. Sensors 14(12):24116–24131

Giardino C, Bresciani M, Braga F, Cazzaniga I, De Keukelaere L, Knaeps E, Brando VE (2017) Bio-optical modeling of total suspended solids. In: Mishra DR, Ogashawara I (eds) Bio-optical modeling and remote sensing of inland waters. Elsevier, Amsterdam, pp 129–156

Gitelson AA, Schalles JF, Hladik CM (2007) Remote chlorophyll-a retrieval in turbid, productive estuaries: Chesapeake bay case study. Remote Sens Environ 109:464–472

Goetz AF (2009) Three decades of hyperspectral remote sensing of the earth: a personal view. Remote Sens Environ 113:S5–S16

Gordon HR, McCluney WR (1975) Estimation of the depth of sunlight penetration in the sea for remote sensing. Appl Opt 14(2):413–416

Gordon HR, Brown OB, Jacobs MM (1975) Computed relationships between the inherent and apparent optical properties of a flat, homogeneous ocean. Appl Opt 14:417–427

Guanter L, Brell M, C-W Chan J, Giardino C, Gomez-Dans J, Mielke C, Morsdorf F, Segl K, Yokoya N (in press) Synergies of spaceborne imaging spectroscopy with other remote sensing techniques, Surv Geophys

Hartmann DL, Tank AMK, Rusticucci M, Alexander LV, Brönnimann S, Charabi YAR et al (2013) Observations: atmosphere and surface. In: Change Climate (ed) Climate change 2013 the physical science basis: working group I contribution to the fifth assessment report of the intergovernmental panel on climate change. Cambridge University Press, Cambridge

Hedley JD, Harborne AR, Mumby PJ (2005) Simple and robust removal of sun glint for mapping shallow-water benthos. Int J Remote Sens 26(10):2107–2112

Hedley J, Roelfsema C, Koetz B, Phinn S (2012) Capability of the Sentinel 2 mission for tropical coral reef mapping and coral bleaching detection. Remote Sens Environ 120:145–155

Hedley JD, Roelfsema CM, Chollett I, Harborne AR, Heron SF et al (2016) Remote sensing of coral reefs for monitoring and management: a review. Remote Sens 8(2):118

Heege T, Bogner A, Pinnel N (2003) Mapping of submerged aquatic vegetation with a physically based process chain. In: Kramer E (Ed.), Remote sensing of the ocean and sea ice, September 8–12, 2003. SPIE, The International Society for Optical Engineering, vol 5233, Barcelona, Spain

Hestir EL, Khanna S, Andrew ME, Santos MJ, Viers JH, Greenberg JA, Rajapakse SS, Ustin SL (2008) Identification of invasive vegetation using hyperspectral remote sensing in the California Delta ecosystem. Remote Sens Environ 112(11):4034–4047

Hestir EL, Brando VE, Bresciani M, Giardino C, Matta E, Villa P, Dekker AG (2015) Measuring freshwater aquatic ecosystems: the need for a hyperspectral global mapping satellite mission. Remote Sens Environ 167:181–195

Hochberg EJ, Atkinson MJ (2000) Spectral discrimination of coral reef benthic communities. Coral Reefs 19:164–171

Hochberg EJ, Atkinson MJ (2003) Capabilities of remote sensors to classify coral, algae and sand as pure and mixed spectra. Remote Sens Environ 85(2):174–189

Hochberg EJ, Andréfouët S, Tyler MR (2003a) Sea surface correction of high spatial resolution Ikonos images to improve bottom mapping in near-shore environments. IEEE Trans Geosci Remote 41(7):1724–1729

Hochberg EJ, Atkinson MJ, Andréfouët S (2003b) Spectral reflectance of coral reef bottom-types worldwide and implications for coral reef remote sensing. Remote Sens Environ 85:159–173

Hochberg EJ, Atkinson MJ, Apprill A, Andréfouët S (2004) Spectral reflectance of coral. Coral Reefs 23(1):84–95

Hoogenboom HJ, Dekker AG, De Haan JF (1998) Retrieval of chlorophyll and suspended matter in inland waters from CASI data by matrix inversion. Can J Remote Sens 24(2):144–152

Huang CC, Shi K, Yang H, Li YM, Zhu AX, Sun DY, Xu LJ, Zou J, Chen X (2015) Satellite observation of hourly dynamic characteristics of algae with Geostationary Ocean Color Imager (GOCI) data in Lake Taihu. Remote Sens Environ 159:278–287
Hunter PD, Gilvear DJ, Tyler AN, Willby NJ, Kelly A (2010) Mapping macrophytic vegetation in shallow lakes using the Compact Airborne Spectrographic Imager (CASI). Aquat Conserv 20:717–727
Hunter PD, Matthews MW, Kutser T, Tyler AN (2016) Remote sensing of cyanobacterial blooms in inland, coastal, and ocean waters. Handb Cyanobacterial Monit Cyanotoxin Anal 18:89–99
Jackson T, Sathyendranath S, Melin F (2017) An improved optical classification scheme for the ocean colour essential climate variable and its applications. Remote Sens Environ 203:152–161
Jeffrey S, Haxo F (1968) Photosynthetic pigments of symbiotic dinoflagellates (zooxanthellae) from corals and clams. Biol Bull 135(1):149–165
Jerlov NG (1976) Marine optics, 2nd edn. Elsevier, Amsterdam
Kalle K (1938) Zum Problem der Meereswasserfarbe. Ann Hydrol Mar Mitt 66:1–13
Keller PA (2001) Comparison of two inversion techniques of a semi-analytical model for the determination of lake water constituents using imaging spectrometry data. Sci Total Environ 14(268):189–196
Khan MF, Maulud KNA, Latif MT, Chung JX, Amil N, Alias A, Nadzir MSM, Sahani M, Mohammad M, Jahaya MF, Hassan H (2018) Physicochemical factors and their potential sources inferred from long-term rainfall measurements at an urban and a remote rural site in tropical areas. Sci Total Environ 613:1401–1416
Kirk JTO (2011) Light and photosynthesis in aquatic ecosystems, 3rd edn. Cambridge University Press, New York
Koponen S, Pulliainen J, Servomaa H, Zhang Y, Hallikainen M, Kallio K, Vepsäläinen J, Pyhälahti T, Hannonen T (2001) Analysis on the feasibility of multi-source remote sensing observations for chl-a monitoring in Finnish lakes. Sci Total Environ 268(1–3):95–106
Korshin GV, Chi-Wang L, Benjamin MM (1997) Monitoring the properties of natural organic matter through UV spectroscopy: a consistent theory. Water Res 31:1787–1795
Kutser T, Pierson DC, Kallio KY, Reinart A, Sobek S (2005) Mapping lake CDOM by satellite remote sensing. Remote Sens Environ 94:535–540
Kutser T, Miller I, Jupp DLB (2006) Mapping coral reef benthic substrates using hyperspectral space-borne images and spectral libraries. Estuar Coast Shelf S 70:449–460
Kutser T, Paavel B, Verpoorter C, Ligi M, Soomets T, Toming K, Casal G (2016) Remote sensing of black lakes and using 810 nm reflectance peak for retrieving water quality parameters of optically complex waters. Remote Sens 8(6):497
Laanen M (2007) Yellow matters—improving the remote sensing of coloured dissolved organic matter in inland freshwaters. Ph.D. dissertation. Vrije Universiteit Amsterdam, pp 267
Lal R (2003) Soil erosion and the global carbon budget. Environ Int 29(4):437–450
Lee Z-P, Carder KL (2002) Effect of spectral band numbers on the retrieval of water column and bottom properties from ocean color data. Appl Opt 41(12):2191–2201
Lee Z-P, Carder KL, Mobley CD, Steward RG, Patch JS (1998) Hyperspectral remote sensing for shallow waters: 1. A semianalytical model. Appl Opt 37:6329–6338
Lee Z-P, Carder KL, Mobley CD, Steward RG, Patch JS (1999) Hyperspectral remote sensing for shallow waters: 2. Deriving bottom depths and water properties by optimization. Appl Opt 38:3831–3843
Lee Z-P, Carder KL, Arnone RA (2002) Deriving inherent optical properties from water color: a multiband quasi-analytical algorithm for optically deep waters. Appl Opt 41:5755–5772
Lee Z-P, Carder KL, Arnone RA, He M (2007) Determination of primary spectral bands for remote sensing of aquatic environments. Sensors 7(12):3428–3441
Lee Z-P, Weidemann A, Arnone R (2013) Combined Effect of reduced band number and increased bandwidth on shallow water remote sensing: the case of worldview 2. IEEE Trans Geosci Remote 51(5):2577–2586
Lindell T, Pierson D, Premazzi G, Zilioli E (1999) Manual for monitoring European lakes using remote sensing techniques. EUR report no.18665 EN. Luxembourg: Office for Official Publications of the European Communities
Liu Y, Roettgers R, Ramírez-Pérez M, Dinter T, Steinmetz F, Noethig E-M, Hellmann S, Wiegmann S, Bracher A (2018) Underway spectrophotometry in the Fram Strait (European Arctic Ocean): a highly resolved chlorophyll a data source for complementing satellite ocean color. Optics Express: ID 313439
Lobo FL, Costa MP, Novo EM (2015) Time-series analysis of Landsat-MSS/TM/OLI images over Amazonian waters impacted by gold mining activities. Remote Sens Environ 157:170–184

Malthus M (2017) Bio-optical modeling and remote sensing of aquatic macrophytes. In: Mishra DR, Ogashawara I (eds) Bio-optical modeling and remote sensing of inland waters. Elsevier, Amsterdam, pp 263–308
Manzo C, Bresciani M, Giardino C, Braga F, Bassani C (2015) Sensitivity analysis of a bio-optical model for Italian lakes focused on Landsat-8, Sentinel-2 and Sentinel-3. Eur J Remote Sens 48(1):17–32
Markelin L, Simis SG, Hunter PD, Spyrakos E, Tyler AN, Clewley D, Groom S (2016) Atmospheric correction performance of hyperspectral airborne imagery over a small eutrophic lake under changing cloud cover. Remote Sens 9(1):2
Matthews MW (2017) Bio-optical modeling of phytoplankton chlorophyll-a. In: Mishra DR, Ogashawara I (eds) Bio-optical modeling and remote sensing of inland waters. Elsevier, Amsterdam, pp 157–182
Matthews MW, Odermatt D (2015) Improved algorithm for routine monitoring of cyanobacteria and eutrophication in inland and near-coastal waters. Remote Sens Environ 156:374–382
Mazel CH (1990) Spectral transformation of downwelling radiation by autofluorescent organisms in the sea. In: Ocean Optics X, vol 1302. International Society for Optics and Photonics, pp 320–328
Mazel CH (1995) Spectral measurements of fluorescence emission in Caribbean cnidarians. Mar Ecol Prog Ser 120:185–191
Mazel CH (1996) Coral fluorescence characteristics: excitation-emission spectra, fluorescence efficiencies, and contribution to apparent reflectance. SPIE Vol 2963 Ocean Opt XIII 1:240–245
Mazel CH, Fuchs E (2003) Contribution of fluorescence to the spectral signature and perceived color of corals. Limnol Oceanogr 48:390–401
Mazel CH, Lesser MP, Gorbunov MY, Barry TM, Farrel JH, Wyman KD et al (2003) Green-fluorescent proteins in Caribbean corals. Limnol Oceanogr 48:402–411
Mobley CD (1994) Light and water. Academic press, San Diego
Moore TS, Campbell JW, Dowell MD (2009) A class-based approach to characterizing and mapping the uncertainty of the MODIS ocean chlorophyll product. Remote Sens Environ 113:2424–2430
Morel A (2001) Bio-optical models. In: Thorpe SA (eds) Encyclopedia of ocean sciences. Academic press, San Diego, pp 317–326. https://doi.org/10.1006/rwos.2001.0407
Morel A, Maritorena S (2001) Bio-optical properties of oceanic waters: a reappraisal. J Geophys Res 106:7163–7180
Moses WJ, Sterckx S, Montes MJ, De Keukelaere L, Knaeps E (2017) Atmospheric correction for inland waters. In: Mishra DR, Ogashawara I (eds) Bio-optical modeling and remote sensing of inland waters. Elsevier, Amsterdam, pp 69–100
Mouw CB, Greb S, Aurin D, DiGiacomo PM, Lee Z, Twardowski M et al (2015) Aquatic color radiometry remote sensing of coastal and inland waters: challenges and recommendations for future satellite missions. Remote Sens Environ 160:15–30
Mouw C, Hardman-Montford N, Alvain S, Bracher A, Brewin R, Bricaud A, Ciotti A, Devred E, Fujiwara A, Hirata T, Hirawake T, Kostadinov T, Roy S, Uitz J (2017) A consumer's guide to satellite remote sensing of multiple phytoplankton groups in the global ocean. Front Mar Sci Res Topic CLEO. https://doi.org/10.3389/fmars.2017.00041
Muller-Karger FE, Hestir E, Ade C, Turpie K, Roberts D, Siegel D, Miller R et al (2018) Satellite sensor requirements for monitoring essential biodiversity variables of coastal ecosystems. Ecol Appl. https://doi.org/10.1002/eap.1682
Nair A, Sathyendranath S, Platt T, Morales J, Stuart V, Forget M-H, Devred E, Bouman H (2008) Remote sensing of phytoplankton functional types. Remote Sens Environ 112:3366–3375
Nechad B, Ruddick KG, Park Y (2010) Calibration and validation of a generic multisensor algorithm for mapping of Total Suspended Matter in turbid waters. Remote Sens Environ 114(4):854–866
Odermatt D, Gitelson A, Brando VE, Schaepman M (2012) Review of constituent retrieval in optically-deep and complex waters from satellite imagery. Remote Sens Environ 118:116–126
Olmanson LG, Bauer ME, Brezonik PL (2008) A 20-year Landsat water clarity census of Minnesota's 10,000 lakes. Remote Sens Environ 112(11):4086–4097
Olmanson LG, Brezonik PL, Bauer ME (2015) Remote sensing for regional lake water quality assessment: capabilities and limitations of current and upcoming satellite systems. In: Younos T, Parece TE (eds) Advances in watershed science and assessment. Springer, Cham, pp 111–140
Ozesmi SL, Bauer ME (2002) Satellite remote sensing of wetlands. Wetl Ecol Manag 10(5):381–402
Pahlevan N, Lee Z, Wei J, Schaaf CB, Schott JR, Berk A (2014) On-orbit radiometric characterization of OLI (Landsat-8) for applications in aquatic remote sensing. Remote Sens Environ 154:272–284
Paringit EC, Nadaoka K (2012) Simultaneous estimation of benthic fractional cover and shallow water bathymetry in coral reef areas from high-resolution satellite images. Int J Remote Sens 33:3026–3047

Pierson DC, Strömbeck N (2001) Estimation of radiance reflectance and the concentrations of optically active substances in Lake Malaren, Sweden, based on direct and inverse solutions of a simple model. Sci Total Environ 268:171–188
Pinnel N (2007) A method for mapping submerged macrophytes in lakes using hyperspectral remote sensing. Limnologische Station, WZW, Department für Ökologie, Dissertation, p 164
Pozdnyakov D, Lyaskovsky A, Grassl H, Pettersson L (2002) Numerical modelling of transspectral processes in natural waters: implications for remote sensing. Int J Remote Sens 22:1581–1607
Preisendorfer RW (1961) Application of radiative transfer theory to light measurements in the sea. Union Geod Geophys Inst Monogr 10:11–30
Prezelin BB (1987) Photosynthetic physiology of dinoflagellates. In: Taylor FJF (ed) The biology of dinoflagellates. Blackwell Scientific Publications, Oxford, pp 174–223
Roelfsema CM, Phinn SR, Dennison WC (2002) Spatial distribution of benthic microalgae on coral reefs determined by remote sensing. Coral Reefs 21:264–274
Sadeghi A, Dinter T, Vountas M, Taylor B, Peeken I, Altenburg Soppa M, Bracher A (2012) Improvements to the PhytoDOAS method for identification of coccolithophores using hyper-spectral satellite data. Ocean Sci 8:1055–1070
Salem SI, Higa H, Kim H, Kazuhiro K, Kobayashi H, Oki K, Oki T (2017) Multi-algorithm indices and look-up table for chlorophyll-a retrieval in highly turbid water bodies using multispectral data. Remote Sens 9(6):556
Salih A, Larkum A, Cox G, Kühl M, Hoegh-Guldberg O (2000) Fluorescent pigments in corals are photoprotective. Nature 408:850–853
Santini F, Alberotanza L, Cavalli RM, Pignatti S (2010) A two-step optimization procedure for assessing water constituent concentrations by hyperspectral remote sensing techniques: an application to the highly turbid Venice lagoon waters. Remote Sens Environ 114(4):887–898
Sathyendranath S, Aiken J, Alvain S, Barlow R, Bouman H, Bracher A, Brewin R, Bricaud A, Brown CW, Ciotti AM, Clementson L, Craig SE, Devred E, Hardman-Mountford N, Hirata T, Hu C, Kostandinov TS, Lavender S, Loisel H, Moore TS, Morales J, Moulin C, Mouw CB, Nair A, Raitsos D, Roesler C, Shutler JD, Sosik H, Soto I, Stuart V, Subramaniam A, Uitz J (2014) Phytoplankton types from space. In: Sathyendranath S (ed) IOCCG report 15. Reports and Monography of the International Ocean Colour Coordination Group. http://www.ioccg.org/reports/IOCCG_Report_15_2014.pdf
Sathyendranath S, Brewin RJ, Jackson T, Mélin F, Platt T (2017) Ocean-colour products for climate-change studies: what are their ideal characteristics? Remote Sens Environ 203:125–138
Schaepman-Strub G, Schaepman ME, Painter TH, Dangel S, Martonchik JV (2006) Reflectance quantities in optical remote sensing—definitions and case studies. Remote Sens Environ 103:27–42
Schalles JF, Yacobi YZ (2000) Remote detection and seasonal patterns of phycocyanin, carotenoid and chlorophyll pigments in eutrophic waters. Arch Hydrobiol Spec Issues Adv Limnol 55:153–168
Schroeder T, Schaale M, Fischer J (2007) Retrieval of atmospheric and oceanic properties from MERIS measurements: a new Case-2 water processor for BEAM. Int J Rem Sens 28(24):5627–5632
Schwarz JN, Kowalczuk P, Kaczmarek S, Cota GF, Mitchell BG, Kahru M, Chavez FP, Cunningham A, McKee D, Gege P, Kishino M, Phinney DA, Raine R (2002) Two models for absorption by coloured dissolved organic matter (CDOM). Oceanologia 44:209–241
Shen F, Verhoef W, Zhou Y, Salama MS, Liu X (2010) Satellite estimates of wide-range suspended sediment concentrations in Changjiang (Yangtze) estuary using MERIS data. Estuaries Coasts 33(6):1420–1429
Simis SG, Olsson J (2013) Unattended processing of shipborne hyperspectral reflectance measurements. Remote Sens Environ 135:202–212
Sipelgas L, Arst H, Kallio K, Erm A, Oja P, Soomere T (2003) Optical properties of dissolved organic matter in Finnish and Estonian lakes. Nord Hydrol 34:361–386
Sokoletsky LG, Shen F (2014) Optical closure for remote-sensing reflectance based on accurate radiative transfer approximations: the case of the Changjiang (Yangtze) River Estuary and its adjacent coastal area, China. Int J Remote Sens 35:4193–4224
Sterckx S, Knaeps E, Bollen M, Trouw K, Houthuys R (2007) Retrieval of suspended sediment from advanced hyperspectral sesnor data in the scheldt estuary at different stages in the tidal cycle. Mar Geod 30:1–12
Sterckx S, Knaeps E, Ruddick K (2011) Detection and correction of adjacency effects in hyperspectral airborne data of coastal and inland waters: the use of the near infrared similarity spectrum. Int J Rem Sens 32(21):64796505
Strong AE (1974) Remote sensing of algal blooms by aircraft and satellite in Lake Erie and Utah Lake. Remote Sens Environ 3(2):99–107

Stumpf RP, Wynne TT, Baker DB, Fahnenstiel GL (2012) Interannual variability of cyanobacterial blooms in Lake Erie. PLoS ONE 7(8):e42444
Toming K, Kutser T, Laas A, Sepp M, Paavel B, Nõges T (2016) First experiences in mapping lake water quality parameters with Sentinel-2 MSI imagery. Remote Sens 8(8):640
Tranvik LJ, Downing JA, Cotner JB, Loiselle SA, Striegl RG, Ballatore TJ, Dillon P et al (2009) Lakes and reservoirs as regulators of carbon cycling and climate. Limnol Oceanogr 54(6(2)):2298–2314
Tyler AN, Hunter PD, Spyrakos E, Groom S, Constantinescu AM, Kitchen J (2016) Developments in Earth observation for the assessment and monitoring of inland, transitional, coastal and shelf-sea waters. Sci Total Environ 572:1307–1321
Vanhellemont Q, Ruddick K (2014) Turbid wakes associated with offshore wind turbines observed with Landsat 8. Remote Sens Environ 145:105–115
Veron JEN (2000) Corals of the world. Australian Institute of Marine Science, Townsville
Verpoorter V, Loisel H, Dessailly D, Mauriaux X (2012) Optical classification of contrasted coastal waters. Remote Sens Environ 123:306–323
Verpoorter V, Kutser T, Seekel D, Tranvik L (2014) A global inventory of lakes based on high-resolution satellite imagery. Geol Rese Letters 41(18):6396–6402
Villa P, Bresciani M, Braga F, Bolpagni R (2014) Comparative assessment of broadband vegetation indices over aquatic vegetation. IEEE J STARS 7(7):3117–3127
Wang M, Shi W (2008) Satellite observed algae blooms in China's Lake Taihu. Eos Trans Am Geophys Union 89:201–202
Werdell PJ, Franz BA, Bailey SW, Feldman GC, Boss E, Brando VE, Dowell M, Hirata T, Lavender SJ, Lee Z, Loisel H (2013) Generalized ocean color inversion model for retrieving marine inherent optical properties. Appl Opt 52(10):2019–2037
Wolanin A, Soppa MA, Bracher A (2016) Investigation of spectral band requirements for improving retrievals of phytoplankton functional types. Remote Sens 8:871. https://doi.org/10.3390/rs8100871
Ylöstalo P, Seppälä J, Kaitala S, Maunula P, Simis S (2016) Loadings of dissolved organic matter and nutrients from the Neva River into the Gulf of Finland-Biogeochemical composition and spatial distribution within the salinity gradient. Mar Chem 186:58–71
Zepp RG, Schlotzhauer PF (1981) Comparison of photochemical behaviour of various humic substances in water: III. Spectroscopic properties of humic substances. Chemosphere 10:479–486

Surveys in Geophysics (2019) 40:431–470
https://doi.org/10.1007/s10712-019-09523-1

# Imaging Spectroscopy for the Detection, Assessment and Monitoring of Natural and Anthropogenic Hazards

**C. Ong[1] · V. Carrère[2] · S. Chabrillat[3] · R. Clark[4] · T. Hoefen[5] · R. Kokaly[5] · R. Marion[6] · C. R. Souza Filho[7] · G. Swayze[5] · D. R. Thompson[8]**

Received: 15 January 2018 / Accepted: 23 February 2019 / Published online: 18 March 2019


**Abstract**

Natural and anthropogenic hazards have the potential to impact all aspects of society including its economy and the environment. Diagnostic data to inform decision-making are critical for hazard management whether for emergency response, routine monitoring or assessments of potential risks. Imaging spectroscopy (IS) has unique contributions to make via the ability to provide some key quantitative diagnostic information. In this paper, we examine a selection of key case histories representing the state of the art to gain an insight into the achievements and perspectives in the use of visible to shortwave infrared IS for the detection, assessment and monitoring of a selection of significant natural and anthropogenic hazards. The selected key case studies examined provide compelling evidence for the use of the IS technology and its ability to contribute diagnostic information currently unattainable from operational spaceborne Earth observation systems. User requirements for the applications were also evaluated. The evaluation showed that the projected launch of spaceborne IS sensors in the near-, mid and long term future, together with the increasing availability, quality and moderate cost of off the shelf sensors, the possibilities to couple unmanned autonomous systems with miniaturized sensors, should be able to meet these requirements. The challenges and opportunities for the scientific community in the future when such data become available will then be ensuring consistency between data from different sensors, developing techniques to efficiently handle, process, integrate and deliver the large volumes of data, and most importantly translating the data to information that meets specific needs of the user community in a form that can be digested/understood by them. The latter is especially important to transforming the technology from a scientific to an operational tool. Additionally, the information must be independently validated using current trusted practices and uncertainties quantified before IS derived measurement can be integrated into operational monitoring services.

**Keywords** Hazards · Imaging spectroscopy · Hyperspectral · Acid mine drainage · Dust · Hydrocarbon · Atmospheric emissions · Aerosols · Asbestos · Swelling clay

***Disclaimers***: Any use of trade, product or firm names in this chapter is for descriptive purposes only and does not imply endorsement by the US Government.

✉ C. Ong
cindy.ong@csiro.au

Extended author information available on the last page of the article

## 1 Introduction

A hazard is broadly defined as a process, phenomenon or human activity that may cause loss of life, injury or other health impacts, property damage, social and economic disruption or environmental degradation (United Nations Office for Disaster Risk Reduction 2017). Hazards may be classified as active where the condition is certain to cause harm, or the hazards are classified as latent where they represent future threats. Most hazards are dormant with only a theoretical risk of harm. When the level of risk/threat and vulnerability elevates to cause the hazards to be severe and large scale, they are classified as disasters (Jha 2010).

Natural hazards are those that are associated with natural processes and phenomena including but not limited to wildland fires, dust storms, volcanic eruptions, hurricanes, floods, earthquakes, swelling clays (when exposed to excess water) and acid sulphate soils. Natural hazards have significant societal and environmental impacts. Anthropogenic hazards are those that are induced by human forces and encompass those that are both accidental and intentional in nature. Anthropogenic hazards can occur from activities such as non-renewable resource exploitation, industrial processing, agriculture, urban development, transportation, and shipping. Lastly, some hazards are classified as socio-natural, in that they are associated with a combination of natural and anthropogenic factors including environmental degradation and climate change.

Passive and active satellite Earth observations have been used for decades to inform decision-making on natural and anthropogenic hazards (Hong et al. 2007; Joyce et al. 2009; Lentile et al. 2006; Metternicht and Zinck 2003; Tralli et al. 2005) and are now generally accepted operational tools for this purpose. Until recently, imaging spectroscopy (IS), particularly from satellite platforms, has not been operationally available and hence is not yet as widely or operationally used. IS is defined as the simultaneous acquisition of spatially co-registered images in many spectrally contiguous bands (>50), measured in calibrated radiance units, from a remotely operated platform (Schaepman et al. 2009; Goetz 2009). IS has unique contributions to make via its ability to provide quantitative diagnostic information to better inform assessments and monitor hazards and their related threats. In the last two decades, with the increasing availability of IS particularly in the visible (VIS) to shortwave infrared (SWIR) (400–2500 nm) range mainly from airborne platforms, some convincing studies have demonstrated its remarkable effectiveness (for acid drainage see Buzzi et al. 2016; Davies and Calvin 2017; Farrand 1997; Farrand and Harsanyi 1997; Fenstermaker and Miller 1994; Kemper and Sommer 2004; King et al. 1995; Kopačková 2014; Kruse et al. 1989; Lopez-Pamo et al. 1999; Ong and Cudahy 2014; Pearson et al. 1997; Peters et al. 1995; Shi et al. 2014a, b; Swayze et al. 1996, 2000; Rockwell et al. 2005; Zabcic et al. 2014; for fugitive dust see Clark et al. 2006; Chudnovsky et al. 2009, 2011; Ong et al. 2003a, b, c, 2008; Ong 2013; Pascucci et al. 2012; for hydrocarbon contamination see Beland et al. 2016; Clark et al. 2010; Khanna et al. 2013; Kokaly et al. 2013; Peterson et al. 2015; for atmospheric emissions see Bradley et al. 2011; Dennison et al. 2013; Deschamps et al. 2013; Franke et al. 2009; Frankenberg et al. 2016; Frassy et al. 2014; Krautwurst et al. 2017; Leifer et al. 2012; Marion et al. 2004; Roberts et al. 2010; Spinetti et al. 2008; Thompson et al. 2105b; Thorpe et al. 2012, 2013; and, for naturally occurring hazardous soils and rocks, see Bonifazi and Serranti 2016; Chabrillat et al. 2002, Chabrillat and Goetz 2006; Clark et al. 2001, 2003b; Marino et al. 2000; Massarelli et al. 2017; Swayze et al. 2004, 2009; Swayze and Kokaly 2016, and many others).

In this paper, we examine a selection of significant anthropogenic and natural hazards and show, through some key case histories, how visible (VIS)- shortwave (SWIR) IS has been used to enhance understanding and improve decision-making and management of these challenges. Finally, we explore the limitations of current systems and requirements for future operational systems.

## 2 Theory and Concepts

Most of the studies conducted thus far on the use of IS for natural and anthropogenic hazards exploit the fact that the hazardous material (be it soil, water or air) has compositional characteristics that are spectrally distinct. In contrast, the surrounding non-hazardous material may have different or no diagnostic spectral features that can be used to distinguish it from the hazardous material. Therefore, the use of IS for hazards is essentially the use of spectroscopy for the detection and identification of materials be they solid, liquid or gas.

Spectroscopy deals with the interpretation, as a function of wavelength, of light reflected, emitted, transmitted or scattered, resulting from the interaction of electromagnetic radiation with matter (atoms and molecules). Absorption and scattering are the two main physical properties that can be used to identify and quantify constituents from both the surface and the atmosphere.

Figure 1 illustrates the driving mechanisms for spectral absorption features, the significance of the wavelength location and shape of spectral features for identification and quantification of the material of interest. As most of the key components of the anthropogenic hazards examined for this paper were characterized by their mineralogical composition, this figure is illustrated with two minerals. The principles described here are similarly applicable to other materials. In the ultraviolet (UV)–VIS wavelength range, the energy of a photon is used to excite electrons to a higher energy level (transition). The energy transitions for the electrons of each element are unique and are distinct from one another. For minerals, absorption features in the UV–VIS–nearinfrared (VNIR) spectral range are generally broad and caused by electronic transitions, crystal field effects and charge transfers that can be used to distinguish the observed material. For example, for the jarosite (an iron-bearing mineral) spectrum in Fig. 1, the electronic transitions and crystal field effects at wavelengths around 430, 550 and 900 nm allow for identification of materials containing iron ($Fe^{+3}$).

At longer wavelengths beyond the NIR towards the shortwave (SWIR) and long-wave infrared (LWIR), the energy of a photon produces changes within the atoms and molecules, more specifically vibrations of atomic or molecular bonds. As a result, the specific wavelengths of light are absorbed, resulting in narrow diagnostic absorption features that can be used to uniquely identify the observed material. Molecular vibrational features at wavelengths between 1000 and 2500 nm are diagnostic of materials containing anion groups, such as Al–OH, Mg–OH, Fe–OH, Si–OH, $CO_3$, $NH_4$ and $SO_4$ as illustrated by the Al–OH and Fe–OH features in Fig. 1 diagnostic of alunite and jarosite, respectively. Often, the various diagnostic absorption features of materials (such as the crystal field and Fe–OH of jarosite) are used in unison to interpret the spectrum, which explains why the complete VIS-SWIR is required. Small differences in the absorption band wavelength position and shape are correlated with both material composition and variability.

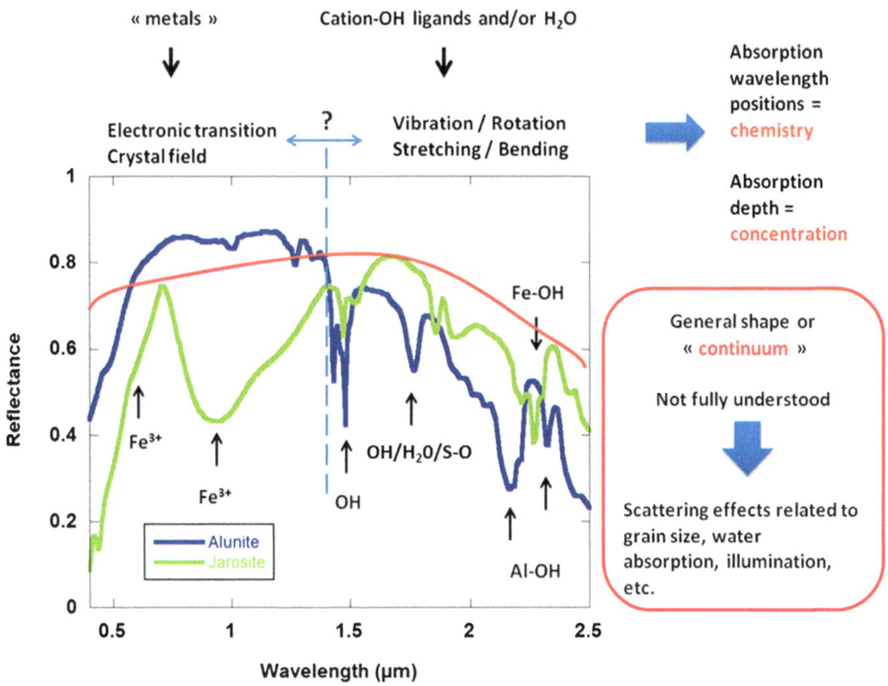


**Fig. 1** Illustration of diagnostic spectral absorption features and the processes that cause them, and the relevance of the wavelength positions and shape for discriminating the material. The red line represents the hull or continuum of the spectrum

Additionally, absorption strength (or depth) is controlled by the optical properties, grain size and concentration of the chemical constituent (mineral, liquid or gas), according to Beer–Lambert's law (Swinehart 1962). The general shape of a reflectance spectrum (also called continuum) is mostly influenced by the physical properties of the medium. Photons can be reflected from grain surfaces or refracted through a particle and thus be scattered. The intensity of the scattering process is controlled by the size of the grains, their optical properties and the energy of the photons. The amount of light scattered and absorbed by a grain is dependent on grain size, more specifically on the surface-to-volume ratio. The larger the grain, the larger the internal path and the higher the probability for the photons to be absorbed. Conversely, the smaller the grain, the more the surfaces of reflection and the more the scattering. As a consequence, the reflectance decreases with increasing grain size. The shape of the continuum may also be due to larger absorption features such as the wing of the 2800-nm fundamental water absorption in the SWIR.

In addition to mineralogical compositions, the other key component used to characterize the hazards examined was gaseous constituents. Gases and aerosols also have diagnostic absorption or emission features in the VIS-SWIR from which their compositions can be distinguished, which are driven by electronic transitions in the UV-SWIR region and, by vibrational–rotational transitions in the VIS-SWIR. For example, $CH_4$ has diagnostic absorption features around 1700 nm, which is the first overtone of the C–H stretch fundamentals around 3400 nm, and, has an additional diagnostic absorption feature at 2300 nm likely caused by multiple combinations and overtones (Clark et al. 2009) (see spectrum of $CH_4$ in Fig. 2).

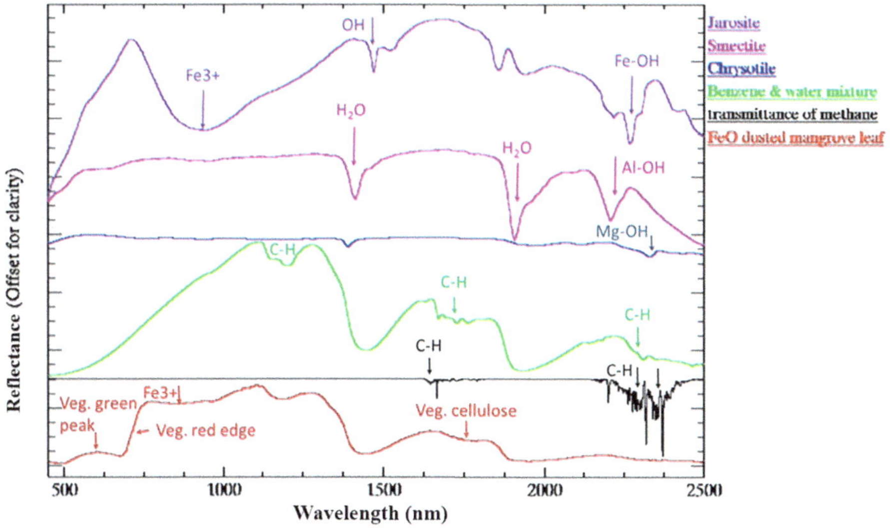


**Fig. 2** Spectra of a selection of key minerals, organic compounds and hydrocarbons related to natural and anthropogenic hazards (sourced from a combination of Kokaly et al. 2017 and Ong 2013). Specifically, the spectra shown are of: jarosite, a key secondary mineral resulting from acid mine drainage (AMD); smectite, a mineral used to identify swelling clay; chrysotile, a mineral with asbestiform crystal habits; benzene and water mixture; methane, the main composition of natural gas and an iron oxide-dusted mangrove leaf

When spectral data are acquired remotely such as the case of IS, the signal recorded by IS includes the atmospheric column between the sensor and the ground surface, which must be accounted for and removed, particularly for spatially varying constituents such as water vapour. Therefore, investigators now routinely estimate atmospheric gaseous constituents based on their absorption features, via a process called atmospheric correction as an intermediate step before the retrieval of surface reflectance spectra (Frouin et al. 1990; Gao and Goetz 1990; Kaufman and Gao 1992; Carrère and Conel 1993; Schläpfer et al. 1998; Qu and Kindel 2003; Cairns et al. 2003; Thompson et al. 2015). The retrieval of gaseous constituents for gaseous hazard monitoring is analogous. In addition, scattering by air molecules (Rayleigh and Raman scattering), scattering and absorption by aerosol and cloud particles (Mie scattering), refraction due to the density gradient in the atmosphere and surface reflection attenuate these absorption or emission features of gaseous compositions and must be accounted for (Gottwald et al. 2006).

In summary, spectroscopy provides unique diagnostic capabilities for identifying and quantifying solid, liquid and gaseous materials giving IS these added diagnostic capabilities in addition to providing locational and spatial context. The theory, principles and concepts of spectroscopy of rocks, gases and other materials are comprehensively described by Farmer (1974), Burns (1993), Clark (1999), Hunt and Ashley (1979), Hunt (1977, 1980), Gottwald et al. (2006), and others. In addition, the theoretical background concepts on gases are described in more detail in Thompson et al. (2019) (this issue) and Chabrillat et al. (2019) (this issue) provide further supplementary details specifically on soils.

# 3 Applications to Anthropogenic and Natural Hazards

The use of Earth observation for natural and anthropogenic hazards have now been comprehensively proven and is generally accepted as an operational tool (Hong et al. 2007; Joyce et al. 2009; Lentile et al. 2006; Metternicht and Zinck 2003; Tralli et al. 2005). The advantages of Earth observation include the ability to provide non-contact, timely, temporal and spatial context useful for assessments of risks, characterization of hazardous events and monitoring post-event remediation efforts. IS, similarly, has these capabilities, but IS is augmented with unique diagnostic capabilities, allowing more accurate identification and quantification of risks, to improve targetting of emergency responses and, enhanced the understanding of the progress of remediation efforts.

In this section, we aim to show the augmented diagnostic capabilities of IS by emphasizing selected case histories that demonstrate the added values of IS. Additionally, as much as possible we focus on case histories where the full VIS-SWIR wavelength ranges are critical for retrieving accurate and quantitative information. This was particularly done to understand the value of future spaceborne IS which will be operating in those ranges.

Finally, we provide a synopsis on the case histories examined, describing the value of the diagnostic capabilities of IS, the spatial requirements, the status of the processing and analysis methods and the findings in the context of their benefit for improving identification and quantification of risks, improving targeting of emergency responses and enhancing the understanding of the progress of remediation efforts.

## 3.1 Acid Drainage

The application of IS in acid mine drainage (AMD) management is one of the most well-documented uses of IS for studying anthropogenic hazards in the last two decades (see Table 1). AMD is an anthropogenic hazard that has been recognized as one of the most significant environmental challenges of the mineral industry (Akcil and Koldas 2006; Harries 1997). AMD is commonly defined as water seepage, with a pH less than 5, originating from a tailings dam, a waste rock pile or exposed sulfide-rich rock (Shaw et al. 1998). In most cases, acidic seepage comes from the oxidation of iron sulphide or pyrite ($FeS_2$). The oxidation process may be induced by a variety of geochemical agents but is usually associated with Fe-oxidizing thiobacilli organisms, the most common of which is *Thiobacillus ferrooxidans*, at pH values below 4.5 (Bigham et al. 1996).

The use of VIS-SWIR IS for characterizing AMD is advantageous because many of the secondary minerals formed by the oxidation of pyrite and other iron sulfide minerals in mine waste and mineralized rocks have diagnostic spectral features detectable in the VIS-SWIR range. Specifically, many of the secondary minerals such as copiapite, jarosite, schwertmannite, ferrihydrite, goethite and haematite are iron-rich and show spectrally diagnostic features around 430, 550 and 900 nm related to ferric iron crystal field transition (see jarosite spectrum in Fig. 2), and many are hydroxyl-bearing, showing diagnostic absorption features around 2265 nm related to Fe–O–H vibrational mode as shown in the jarosite spectrum in Fig. 2 and therefore having diagnostic spectral reflectance signatures (Cloutis et al. 2006; Swayze et al. 2000; Crowley et al. 2003).

The speciations of these secondary minerals are controlled by pH, the degree of oxidation, moisture content and solution composition (Swayze et al. 2000). A prolonged period of drying of these acidic waters can produce a suite of efflorescent salts (many of which

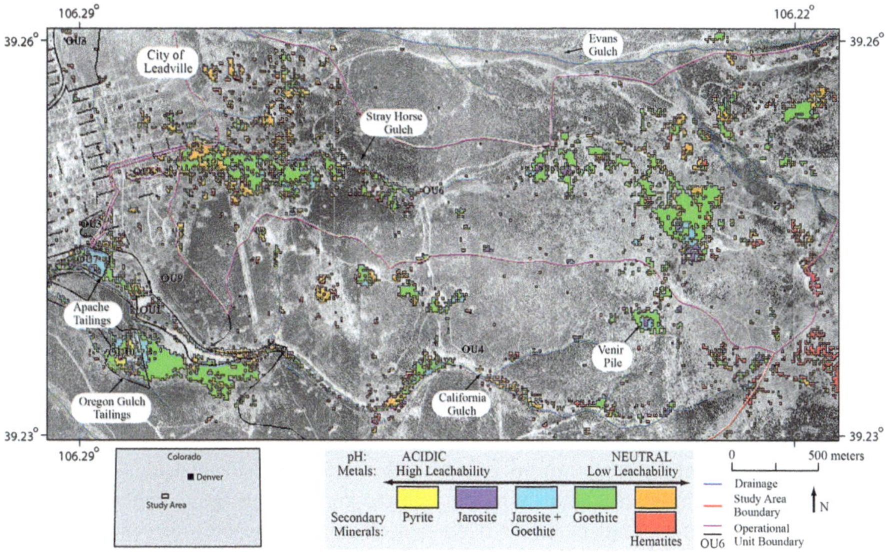


**Fig. 3** Secondary minerals related to the AMD process in the Leadville Mining District, Colorado, USA, mapped using IS data acquired using the airborne AVIRIS-C sensor. An idealized bull's eye model of spectrally detectable secondary mineral zones and their relation to acidity and metal leachability is shown to ease "translation" of the maps into parameters such as pH levels that are usually better known to environmental practitioners (adapted from Swayze et al. 2000)

also have spectrally diagnostic features in the VIS-SWIR range) and when dissolved can rapidly decrease the pH of acid waters and can result in short-term catastrophic effects on metal loadings and harm aquatic ecosystems (Hammarstrom et al. 2005). The leachate acidity is also capable of releasing heavy metals such as As, Pb, Cu, Co, Zn, Cd which are substituted with iron in the secondary minerals. AMD, therefore, can adversely impact the quality of drinking water and the health of riparian ecosystems. The task of identifying the sources, measuring the extent and determining the severity of AMD with traditional field methods can be costly and time-consuming.

Pioneering work by Swayze et al. (1996, 2000) at a test site near Leadville, Colorado, USA, in the district historically mined for gold, silver, lead and zinc since 1859 (Emmons et al. 1927) demonstrated that spectrally dominant iron minerals (pyrite, jarosite, a jarosite and goethite mixture, and haematite) exposed on the surface of the waste rock and tailings piles at Leadville can be identified and mapped using the Fe–O–H and $Fe^{3+}$ diagnostic spectral absorption features related to those secondary minerals (Fig. 2), used from Airborne Visible/Infrared Imaging Spectrometer Classic (AVIRIS-C) (Vane et al. 1993) imaging spectrometer acquired from an airborne platform (Fig. 3).

In the Leadville study, secondary minerals resulting from AMD were mapped using Tetracorder, a spectral shape-matching system (Clark et al. 2003a) applied to atmospherically corrected and calibrated AVIRIS-C data. Tetracorder is a modified least squares shape-matching algorithm that compares the continuum-removed (Clark and Roush 1984) absorption features of spectra of a comprehensive reference library to those of an observed spectrum (in this case the IS data). The continuum removal for each spectrum in the reference library spectra is tuned to the wavelength region around key diagnostic absorption features. Before comparison of each observed spectrum, it is first normalized against its continuum around the same wavelength region of the diagnostic absorption feature of the reference library spectrum and then scaled by a multiplicative constant so that its depth matches that of the corresponding reference absorption feature. Least squares correlation coefficients are then calculated and weighted according to the area under the curve and/or optional user-defined weights. Sets of absorption features for specific materials are compared based on a prescribed set of rules. An identification is then issued based on the application of these rules to the calculated values.

Validation of these mineral maps with a comprehensive suite of ground-based spectral measurements analysed using Tetracorder and independently cross-validated with soil samples collected from the site which were laboratory-analysed for leachate pH, leachate-specific conductance, bulk metal composition and metal leachability (see Swayze et al. 2000 for details) indicated that, the correlation between the AVIRIS-C-derived map and the ground traverse was very good except near man-made structures with rusted metal roofs and walls, which have a goethite spectral signature. The leachate chemical analyses indicate excellent correlation between mine waste pH and spectral zones mapped using IS, and the leachate trace metal concentrations generally showed a good correlation to the spectral mineral zones with metal leachability in the goethite zone with a small fraction of that in the other spectral zones (except for Mn). In fact, leachate concentrations of Pb, Cu, As and Co vary over three orders of magnitude across the spectral zones. This is significant because it indicates that the mineralogy in the spectral zones exerts a powerful control on metal leachability (e.g., the susceptibility of metals to leach into surface run-off) and hence the mapping of these minerals will be a good surrogate for leachability.

In summary, the maps produced from the Leadville study provided identification, location, spatial distribution and quantification of risks contributing towards improved targeting of remediation efforts which would otherwise be time-consuming and costly to map with traditional methods. In this regard, IS data could represent considerable savings in environmental remediation efforts.

Following the Leadville work, many other similar studies were conducted demonstrating the use of imaging spectrometers for mapping secondary minerals related to acid drainage (Kruse et al. 1989; Fenstermaker and Miller 1994; King et al. 1995; Peters et al. 1995; Farrand 1997; Farrand and Harsanyi 1997; Lopez-Pamo et al. 1999; Rockwell et al. 2005; Pearson et al. 1997; Kemper and Sommer 2004 and many others). Of particular interest, are the studies conducted by Zabcic et al. (2014) and Buzzi et al. (2016) who carried out the mapping of efflorescent salts, many of which have spectrally diagnostic features. For example, malenterite has a broad double crystal field transition absorption feature at 892 and 1150 nm related to ferrous iron and a smaller absorption feature at 434 nm related to $Fe^{3+}$–O charge transfer feature (Crowley et al. 2003). These efflorescent salts, which are often difficult to detect remotely because of their small size, were detected using HyMap (Cocks et al. 1998) airborne hyperspectral data acquired at the Sotiel Migollas Mine in the Iberian Pyrite Belt, Spain. The detection followed several years of drought allowing large spatial distributions of these salts at the tailings holding ponds and to a lesser extent on riverbeds to be formed. Here, the mapping of these salts by Zabcic et al. (2014) was done using a combination of the spatial-spectral endmember extraction (SSEE) algorithm (Rogge et al. 2006) for extracting spectral endmembers and iterative linear spectral unmixing analysis (ISMA) (Rogge et al. 2006) to derive the mineral maps, and, by Buzzi et al. (2016) using the Spectral Angle Mapper (SAM) (Kruse and Dwyer 1993) to perform the mineral mapping. The ability to detect these salts is significant as these salts are one of the most potent sources of metals, providing a ready source of acids and metals (Alpers et al.

1994; Plumlee et al. 1999). Specifically for the Sotiel Migollas Mine, the maps showing the abundance of these efflorescent salts is potentially useful for estimating the concentration of heavy metals and other hazardous metals that could be liberated and hence infer the risks of this abandoned mine where the infrastructure is thousands of years old, where an accidental spill or breach of the tailings pond following the next heavy rainfall event would have significant implications as in the case of the Aznalcollar accident (Achterberg et al. 1999; van Geen and Chase 1998). In fact, the use of IS for determining heavy metal loads was demonstrated by Kemper and Sommer (2004) following the Aznalcollar incident providing useful data for understanding the distribution and severity of the hazard and contributed towards better targeting of the clean-up process.

Although mapping of AMD conditions is crucial for characterizing AMD hazards, assessing its risks and targeting remediation work, monitoring is equally essential after remediation work commences. This monitoring aspect is demonstrated by Zabcic et al. (2014), Ong and Cudahy (2014) and Buzzi et al. (2016) using multi-temporal airborne IS data. By exploiting the relation between the minerals generated as a result of AMD and pH, Zabcic et al. (2014) and Ong and Cudahy (2014) extended the mapping of AMD mineral from IS to pH maps using Partial Least Squares (PLS) (Haaland and Thomas 1988). Such translated maps are useful for downstream users such as on-site environmental practitioners to understand the temporal evolution of the acid-producing material and to target and assess the effectiveness of remediation efforts.

Acid sulphate soil (ASS) is a socio-natural hazard where acidic soils or sediments that contain iron sulphides, mainly pyrite, and sulphuric acid are exposed either via natural or anthropogenic activities such as urban development. ASS is widely distributed around the world and has raised environmental concerns because of its the soil's capability to produce strong acidity and consequent trace metal release. The mechanism and oxidation process is similar to AMD, hence the methods used for AMD are equally relevant. As it is challenging to detect the occurrence and severity of ASS by conventional methods, which include time-consuming intensive field surveys followed by chemical measurements, mapping the location and severity of ASS from IS is an attractive prospect. Studies were conducted by Shi et al. (2014a, b), which showed that secondary minerals and associated pH related to ASS surrounding an estuary in Western Australia could be mapped using similar spectral feature fitting and PLS analysis methods as the AMD studies described above. Such maps help to improve targeting of remediation efforts and inform land-use planning. However, ASS generally is spatially small, posing cost challenges for current IS.

In terms of understanding the capabilities for future spaceborne IS, Davies and Calvin (2017) conducted a study to simulate the ability of future spaceborne IS and found that identification of key minerals related to AMD were possible at all spatial resolutions (2–60 m) albeit the finest spatial resolution images provided the best result with the most endmembers with more accurate spatial boundaries, and the lowest spatial resolution lost the ability to identify smaller features. Specifically, the areas mapped were reduced by one half when the spatial resolution was reduced to 15 m, but from 15 to 30 m, no further reduction in the area mapped was seen. Further, they found that near-future spaceborne IS which will predominantly be at a spatial resolution of 30 m, will be most relevant to identifying potentially hazardous AMD that cover large areas of at least 75 m in width.

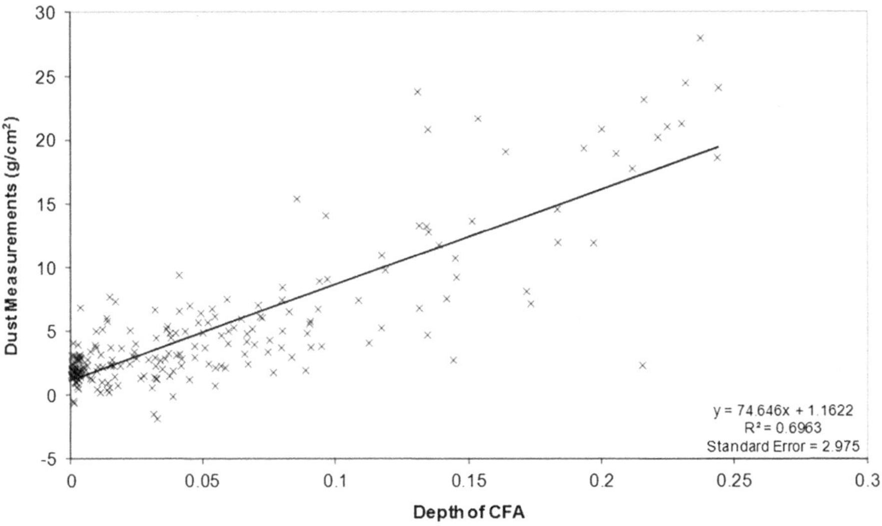


**Fig. 4** Scatterplot of the crystal field absorption (CFA) feature near 900 nm for ferric iron oxide (x) and the mass per unit area of dust on the leaves (y) from which an empirical line relation was developed to estimate iron oxide dust quantities using airborne IS data (sourced from Ong 2013). $R^2$, coefficient of determination

## 3.2 Fugitive Dust

The detection and diagnosis of hazardous fugitive dust, whether naturally occurring or anthropogenic, is based on spectroscopic principles where spectrally diagnostic features of the dust deposited on surfaces of leaves, soils or other background materials are used as markers/fingerprints for detecting the distribution and quantification of the concentration. Also, the underlying surfaces may have spectrally diagnostic features that may be used to identify the material where the dust is deposited. An example of this is the case of the spectrum of an iron oxide-dusted leaf in Fig. 2, where the ferric iron crystal field absorption (CFA) feature around 900 nm is evident together with features related to the chlorophyll vegetation such as the green peak at 550 nm, the red edge between 650 and 750 nm (Filella and Peñuelas 1994 and references therein) and cellulose feature around 1700 nm.

Although the use of IS to manage fugitive dust is not as comprehensively studied as AMD, Ong et al. (2003a, b, c), Ong et al. (2008) and Ong (2013) undertook an extensive study related to mineral resource development and the associated operations at the Port Hedland handling facility in Western Australia to better understand the impacts of fugitive dust originating from the handling of iron ore on the surrounding mangrove ecosystem. The method employed relied on an extensive calibration dataset of field samples to correlate the ferric iron oxide CFA feature around 900 nm to laboratory measurements of surficial dust quantities. Specifically, an empirical line relation between the continuum-removed 900-nm CFA depth and dust concentrations (see Fig. 4) was developed that was then applied to atmospherically corrected and continuum-removed airborne HyMap data collected over a 10-year period.

The resulting dust maps shown in Fig. 5 demonstrate that it is possible to positively identify the source of the fugitive dust and quantitatively identify, measure, map and monitor the spatial distribution of fugitive iron oxide dust through time. The ability to

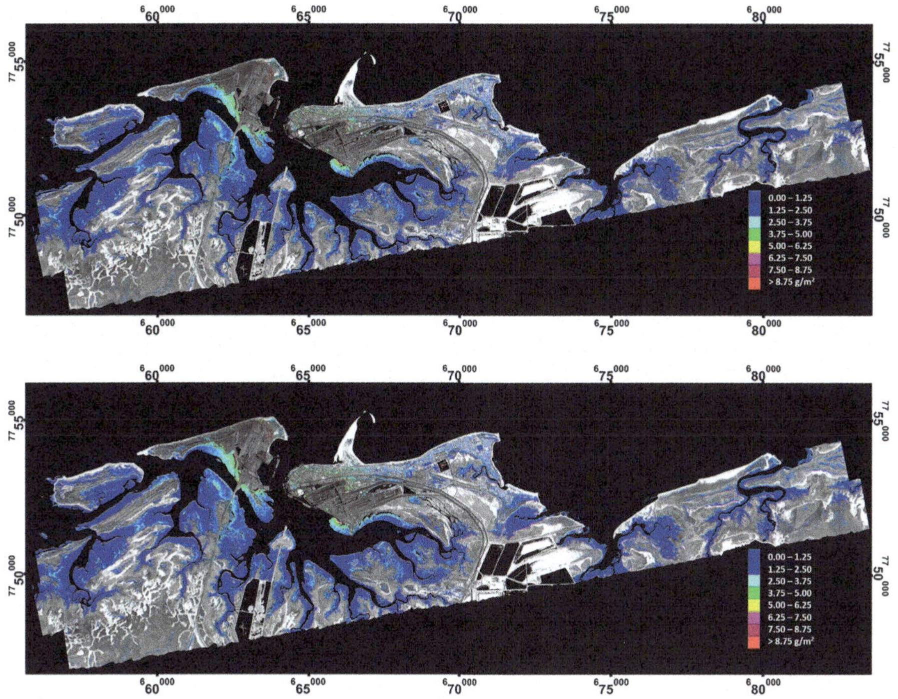


**Fig. 5** Maps of ferric iron oxide dust levels on mangroves and other vegetation surrounding Port Hedland harbour, Western Australia, generated from remotely sensed IS data acquired on 20 October 2002 (top) and 24 August 2006 (bottom). The color coding is such that cool colors (starting with dark blue) denote low fugitive dust levels and hot colors (ending with red) denote high fugitive dust levels. The dust measurements are in units of g/m$^2$. Image sourced from Ong et al. (2008) and Ong (2013)

provide identification and quantification is important in this case as the mining company is obliged via a ministerial agreement to monitor their impact. Additionally, the spectral-derived maps provided new insights into the spatial distribution of fugitive dust across the landscape where it was not possible with previous methods, which, had acquired measurements at a few discrete points and did not allow for rapid identification of the composition of the dust. The spectral-derived maps generated from the Ong (2013) study were comprehensively validated paving the way for the method to potentially be used operationally, especially after the launch of future spaceborne IS sensors such as Environmental Mapping and Analysis Program (EnMAP, Guanter et al. 2015), Hyperspectral Imager Suite (HISUI, Matsunaga et al. 2018), Surface Biology and Geology (SBG, Green 2018, previously Hyperspectral InfraRed Imager (HyspIRI, Lee et al. 2015), Hyperspectral Imaging Mission for the Environment (CHIME, Nieke and Rast 2018) and Hyperspectral Precursor of the Application Mission (Loizzo et al. 2018). Importantly, the Ong (2013) study was undertaken to quantify the uncertainties in the dust products generated using the method. Quantifying these uncertainties is important when such measurements are used to meet regulatory requirements. For example, it was determined that the wavelength miscalibration of up to 2.1 nm could potentially increase the uncertainties up to 34% in the determination of fugitive dust levels.

Further to this work, Ong et al. (2008) and Pascucci et al. (2012) provide supporting evidence of the potential to map and characterize other fugitive dusts based on the spectrally diagnostic features of the hazardous material at other mining operations.

Often large amounts of fugitive dust can be generated as a result of accidental spills or human malice. In some cases, the dust can be hazardous. Therefore, the deployment of IS for the task of remotely and rapidly detecting and locating the deposition of fugitive dust after such a hazardous event presents an important tool for emergency responders and those involved in the clean-up especially where there is a safety concern for those undertaking the clean-up process. IS from AVIRIS-C sensor was deployed for such an application where potentially hazardous dust was discriminated following the 11 September 2001 World Trade Centre (WTC) attack (Clark et al. 2006). Specifically, the AVIRIS-C data were used to map the abundance and extent of dust and debris resulting from the collapse of the WTC using Tetracorder. Of immediate concern was the potential presence of asbestos-bearing materials in the dust, and hence the data were used to determine the presence and distribution of asbestos materials using the Mg–O–H spectral absorption feature present in chrysotile (Figs. 2, 6). Although spectroscopy is only able to measure minerals related to asbestos and will not discriminate between asbestiform or non-asbestiform particles, the AVIRIS-C data were able to identify pockets (or small clusters of pixels) of serpentine, indicating potential asbestos levels of a few per cent, up to 10 and 20%. Such data were valuable as guidance towards determining the risks to those involved in the clean-up operations.

Dust storms are natural atmospheric events that are typically associated with deserts, and they are known to cause significant impacts on the physical environment and human health (Goudie and Middleton 2006). Often dust storms carry mineral dust from eroded exposed soil surfaces.

Chudnovsky et al. (2009, 2011) employed spectral matching to known mineralogical spectral features (Duckworth 2004) and SAM (Kruse and Dwyer 1993) using spaceborne IS data acquired by National Aeronautics and Space Administration (NASA) EO-1 Hyperion sensor to map the mineral composition of the dust at the Bodélé Depression of Northern Chad, considered to be one of the world's largest sources of atmospheric mineral dust. In the Chudnovsky et al. (2009) case study, they demonstrated that it was possible to detect key minerals from suspended dust particles from the remotely sensed spectral data. As the composition of the deposited dust can have varying impacts on the receptor environment (Jickells et al. 2005; Westrich et al. 2016), the ability to determine the composition of the dust is important for quantifying their impacts and predicting their associated risks. An extension of the case study (Chudnovsky et al. 2011) indicated that IS data can be used to map the distribution and composition of the dust via their mineralogical features, which in turn has the potential to be used to track the transportation routes of the dust. As this case study was conducted at similar spatial resolutions as future global and temporal spaceborne IS missions such as the SBG mission (Green 2018; Lee et al. 2015), the next-generation spaceborne IS missions would make such applications viable allowing a better understanding of dust on climate radiative forces.

**Fig. 6** Serpentine and amphibole minerals map of the lower Manhattan area, New York City, New York, USA. Image sourced from Clark et al. (2006)

### 3.3 Hydrocarbon Contamination in Aquatic and Terrestrial Ecosystems

An oil spill is oil, discharged accidentally or intentionally, that floats on the surface of water bodies as a discrete mass and is carried by the wind, currents and tides (United Nations Statistical Division 2017). The term oil spill usually refers to marine oil spills, where oil is released into the ocean or coastal waters, but spills may also occur on land. The largest oil spill in the US history occurred in the Gulf of Mexico on 20 April 2010, when an explosion on the Deepwater Horizon (DWH) offshore drilling rig released ~ 780,000 $m^3$ of crude oil into the Gulf before being capped on 15 July 2010 (Lehr et al. 2010).

The detection of oil spills, whether manifested on the ground or surfaces of rivers, lakes or oceans, from IS depends on the presence of diagnostic hydrocarbon spectral absorptions centered near 1200, 1700 and 2300 nm which arise from overtones and combinations of the fundamental absorptions of C–H bonds (see the green-coloured spectrum of benzene and water mixture in Fig. 2).

During the DWH disaster, Clark et al. (2010) and Leifer et al. (2012) used a combination of airborne AVIRIS-C and multi-spectral Moderate Resolution Imaging Spectroradiometer (MODIS, Salomonson et al. 1989) data to estimate the spatial distribution and thickness of the oil spill on the ocean. Airborne IS data acquired on 17 May 2010 for approximately 30% of the core spill area were used to estimate the composition and thickness of the oil based on diagnostic hydrocarbon spectral features. Specifically, the AVIRIS-C data were processed with Tetracorder using 22 reference spectra of laboratory-constructed emulsions measured at oil thicknesses ranging from 0.025 to 28 mm. Using these spectral-derived maps, the oil volumes within the top few millimeters of the ocean surface were estimated to be 19,000 (conservative) to 34,000 (aggressive) barrels of oil (Clark et al. 2010). The conservative and aggressive volume estimates are based on choosing either 2 mm (conservative) or 4 mm (aggressive) for AVIRIS-C spectra of sea surface emulsions that spectrally resemble similar-looking reference emulsions with those thicknesses.

As the AVIRIS-C data did not cover the entire extent of the oil spill, MODIS data were used to define the spatial boundaries and internal structure of the oil spill to extrapolate the estimation of the oil thickness to the larger spatial distribution of the oil spill. If the core spill areas not covered by AVIRIS-C contained similar amounts of oil and water-in-oil emulsions, then extrapolation to the entire core spill area defined by the MODIS image indicated a minimum of 66,000–120,000 barrels of oil was floating on the surface (Clark et al. 2010). These surface volumes form the basis of the flow rate estimation (Labson et al. 2010) contributing to the clean-up process. However, these were conservative estimates as the calculation was made assuming thick surface oil and did not include oil in sheens, oil under the surface, oil washed onto land, or oil burned, evaporated, or bio-degraded. Indeed, field observations of the oil emulsions indicated that the thickness of oil could sometimes exceed 20 mm. If this larger thickness was used to estimate the volume of oil, the estimation could potentially be as high as 150,000 barrels within the AVIRIS-C coverage. If this volume was extrapolated to the entire spill estimated by MODIS, the estimated volume for the core spill would be approximately 500,000 barrels of thick oil remaining on the ocean surface.

In addition to the oil on the surface of the ocean, the DWH oil spill caused oil to be washed onto approximately 796 km of shoreline comprised of intertidal marshes, disproportionally impacting salt marshes along the coastline of Louisiana, the USA (Michel et al. 2013). Using IS to understand the status and progress of rehabilitation of the impacted area was the subject of several studies.

Kokaly et al. (2013) applied spectral feature analysis to the AVIRIS-C data collected during and after the DWH oil spill to delineate the distribution of oil-damaged canopies in the marshes of Barataria Bay, Louisiana. Spectral feature analysis based on Material Identification and Characterization Algorithm (MICA), a shape-matching algorithm (Kokaly 2011), was used to compare the AVIRIS-C data to reference spectra of oiled marsh collected by the researchers during field surveys. The spectra fitting relied on both the 1700 and 2300 nm C-H absorption features. A composite map of oiled areas at 3.5-m grid spacing (Fig. 7), accumulated over three dates, was found to be 93.4% accurate in detecting oiled shorelines.

Similarly, Peterson et al. (2015) applied multiple endmember spectral mixture analysis (MESMA) (Franke et al. 2009) to the same AVIRIS-C data for Barataria Bay to detect oil on the coastal marsh. Peterson et al. (2015) were able to obtain accuracies ranging

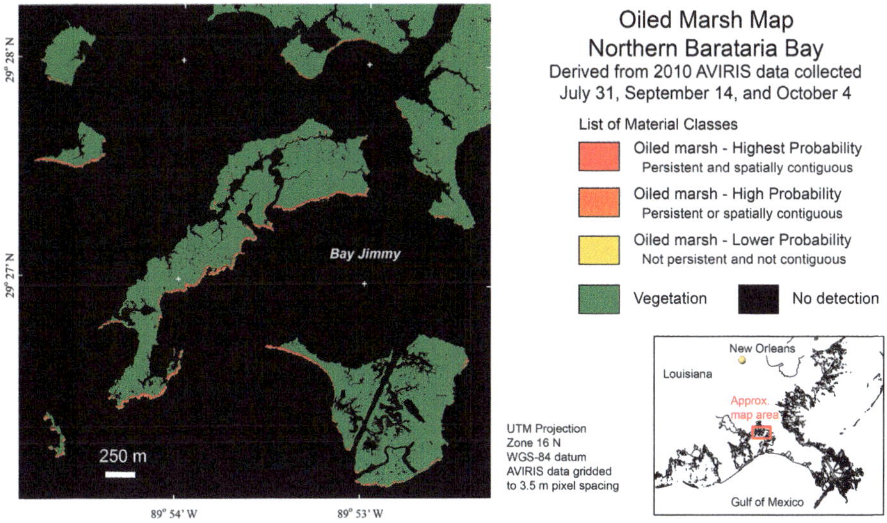


**Fig. 7** Hydrocarbon contamination detected in AVIRIS-C data for the Deepwater Horizon oil spill in a portion of Barataria Bay, Louisiana, the USA. Figure adapted from Kokaly et al. (2013)

from 87.5 to 93.3% by using a nine-band subset emphasizing the hydrocarbon absorption regions.

AVIRIS-C data collected during the DWH oil spill and 1 year afterwards (2011) were used by Khanna et al. (2013) to show that vegetation stress, measured spectrally using the chlorophyll spectral absorption feature as indicated in Fig. 2, was restricted to the tidal zone extending 14 m inland from the shoreline in September 2010. The stress indices used showed that the highest values were located on pixels next to the shoreline and decreased with distance inland from the shore. The later imagery acquired in 2011 indicated that the stress index values were no longer correlated with proximity to oil, suggesting that the marsh on the Barataria Bay was on its way to recovery. One year after the oil spill, the data indicated that areas denuded of vegetation experienced varying degrees of revegetation and that recovery of plants was poorest in the first three pixels (~ 11 m) adjacent to the shoreline.

Beland et al. (2016) evaluated changes in dominant species cover from 2010 to 2012 for oiled and non-oiled shorelines. They used field-referenced image endmembers and canonical discriminant analysis to classify 21 AVIRIS-C images acquired during the autumns of 2010, 2011 and 2012. Marshes that were heavily contaminated with oil exhibited variable responses from 2010 to 2012. Some marsh area was converted to open water along oiled and non-oiled shorelines that were similarly situated in the landscape. Marsh loss along oil-contaminated shorelines was double that of non-oiled shorelines. Analysis of dominant species showed that only *Spartina alterniflora*-dominated marshes were extensively degraded, losing 15% (354,604 $m^2$) cover in oiled shoreline zones, suggesting that *S. alterniflora* marshes may have been more vulnerable to shoreline erosion following hydrocarbon stress, due to their landscape position.

This case study from DWH demonstrated the use of IS not only for locating the oil spill but for identifying the different oil chemistries, including its weathering state and thickness. The provision of such diagnostic data is potentially beneficial for directing recovery efforts to areas where they are most effective. Also, IS data were demonstrated to be sufficient for identifying hydrocarbon contamination on marshes, and multi-temporal IS data were found to be effective for monitoring the impact and the recovery process of the vegetation.

### 3.4 Atmospheric Emissions

Atmospheric emission is a global issue projected by the Organization for Economic Cooperation and Development (OECD) to cause 6 and 9 million premature deaths in 2060 at the cost of USD18–25 trillion (OECD 2016). Much of this relates to air quality issues such as particulates and aerosols, especially in non-OECD (low income, low human development index, less developed) countries. However, in first-world nations, the greatest atmospheric emission issues by far are those that affect the climate—which in turn relates to modern society's heavy reliance on energy and the use of fossil fuels resulting in increased levels of greenhouse gases (OECD 2016). Studies demonstrating the use of IS for the detection of atmospheric gas relied on the fact that the atmospheric gas hazard was either spectrally distinct or was sufficiently enhanced from ambient conditions where the gas was a composition of ambient air. The former may be the case for industrial gases such as hexasulphur ($S_6$), while the latter may be the case with fugitive methane ($CH_4$) emissions from a coal mine or oil production equipment (see transmittance spectrum of methane in Fig. 2).

A long history of sounding[1] missions has measured the atmosphere using absorption of reflected solar light (Rodgers 2000). Such instruments typically have coarse (> 500 m) spatial resolution. In contrast, IS investigations have historically focused on higher spatial resolution mapping of Earth's surface at resolutions of 30–100 m (Schaepman et al. 2009). However, the VIS-SWIR region observes vibrational transitions by trace gases such as $CH_4$ and $CO_2$ and hence IS can be used to map these gases via the diagnostic atmospheric spectral features. The use of solar energy precluded the need for a scene temperature contrast, making the detection strength a predictable function of known illumination and simplifying quantitative analysis. Indeed, investigators routinely estimate atmospheric gas absorption as an intermediate step in atmospheric correction. More recently, advances in instruments and algorithms have made imaging of atmospheric composition a subject of study in its own right. IS has increasingly been deployed to image local atmospheric anomalies such as pollutants, aerosols and enhancements of trace gases such as carbon dioxide ($CO_2$), methane ($CH_4$) and sulphur dioxide ($SO_2$). While trace gas characterization has more often used sounding spectrometers (Gerilowski et al. 2011) or in situ airborne sensors (Conley et al. 2016), imaging spectrometers bring unique capabilities. IS can survey wide areas, detecting spatially small sources that had not been identified previously. The higher spatial resolutions afforded by IS provide the ability to improve localization and attribution to specific point sources. Finally, IS can spatially characterize the structure of extended enhancements, such as volcanic plumes, to better constrain spatial extent and transport.

In recent times, AVIRIS-C has been successfully used to find local gas enhancements in the VIS-SWIR (Thorpe et al. 2012, 2013). Researchers generated images of industrial stack plumes (Dennison et al. 2013) as well as natural seeps such as coal oil point (Roberts et al. 2010; Bradley et al. 2011). Promising results have also been obtained for mapping nitrogen dioxide ($NO_2$) sources, such as the Zurich (Switzerland) airport, a waste incinerator

[1] A sounding system measures the intensity of radiation to provide information about a particular property as a function of height in the atmosphere.

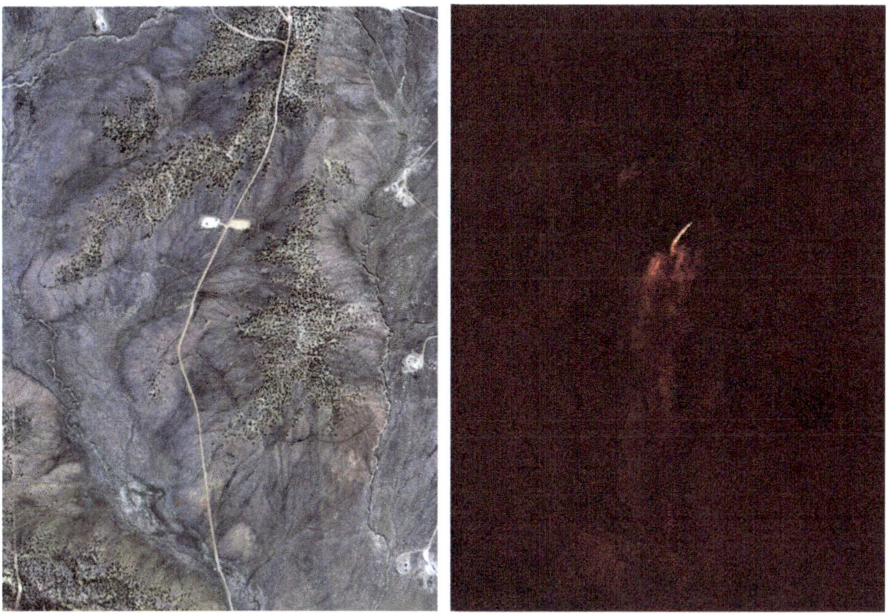

**Fig. 8** (Left) True color segment of an AVIRIS-NG flight line acquired in 2014 from the Four Corners Campaign with spectroscopic data available from http://avirisng.jpl.nasa.gov. (Right) A $CH_4$ retrieval using the technique of Thompson et al. (2015b) shows a plume emanating from a gas well pad. Brighter pixels correspond to higher local enhancements

and motorways, from the Airborne Prism Experiment (APEX) imaging spectrometer (Popp et al. 2012). AVIRIS-C overflew the Hawaiian volcano at Kilauea (USA), providing the first ever map of $CO_2$ concentrations in a volcanic plume (Spinetti et al. 2008). The study of an AVIRIS-C image of a biomass plume from a prescribed fire in Quinault, Washington, USA, by Marion et al. (2004, 2006), Alakian et al. (2008, 2009) and Deschamps et al. (2012, 2013) and highlighted the need to correct the radiative effect of aerosols in the $CO_2$ absorption band at 2000 nm to accurately derive $CO_2$ concentration maps in strong optically thick smoke plume. The next-generation AVIRIS (AVIRIS-NG) instrument (Hamlin et al. 2011) offered greater sensitivity in comparison to AVIRIS-C thanks to higher signal-to-noise ratios (SNRs) and finer spectral sampling. Controlled release experiments have characterized the sensitivity of AVIRIS-NG at multiple wind speeds and altitudes (Thorpe et al. 2016).

In recent years, VIS-SWIR IS has been increasingly used for wide-area surveys to locate, characterize and quantify gaseous leaks related to the oil and gas industry. This is further assisted by the recent incorporation of detection algorithms into real-time analysis on-board the aircraft, permitting radiometric calibration, detection, quantification, mapping and finally geolocation by ray tracing with digital elevation maps and camera models (Thompson et al. 2105b). These operations execute immediately, generating results to a real-time display so that operators can adjust future flight lines and guide field teams that deploy instruments in situ. This capability was introduced during an AVIRIS-NG campaign in 2014, demonstrated the ability to map oil- and gas-related emissions (Thompson et al. 2105b) and observed plumes from a landfill in the Los Angeles Basin, California, USA

(Krautwurst et al. 2017). After that, AVIRIS-NG was acquired jointly with the HyTES instrument operating in thermal ranges (Hulley et al. 2016), and other National Oceanic and Atmospheric Administration (NOAA) and university-sponsored teams, to characterize the Four Corners area of the USA. This area had been identified as a $CH_4$ emission "hot spot," but the attribution of this emission to specific sources was not well understood (Frankenberg et al. 2015). Two aircrafts mapped a large fraction of this region, identifying over 200 distinct anthropogenic sources (Fig. 8) including gas well pad unloading events, drilling operations, storage venting and emissions from processing facilities (Frankenberg et al. 2016) based on methane spectral absorption features (Fig. 2) from AVIRIS-NG (Fig. 7). In addition, the IS data detected multiple underground pipeline leaks from transmission lines. These measurements demonstrated the use of IS to provide a reliable statistical characterization of emission sources and accounted for a large fraction of the overall emissions observed in the region.

After the Four Corners survey in 2014, the number of $CH_4$ investigations using imaging spectrometers in the USA has grown considerably in a short time. VIS-SWIR and thermal sensors viewed the Aliso Canyon, Southern California, USA, event, a large accidental release of methane from an underground storage facility in autumn 2015–winter 2016. The Aliso Canyon event was the largest single $CH_4$ point-source-emitter detected to date and was successfully observed by the NASA EO-1 Hyperion (Pearlman et al. 2001) instrument on-board the Earth-orbiting EO-1 spacecraft (Thompson et al. 2016). The EO-1 spacecraft was only able to image the plume under sub-optimal illumination conditions, resulting in an extremely low signal-to-noise ratio (SNR). Nevertheless, it observed the plume morphology on three occasions to achieve the first ever detection and attribution of a single $CH_4$ point source from orbit. In 2016, the state of California began an extended campaign by using AVIRIS-NG to characterize $CH_4$ point sources in key areas across the state (Duren et al. 2017). This represented the first systematic, multi-sector effort involving remote sensing technology. Through these efforts, spectroscopic $CH_4$ imaging has continued to evolve from a purely scientific method to practical policymaking support tool, and is an important component of a multi-tier $CH_4$-accounting regime (Hopkins et al. 2016).

### 3.5 Naturally Occurring Hazardous Soils and Rocks

Expansive or swelling clays and clay shales are soils that have the potential to significantly change their volume with changes in moisture, hence, with wet and dry cycles, these volumetric changes can cause extreme pressure in the soil profile resulting in severe engineering problems (building and road cracks, foundation deformation), agricultural problems (root cutting), as well as environmental problems (drying up of the sub-surface horizon, floods, landslides) causing billions of dollars of damage worldwide every year, more than all other natural hazards combined (Jones and Jefferson 2012). Expansion of soils is a result of a combination of minerals present in the matrix (mainly clay mineral species and content), water quality (fresh or saline), organic matter content, availability of Ca ions, as well as the origin and content of free iron oxides. In particular, aluminium-rich smectite (montmorillonite) (see Fig. 2 for spectrum) has been shown to be the main culprit for the expansive potential hazard. IS with its capabilities for direct identification of the key culprit constituent minerals in soils is potentially an attractive tool for rapid and inexpensive regional and local mapping to augment standard engineering and geological methods which are time-consuming and expensive, and do not provide spatial coverage.

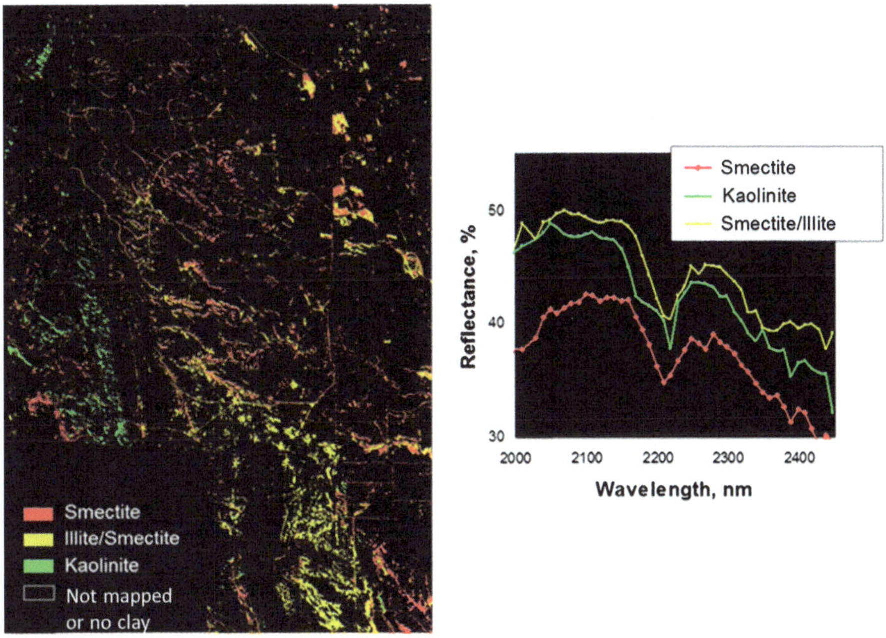


**Fig. 9** Regional map for the Front Range Urban corridor near Denver, Colorado, USA of expansive clays exposed at the surface and of swelling soils hazards (red: high, yellow: medium, green: low/none) based on airborne AVIRIS-C IS data. Figure modified from Chabrillat and Goetz (2006)

Pioneering work from Chabrillat et al. (2002) demonstrated the potential of IS to detect and map swelling clays with a study along the Front Range Urban corridor near Denver, Colorado, USA, a rapidly growing urban area that was severely affected by swelling soils. This was due to the heaving bedrock hazard associated with the coincidence of steeply dipping layers of sedimentary bedrock having dip angles of greater than 30°, and zones of expansive bedrock that swell in volume when excess moisture was introduced. Chabrillat et al. (2002) showed that VIS-SWIR IS from AVIRIS-C can be used to discriminate pure smectite from mixed layers of illite/smectite samples (associated with decreasing swelling potential), and thus can be used to produce a map of hazards for the studied area (e.g., Chabrillat et al. 2002; Chabrillat and Goetz 2006). Spectral discrimination and identification of variable clay mineralogy (smectite, illite/smectite and kaolinite), related to variable swelling potential, was possible from airborne IS data at high spatial resolution (<5 m), and on a regional basis with slightly lower spatial resolution (20 m) (Fig. 9). Spectral feature fitting was used to discriminate the swelling soils, and the level of success was dependent on the level of the exposures and the variable mineralogy, and the presence of partial vegetation cover.

Subsequently, many studies extended the concept and extended the knowledge on swelling soils' optical properties (e.g., Kariuki et al. 2004; Bourguignon et al. 2007; Yitegasu et al. 2009; Dufrechou et al. 2016) and their link to potential swelling hazard. Nowadays, IS is a proven method for the identification and mapping of the occurrences of swelling soils, as demonstrated in many regions around the world (the USA, Kenya and France). Such optical remote sensing methods, coupled with a reduced amount of field and laboratory

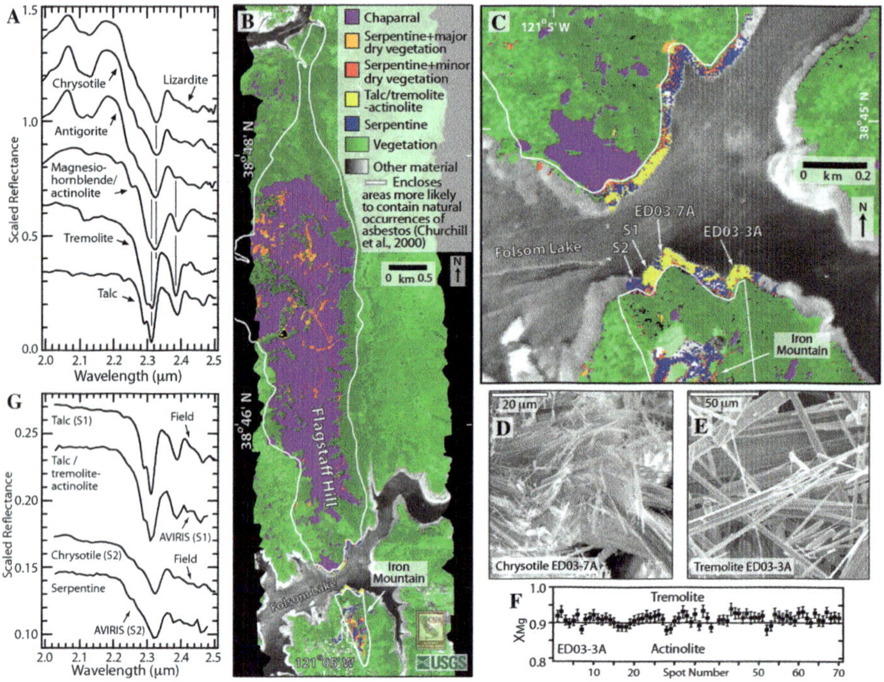


**Fig. 10** **A** Reflectance spectra of some minerals associated with serpentine deposits in Calvaras County in California, USA. **B**, **C** Potentially asbestos-bearing minerals and vegetation preferentially growing on asbestos-bearing rocks in western El Dorado County in the Sierra Nevada of California, USA, mapped using IS data acquired using the AVIRIS-C. **D**, **E** Scanning electron micrograph of chrysotile fiber bundles. **F** Comparison of spectra extracted from AVIRIS-C to spectra collected in the field collected over the Red Hill Complex, in the foothills of the Sierra Nevada, California, USA. Figure sourced from Swayze et al. (2009)

analyses, can offer a practical tool for town planners, as long as the swelling soils are adequately exposed at the surface and the instrument has a high enough SNR particularly in the SWIR.

Hazards related to asbestos exposure and their detrimental implications for human health are well documented (International Agency for Research on Cancer 2012; van Oss et al. 1999). Asbestos is a morphological term describing mineral fibres with lengths that are much greater than their widths. Consequently, the asbestos name does not always correspond to a mineral name. A variety of minerals have asbestiform crystal habits (Ross 1999). The two main categories used commercially for asbestos insulation are chrysotile, the only fibrous member of the serpentine mineral group, and several fibrous amphibole minerals comprised of anthophyllite, amosite, crocidolite, tremolite and actinolite. IS can be used to identify some of these minerals such as chrysotile (see spectrum and related diagnostic feature in Fig. 2), and indeed, there have been a handful of case studies on the use of the technology for identifying asbestos-bearing rocks and man-made materials (e.g., Marino et al. 2000; Clark et al. 2001, 2003b; Swayze et al. 2004, 2009; Frassy et al. 2014; Massarelli et al. 2017; Bonifazi and Serranti 2016).

Although there is high public awareness of the hazards related to insulation and other manufactured goods, and even legacy mines, issues related to naturally occurring

**Table 1** Summary of anthropogenic and natural hazards examined, sensors used and associated spectral and spatial specifications, and the spatial extent of the studies examined

| Hazard examined | Sensors | Spectral specifications | Spatial resolution | Area covered |
|---|---|---|---|---|
| A AD, A AE, N AE, NOHSR, HC | AVIRIS-C | Range 360–2500 nm, 224 bands<br>Resolution: 6.7 nm (360–670 nm), 9.5 nm (660–1280 nm), 10.0 nm (1260–2500 nm) | 5–20 m | 5–12.5 $km^2$ |
| A AE | AVIRIS-NG | Range 380–2520 nm<br>Resolution: 5 nm | 0.4–3.5 m | Variable |
| A AD S-N AD, A FD | HyMap | Range 450–24,800 nm<br>Resolution: 15–16 nm (450–1800 nm), 18–20 nm (1950–2480 um) | 1.5–10 m | 7.5–100 $km^2$ |
| A AD | ProspectTIR | Range 398–2455 nm<br>Resolution: 5 nm | 2 m | |
| N FD | Hyperion* | Range 400–2500 nm<br>Resolution: 10 nm | 30 m | Typically 750 $km^2$ |
| A AD | MIVIS | Range 430–830, 1150–1550, 1983–2478, 8180–12,700 nm<br>Resolution: 20 nm (VIS), 50 nm (NIR), 90 nm (SWIR), 340–350 nm (TIR) | 3.5 m | ~32 $km^2$ |

* denotes spaceborne

The hazards examined were anthropogenic acid drainage (AD), socio-natural acid drainage (S-N AD), anthropogenic fugitive dust (A FD), natural fugitive dust (N FD), anthropogenic atmospheric emission (A AE), hydrocarbon contamination (HC), natural atmospheric emission (N AE), naturally occurring hazardous soils and rocks (NOHSR)

**Table 2** Summary of the selected case studies examined, analysis employed, calibration and validation methods employed, and key findings

| Hazard | Reference of detailed investigation | Analysis methods | Calibration and validation information | Key findings |
|---|---|---|---|---|
| A AD | Swayze et al. (1996, 2000), Kruse et al. (1989), Fenstermaker and Miller (1994), King et al. (1995), Peters et al. (1995), Farrand (1997), Farrand and Harsanyi (1997), Lopez-Pamo et al. (1999), Rockwell et al. (2005), Pearson et al. (1997) and Kemper and Sommer (2004) | Tetracorder<br>(Clark et al. 2003a, b) mapping distinct diagnostic spectral features of AMD secondary mineral.<br>SAM (Kruse and Dwyer 1993)<br>Variable multiple endmember spectral mixture analysis (VMESMA) (García-Haro et al. 2005) | Field spectroscopy and laboratory chemical analysis (X-ray diffraction (XRD), XRF, pH, specific conductance, bulk metal composition and metal leachability) | Key AMD minerals can be detected using various spectral feature fitting and unmixing methods<br>Mineral maps useful to quantify acidity and metal leaching potentially useful for prioritizing remediation efforts |
| A AD | Zabcic et al. (2014), Buzzi et al. (2016), Ong et al. (2003a, b, c), Ong and Cudahy (2014) and Davies and Calvin (2017) | SSEE (Rogge et al. 2006) to extract spectral endmember and ISMA (Rogge et al. 2006) to map minerals<br>PLS (Haaland and Thomas 1988) to infer pH from the spectral data<br>SAM | Field spectroscopy and laboratory chemical analysis (XRD, pH) | Extension beyond detection and mapping to monitoring demonstrated with mineral and pH maps from multiple dates<br>Proper atmospheric correction and calibration essential for temporal work<br>Efflorescent salts can be detected, and multi-temporal maps can be used to monitor AMD processes and infer risks of significant trace metal leaching potential especially at ancient mine infrastructure<br>pH maps were useful for understanding the evolution of the acid-producing material, target areas for remediation and monitoring remediation efforts<br>Simulation of future satellite systems indicates that secondary minerals can be detected for AMD impacts on areas larger than 75 m |

**Table 2** (continued)

| Hazard | Reference of detailed investigation | Analysis methods | Calibration and validation information | Key findings |
|---|---|---|---|---|
| S-N AD | Shi et al. (2014a, b) | PLS to infer pH from the spectral data<br>Multi-spectral feature fitting (MSFF) similar to Clark et al. 2003a, b | Field spectroscopy and laboratory chemical analysis (XRD, scanning electron microscope (SEM), energy-dispersive X-ray analysis (EDXRF)) | ASS minerals detectable with present imaging spectrometers but constrained by spatial resolution<br>pH maps can be used to establish the spatial extent and severity of soil acidity |
| A FD | Ong et al. (2003a, b, c), Ong et al. (2008), Ong (2013) and Pascucci et al. (2012) | Tailored empirical relationship between spectral feature depth and dust quantities | Field spectroscopy, laboratory analysis of weight of dust<br>XRD and EDXRF | Source of dust and receptor can be uniquely identified;<br>Dust concentration can be quantified;<br>Initial map provided an understanding of spatial distribution and extent of the impact;<br>Temporal maps were used to monitor dust levels and dust suppression efforts |
| A FD | Clark et al. (2006) | Tetracorder | Field spectroscopy and laboratory chemical analysis (XRD, XRF) | Fingerprinting of dust fallout to identify hazardous areas used to aid clean-up operations |
| N FD | Chudnovsky et al. (2009, 2011) | Spectral matching to known mineralogical features (Duckworth, 2004)<br>SAM | Field spectroscopy and laboratory chemical analysis (XRD, XRF, free iron, granulometry) | The composition of the deposited dust was used to understand impacts on the receptor environment<br>Potential to track the transportation routes of the dust |

**Table 2** (continued)

| Hazard | Reference of detailed investigation | Analysis methods | Calibration and validation information | Key findings |
|---|---|---|---|---|
| A HCC | Clark et al. (2010), Leifer et al. (2012), Kokaly et al. (2013), Peterson et al. (2015), Khanna et al. (2013) and Beland et al. (2016) | Tetracorder<br>MICA shape-matching algorithm (Kokaly 2011)<br>Multiple endmember spectral mixture analysis (MESMA) (Franke et al. 2009) | Field spectroscopy and laboratory synthetic oil mixtures | Quantification of surface volumes based on diagnostic hydrocarbon spectral features contributed towards the flow rate estimation for the clean-up process<br>IS was able to delineate the distribution of oil-damaged canopies in the marshes<br>Comparison of IS data before and 1 year after oil spill indicated that areas denuded of vegetation experienced varying degrees of revegetation and recovery of plants was poorest adjacent to the shoreline<br>Data from 3 consecutive years were used to identify progress changes in dominant species at the marsh |
| A AE | Bradley et al. (2011), Dennison et al. (2013), Deschamps et al. (2013), Franke et al. (2009), Frankenberg et al. (2016), Frassy et al. (2014), Krautwurst et al. (2017), Leifer et al. (2012), Marion et al. (2004); Roberts et al. (2010), Spinetti et al. (2008), Thompson et al. (2015b), Thorpe et al. (2012, 2013), 2016) and Thorpe (2016) | MODTRAN modelling<br>Cluster-tuned matched filter (CTMF) (Funk et al. 2001, Thorpe et al. 2013)<br>Classical matched filter (Manolakis et al. 2014) | Controlled release of gas<br>Field gas sampling | $CH_4$, $NO_2$, $CO_2$ absorption in the VIS-SWIR was measured using imaging spectrometer data<br>Over 200 $CH_4$ plume sources were identified across the gas producing areas<br>Real-time analysis consisting of radiometric calibration, detection, quantification, mapping and geolocation demonstrated<br>Large $CH_4$ sources mapped using spaceborne IS |

**Table 2** (continued)

| Hazard | Reference of detailed investigation | Analysis methods | Calibration and validation information | Key findings |
| --- | --- | --- | --- | --- |
| NOHSR | Chabrillat et al. (2002) and Chabrillat and Goetz (2006)<br>Marino et al. (2000), Clark et al. (2001, 2003b), Swayze et al. (2004), Frassy et al. (2014), Massarelli et al. (2017), Bonifazi and Serranti (2016), Swayze et al. (2009) and Swayze and Kokaly (2016) | Spectral feature fitting<br>Tetracorder | Field spectroscopy and laboratory chemical analysis (XRD, XRF) | Key minerals driving swelling clay can be mapped using data at 5-20-m spatial resolution<br>Mapping of minerals related to asbestos achieved with field validation shows good agreement with the map<br>Asbestiform crystal habits |

asbestos (NOA) are not as commonly known even though it can potentially be a hazard for people living near its sources (Pan et al. 2005). For example, in central and northern California, USA, chrysotile and fibrous amphiboles like tremolite and actinolite are dominantly, but not exclusively, associated with serpentinite and ultramafic rocks, which are common in the Sierra Nevada, Coast Ranges and Klamath Mountains. Consequently, it is important that these rock types be identified to help reduce the public's exposure to potentially NOA-bearing dust generated during building and outdoor activities.

A study using data collected by AVIRIS-C in western El Dorado County in the Sierra Nevada of California, USA, demonstrated the use of IS for mapping the distribution of these potentially NOA-bearing serpentinite and ultramafic rocks (Swayze et al. 2009; Swayze and Kokaly 2016). Here, direct detection of potentially NOA-bearing minerals exposed in rock outcrops and mapping of vegetation cover that preferentially grows on these rocks as a surrogate of NOA-bearing minerals was achieved using Tetracorder as shown in Fig. 10 where the fibrous chrysotile was mapped together with other non-fibrous serpentine minerals and the potentially fibrous tremolite and actinolite were combined into a single spectral category called talc/tremolite-actinolite (T/TA). Compared to traditional field mapping, nearly all of the high-density clusters of pixels mapped as potentially NOA: that is serpentine, serpentine + minor dry vegetation and T/TA, either fall inside or within a few hundred metres of the boundaries identified by Churchill and Hill (2000) and Churchill et al. 2000 (i.e. shown by white polygons in Fig. 10B, C) except where transported by human activities or exposed by low-water condition of Folsom Lake. Field validation work verified the exposures of NOA along the Folsom Lake shoreline as seen in Fig. 10 C to be chrysotile-bearing serpentinite and tremolite-talc schist. Further, there is potential to identify serpentine by mapping chaparral as these species of vegetation have a local affinity for serpentine (Kruckeberg 1984).

Spectral maps that integrate information from substrate-dependent vegetative classification and lithologically independent mineral identification can be used as a tool for delineating areas more likely to contain NOA. These maps can also be used to identify roads surfaced with serpentine aggregate, locate areas in need of dust control or remediation and help fill gaps in geologic mapping where access is limited.

### 3.6 Summary Discussions

Use cases on the application of IS for anthropogenic acid drainage (A AD), socio-natural acid drainage (S-N AD), anthropogenic fugitive dust (A FD), natural fugitive dust (N FD), anthropogenic atmospheric emission (A AE), hydrocarbon contamination (HC C), natural atmospheric emission (N AE), naturally occurring hazardous soils and rocks (NOHSR) (reference shown in Table 2) were examined as part of this paper.

All the case studies examined showed that diagnostic spectral features in the VIS-SWIR range were crucial for providing identification and quantification of the hazards. Specifically, the hazardous materials for A AD, S-N AD, A FD, N FD and NOHSR all have unique mineralogical spectral features that can be discriminated across the VIS-SWIR. In addition, hazardous A AE and HC C contamination also has spectrally distinguishing features across the VIS-SWIR region. These features and their associated shapes/geometries were key to accurate identification and quantification of the hazardous materials. These identification and quantification capabilities then translated to the enabling of better assessments of risks as shown for all case histories, improved targeting of emergency responses

**Table 3** User requirements for natural and anthropogenic hazard applications examined as part of this review

| User objective | Spatial requirements ground sample distance (GSD) | Temporal requirements | Capture to delivery time | IS capabilities |
|---|---|---|---|---|
| Anthropogenic hazards emergency response | < 2–5 m | 1–3 days aligned with remediation efforts | Real time | Target mission |
| Anthropogenic hazard routine monitoring | < 2–10 m | Dependent on regulatory requirements | Days to weeks | Target mission |
| Anthropogenic baseline risk assessment | < 5 m | Once off to be followed by monitoring | Weeks | Target mission |
| Natural hazards regional-scale routine monitoring | < 20–30 m | Dependent on climate, data frequencies of 30–100 days may be sufficient for most applications | Weeks | Mapping mission |
| Natural hazards local-scale (e.g., ASS) routine assessment and monitoring | < 5 m | Dependent on climate, capturing events as required and possibly biannual to annual assessments | Weeks | Target mission |

demonstrated by case histories in A FD and HC C, and enhanced understanding of the progress of remediation efforts shown for monitoring A AD, A FD and HC C hazards.

It is important to note that for all the case studies examined, the data had to be processed to remove atmospheric attenuation. In addition, recalibration (wavelength and radiometric) was also essential for the retrieval of accurate data particularly for multi-temporal monitoring. There are now operational techniques for this process (Richter and Schläpfer 2002; FLAASH, Adler-Golden et al. 1998; SODA, Rodger 2011; CSES 1997; Berk et al. 1998; Berk 2005; Carrère and Conel 1993). To encourage the uptake of the technology, it may be important for data providers to consider the delivery of data with this level of processing. In addition, for spaceborne IS, international networks such as the Radiometric Calibration Network (RadCalNet) (RadCalNet 2018) should prove to be valuable for radiometric calibration.

A variety of techniques have been developed to enable the detection and quantification of these natural and anthropogenic hazards starting with simple ratios, unmixing algorithms including different variants of matched filters (Stocker et al. 1990; Manolakis et al. 2009; Funk et al. 2001), spectral feature matching (Chabrillat et al. 2016; Clark et al. 2003a), spectral deconvolution approaches (Sunshine et al. 1990, Brossard et al. 2016) and the employment of chemometric methods such as PLS (Haaland and Thomas 1988) (see more details in Table 2). A notable development relevant to emergency response is the work by Thompson et al. (2105b) which provides near-real-time capability across the complete chain of processing from radiometric correction to information products. Such capabilities would enable IS technologies to be more readily deployed operationally. If such systems were to be developed further, it would extend the potential for future spaceborne IS to real-time decision-making in disaster management. In the area of risk assessments and monitoring linked to regulatory requirements, the method developed by Ong (2013) included an assessment of the uncertainties in the method essential for meeting the compliance requirements and hence is at a state where it is ready to be operationally deployed.

Associated with the development of these analysis techniques, spectral libraries and associated corroborative information were compiled to better understand the material, background, and mixtures. This has resulted in quite a few new and extended reference spectral libraries (e.g., for AMD secondary minerals and asbestos (Kokaly et al. 2017; Crowley et al. 2003); for oils of varying composition and oil emulsions of different oil:water ratios and thicknesses (Lammoglia and Souza Filho 2011, 2012; Clark et al. 2010; Kokaly et al. 2017); for mixtures of oils with a variety of minerals and soils (Scafutto and Souza Filho 2016; Scafutto et al. 2016, 2017)). These techniques and spectral libraries provide a strong basis for the application of IS for natural and anthropogenic hazards detection, assessment and monitoring.

The data used were acquired from a range of VIS-SWIR imaging spectrometers (see Table 1) mainly from airborne platforms except for EO-1 Hyperion. Note that no unmanned autonomous systems (UAS) were examined as there are insufficiently documented case histories deploying the full VIS-SWIR. The spatial resolutions used the range from 0.4 to 20 m, but most of them were less than 5 m and the areas covered was generally small with the largest being 100 km$^2$. Generally, the spatial resolution chosen was governed by the size of the impacted area and the ability to discriminate the spectral feature. Therefore, the choice of the optimal spatial resolution cannot be generalized for any application.

## 4 User and Observational Requirements

A diverse range of applications of IS for natural and anthropogenic hazards (see Table 2) for risk assessments, emergency response, targeting clean-up/remediation and monitoring remediation efforts have been examined in this paper. The area of impact varied as well as the response time required to attend to the hazard and the remediation impacts to be seen. The observational requirements which vary according to the needs of the hazard and user community, are summarized in Table 3 and are based on the use case examined.

In general, the temporal requirements depend on whether the data would be used for emergency responses, routine monitoring or baseline assessments. If the applications are for emergency responses, spaceborne pointing capabilities would be ideal allowing capture immediately following an accidental oil spill, tailings dam breach or gas/chemical release, earthquake, flood, wildfire or terrorist attack. Repeat observations every several days aligned with the schedule of the clean-up process would be useful to help to guide and focus the remediation effort. Such applications would require the data to be provided to the clean-up team in real time or at least over a very short time frame (that is, hours). Strategies such as on-board processing and fast downlink are critical so that data can be incorporated into the clean-up framework. After containment of the pollutant, the frequency of acquisition likely could be reduced to once per a couple of months depending on the progress of remediation. Regarding spatial requirements, most emergency responses require spatial resolutions of 5 m or finer. For example, the data captured for the WTC attack were at 2 m spatial resolution.

If the user requirement is for routine monitoring, in general, a moderate repeat time of 30 days will suffice for most applications. For anthropogenic hazards, the temporal frequency of data acquisition is usually determined by the regulatory requirements; whereas for natural hazards, the temporal frequency may be determined by the need to capture critical climatic events or changes as they occur. For example, in the monitoring of dust resulting from iron ore handling, current routine ground monitoring using dust depositional gauges is conducted monthly, and hence the availability of remotely sensed dust maps in line with this schedule would be sufficient. Further, for hydrocarbon monitoring, continuous, natural releases of hydrocarbons are not constrained within a specific collection window, but if pollutants are on the water surface, wind conditions may constrain data collection. Therefore, a moderate 30-day repeat or longer revisit cycle, which takes surface conditions into account, may be sufficient for their detection, assuming spatial resolution is adequate.

With regard to spatial resolution, anthropogenic hazard applications usually require finer spatial resolutions than natural hazards as monitoring is usually conducted at the local scale, for example within a mineral lease area. Most of the examples examined captured data with the spatial resolutions of 2–5 m. A pixel resolution of 20 m appears to be sufficient for most natural hazards monitoring applications. However, there are some natural hazard applications such as mapping acid sulphate soils where finer spatial resolutions of less than 5 m are required as the hazards are usually localized point sources.

One of the most comprehensive studies on spectral requirements was conducted by Swayze et al. (2003). In this study, the effects of spectrometer bandpass, sampling and SNR on the spectral identification of a range of surface materials relevant to anthropogenic and natural hazards was examined. The study results indicated that when the sampling interval (SI) is broad, it is the main factor that drives the ability to identify a material because of spectral aliasing, a condition where absorption features of different materials can resemble one another. At narrow SI values relative to the natural full width at half maximum of absorptions, the SNR drives the identification of materials. The bandpass (BP) is less critical to spectral identification than the SI or SNR because broadening the BP does not induce spectral aliasing. It is possible to use the results from studies such as Swayze et al. (2003) to define the optimum combination of BP, SI and SNR for particular applications. Therefore, if we assume the materials examined in Swayze et al. (2003) are fair representations of surface materials encountered for natural and anthropogenic hazards, then the lowest SI to achieve correct spectral identification at 50% of the time for the electronic region (400–1350 nm) is around 13 nm, and 10 nm for the vibrational region (1350–2500 nm). If higher accuracy is desired finer sampling will be required. These values are in the range of the specifications of future spaceborne IS missions such as Environmental Mapping and Analysis Program (EnMAP, Guanter et al. 2015), Hyperspectral Imager Suite (HISUI, Matsunaga, et al. 2018), Surface Biology and Geology (SBG, Green 2018, previously Hyperspectral InfraRed Imager (HyspIRI), Lee et al. 2015), Hyperspectral Imaging Mission for the Environment (CHIME, Nieke and Rast 2018) and Hyperspectral Precursor of the Application Mission (Loizzo et al. 2018). The SNR specifications for future spaceborne missions are also higher than the SNR required for the identification of the different materials examined.

It is important to note that the Swayze et al. (2003) study examined pure materials, and in real-world situations usually mixtures of materials are encountered; therefore, further studies to examine these implications are required. Additionally, Swayze et al. (2003) did not cover hydrocarbons, gaseous emissions and aerosols. It would be useful to conduct studies to examine the requirements for spectral identification of these materials.

## 5 Challenges and Future Directions

This paper examined the use of VIS-SWIR IS to inform decision-making on natural and anthropogenic hazards. Specifically, several case histories for management of A AD, S-N AD, A FD, N FD, A AE, HC C, NOHSR representative of some significant societal and global challenges were examined. These case histories were chosen predominantly to highlight the distinguishing capability of VIS-SWIR IS—the ability to provide quantitative diagnostic information. These diagnostic data generally cannot be obtained at the quantification level, speed, spatial comprehension or resolution from current operational remote sensing systems.

Although the case studies provided compelling evidence for the use of IS, the technology is currently still generally considered to be a scientific tool because of the lack of operational systems, the perceived complexity when using the data and the high costs of acquisitions. The projected launch of spaceborne IS sensors in the near future should fill a proportion of this gap especially for applications where the areas of impact is large as found by Davies and Calvin (2017). However, for many of the case histories examined, the small areas of impact preclude the use of these near term spaceborne IS, which generally has spatial resolution of 30 m. For these cases, in the midterm, there are a small number of high spatial resolution spaceborne IS systems being suggested. Additionally, high quality and moderate cost of off the shelf sensors are becoming more accessible. These next-generation technologies may have the potential to fill this gap. With the latter, there is still the issue of operational costs with airborne systems. Here, the possibilities to couple UAS with miniaturized systems operating across the complete VIS-SWIR ranges, albeit not yet currently available, are a long-term future opportunity to overcome some of the challenges with current operational systems.

With regard to the availability of scientific tools and knowledge to handle the data, the associated studies accompanying many spaceborne missions and the availability of commercial off the shelf sensors have seen an incremental increase in the tools being made publically available, such as those provided by the EnMAP mission (van der Linden et al. 2015). The van der Linden et al. (2015) study shows that many analysis methods have been developed and many of them have also been verified. There is an opportunity now to capture these verified methods in analysis packages in readiness for future spaceborne IS so that downstream products can be quickly accessed enabling faster uptake of the technology.

The challenges and opportunities for the scientific community when future spaceborne IS and next-generation sensors from airborne or UAS platforms become available will be ensuring consistency between data from different sensors, handling, processing, integrating and delivering the large volumes of data and most importantly translating the data to information that meets specific unmet needs of the user community in a form that can be digested/understood by them. The latter is especially important for transferring the technology from a scientific to an operational tool. Good progress has been made by the airborne community in this area especially by organizations such as the European Facility for Airborne Research (EUFAR, http://www.eufar.net/tools/) which has developed an extensive variety of tools for these purposes. These tools could potentially be a good basis for future spaceborne IS missions.

In addition to the consistency of products, the downstream information products must be independently validated using current trusted practices and uncertainties quantified before the IS derived measurement can be integrated into operational monitoring services. Most of the case histories captured in this paper demonstrated this aspect where the products have been independently validated.

**Acknowledgements** This paper is an outcome of a Workshop on Requirements, capabilities and directions in spaceborne IS held at the International Space Science Institute (ISSI) in Bern, Switzerland, in November 2016. The support of ISSI is gratefully acknowledged. A portion of this research was carried out at the Jet Propulsion Laboratory, California Institute of Technology, under contract with the National Aeronautics and Space Administration (NASA). US Government support is acknowledged.



## References

Achterberg EP, Braungardt C, Morley NH, Elbaz-Poulichet F, Leblanc M (1999) Impact of Los Frailes mine spill on riverine, estuarine and coastal waters in southern Spain. Water Res 33:3387–3394. https://doi.org/10.1016/S0043-1354(99)00282-1

Adler-Golden S, Berk A, Bernstein LS, Richtsmeier S (1998) FLAASH, A Modtran4 Atmospheric correction package for hyperspectral data summaries of the 7th annual JPL Airborne earth science workshop, Jet Propulsion Laboratory Publication 98-21 vol 1, AVIRIS workshop, Jet Propulsion Laboratory, California Institute of Technology, Pasadena

Akcil A, Koldas S (2006) Acid mine drainage (AMD): causes, treatment and case studies. J Clean Prod 14:1139–1145. https://doi.org/10.1016/j.jclepro.2004.09.006

Alakian A, Marion R, Briottet X (2008) Remote sensing of aerosol plumes: a semi-analytical model. Appl Opt 47:1851–1866

Alakian A, Marion R, Briottet X (2009) Retrieval of microphysical and optical properties in aerosol plumes with hyperspectral imagery: l-APOM method. Remote Sens Environ 113:781–793
Alpers CN, Blowes DW, Nordstrom DK, Jambor JL (1994) Secondary minerals and acid mine-water chemistry. In: Blowes DW, Jambor JL (eds) The environmental geochemistry of sulfide mine-wastes, short course handbook, vol 2. Waterloo, Mineralogical Association of Canada, pp 247–270
Beland M, Roberts DA, Peterson SH, Biggs TW, Kokaly RF, Piazza S, Ustin SL (2016) Mapping changing distributions of dominant species in oil-contaminated salt marshes of Louisiana using imaging spectroscopy. Remote Sens Environ 182:192–207. https://doi.org/10.1016/j.rse.2016.04.024
Berk A, Bernstein LS, Anderson GP, Acharya PK, Robertson DC, Chetwynd JH, Adler-Golden SM (1998) MODTRAN cloud and multiple scattering upgrades with application to AVIRIS. Remote Sens Environ 65:367–375
Berk A et al (2005) MODTRAN 5: a reformulated atmospheric band model with auxiliary species and practical multiple scattering options: update. In: Defense and security, SPIE, vol 5806
Bigham JM, Schwertmann U, Pfab G (1996) Influence of pH on mineral speciation in a bioreactor simulating acid mine drainage. Appl Geochem 11:845–849
Bonifazi G, Serranti S (2016) Hyperspectral imaging for asbestos identification in cement products. SPIE 1:1. https://doi.org/10.1117/2.1201608.006645
Bourguignon A, Delpont G, Chevrel S, Chabrillat S (2007) Detection and mapping of shrink-swell clays in SW France using ASTER imagery. In: Teeuw RM (ed) Mapping hazardous terrain using remote sensing, vol 283. Geological Society, Special Publications, London, pp 117–124
Bradley ES, Leifer I, Roberts DA, Dennison PE, Washburn L (2011) Detection of marine methane emissions with AVIRIS band ratios. Geophys Res Lett 38:10
Brossard M, Marion R, Carrère V (2016) Deconvolution of SWIR reflectance spectra for automatic mineral identification in hyperspectral imaging. Remote Sens Lett 7(6):581–590. https://doi.org/10.1080/2150704X.2016.1168946
Burns RG (1993) Mineralogical applications of crystal field theory, 2nd edn. Cambridge Topics in Mineral Physics and Chemistry, Cambridge
Buzzi J, Riaza A, García-Meléndez E, Carrère V, Holzwarth S (2016) Monitoring of river contamination derived from acid mine drainage using airborne imaging spectroscopy (HyMap) data. South West Spain River Res Appl 32:125–136. https://doi.org/10.1002/rra.2849
Cairns B, Carlson BE, Ying R, Lacis AA, Oinas V (2003) Atmospheric correction and its application to an analysis of Hyperion data. IEEE Trans Geosci Remote Sens 41(6):1232–1245
Carrère V, Conel JE (1993) Recovery of atmospheric water vapor total column abundance from imaging spectrometer data around 940 nm—sensitivity analysis and application to airborne visible/infrared imaging spectrometer (AVIRIS) data. Remote Sens Environ 44:179–204
Chabrillat S, Goetz AFH (2006) Remote sensing of expansive soils: use of hyperspectral methodology for clay mapping and hazard assessment. In: Ali Al-Rawas A, Goosen MFA (eds) Expansive soils: recent advances in characterization and treatment. Taylor & Francis, London, pp 187–209
Chabrillat S, Goetz AFH, Olsen HW, Krosley L (2002) Use of hyperspectral images in the identification and mapping of expansive clay soils and the role of spatial resolution. Remote Sens Environ 82(2–3):431–445
Chabrillat S, Guillaso S, Rabe A, Foerster S, Guanter L (2016) From HYSOMA to ENSOMAP - A new open source tool for quantitative soil properties mapping based on hyperspectral imagery from airborne to spaceborne applications. In: EGU General Assembly 2016, Vienna, Austria
Chabrillat S, Ben-Dor E, Cierniewski J, Gomez C, Schmid T, van Wesemael B (2019) Imaging spectroscopy for soil mapping and monitoring. Surv in Geophys. https://doi.org/10.1007/s10712-019-09524-0 (this issue)
Chudnovsky A, Ben-Dor E, Kostinski AB, Koren I (2009) Mineral content analysis of atmospheric dust using hyperspectral information from space. Geophys Res Lett. https://doi.org/10.1029/2009gl037922
Chudnovsky A, Kostinski A, Herrmann L, Koren I, Nutesku G, Ben-Dor E (2011) Hyperspectral spaceborne imaging of dust-laden flows: anatomy of Saharan dust storm from the Bodele depression. Remote Sens Environ 115:1013–1024. https://doi.org/10.1016/j.rse.2010.12.006
Churchill RK, Hill RL (2000) A general location guide for ultramafic rocks in California: areas more likely to contain naturally occurring asbestos. California Department of Conservation, Division of Mines and Geology Open-File Report 2000–19, Sacramento
Churchill RK, Higgins CT, Hill B (2000) Areas more likely to contain natural occurrences of asbestos in western El Dorado County, California. California Department of Conservation, Division of Mines and Geology, Open-File Report 2000–002, Sacramento
Clark RN (1999) Spectroscopy of rocks and minerals, and principles of spectroscopy. In: Rencz A (ed) Manual of remote sensing. Wiley, New York, pp 3–58

Clark RN, Roush TL (1984) Spectral properties of mixtures of montmorillonite and dark carbon grains: implications for remote sensing minerals containing chemically and physically adsorbed water. J Geophys Res 89:6329–6340
Clark RN, Green RO, Swayze GA, Meeker G, Sutley S, Hoefen TM, Livo KE, Plumlee G, Parvi B, Sarture C, Wilson S, Hageman P, Lamoth P, Vance JS, Boardman J, Brownfield I, Gent C, Morath LC, Taggart J, Theodorakos PM, Adams M (2001) Environmental studies of the world trade center area after the September 11, 2001 attack. U.S. Geological Survey Open-File Report 01-0429
Clark RN, Swayze GA, Livo KE, Kokaly RF, Sutley SJ, Dalton JB, McDougal RR, Gent CA (2003a) Imaging spectroscopy: earth and planetary remote sensing with the USGS tetracorder and expert systems. J Geophys Res 108(E12):5131. https://doi.org/10.1029/2002JE001847
Clark RN, Hoefen TM, Swayze GA, Livo KE, Meeker GP, Sutley SJ, Wilson S, Brownfield IK, Vance JS (2003b). Reflectance spectroscopy as a rapid assessment tool for the detection of amphiboles from the Libby, Montana region. U.S. Geological Survey Open-File Report 03-128
Clark RN, Swayze GA, Hoefen TM, Green RO, Livo KE, Meeker G, Sutley S, Plumlee G, Pavri B, Sarture C, Boardman J, Brownfield I, Morath LC (2006) Chapter 4: environmental mapping of the World Trade Center area with imaging spectroscopy after the September 11, 2001 attack: in Urban Aerosols and Their Impacts: Lessons Learned from the World Trade Center Tragedy vol American Chemical Society, Symposium Series 919. Oxford University Press
Clark RN, Curchin JM, Hoefen TM, Swayze GA (2009) Reflectance spectroscopy of organic compounds: 1. Alkanes J Geophys Res. https://doi.org/10.1029/2008je003150
Clark RN, Swayze GA, Leifer I, Livo KE, Kokaly RF, Hoefen T, Lundeen S, Eastwood M, Green RO, Pearson N, Sarture C, McCubbin I, Roberts D, Bradley E, Steele D, Ryan T, Dominguez R (2010) A method for quantitative mapping of thick oil spills using imaging spectroscopy: U.S. Geological Survey Open-File Report 2010–1167, p 51
Cloutis EA, Hawthorne FC, Mertzman SA, Krenn K, Cairg MA, Marcino D, Methot M, Strong J, Mustard JF, Blaney DL, Bell JF III, Vilas F (2006) Detection and discrimination of sulfate minerals using reflectance spectroscopy. Icarus 184:121–157
Cocks T, Jenssen R, Stewart A, Wilson I, Shields T (1998) The Hymap Airborne hyperspectral sensor: the system, Calibration and performance. In: Proceedings of 1st EARSEL workshop on imaging spectroscopy, Zurich, Switzerland, October 1998. EARSEL, pp 37–42
Conley S, Franco G, Faloona I, Blake DR, Peischl J, Ryerson TB (2016) Methane emissions from the 2015 Aliso Canyon blowout in Los Angeles, CA. Science 351(6279):1317–1320
Crowley JK, Williams DE, Hammarstrom JM, Piatak N, Chou I-M, Mars JC (2003) Spectral reflectance properties (0.4–2.5 μm) of secondary Fe-oxide, Fe-hydroxide, and Fe-sulphate-hydrate minerals associated with sulphide-bearing mine waste. Geochem Explor Environ Anal 3:219–228
CSES (1997) ATmosphere REMoval Program (ATREM) User's Guide Version 3.0 July 1997. University of Colorado, Boulder, Colorado
Davies G, Calvin W (2017) Mapping acidic mine waste with seasonal airborne hyperspectral imagery at varying spatial scales. Environ Earth Sci 76:1–14. https://doi.org/10.1007/s12665-017-6763-x
Dennison PE, Thorpe AK, Pardyjak ER, Roberts DA, Qi Y, Green RO, Bradley ES, Funk CC (2013) High spatial resolution mapping of elevated atmospheric carbon dioxide using airborne imaging spectroscopy: radiative transfer modeling and power plant plume detection. Remote Sens Environ 139(2013):116–129
Deschamps A, Marion R, Foucher P-Y, Briottet X (2012) Estimation of aerosol type from airborne hyperspectral data: a new technique designed for industrial plume characterization. In: Proceedings of SPIE, 8534. Edinburgh
Deschamps A, Marion R, Briottet X, Foucher P-Y (2013) Simultaneous retrieval of $CO_2$ and aerosols in a plume from hyperspectral imagery: application to the characterization of forest fire smoke using AVIRIS data. Int J Remote Sens 34(19):6837–6864. https://doi.org/10.1080/01431161.2013.809499
Duckworth J (2004) Mathematical data processing. In: Roberts CA, Workman J, Reeves JB III (eds) Near-infrared reflectance spectroscopy, Agronomy Monograph, vol 44. American Society of Agronomy, Madison, pp 115–132
Dufrechou G, Hohmann A, Bourguignon A, Grandjean G (2016) Targeting and mapping expansive soils (Loiret, France): geometrical analysis of laboratory soil spectra in the short-wave infrared domain (1100–2500 nm). Bull Soc Géol France 187(3):169–181
Duren R, Thorpe A, Sanders A (2017) Carlifornia baseline methane survey. Interim phase 1 report. Jet Propulsion Laboratory, California
Emmons SF, Irving JD, Loughlin GF (1927) Geology and ore deposits of the Leadville Mining District, Colorado. U.S. Geological Survey professional paper 148

Farmer VC (1974) The infrared spectra of minerals. Mineralogical Society, London, p 539

Farrand WH (1997) Identification and mapping of ferric oxide and oxyhydroxide minerals in imaging spectrometer data of Summitville, Colorado, U.S.A., and the surrounding San Juan Mountains. Int J Remote Sens 18:1543–1552

Farrand WH, Harsanyi JC (1997) Mapping the distribution of mine tailings in the Coeur d = Alene River Valley, Idaho, through the use of a constrained energy minimization technique. Remote Sens Environ 59:64–76

Fenstermaker LK, Miller JR (1994) Identification of fluvially redistributed mill tailings using high spectral resolution aircraft data. Photogramm Eng Remote Sens 60:989–995

Filella I, Peñuelas J (1994) The red edge position and shape as indicators of plant chlorophyll content, biomass and hydric status Int. J Remote Sens 15(7):1459–1470

Franke J, Roberts DA, Halligan K, Menz G (2009) Hierarchical multiple endmember spectral mixture analysis (MESMA) of hyperspectral imagery for urban environments. Remote Sens Environ 113:1712–1723. https://doi.org/10.1016/j.rse.2009.03.018

Frankenberg C, Thorpe AK, Hook SJ, Green RO, Thompson DR, Kort EA, Hulley GC, Vance N, Bue BD, Aubrey AD (2015) The 4-Corners methane hotspot: mapping CH4 plumes at 60 km through 1 m resolution using space-and airborne spectrometers. In: AGU fall meeting abstracts

Frankenberg C, Thorpe AK, Thompson DR, Hulley G, Kort EA, Vance N, Borchardt J et al (2016) Airborne methane remote measurements reveal heavy-tail flux distribution in Four Corners region. In: Proceedings of the national academy of sciences: 201605617

Frassy F, Candiani G, Rusmini M, Maianti P, Marchesi A, Rota Nodari F, Dalla Via G, Albonico C, Gianinetto M (2014) Mapping asbestos-cement roofing with hyperspectral remote sensing over a large mountain region of the Italian Western Alps. Sensors 14:15900–15913

Frouin R, Deschamps A, Lecomte P (1990) Determination from space of atmospheric total water vapor amounts by differential absorption near 940 nm: theory and airborne verification. J Appl Meteorol 29:448–460

Funk CC, Theiler J, Roberts DA, Borel CC (2001) Clustering to improve matched filter detection of weak gas plumes in hyperspectral thermal imagery. IEEE Trans Geosci Remote Sens 39(7):1410–1420

Gao B-C, Goetz AFH (1990) Column atmospheric water vapor and vegetation liquid water retrievals from airborne imaging spectrometer data. J Geophys Res 95(D4):3549–3564

García-Haro FJ, Sommer S, Kemper T (2005) A new tool for variable multiple endmember spectral mixture analysis (VMESMA). Int J Remote Sens 26:2135–2162. https://doi.org/10.1080/01431160512331337817

Gerilowski K, Tretner A, Krings T, Buchwitz M, Bertagnolio PP, Belemezov F, Erzinger J, Burrows JP, Bovensmann H (2011) MAMAP-a new spectrometer system for column-averaged methane and carbon dioxide observations from aircraft: instrument description and performance analysis. Atmos Meas Tech 4(2):215

Goetz AFH (2009) Three decades of hyperspectral remote sensing of the earth: a personal view. Remote Sens Environ 113:S5–S16. https://doi.org/10.1016/j.rse.2007.12.014

Gottwald M, Bovensmann H, Lichtenberg G, Noël S, von Bargen A, Slijkhuis S, Piters A, Hoogeveen R, von Savigny C, Buchwitz M, Kokhanovsky A, Richter A, Rozanov A, Holzer-Popp T, Bramstedt K, Lambert J-C, Skupin J, Wittrock F, Schrijver H, Burrows JP (2006) SCIAMACHY, monitoring the changing earth's atmosphere. DLR, Institut für Methodik der Fernerkundung (IMF), Weßling

Goudie A, Middleton NJ (2006) Desert dust in the global system. Springer, Berlin

Green RO (2018) Global VSWIR imaging spectroscopy and the 2017 decadal survey. In: IGARSS 2018–2018 IEEE international geoscience and remote sensing symposium, Valencia, pp 183–185. https://doi.org/10.1109/igarss.2018.8518744

Guanter L, Kaufmann H, Segl K, Foerster S, Rogass C, Chabrillat S, Kuester T, Hollstein A, Rossner G, Chlebek C, Straif C, Fischer S, Schrader S, Storch T, Heiden U, Mueller A, Bachmann M, Mühle H, Müller R, Habermeyer M, Ohndorf A, Hill J, Buddenbaum H, Hostert P, van der Linden S, Leitão PJ, Rabe A, Doerffer R, Krasemann H, Xi H, Mauser W, Hank T, Locherer M, Rast M, Staenz K, Sang B (2015) The environmental mapping and analysis program (EnMAP) spaceborne imaging spectroscopy mission for. Earth Obs Remote Sens 7:8830–8857. https://doi.org/10.3390/rs70708830

Haaland DM, Thomas EV (1988) Partial least-squares methods for spectral analyses. 1. Relation to other quantitative calibration methods and the extraction of qualitative information. Anal Chem 60:1193–1200

Hamlin L, Green RO, Mouroulis P, Eastwood M, Wilson D, Dudik M, Paine C (2011) Imaging spectrometer science measurements for terrestrial ecology: AVIRIS and new developments. In: 2011 Aerospace conference, pp 1–7

Hammarstrom JM, Seal RR II, Meier AL, Kornfeld JM (2005) Secondary sulfate minerals associated with acid drainage in the eastern US: recycling of metals and acidity in surficial environments. Chem Geol 215:407–431

Harries J (1997) Acid mine drainage in Australia: its extent and potential future liability. Kingston, ACT: Supervising Scientist

Hong Y, Adler R, Huffman G (2007) Use of satellite remote sensing data in the mapping of global landslide susceptibility. Nat Hazards 43:245–256. https://doi.org/10.1007/s11069-006-9104-z

Hopkins FM, Ehleringer JR, Bush SE, Duren RM, Miller CE, Lai C-T, Hsu Y-K, Carranza V, Randerson JT (2016) Mitigation of methane emissions in cities: how new measurements and partnerships can contribute to emissions reduction strategies. Earth's Future 4(9):408–425

Hulley GC, Duren RM, Hopkins FM, Hook SJ, Johnson WR, Eng BT, Mihaly JM et al (2016) High spatial resolution imaging of methane and other trace gases with the airborne hyperspectral thermal emission spectrometer (HyTES). Atmos Meas Tech 9(5):2393

Hunt GR (1977) Spectral signatures of particulate minerals, in the visible and near-infrared. Geophysics 42:501–513

Hunt GR (1980) Spectroscopic properties of rocks and minerals. In: Carmichael RS (ed) Handbook of physical properties of rocks. CRC Press, Boca Raton

Hunt GR, Ashley RP (1979) Spectra of altered rocks in the visible and near infrared. Econ Geol 74:1613–1629

International Agency for Research on Cancer (2012) Arsenic, metals, fibres and dusts. In: A review of human carcinogens. International Agency for Research on Cancer Monographs on the Evaluation of Carcinogenic Risks to Humans, vol 100 C. International Agency for Research on Cancer, Lyon, France

Jha MK (2010) Natural and anthropogenic disasters: an overview. In: Jha MK (ed) Natural and anthropogenic disasters: vulnerability, preparedness and mitigation. Springer, Dordrecht, pp 1–16. https://doi.org/10.1007/978-90-481-2498-5_1

Jickells TD, An ZS, Andersen KK, Baker AR, Bergametti G, Brooks N, Cao JJ, Boyd PW, Duce RA, Hunter KA, Kawahata H, Kubilay N, laRoche J, Liss PS, Mahowald N, Prospero JM, Ridgwell AJ, Tegen I, Torres R (2005) Global iron connections between desert dust. Ocean Biogeochem Clim Sci 308:67–71. https://doi.org/10.1126/science.1105959

Jones L, Jefferson I (2012) Expansive soils. In: Burland J (ed) ICE manual of geotechnical engineering. Geotechnical engineering principles, problematic soils and site investigation, vol 1. ICE Publishing, pp 413–441

Joyce KE, Belliss SE, Samsonov SV, McNeill SJ, Glassey PJ (2009) A review of the status of satellite remote sensing and image processing techniques for mapping natural hazards and disasters. Prog Phys Geogr 33:183–207. https://doi.org/10.1177/0309133309339563

Kariuki PC, Woldai T, van der Meer F (2004) Effectiveness of spectroscopy in identification of swelling indicator clay minerals. Int J Remote Sens 25:455–469

Kaufman YJ, Gao B-C (1992) Remote sensing of water vapor in the near IR from EOS/MODIS. IEEE Trans Geosci Remote Sens 30(5):871–884

Kemper T, Sommer S (2004) Use of airborne hyperspectral data to estimate residual heavy metal contamination and acidification potential in the Guadiamar floodplain Andalusia, Spain after the Aznacollar mining accident. In: Proceedings of the SPIE 5574, remote sensing for environmental monitoring, GIS applications, and geology IV, pp 224

Khanna S, Santos MJ, Ustin SL, Koltunov A, Kokaly RF, Roberts DA (2013) Detection of salt marsh vegetation stress and recovery after the Deepwater Horizon oil spill in Barataria Bay, Gulf of Mexico using AVIRIS data. PLoS ONE 8(11):e78989. https://doi.org/10.1371/journal.pone.0078989

King TVV, Clark RN, Ager C, Swayze GA (1995). In: Posey HH, Pendelton JA, Van Zyl D (eds) Proceedings: summitville forum = 95. Colorado Geological Survey Special Publication 38, pp 59–63

Kokaly RF (2011) PRISM: processing routines in IDL for spectroscopic measurements (installation manual and user's guide, version 1.0). U.S. Geological Survey Open-File Report 2011–1155. https://pubs.usgs.gov/of/2011/1155/

Kokaly RF, Couvillion BR, Holloway JM, Roberts DA, Ustin SL, Peterson SH, Khanna S, Piazza SC (2013) Spectroscopic remote sensing of the distribution and persistence of oil from the Deepwater Horizon spill in Barataria Bay marshes. Remote Sens Environ 129:210–230. https://doi.org/10.1016/j.rse.2012.10.028

Kokaly RF, Clark RN, Swayze GA, Livo KE, Hoefen TM, Pearson NC, Wise RA, Benzel WM, Lowers HA, Driscoll RL, Klein AJ (2017) USGS spectral library version 7. U.S. Geological Survey Data Series 1035. https://doi.org/10.3133/ds1035

Kopačková V (2014) Using multiple spectral feature analysis for quantitative pH mapping in a mining environment. Int J Appl Earth Obs Geoinf 28:28–42. https://doi.org/10.1016/j.jag.2013.10.008
Krautwurst S, Gerilowski K, Jonsson HH, Thompson DR, Kolyer RW, Thorpe AK, Horstjann M, Eastwood M, Leifer I, Vigil S, Krings T, Borchardt J, Buchwitz M, Fladeland MM, Burrows JP, Bovensmann H (2017) Methane emissions from a Californian landfill, determined from airborne remote sensing and in situ measurements. Atmos Meas Tech 10:3429–3452
Kruckeberg AR (1984) California serpentines: flora, vegetation, geology, soils, and management problems. University of California Press, Los Angeles
Kruse FA, Dwyer JL (1993) The effects of AVIRIS atmospheric calibration methodology on identification and quantitative mapping of surface mineralogy, Drum Mtns., Utah Summaries of the Fourth JPL Airborne Geoscience Workshop JPL Publication, vol 1. Pasadena, California, pp 93–26
Kruse FA, Hauff P, Dietz J, Brock JC, Hampton L (1989) Final report contract no. 412.DEN.63786.FM (CH2 M HILL). Center for the Study of Earth from Space, University of Colorado, Boulder
Labson VF, Clark RN, Swayze GA, Hoefen TM, Kokaly R, Livo KE, Powers MH, Plumlee GS, Meeker GP (2010) Estimated lower bound for leak rates from the Deepwater Horizon spill—Interim report to the Flow Rate Technical Group from the Mass Balance Team. U.S. Geological Survey Open-File Report 2010–1132. http://pubs.usgs.gov/of/2010/1132/
Lammoglia T, Souza Filho CR (2011) Spectroscopic characterization of oils yielded from Brazilian offshore basins: potential applications of remote sensing. Remote Sens Environ 115:2525–2535
Lammoglia T, Souza Filho CR (2012) Mapping and characterization of the API gravity of offshore hydrocarbon seepages using multispectral ASTER data. Remote Sens Environ 123:381–389
Lee CM, Cable ML, Hook SJ, Green RO, Ustin SL, Mandl DJ, Middleton EM (2015) An introduction to the NASA Hyperspectral InfraRed Imager (HyspIRI) mission and preparatory activities. Remote Sens Environ 167:6–19. https://doi.org/10.1016/j.rse.2015.06.012
Lehr B, Sky B, Possolo A et al (2010) Oil budget calculator deepwater horizon: a report to the National Incident Command. NOAA
Leifer I, Lehr WJ, Simecek-Beatty D, Bradley E, Clark R, Dennison P, Hu Y, Matheson S, Jones CE, Holt B, Reif M, Roberts DA, Svejkovsky J, Swayze G, Wozencraft J (2012) State of the art satellite and airborne marine oil spill remote sensing: application to the BP Deepwater Horizon oil spill. Remote Sens Environ 124:185–209
Lentile LB, Holden ZA, Smith AMS, Falkowski MJ, Hudak AT, Morgan P, Lewis SA, Gessler PE, Benson NC (2006) Remote sensing techniques to assess active fire characteristics and post-fire effects. Int J Wildland Fire 15:319–345. https://doi.org/10.1071/WF05097
Loizzo R, Guarini R, Longo F, Scopa T, Formaro R, Facchinetti C, Varacalli G (2018) Prisma: the Italian hyperspectral mission. IGARSS 2018–2018 IEEE international geoscience and remote sensing symposium, Valencia, pp 175–178. https://doi.org/10.1109/igarss.2018.8518512
Lopez-Pamo E, Barettino D, Anton-Pacheco C, Ortiz G, Arranz JC, Gumiel JC, Martinez-Pledel B, Aparico M, Montouto O (1999) The extent of the Aznalcollar pyretic sludge spill and its effects on soils. Sci Total Environ 242:57–88
Manolakis DG, Lockwood R, Cooley T, Jacobson J (2009) Is there a best hyperspectral detection algorithm. In: Proceedings of the SPIE 7334, p. 733402, https://doi.org/10.1117/12.816917. c© 2009 SPIE
Manolakis D, Truslow E, Pieper M, Cooley T, Brueggeman M (2014) Detection algorithms in hyperspectral imaging systems: an overview of practical algorithms. IEEE Signal Proc Mag 31:24–33
Marino CM, Panigada C, Busetto L, Galli A (2000) Environmental applications of airborne hyperspectral remote sensing: asbestos concrete sheeting identification and mapping. In: Proceedings of the fourteenth international conference: applied geologic remote sensing, Las Vagas, November 6–8th. Veridian ERIM International, pp 607–609
Marion R, Michel R, Faye C (2004) Measuring trace gases in plumes from hyperspectral remotely sensed data. IEEE Trans Geosci Remote Sens 42(4):854–864
Marion R, Michel R, Faye C (2006) Atmospheric correction of hyperspectral data over dark surfaces via simulated annealing. IEEE Trans Geosci Remote Sens 44(6):1566–1574
Massarelli C, Matarrese R, Uricchio VF, Muolo MR, Laterza M, Ernesto L (2017) Detection of asbestos-containing materials in agro-ecosystem by the use of airborne hyperspectral CASI-1500 sensor including the limited use of two UAVs equipped with RGB cameras. Int J Remote Sens 38:2135–2149
Matsunaga T, Iwasaki A, Tsuchida S, Iwao K, Tanii J, Nakamura R, Yamamoto H, Kato S, Obata K, Kashimura O, Tanii J, Mouri K and Tachikawa T (2018) HISUI status toward FY2019 launch. In: IGARSS 2018–2018 IEEE international geoscience and remote sensing symposium, Valencia, pp 160–163. https://doi.org/10.1109/igarss.2018.8518639
Metternicht GI, Zinck JA (2003) Remote sensing of soil salinity: potentials and constraints. Remote Sens Environ 85:1–20. https://doi.org/10.1016/S0034-4257(02)00188-8

Michel J, Owens EH, Zengel S, Graham A, Nixon Z, Allard T, Holton W, Reimer PD, Lamarche A, White M, Rutherford N, Childs C, Mauseth G, Challenger G, Taylor E (2013) Extent and degree of shoreline oiling: deepwater horizon oil spill, gulf of Mexico, USA. PLOS ONE 8:e65087
Nieke J, Rast M (2018) Towards the copernicus hyperspectral imaging mission for the environment (CHIME). In: IGARSS 2018–2018 IEEE international geoscience and remote sensing symposium, 22–27 July 2018. pp 157–159. https://doi.org/10.1109/igarss.2018.8518384
OECD (2016) The economic consequences of outdoor air pollution. OECD Publishing, Paris
Ong CCH (2013) Mapping and monitoring the environmental impacts of mining using hyperspectral data. Curtin University of Technology, Bentley
Ong CCH, Cudahy TJ (2014) Mapping contaminated soils: using remotely-sensed hyperspectral data to predict pH. Eur J Soil Sci 65:897–906. https://doi.org/10.1111/ejss.12160
Ong C, Swayze G, Clark R (2003a) An investigation of the Tetracorder expert system for multi-temporal mapping of acid drainage-related minerals using airborne hyperspectral data. In: Proceedings of the 3rd EARSel workshop on imaging spectroscopy, Herrsching, Germany, 13–16 May 2003, pp 357–362
Ong CCH, Cudahy TJ, Caccetta MS, Piggott MS (2003b) Deriving quantitative dust measurements related to iron ore handling from airborne hyperspectral data. Min Technol 112:A158–A163
Ong C, Cudahy TJ, Swayze G (2003c) Predicting acid drainage related physicochemical measurements using hyperspectral data. In: Proceedings of the 3rd EARSeL workshop on imaging spectroscopy, Herrsching, Germany, 13–16 May 2003, pp 363–373
Ong C, Lau I, Hewson R, Tongway D, Eisele A (2008) MERIWA project M385: the development of a hyperspectral environmental measurement tool for monitoring mining related infrastructure and rehabilitation: CSIRO exploration and mining report no. 271, p 122
Pan XL, Day HW, Wang W, Beckett LA, Schenker MB (2005) Residential proximity to naturally occurring asbestos and mesothelioma risk in California. Am J Respir Crit Care Med 172:1019–1025
Pascucci S, Belviso C, Cavalli RM, Palombo A, Pignatti S, Santini F (2012) Using imaging spectroscopy to map red mud dust waste: the Podgorica aluminum complex case study. Remote Sens Environ 123:139–154. https://doi.org/10.1016/j.rse.2012.03.017
Pearlman J, Carman S, Segal C, Jarecke P, Clancy P, Browne W (2001) Overview of the hyperion imaging spectrometer for the NASA EO-1 mission. In: IGARSS 2001. scanning the present and resolving the future. Proceedings. IEEE 2001 international geoscience and remote sensing symposium (Cat. No.01CH37217), 2001 2001, vol 3037. pp 3036–3038. https://doi.org/10.1109/igarss.2001.978246
Pearson RM, Swayze GA, Wiltshire R, Serina D, Clark RN (1997) Using AVIRIS imagery mineral mapping for identification of acid rock drainage sources at the California Gulch NPL Site and fluvial tailings deposits on the Arkansas River, Leadville, Colorado, USA. In: Proceedings of the 3rd international airborne remote sensing conference, Copenhagen, Denmark, pp I-102–I-109
Peters DC, Hauff PL, Livo KE (1995) Remote sensing for mine waste discrimination and characterization. In: Curran PJ, Roberston YC (eds) Proceedings of RSS 95: remote sensing in action, September 11–14, 1995, Southhampton. The Remote Sensing Society, pp 866–877
Peterson SH, Roberts DA, Beland M, Kokaly RF, Ustin SL (2015) Oil detection in the coastal marshes of Louisiana using MESMA applied to band subsets of AVIRIS data. Remote Sens Environ 159:222–231. https://doi.org/10.1016/j.rse.2014.12.009
Plumlee GS, Smith KS, Montour MR, Ficklin WH, Mosier EL (1999) Geologic controls on the composition of mine waters and natural waters draining diverse mineral-deposit types. In: Filipek LH, Plumlee GS (eds) Reviews in economic geology, part B: case studies and research topics. Society of Economic Geologists, 6B, Littleton, pp 373–432
Popp C, Brunner D, Damm A, Van Roozendael M, Fayt C, Buchmann B (2012) High-resolution $NO_2$ remote sensing from the Airborne Prism EXperiment (APEX) imaging spectrometer. Atmos Meas Tech 5:2211–2225
Qu Z, Kindel BC, Goetz AFH (2003) The high accuracy atmospheric correction for hyperspectral data (HATCH) model. IEEE Trans Geosci Remote Sens 41(6):1223–1231
RadCalNet (2018) Radiometric Calibration Network. https://www.radcalnet.org/#!/. Accessed 2 Feb 2019
Richter R, Schläpfer D (2002) Geo-atmospheric processing of airborne imaging spectrometry data: part 2. Atmospheric/topo-graphic correction. Int J Remote Sens 23:2631–2649
Roberts DA, Bradley ES, Cheung R, Leifer I, Dennison PE, Margolis JS (2010) Mapping methane emissions from a marine geological seep source using imaging spectrometry. Remote Sens Environ 114(3):592–606
Rockwell BW, McDougal RR, Gent CA (2005) Remote sensing for environmental site screening and watershed evaluation in Utah mine lands—East Tintic Mountains, Oquirrh Mountains, and Tushar Mountains. U.S. Geological Survey Scientific Investigations Report 2004–5241

Rodger A (2011) SODA: a new method of in-scene atmospheric water vapor estimation and post-flight spectral recalibration for hyperspectral sensors application to the HyMap sensor at two locations. Remote Sens Environ 115(2):536–547

Rodgers CD (2000) Inverse methods for atmospheric sounding: theory and practice, vol 2. World Scientific, New York

Rogge DM, Rivard B, Zhang J, Feng J (2006) Iterative spectral unmixing for opti-mizing per-pixel endmember sets. IEEE Trans Geosci Remote Sens 44(12):3725–3736

Ross M (1999) The health effects of mineral dusts. In: Plumlee GS, Logsdon MJ (eds) Reviews in economic geology, part a: the environmental geochemistry of mineral deposits: processes, techniques, and health issues, 6A. Society of Economic Geologists, Littleton, pp 339–356

Salomonson VV, Barnes WL, Maymon PW, Montgomery HE, Ostrow H (1989) MODIS: advanced facility instrument for studies of the Earth as a system. IEEE Trans Geosci Remote Sens 27:145–153. https://doi.org/10.1109/36.20292

Scafutto RDM, Souza Filho CR (2016) Quantitative characterization of crude oils and fuels in mineral substrates using reflectance spectroscopy: implications for remote sensing. Int J Appl Earth Obs Geoinf 50:221–242

Scafutto RDM, Souza Filho CR, Rivard B (2016) Characterization of mineral substrates impregnated with crude oils using proximal infrared hyperspectral imaging. Remote Sens Environ 179:116–130

Scafutto RDM, Souza Filho CR, de Oliveira WJ (2017) Hyperspectral remote sensing detection of petroleum hydrocarbons in mixtures with mineral substrates: implications for onshore exploration and monitoring. ISPRS J Photogramm Remote Sens 128:146–157

Schaepman ME, Ustin SL, Plaza AJ, Painter TH, Verrelst J, Liang S (2009) Earth system science related imaging spectroscopy—an assessment. Remote Sens Environ 113:S123–S137

Schläpfer D, Borel CC, Keller J, Itten KI (1998) Atmospheric pre-corrected differential absorption technique to retrieve columnar water vapor. Remote Sens Environ 65:353–366

Shaw SC, Groat LA, Jambor JL, Blowes DW, Hanton-Fong CJ, Stuparyk RA (1998) Mineralogical study of base metal tailings with various sulfide contents, oxidized in laboratory columns and field lysimeters. Environ Geol 33(2/3):209–217

Shi X-Z, Lau IC, Aspandiar M (2014a) Comparison of PLSR modelling and indicative mineral mapping of airborne hyperspectral imagery for acid sulphate soil assessment. AU Int J Remote Sens 35:1309–1330. https://doi.org/10.1080/01431161.2013.876121

Shi X, Aspandiar M, Lau IC, Oldmeadow D (2014b) Assessment of acid sulphate soil both on surface and in subsurface using hyperspectral data. Can J Remote Sens 39(6):468–480

Spinetti C, Carrère V, Buongiorno MF, Sutton AJ, Tamar Elias T (2008) Carbon dioxide of PuuOo volcanic plume at Kilauea retrieved by AVIRIS hyperspectral data. Remote Sens Environ 112(6):3192–3199

Stocker AD, Reed IS, Yu X-L (1990) Multidimensional signal processing for electro-optical target detection. In: Proceedings of the SPIE 1305, signal and data processing of small targets 1990, 218 (October 1, 1990). https://doi.org/10.1117/12.21593

Sunshine JM, Pieters CM, Pratt SF (1990) Deconvolution of mineral absorption bands: an improved approach. J Geophys Res 95:6955–6966. https://doi.org/10.1029/jb095ib05p06955

Swayze GA, Kokaly RF (2016) Detection of potentially asbestos-bearing rocks using imaging spectroscopy, developing integrated methods to address complex resource and environmental issues. In: Smith KS, Phillips JD, McCafferty AE, Clark RN (eds) Developing integrated methods to address complex resource and environmental issues. U.S. Geological Survey Circular 1413, pp 95–100

Swayze GA, Clark RN, Pearson RM, Livo KE (1996) Mapping acid-generating minerals at the California Gulch Superfund Site in Leadville, Colorado using imaging spectroscopy. In: Summaries of the Sixth Annual JPL Airborne Earth Science Workshop, Jet Propulsion Laboratory Publication 96-4, 1, pp 231–234

Swayze GA, Smith KS, Clark RN, Sutley SJP, Pearson RM, Vance JS, Hageman PL, Briggs PH, Meier AL, Singleton MJ, Roth S (2000) Using imaging spectroscopy to map acidic mine waste. Environ Sci Technol 34:47–54

Swayze GA, Clark RN, Goetz AFH, Chrien TG, Gorelick NS (2003) Effects of spectrometer band pass, sampling, and signal-to-noise ratio on spectral identification using the Tetracorder algorithm. J Geophys Res Planets 108:5105. https://doi.org/10.1029/2002je001975

Swayze GA, Kokaly RF, Higgins CT, Clark RN, Meeker GP, Clinkenbeard JP, Sutley SJ (2004) Using imaging spectroscopy to map ultramafic rocks, serpentinites, and tremolitic rocks in El Dorado and Plumas Counties, California. U.S. Geological Survey Open-File Report 2004-1304

Swayze GA, Kokaly RF, Higgins CT, Clinkenbeard JP, Clark RN, Lowers HA, Sutley SJ (2009) Mapping potentially asbestos-bearing rocks using imaging spectroscopy. Geology 37(8):763–766

Swinehart DF (1962) The Beer-Lambert law. J Chem Educ 39(7):333

Thompson DR, Gao B-C, Green RO, Roberts DA, Dennison PE, Lundeen SR (2015) Atmospheric correction for global mapping spectroscopy: ATREM advances for the HyspIRI preparatory campaign. Remote Sens Environ 167:64–77
Thompson DR, Leifer I, Bovensmann H, Eastwood M, Fladeland M, Frankenberg C, Gerilowski K et al (2105) Real-time remote detection and measurement for airborne imaging spectroscopy: a case study with methane. Atmos Meas Tech 8(10):4383–4397
Thompson DR, Thorpe AK, Frankenberg C, Green RO, Duren R, Guanter L, Hollstein A, Middleton E, Ong L, Ungar S (2016) Space-based remote imaging spectroscopy of the Aliso Canyon CH4 super-emitter. Geophys Res Lett 43(12):6571–6578
Thompson DR, Guanter L, Berk A, Gao B-C, Richter R, Schläpfer D, Thome KJ (2019) Retrieval of atmospheric parameters and surface reflectance from visible and shortwave infrared imaging spectroscopy data. Surv Geophys. https://doi.org/10.1007/s10712-018-9488-9 **(this issue)**
Thorpe AK, Roberts DA, Dennison PE, Bradley ES, Funk CC (2012) Point source emissions mapping using the airborne visible/infrared imaging spectrometer (AVIRIS). In: SPIE defense, security, and sensing. International Society for Optics and Photonics, pp 839013–839013
Thorpe AK, Roberts DA, Bradley ES, Funk CC, Dennison PE, Leifer I (2013) High resolution mapping of methane emissions from marine and terrestrial sources using a cluster-tuned matched filter technique and imaging spectrometry. Remote Sens Environ 134:305–318
Thorpe AK, Frankenberg C, Aubrey AD, Roberts DA, Nottrott AA, Rahn TA, Sauer JA, Dubey MK, Costigan KR, Arata C, Steffke AM, Hills S, Haselwimmer C, Charlesworth D, Funk CC, Green RO, Lundeen SR, Boardman JW, Eastwood ML, Sarture CM, Nolte SH, Mccubbin IB, Thompson DR, McFadden JP (2016) Mapping methane concentrations from a controlled release experiment using the next generation airborne visible/infrared imaging spectrometer (AVIRIS-NG). Remote Sens Environ 179:104–115
Tralli DM, Blom RG, Zlotnicki V, Donnellan A, Evans DL (2005) Satellite remote sensing of earthquake, volcano, flood, landslide and coastal inundation hazards. ISPRS 59:185–198. https://doi.org/10.1016/j.isprsjprs.2005.02.002
United Nations Office for Disaster Risk Reduction (2017) Terminology. United Nations Office for Disaster Risk Reduction. http://www.unisdr.org/we/inform/terminology. Accessed 08 Oct 2017
van der Linden S, Rabe A, Held M, Jakimow B, Leitão P, Okujeni A, Schwieder M, Suess S, Hostert P (2015) The EnMAP-Box—a toolbox and application programming interface for EnMAP data processing. Remote Sens 7:11249
van Geen A, Chase Z (1998) Recent mine spill adds to contamination of southern Spain Eos. Trans Am Geophys Union 79:449–455. https://doi.org/10.1029/98EO00340
Van Oss CJ, Naim JO, Costanzo PM, Giese RF Jr, Wu W, Sorling AF (1999) Impact of different asbestos species and other mineral particles on pulmonary pathogenesis. Clays Clay Miner 47(6):697–707
Vane G, Green RO, Chrien TG, Enmark HT, Hansen EG, Porter WM (1993) The airborne visible/infrared imaging spectrometer (AVIRIS). Remote Sens Environ 44:127–143. https://doi.org/10.1016/0034-4257(93)90012-M
Westrich JR, Ebling AM, Landing WM, Joyner JL, Kemp KM, Griffin DW, Lipp EK (2016) Saharan dust nutrients promote Vibrio bloom formation in marine surface waters. Proc Natl Acad Sci 113:5964–5969. https://doi.org/10.1073/pnas.1518080113
Yitegasu FA, Van der Meer F, van der Werff H, Zigterman W (2009) Quantifying engineering parameters of expansive soils from their reflectance spectra. Eng Geol 105(3–4):151–160
Zabcic N, Rivard B, Ong C, Mueller A (2014) Using airborne hyperspectral data to characterize the surface pH and mineralogy of pyrite mine tailings. Int J Appl Earth Obs Geoinf 32:152–162. https://doi.org/10.1016/j.jag.2014.04.008

**Publisher's Note** Springer Nature remains neutral with regard to jurisdictional claims in published maps and institutional affiliations.

## Affiliations

**C. Ong[1] · V. Carrère[2] · S. Chabrillat[3] · R. Clark[4] · T. Hoefen[5] · R. Kokaly[5] · R. Marion[6] · C. R. Souza Filho[7] · G. Swayze[5] · D. R. Thompson[8]**

[1] Commonwealth Scientific Industrial Research Organisation (CSIRO), 26 Dick Perry Avenue, Kensington, WA 6151, Australia

[2] Faculté des Sciences et des Techniques UMR-CNRS 6112 - Laboratoire Planetologie et Geodynamique, Université de Nantes, 2 rue de la Houssinière, BP 92208, 44322 Nantes Cedex 3, France

[3] German Research Centre for Geosciences (GFZ), Telegrafenberg, 14473 Potsdam, Germany

[4] Planetary Science Institute, 1546 Cole Blvd #120, Lakewood, CO 80401, USA

[5] United States Geological Survey (USGS), Bld. 20 Rm C2019, Denver Federal Center, Kipling and 6th, Denver, CO 80225, USA

[6] Laboratoire Télédétection, Surveillance, Environnement Commissariat à l'Energie Atomique et aux énergies alternatives Centre DAM/Ile de France, Bruyères-le-Châtel, 91297 Arpajon Cedex, France

[7] Institute of Geosciences, State University of Campinas (UNICAMP), Rua Carlos Gomes, 250, Campinas, SP CEP 13083-855, Brazil

[8] Jet Propulsion Laboratory, California Institute of Technology, 4800 Oak Grove Dr, Pasadena, CA 91109, USA

Surveys in Geophysics (2019) 40:471–488
https://doi.org/10.1007/s10712-018-9486-y

# Imaging Spectroscopy of Urban Environments

**S. van der Linden[1,2] · A. Okujeni[1] · F. Canters[3] · J. Degerickx[4] · U. Heiden[5] · P. Hostert[1,2] · F. Priem[3] · B. Somers[4] · F. Thiel[1]**

Received: 2 October 2017 / Accepted: 5 July 2018 / Published online: 16 July 2018



## Abstract

Future spaceborne imaging spectroscopy data will offer new possibilities for mapping ecosystems globally, including urban environments. The high spectral information content of such data is expected to improve accuracies and thematic detail of maps on urban composition and urban environmental condition. This way, urgently needed information for environmental models will be provided, for example, for microclimate or hydrological models. The diverse vertical structures, highly frequent spatial change and a great variety of materials cause challenges for urban environmental mapping with Earth observation data, especially at the 30 m spatial resolution of data from future spaceborne imaging spectrometers. This paper gives an overview of the state-of-the-art in urban imaging spectroscopy considering decreasing spatial resolution, the related user requirements and existing knowledge gaps, as well as expected future directions for the work with new data sets.


**Keywords** Imaging spectroscopy · Hyperspectral · Urban · Unmixing · Spatial resolution · Environmental Mapping and Analysis Program (EnMAP)

## 1 Introduction

In 2018, 55% of the world's population is estimated to live in cities and the numbers of urban dwellers and of mega cities will continue to increase rapidly (UN 2018). The twenty-first century is therefore referred to as the *Century of the City* (Seto et al. 2010).

✉ S. van der Linden
sebastian.linden@geo.hu-berlin.de

1 Geography Department, Humboldt-Universität zu Berlin, Unter den Linden 6, 10099 Berlin, Germany

2 Integrative Research Institute on Transformations of Human-Environment Systems, Humboldt-Universität zu Berlin, Unter den Linden 6, 10099 Berlin, Germany

3 Department of Geography, Cartography and GIS Research Group, Vrije Universiteit Brussel, Pleinlaan 2, 1050 Brussels, Belgium

4 Division of Forest, Nature and Landscape, KU Leuven, Celestijnenlaan 200e, 3001 Leuven, Belgium

5 Department of Land Surface Dynamics, German Remote Sensing Data Center, German Aerospace Center, Münchner Str. 20, 82234 Weßling, Germany

Urbanization constitutes a demographic, socioeconomic and biophysical process that substantially contributes to global environmental change (Lambin et al. 2001; Pickett et al. 2011). The expansion of impervious areas, the loss of natural environments as well as the increased landscape fragmentation along with the concentration of human consumption, production and waste generation alter biogeochemical cycles, climate regimes, hydrological systems and biodiversity at multiple scales (Grimm et al. 2008; Lawler et al. 2014). Globally, urban areas are a major source of $CO_2$ and other greenhouse gas emissions (Kennedy et al. 2009). At the regional and local scale, cities are characterized by the urban heat island, that is, an increased air temperature (Grimmond 2007; Oke 1982), and by air pollution (Fenger 1999; Hatt et al. 2004). Urban ecosystem condition, its functioning and spatial–temporal change influence the quality of life and human health (Harlan et al. 2006; Pauleit et al. 2005; Tan et al. 2010). At the same time, contemporary urbanization has the chance to promote ecological sustainability due to increased environmental awareness and returns from innovation, productivity and efficiency (Grimm et al. 2008; Seto et al. 2010). To ensure that policy sustains benefits of urbanization and anticipates or manages the negative consequences of urban growth (UN-Habitat 2010), the interactions between socioeconomic and environmental processes in urban landscapes need to be better understood (Alberti 2005). Integrated research targeting the diverse aspects and implications of urban systems in times of rapid urbanization is of great importance to face the complexity of urban dynamics.

Remote sensing can contribute in many ways to the analysis of urban environments and provides substantial information on ecosystem characteristics (see Small et al. 2018 for a general overview on urban remote sensing). We define the term *urban environment* in this paper as contiguous areas of anthropogenic, artificial surfaces used for transport, commerce, production, administration or housing, plus the included or adjacent vegetation surfaces that are intensively managed and directly influenced by the artificial cover. Land cover composition as detailed as construction material abundance (Herold et al. 2003; Roessner et al. 2001), vegetation characteristics such as type, structure or condition (Alonzo et al. 2015; Degerickx et al. 2018; Delegido et al. 2014; Pontius et al. 2017), information on vertical structures and surface roughness (Zhou et al. 2005), water run-off potentials (Weng 2001) or spatial–temporal surface temperature changes (Deng and Wu 2013; Imhoff et al. 2010) can be deduced when the full spectrum of remote sensing technology is considered, including multi- and hyperspectral optical sensors, thermal sensors, microwave and lidar systems (Small et al. 2018). This way, ecologically relevant information such as spatial patterns of impervious and vegetated surface, biomass estimates and inputs for microclimatic or hydrological models can be derived (Carlson and Arthur 2000; Heldens et al. 2017; Huang et al. 2013; Ngie et al. 2014; Voogt and Oke 2003).

The contribution from optical sensors to the list of variables that can be derived from remote sensing data is mostly related to the identification of land cover types and materials or the type and biophysical condition of urban vegetation. Particularly, the spectral information from imaging spectroscopy data adds relevant detail to the analysis of the urban environment, for example, to the derivation of ecological indicators (Behling et al. 2015) or urban structure types (Heiden et al. 2012), the characterization of urban greenness (Degerickx et al. 2018; Delegido et al. 2014) or modeling meteorological variables (Brackx et al. 2017). Nevertheless, imaging spectroscopy and optical remote sensing in general also face challenges when urban environments are mapped. On the one hand, mapping is complicated by the high number of spectrally mixed pixels typical of urban imagery with a great variety of surface materials and high frequent spatial change (Small 2003). Phenomena such as shadow, illumination differences and object displacement from vertical structures, on the other hand, hamper the remote sensing analyses at all scales (Adeline et al. 2013; van der Linden and Hostert 2009). The spectral ambiguity between land cover types, due to the presence of construction materials with similar reflectance properties, and the spectral heterogeneity within single land covers, due to the high diversity of materials and species or the high spectral variability of the single materials or species, constitute further challenges (Fig. 1; Herold and Roberts 2005; Herold et al. 2004; Small 2001; van der Linden et al. 2007). Here, imaging spectroscopy data have proven more reliable than data from multispectral instruments. Urban vegetation also marks an especially challenging case. Street trees, for example, live under extreme conditions, often being exposed to higher actinic stress, lower water supplies and higher loads of air pollution (Damm 2008). Their sparse spatial distribution requires very high spatial resolution imagery for accurate mapping (Tigges et al. 2013), while urban street canyons may cause extreme illumination situations and trees are often obscured by surrounding buildings (van der Linden and Hostert 2009). Again, an increased spectral information adds to the quality of results (Alonzo et al. 2014). The combined complexity of material composition and vertical structures also renders the modeling of radiative transfer challenging and processing intensive. Despite the existence of ray-tracing models that can model urban radiative transfer under certain conditions (Gastellu-Etchegorry et al. 2015), studies reporting the use of model inversion for mapping biophysical parameters for different land covers at relevant spatial extent are still missing.

Given the fine-scale spatial patterns and highly frequent spectral variation of urban surfaces, the spatial resolution of remote sensing imagery becomes a key factor in mapping urban environments. Thus, the soon-to-come step from airborne to spaceborne imaging spectroscopy marks a major challenge for urban remote sensing. This overview paper reflects on the state of imaging spectroscopy and the respective potentials and challenges

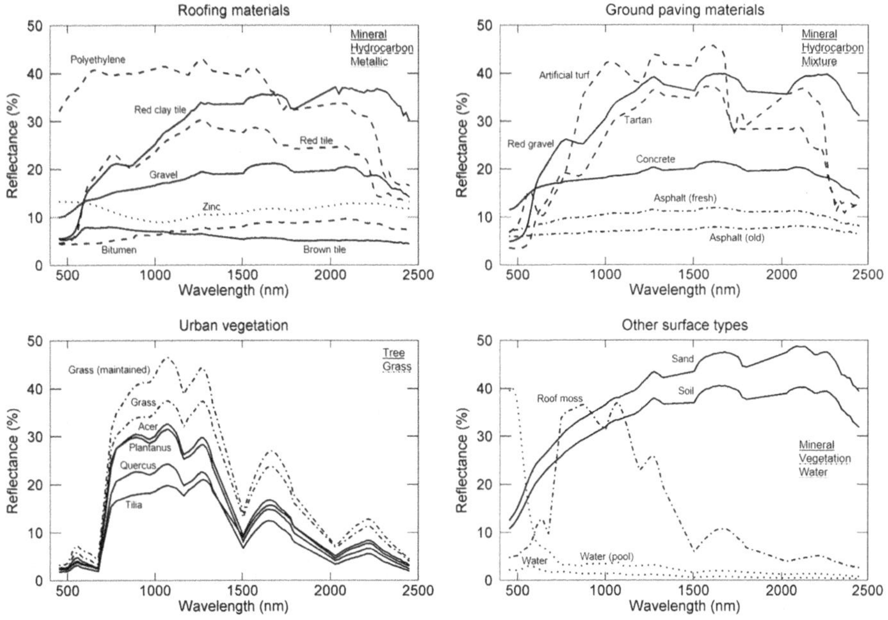


**Fig. 1** Examples for urban surface spectra ordered by land cover type. Plots reveal the spectral heterogeneity within built-up land cover classes and the similarity between them

in the context of future spaceborne imaging spectrometers such as the Environmental Mapping and Analysis Program *EnMAP* (Guanter et al. 2015) or the Hyperspectral Infrared Imager *HyspIRI* (Hochberg et al. 2015).

## 2 State-of-the-Art in Imaging Spectroscopy of Urban Environments

Imaging spectroscopy has been shown to be a powerful tool for remote sensing of urban environments in many studies (Herold et al. 2007; Roberts et al. 2012). Its high spectral information content overcomes the challenges caused by spectral ambiguities in multispectral data to a certain extent (Herold et al. 2004) and allows analysis at different levels of thematic detail (Fig. 2). Airborne imaging spectroscopy data have long been used for detailed urban mapping up to the material level (e.g., Heiden et al. 2007; Herold et al. 2003). However, airborne observations are mostly constrained in spatial coverage (i.e., usually only parts of cities) and temporal frequency (i.e., irregular and usually not multiple annual observations), which so far has led to a limited use in city- to regional-scale urban applications such as urban ecosystem management or urban planning. Relatively few studies include satellite-based imaging spectroscopy data (e.g., Demarchi et al. 2012a; Fan and Deng 2014; Licciardi and Del Frate 2011; Weng et al. 2008) or simulations of future spaceborne systems (e.g., Okujeni et al. 2015; Roberts et al. 2012; Rosentreter et al. 2017), which lack the spatial detail needed for many applications, especially in the context of urban planning. Once upcoming spaceborne sensors provide high-quality full-range imaging spectroscopy data with large area coverage and high temporal frequency, the number of urban environmental applications with imaging spectroscopy data that do not require very high spatial detail is expected to increase, for example, the mapping of physical composition for urban climate models.

The decrease in spatial resolution, when moving from air- to spaceborne observations, leads to the aggregation of different surface types into larger pixels (Fig. 3) and, hence, to severe mixing of spectral signals (Fig. 4). The number of pure spectral signals that mark the vertices of the spectral feature space is quickly reduced when pixels from different surface types are aggregated. At 30 m spatial resolution, only few pure clusters remain along the edges of the typical triangular mixing space described by Small (2005), while the majority of pixels is located in the center of the spectral feature space. With increasing spectral mixing, the distinct spectral characteristics of unique urban materials are lost, which makes classification at the urban material level

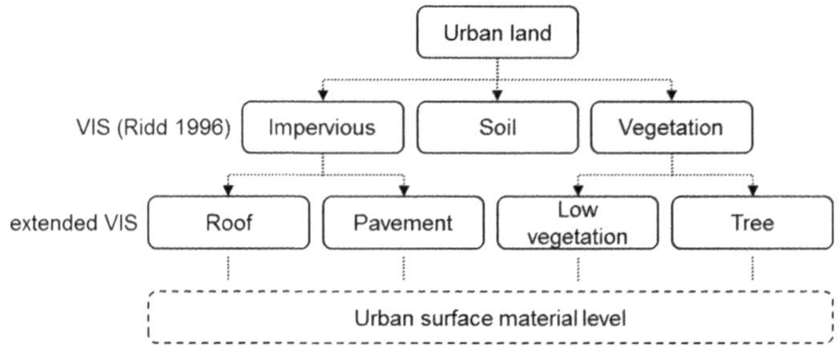


**Fig. 2** Hierarchical land cover categorization scheme adopted for urban mapping. VIS refers to a vegetation-impervious surface-soil (V-I-S) model. From Heiden et al. (2007, modified)

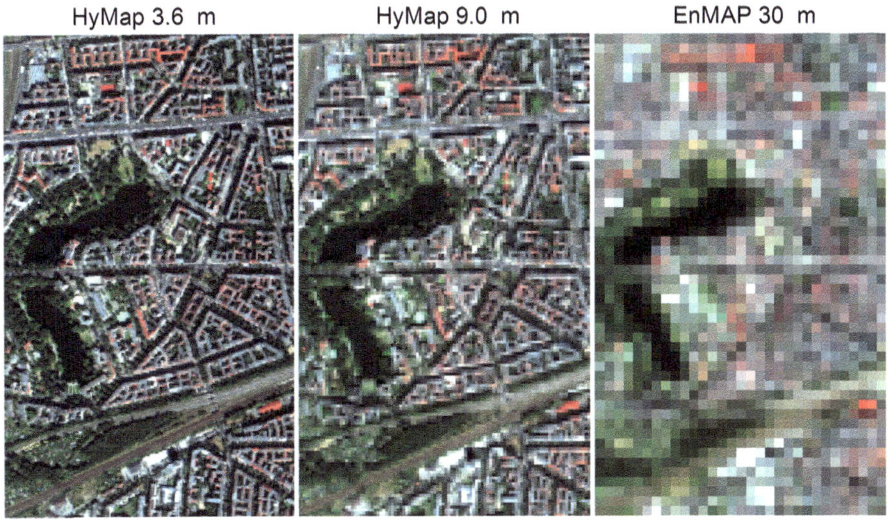


**Fig. 3** True color images from airborne Hyperspectral Mapper (HyMap) data at 3.6 and 9.0 m and from simulated Environmental Mapping and Analysis Program (EnMAP) data at 30 m spatial resolution for Berlin, Germany, showing the loss of spatial and spectral detail caused by spatial aggregation of urban structures

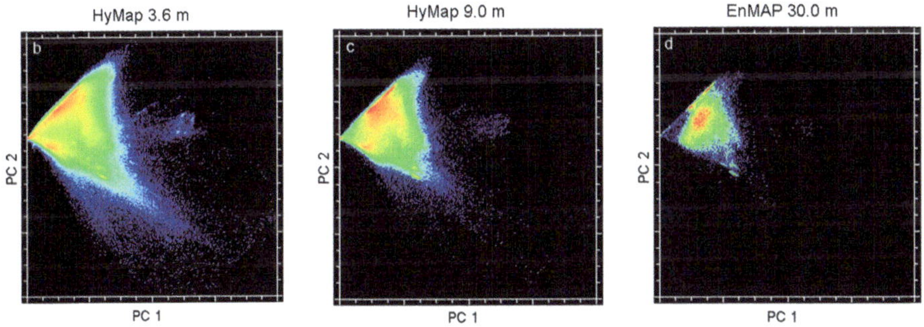


**Fig. 4** Two-dimensional spectral feature space along first and second principal components of data from HyMap at 3.6 and 9 m spatial resolution and from simulated EnMAP data at 30 m. Density is displayed as color range from dark blue over green and yellow to red; scaling is irrelevant, yet identical. Details see text

unrealistic. Urban mapping at coarser resolutions therefore rather focuses on delineating the three main urban cover types vegetation, impervious and soil (VIS) (Ridd 1995), or extended variants of the VIS framework (Fig. 2; e.g., Okujeni et al. 2015; Roberts et al. 2012; Weng et al. 2008). Due to the significant influence of spatial resolution on the level of thematic detail and accuracy of existing studies, the remainder of this section will be structured along the spatial scale, starting with the laboratory and field (indefinitely high spatial resolution) through to the airborne (very high to high spatial resolution) and spaceborne (medium spatial resolution) scale. For each spatial scale, presented literature is organized along the type of application, starting with qualitative analysis, that is, discrete classification into land cover, material or vegetation types, leading over to quantitative analysis, that is, mapping of cover type fractions, and ending with the description of biophysical parameters, if existent at that scale.

*Field and laboratory spectroscopy* allows for the most detailed spectral measurements of urban materials or urban vegetation, given the very high number of spectral bands, for example, 2150 in the case of the full-range Analytical Spectral Devices (ASD) spectroradiometer, indefinitely high spatial detail, and (at least partially) standardized illumination. Multiple spectral measurements from the field or laboratory are usually collected in spectral libraries and analyzed for characteristics that have been acquired in parallel to the measurement, for example, material composition or biophysical variables. However, only few results from urban areas using field- or laboratory-based spectral libraries are reported in the literature. Herold et al. (2004) developed a comprehensive field spectral library of more than 4500 individual spectra for the vicinity of Santa Barbara, USA. In a systematic statistical analysis, they revealed opportunities and limitations for separating urban land cover types and for identifying important wavelength regions for urban remote sensing. Ben-Dor et al. (2001) developed an urban spectral library from a collective spectral database (Price 1995) to explore the visible and near-infrared spectral characteristics of urban cover types typical for Tel-Aviv, Israel. In a follow-up study, the use of the entire visible to shortwave-infrared region based on a spectral library obtained from in situ measurements for the recognition of urban materials was illustrated (Ben-Dor 2001).

Pure image spectra that have been extracted from spectrally homogeneous regions in high spatial resolution airborne imaging spectroscopy data constitute an alternative means to develop or complement urban spectral libraries. For example, Heiden et al. (2007) developed a comprehensive field and image spectral library of more than 21,000 urban materials representative for many German cities. They explored this library to identify unique material-specific spectral features that are robust against spectral overlap between material classes and within-class variability. Findings obtained from field, laboratory or high spatial resolution image spectral libraries mainly contribute to an improved understanding of general spectral properties of urban materials, including their brightness (albedo), slope and shape of the broad continuum, absorption features or separability.

Beyond the description of urban materials, spectral measurements are also used to describe vegetation properties and the relevance of very high spectral detail has, for example, been shown in the context of mapping urban air pollution based on leaf chlorophyll content and fluorescence yield of urban street trees (Brackx et al. 2017; Van Wittenberghe et al. 2013). Even though the mixed pixel challenge is not tackled at this scale, insights into opportunities and limitations of imaging spectroscopy for urban mapping at airborne and spaceborne scales are provided.

*With airborne imaging spectroscopy at* 1–5 m *spatial resolution*, classification of urban land cover has been shown to be highly accurate (e.g., Demarchi et al. 2014; Roessner et al. 2001). The level of thematic detail in classification approaches (compare Fig. 2) is related to the subsequent analyses and reaches from extended variants of VIS (e.g., Roessner et al. 2001; van der Linden et al. 2007), to more material-oriented mapping approaches (Franke et al. 2009; Herold et al. 2003). Priem and Canters (2016) used airborne prism experiment (APEX) data of 2 m spatial resolution and 252 spectral bands for mapping the city center of Brussels, Belgium, with 27 material and land cover classes (Fig. 5). They applied kernel-based support vector machines from the field of machine learning and derived a map of 81% overall accuracy (Fig. 5c). They could reliably separate distinct types of roofs (ceramic tiles, shingle, fiber cement), pavement (asphalt, concrete, cobble stone), soil, trees and grass. The strength of machine learning approaches like support vector classification for the analysis of urban imaging

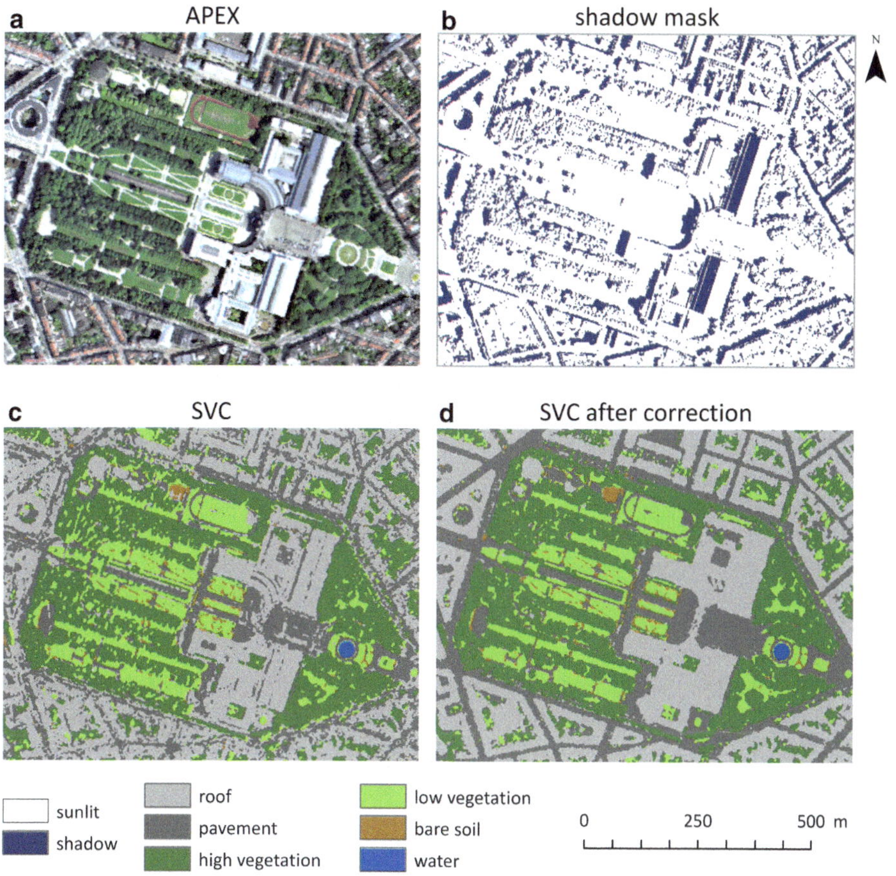


**Fig. 5** True color red/green/blue (RGB) visualization of APEX data from Brussels, Belgium (**a**), shadow mask representing the digital surface model information used for post-classification correction (**b**) and the results from support vector classification (SVC) without (**c**) and with (**d**) inclusion of the digital surface model. From Priem and Canters (2016)

spectroscopy data at very high spatial resolution is also shown in studies aiming at extended VIS classes that include multimodal spectral distributions, typical for urban environments (e.g., van der Linden et al. 2007). An alternative to machine learning classification is classification based on unmixing approaches (Franke et al. 2009) or combinations of unmixing and statistical classification (Roessner et al. 2001). Despite the high level of thematic detail and accuracy that is achieved with very high spatial and spectral resolution imaging spectroscopy data and powerful classifiers, some typical urban challenges like spectral ambiguity and low signal in shadowed areas remain. Priem and Canters (2016), for example, further improved their results from 81 to 88% overall accuracy over Brussels by post-processing of their APEX-based map using information from high-resolution lidar data (Fig. 5d). Studies focusing on classifying vegetation cover types in urban areas come to similar conclusions: very high-resolution airborne visible/infrared imaging spectrometer (AVIRIS) data alone allowed high overall accuracies when mapping, for example, 15 tree species over Santa Barbara, USA, while tree crown mapping was more reliable when including vertical density measures from full waveform lidar data (Alonzo et al. 2013, 2014).

Even at spatial resolutions below 5 m, a high number of spectral mixings occur and single pixels are often composed of more than one spectral *endmember*, that is, a spectral signal that is considered as being the *pure* representation of a cover type at the thematic level of the given research. Wetherley et al. (2017), for example, stated that about half of the pixels in a 4-m resolution AVIRIS Next-Generation (NG) image consisted of two or more endmembers according to model selection in a multiple endmember spectral mixture analysis (MESMA). This finding is confirmed by a multi-scale analysis at the thematic level of extended VIS that was synthesized in the context of this review from existing studies: Approximately 45% of the pixels in 3.6-m resolution data are composed of more than two of the extended VIS land cover types according to detailed reference maps (Fig. 6). The share of mixed pixels and the number of land cover types mixed in single pixels increases with coarser spatial resolution.

To effectively cope with the spectral mixing without information loss, quantitative mapping approaches, that is, approaches estimating land cover fractions within pixels, are used to estimate sub-pixel fractions of urban land cover at the pixel scale. Okujeni et al. (2013) showed that the high information content of airborne imaging spectroscopy data allowed for extended VIS fraction mapping, that is, impervious, the *I*, into roofs and paved surfaces and vegetation, the *V*, into trees and low vegetation, in a support vector regression approach that used synthetically mixed training data from a 3.6-m HyMap data image endmember spectral library. Similar results for an extended VIS framework are reported by Degerickx et al. (2017) on the same data set using MESMA with different endmember selection techniques and Wetherley et al. (2017) using AVIRIS NG data at 4 m with MESMA, as well as by Chen et al. (2017) using reflective optics system imaging spectrometer (ROSIS) data at 1.3 m covering the visible to near-infrared range and a more recent sparse spectral unmixing approach.

Beyond mapping and quantifying of urban land cover, the information in high spatial and spectral resolution airborne data can be used to characterize the state and condition of urban green. Degerickx et al. (2018) employ 2-m resolution APEX data in combination with airborne lidar data to retrieve the chlorophyll content and leaf area index of individual urban trees in the city of Brussels, Belgium, using partial least squares regression. By combining these outcomes into an objective urban tree health assessment, which is shown to

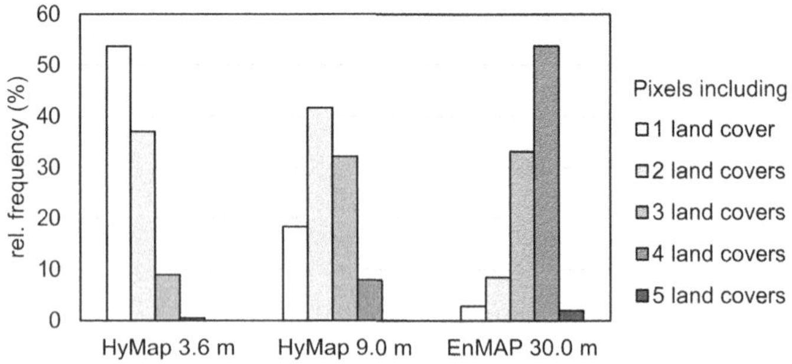


**Fig. 6** Relative frequencies of pixels including different numbers of discrete land cover types in HyMap imagery at 3.6 and 9 m spatial resolution and a simulated EnMAP image at 30 m resolution. High-resolution land cover reference information included the extended VIS classes roof, pavement, low vegetation, trees and soil. Based on data from Okujeni et al. (2013, 2014, 2015)

match the results from a traditional visual tree health assessment, the authors clearly illustrate the potential of imaging spectroscopy for urban tree health monitoring.

*With airborne imagery at coarser spatial resolution of* > 5 m *ground sampling distance*, classification into discrete classes has hardly been used in the urban context and quantitative mapping techniques are becoming even more important. Several authors report on the strength of imaging spectroscopy data in this context. Wetherley et al. (2017) compared results derived from AVIRIS NG at 4 m and from AVIRIS Classic at 18 m using MESMA with different image endmember library variants. The two AVIRIS sensors were flown at different altitudes and could, hence, be processed at native resolutions. For their best fraction estimates, the authors reported similar accuracies for the green vegetation, impervious and pervious (mean $R^2$ of 0.933 at 4 m and 0.900 at 18 m). For spectrally more similar cover types in the *extended VIS* model, that is, turf grass, tree, paved, roof, non-photosynthetic vegetation and soil, they discovered a decrease in accuracies (mean $R^2$ of 0.837 at 4 m and 0.760 at 18 m). Despite this decrease, these findings indicate the possibility to work with more detailed classes when adopting coarser resolution data. Okujeni et al. (2014) transferred their findings from 3.6 m data to 9 m and reported high accuracies for the work with synthetic training data when combined with support vector regression, but also other machine learning approaches. More detailed maps on urban vegetation also require quantitative mapping frameworks. Gu et al. (2015), for example, used non-metric multidimensional scaling ordination (NMDS) to map canopy traits in 11.8-m AVIRIS data. Based on these trait maps, they predicted dominant species composition of urban forest patches. However, studies at spatial scales > 5 m do not achieve the accuracies of studies with higher spatial resolution (Delegido et al. 2014) or studies including structural information from, for example, lidar sensors (Alonzo et al. 2015).

Nowadays, *spaceborne imaging spectrometers* can acquire data at minimal ground sampling distances of 30 m, as will be the case for EnMAP or HyspIRI (Guanter et al. 2015). So far, only few studies in urban environments have used Hyperion imagery, which comes closest to the characteristics of future systems but lacks sufficient radiometric quality in the shortwave-infrared (Kruse et al. 2003). All these studies aim for land cover fractions instead of discrete classification. Weng et al. (2008) extracted the fraction of impervious surface from Hyperion data with an RMSE of about 17% using a VIS approach that separated high and low albedo impervious surfaces as suggested by Wu and Murray (2003) based on multispectral data. Fan and Deng (2014) achieved similar accuracies for impervious surface fractions, while separating tree and non-woody vegetation in an enhanced MESMA approach. The spaceborne compact high-resolution imaging spectrometer (CHRIS) only covers the wavelength range from 410 to 1050 nm and can therefore achieve 18-m ground instantaneous field of view. It is capable of tilting to derive quasi-simultaneous image acquisitions at multiple viewing angles and has been used by various authors in an urban context (e.g., Demarchi et al. 2012a; Duca and Del Frate 2008; Licciardi and Del Frate 2011). For example, Duca and Del Frate (2008) used multi-angular acquisitions from the CHRIS instrument and improved classification accuracies of an extended VIS scheme (dark asphalt, bright asphalt, buildings, among other cover types) by almost 7% compared to a single nadir acquisition. Demarchi et al. (2012b) also showed an improvement in the accuracy of quantitative mapping of imperviousness from CHRIS data when using multi-angle instead of single, close-to-nadir imagery.

The limitations of existing spaceborne imaging spectrometers have led to an increased use of data from end-to-end sensor simulations in the context of mission preparations for EnMAP or HyspIRI (Roberts et al. 2012; Segl et al. 2012). During end-to-end simulation, an at-sensor raw digital number (DN) image is modeled for an area with known reflectance factors, assuming realistic solar illumination and atmospheric conditions. The reflectance factors may stem from high spectral and spatial airborne imagery or from forward modeling of, for example, vegetation canopies. The full radiative path from Sun to the geographic location of the area and to the sensor is simulated, followed by a simulation of the signal's modulations within the sensor and during recording. The resulting raw DN product is afterward preprocessed to level 2a using the standard ground segment workflow. Several authors have worked on urban environments with future Earth observation imaging spectroscopy data from such simulations. Okujeni et al. (2015) used simulated EnMAP data for mapping extended VIS components. The study proved both the capability of imaging spectrometry data to quantify spectrally similar cover types like roofs and pavements, or low and high vegetation at 30 m and the superiority of the higher spectral resolution in comparison to multispectral Landsat data. This is in line with results from Roberts et al. (2012) using simulated HyspIRI data with MESMA and from Rosentreter et al. (2017) who used simulated EnMAP data and a multi-class SVR. They all conclude that mapping class fractions at the thematic detail of extended VIS from spaceborne imaging spectroscopy is worthwhile.

## 3 User and Observational Requirements

The literature overview in the previous section clearly shows that imaging spectroscopy data is mainly used for the identification of materials and the estimation of their associated land cover. Based on the reported state-of-the-art, the following user and observational requirements can be derived.

The requirements in terms of *temporal resolution* strongly depend on the thematic classes to be studied. On the one hand, individual images are usually sufficient to map anthropogenic surfaces. Ideally, the acquisition date and time minimize the share of shadow and the degree of reflectance anisotropies. The combination of leaf-on and leaf-off data from the same year, on the other hand, can help quantifying deciduous tree cover or surface cover underneath canopies. The advantage of multiple images per year is even higher when mapping tree species: information from various dates to better describe phenology stages will increase separability and improve characterization of vegetation structure and vigor (e.g., using multispectral data refer to Pontius et al. 2017; Tigges et al. 2013; Wirion et al. 2017). Denser time series of acquisitions may help characterize phenology and the influences of, for example, neighboring artificial surfaces on microclimate models.

*Radiometric resolution* requirements are not as high as, for example, in aquatic applications (see Giardino et al. 2018, this issue). However, the analysis of images with a high amount of shadow may benefit from high radiometric resolution and better signal-to-noise ratios in order to better differentiate land cover in shaded areas (van der Linden and Hostert 2009).

Looking at *spectral resolution* requirements, the benefits of the high spectral information content have been demonstrated in many studies. Particularly, when classification schemes aim for a high level of detail, such as material composition, additional spectral information leads to significantly higher accuracies at all spatial scales (Herold et al. 2003, Okujeni et al. 2015). The description of vegetation characteristics in urban environments also benefits from high spectral resolution (Alonzo et al. 2014; Delegido et al. 2014; Launeau et al. 2017).

To acquire data with very high spectral resolution from space, however, the *spatial resolution* must be decreased. Maps on urban material composition at more thematic detail than extended VIS schemes appear unrealistic with the near future spaceborne imaging spectrometers at 30 m spatial resolution, due to the high degree of spectral mixing (Fig. 4), while the level of detail as in extended VIS schemes appears feasible to differentiate (Okujeni et al. 2015; Roberts et al. 2012). To summarize the influence of spatial scale as reported in Sect. 2, results from existing studies have been collected and reanalyzed in the context of the present review (data from Okujeni et al. 2013, 2014, 2015). The deficiencies and potential of mapping extended VIS components at scales reaching from 3.6-m spatial resolution airborne data to (simulated) spaceborne resolution are illustrated in Figs. 7 and 8. Despite the significant increase in spectral mixing and loss in spatial detail, Fig. 7 shows general spatial patterns of the extended VIS fraction estimates as derived by support vector regression models that were trained with synthetic library mixtures are well preserved across scales (Fig. 7). Roof, low vegetation, trees and soil fractions can be estimated with a slight decrease in model performance when progressing from < 5 to 30 m spatial resolution (Fig. 8). Pavement fractions are inaccurate and underestimated at all scales, while roofs and soils are often overestimated at very low fractions. This observation points to the general limitation of separating paved areas from roofs and soils in optical remote sensing data even when using imaging spectroscopy for urban mapping. Similarly, low vegetation is underestimated, while trees are slightly overestimated across the entire range of values. Both effects relate to spectral ambiguity of materials, high shadow fractions and high frequencies of spectral mixing along small objects or linear features.

Therefore, there is evidence that data from future spaceborne spectrometers will contribute relevant information to studies at 30-m scale and for subsequent analyses, for example, environmental models, at such spatial scales, despite the loss of spatial detail compared to airborne sensors. The final accuracy and level of detail that can be achieved with actual

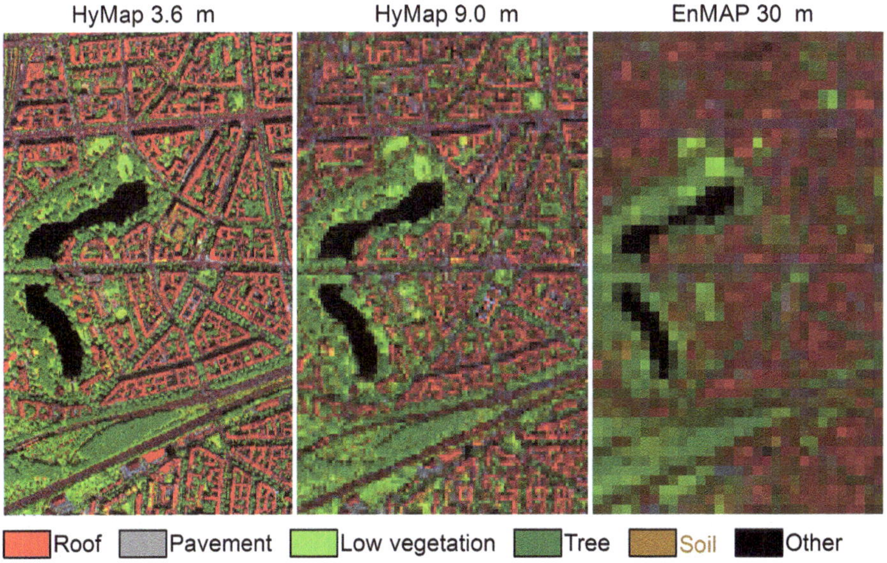


**Fig. 7** Quantitative land cover maps of Berlin using imaging spectroscopy data at different spatial resolutions based on the synthetic unmixing approach by Okujeni et al. (2013). Map colors are weighted averages of class colors. Gray to brown colors, as visible in the 30-m resolution map, depict multi-class mixtures

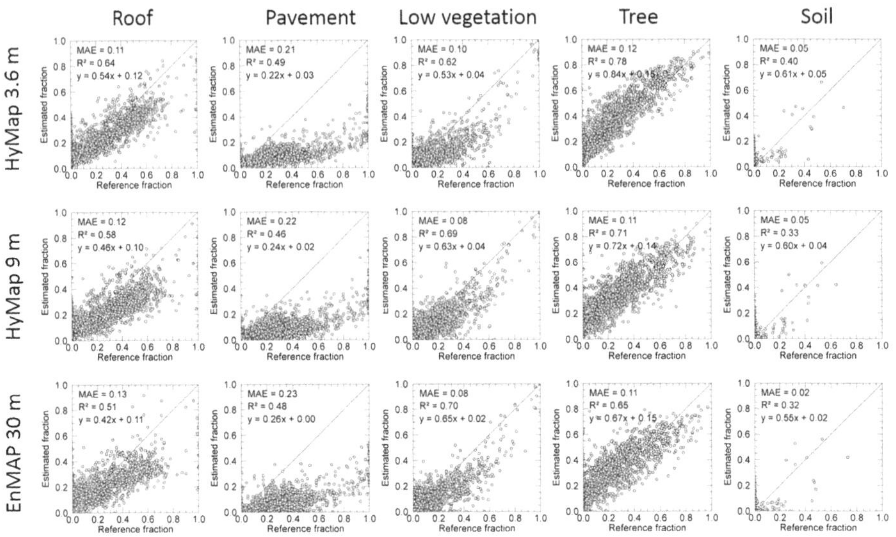


**Fig. 8** Scatter plots and error metrics of fraction estimates for rooftops, paved surfaces, trees, low vegetation and soil based on imaging spectroscopy data at 3.6, 9 and 30 m spatial resolution. Estimates are compared to reference fractions based on very high-resolution vector layers aggregated to the 30-m raster cells of the simulated EnMAP data. Results based on reanalyzed data from Okujeni et al. (2013, 2014, 2017)

data cannot be predicted based on simulated data from end-to-end simulation. The reflectance surfaces underlying such simulations always include artefacts following the complex surface geometry of the urban environment, for example, through object displacement (van der Linden and Hostert 2009), or illumination anisotropies from large view angles in the original airborne data (Schiefer et al. 2006).

In comparison with other environments, the requirements for mapping and monitoring urban areas with Earth observation data may to some extent be similar to those from geological applications (e.g., van der Meer et al. 2012), where material identification and fraction estimates prevail. Only relatively large and homogeneous urban vegetation surfaces, for example, larger forest patches, may require similar approaches as forest mapping outside urban areas (see Hill et al., under review, this issue).

## 4 Challenges and Future Directions

Nowadays, spaceborne multispectral imagery is available from finer than 1 m spatial resolution for the analysis of urban environments. In the near future, imaging spectroscopy data will be acquired at 30 m resolution with sensors such as EnMAP and HyspIRI, as well as the Hyperspectral Imager Suite (Matsunaga et al. 2017) and PRISMA (Guarini et al. 2017). These sensors enable repeated observations almost everywhere on the globe under more standardized acquisition conditions and with operational preprocessing. Still, the selection of data sets by users working in urban environments will always depend on the intended level of spatial detail. Given the technical trade-off in spectral and spatial resolution, spaceborne imaging spectroscopy data alone will hence not automatically become number one choice, for example, by urban planners, due to the small size of most urban objects.

To incorporate the potential of higher spectral resolution, data fusion techniques can be expected to play a pivotal role in fostering the use of such data. For the fusion with secondary optical data, approaches based on Sentinel-2 Multispectral Imager (MSI) data appear promising (Yokoya et al. 2016). The relevance of increased spatial detail in combination with high spectral information content for urban mapping has already been shown using CHRIS data (Demarchi et al. 2012b).

The benefits of including structural information especially from lidar are mentioned in Sect. 2. With the Carnegie Airborne Observatory-2 (Asner et al. 2012), similar multi-sensor constellations have been successfully implemented for airborne observation. Such structural information may be used at the original 30 m or even higher resolutions by downscaling the spectral signal as done by Alonzo et al. (2014) for analyzing single urban trees. With detailed digital surface models being available for many urban areas around the world as well as products from microwave sensors such as TanDEM-X, more fusion approaches become feasible, once imaging spectroscopy is available at higher frequencies. Attempts to achieve this in urban environments with multispectral data exist (Zhu et al. 2012).

For studies aiming at monitoring temporal developments or mapping entire regions as well as for comparative studies on urban areas at continental or even global scale, future spaceborne imaging spectrometers will introduce major advantages with regard to data availability: (1) at least for selected regions, multiple observations per month become possible and (2) data will be globally available under stable sensor characteristics and with standardized preprocessing workflows. These changes offer several opportunities for the work with imaging spectroscopy data. Yet the suite of analysis approaches will have to be adapted at some points, before being able to make ideal use of the new wealth of data. With regard to multi-temporal data sets, for example, time-series analysis has to move beyond the analysis of simple, mostly vegetation-related indices that may also be calculated from multispectral data. Such indices are better derived from instruments such as Sentinel-2 with significantly higher repetition rate and spatial detail. Instead, new spectral indices or feature extractions from waveband regions only present in imaging spectroscopy data are required (Leitão et al. 2015). This way, temporal changes may be incorporated in the analysis of urban areas to observe urban growth, densification or changes in composition.

When the same sensor and preprocessing are used for comparative analysis, the transfer of models, spectral libraries and even expert knowledge, for example, on characteristic spectral features as described in Heiden et al. (2007) will offer new ways for developing biophysically based semiempirical or purely empirical mapping models.

This will, however, require joint activities, for example, the generation of urban spectral libraries including image spectra from cities in all regions that shall be monitored. First attempts in this direction are already being made for foliar traits (e.g., Singh et al. 2015) and may in similar ways be extended for urban areas. The idea of using synthetic mixtures between different surface spectra from spectral libraries (Okujeni et al. 2017) to map surface fractions in imagery of different spectral resolution has been proven useful (Priem et al. 2016, 2017). Similarly, Wetherley et al. (2017) used spectra from images at different spatial resolutions and this way improved results at all scales. In addition, more work is needed to adapt spectral libraries to new image data sets, by pruning techniques (Degerickx et al. 2017) or extension of libraries by automated integration of unknown surface material spectra (Jilge et al. 2017).

Finally, the production of more detailed and accurate maps requires new frameworks that allow several application fields to make use of the same high-quality maps derived from urban imaging spectroscopy data. In the context of land cover mapping, advances were made regarding land cover classification scheme standardization. Most of urban planning activities are based on urban spatial units such as land parcels, urban blocks, urban structure types, urban biotopes or administrative units. However, the nomenclature of such units is highly complex and lack standardization and consistency. This led, for example, to the concept of Functional Urban Areas in the Urban Atlas of the European Copernicus Services (http://land.copernicus.eu/local/urban-atlas). Environmental modelers need more physical-based parameters such as abundance of impervious areas, the building density and the abundance, structure and condition of vegetation. Over the past years, the concept of local climate zone mapping, for example, has become widely recognized even beyond pure climate science. It characterizes regions of uniform surface cover, structure, material and human activity (Bechtel et al. 2015; Stewart and Oke 2012). Based on spaceborne urban surface material mapping, any urban spatial unit can be characterized and thus can support the standardized physical description of cities.

The future availability of spaceborne imaging spectroscopy data will offer room for new analysis approaches of urban environments, yet efforts at several ends are needed for an operational use of the new data.

**Acknowledgements** The authors are grateful to the editors of the special issue and the organizers of the ISSI workshop on *Exploring the Earth's Ecosystems at a Global Scale with Imaging Spectroscopy Data* in Bern, Switzerland, in November 2016, where the present paper was framed. The work of the authors was supported by the German Federal Ministry of Economic Affairs and Energy in the framework of the EnMAP Core Science Team (FKZ 50EE1622) and by the Belgian Science Policy Office in the framework of the Stereo III Project UrbanEARS (SR/00/307).

## References

Adeline KRM, Chen M, Briottet X, Pang SK, Paparoditis N (2013) Shadow detection in very high spatial resolution aerial images: a comparative study. ISPRS J Photogramm Remote Sens 80:21–38. https://doi.org/10.1016/j.isprsjprs.2013.02.003

Alberti M (2005) The effects of urban patterns on ecosystem function. Int Reg Sci Rev 28:168–192

Alonzo M, Roth K, Roberts D (2013) Identifying Santa Barbara's urban tree species from AVIRIS imagery using canonical discriminant analysis. Remote Sens Lett 4:513–521. https://doi.org/10.1080/2150704X.2013.764027

Alonzo M, Bookhagen B, Roberts DA (2014) Urban tree species mapping using hyperspectral and lidar data fusion. Remote Sens Environ 148:70–83. https://doi.org/10.1016/j.rse.2014.03.018

Alonzo M, Bookhagen B, McFadden JP, Sun A, Roberts DA (2015) Mapping urban forest leaf area index with airborne lidar using penetration metrics and allometry. Remote Sens Environ 162:141–153. https://doi.org/10.1016/j.rse.2015.02.025

Asner GP et al (2012) Carnegie Airborne Observatory-2: Increasing science data dimensionality via high-fidelity multi-sensor fusion. Remote Sens Environ 124:454–465. https://doi.org/10.1016/j.rse.2012.06.012

Bechtel B et al (2015) Mapping local climate zones for a worldwide database of the form and function of cities. ISPRS Int J Geoinf 4:199. https://doi.org/10.3390/ijgi4010199

Behling R, Bochow M, Foerster S, Roessner S, Kaufmann H (2015) Automated GIS-based derivation of urban ecological indicators using hyperspectral remote sensing and height information. Ecol Ind 48:218–234. https://doi.org/10.1016/j.ecolind.2014.08.003

Ben-Dor E (2001) Imaging spectrometry for urban applications. In: van der Meer FD, De Jong SM (eds) Imaging spectrometry—basic principles and prospective applications. Remote sensing and digital image processing, vol 4. Springer, Dordrecht, pp 243–281

Ben-Dor E, Levin N, Saaroni H (2001) A spectral based recognition of the urban environment using the visible and near-infrared spectral region (0.4–1.1 μm). A case study over Tel-Aviv, Israel. Int J Remote Sens 22:2193–2218

Brackx M, Van Wittenberghe S, Verhelst J, Scheunders P, Samson R (2017) Hyperspectral leaf reflectance of *Carpinus betulus* L. saplings for urban air quality estimation. Environ Pollut 220:159–167. https://doi.org/10.1016/j.envpol.2016.09.035

Carlson TN, Arthur ST (2000) The impact of land use—land cover changes due to urbanization on surface microclimate and hydrology: a satellite perspective. Glob Planet Change 25:49–65

Chen F, Wang K, Van de Voorde T, Tang TF (2017) Mapping urban land cover from high spatial resolution hyperspectral data: an approach based on simultaneously unmixing similar pixels with jointly sparse spectral mixture analysis. Remote Sens Environ 196:324–342. https://doi.org/10.1016/j.rse.2017.05.014

Damm A (2008) Hyperspektrale Fernerkundung zur Ableitung pflanzenphysiologischer Parameter von Stadtbäumen – Strahlungstransfermodellierung für Berliner Kastanienbestände. Dissertation, Humboldt-Universität zu Berlin

Degerickx J, Okujeni A, Iordache M-D, Hermy M, van der Linden S, Somers B (2017) A novel spectral library pruning technique for spectral unmixing of urban land cover. Remote Sens 9:565

Degerickx J, Roberts D, McFadden JP, Hermy M, Somers B (2018) Urban tree health assessment using airborne hyperspectral and LiDAR imagery. Int J Appl Earth Obs Geoinf 73:26–38

Delegido J et al (2014) Chlorophyll content mapping of urban vegetation in the city of Valencia based on the hyperspectral NAOC index. Ecol Ind 40:34–42. https://doi.org/10.1016/j.ecolind.2014.01.002

Demarchi L, Canters F, Chan JC-W, van de Voorde T (2012a) Multiple endmember unmixing of CHRIS/Proba imagery for mapping impervious surfaces in urban and suburban environments. IEEE Trans Geosci Remote Sens 50:3409–3424. https://doi.org/10.1109/tgrs.2011.2181853

Demarchi L, Chan JC-W, Ma J, Canters F (2012b) Mapping impervious surfaces from superresolution enhanced CHRIS/Proba imagery using multiple endmember unmixing. ISPRS J Photogramm Remote Sens 72:99–112. https://doi.org/10.1016/j.isprsjprs.2012.05.015

Demarchi L, Canters F, Cariou C, Licciardi G, Chan JC-W (2014) Assessing the performance of two unsupervised dimensionality reduction techniques on hyperspectral APEX data for high resolution urban land-cover mapping. ISPRS J Photogramm Remote Sens 87:166–179. https://doi.org/10.1016/j.isprsjprs.2013.10.012

Deng C, Wu C (2013) Examining the impacts of urban biophysical compositions on surface urban heat island: a spectral unmixing and thermal mixing approach. Remote Sens Environ 131:262–274. https://doi.org/10.1016/j.rse.2012.12.020

Duca R, Del Frate F (2008) Hyperspectral and multiangle CHRIS–PROBA images for the generation of land cover maps. IEEE Trans Geosci Remote Sens 46:2857–2866. https://doi.org/10.1109/tgrs.2008.2000741

Fan F, Deng Y (2014) Enhancing endmember selection in multiple endmember spectral mixture analysis (MESMA) for urban impervious surface area mapping using spectral angle and spectral distance parameters. Int J Appl Earth Obs Geoinf 33:290–301. https://doi.org/10.1016/j.jag.2014.06.011

Fenger J (1999) Urban air quality. Atmos Environ 33:4877–4900

Franke J, Roberts DA, Halligan K, Menz G (2009) Hierarchical multiple endmember spectral mixture analysis (MESMA) of hyperspectral imagery for urban environments. Remote Sens Environ 113:1712–1723

Gastellu-Etchegorry J-P et al (2015) Discrete anisotropic radiative transfer (DART 5) for modeling airborne and satellite spectroradiometer and LIDAR acquisitions of natural and urban landscapes. Remote Sens 7:1667

Giardino C et al (2018) Imaging spectrometry of inland and coastal waters: state of the art, achievements and perspectives. Surv Geophy. https://doi.org/10.1007/s10712-018-9476-0

Grimm NB, Faeth SH, Golubiewski NE, Redman CL, Wu JG, Bai XM, Briggs JM (2008) Global change and the ecology of cities. Science 319:756–760. https://doi.org/10.1126/science.1150195

Grimmond S (2007) Urbanization and global environmental change: local effects of urban warming. Geogr J 173:83–88. https://doi.org/10.1111/j.1475-4959.2007.232_3.x

Gu H, Singh A, Townsend PA (2015) Detection of gradients of forest composition in an urban area using imaging spectroscopy. Remote Sens Environ 167:168–180. https://doi.org/10.1016/j.rse.2015.06.010

Guanter L et al (2015) The EnMAP spaceborne imaging spectroscopy mission for Earth observation. Remote Sens 7:8830

Guarini R, Loizzo R, Longo F, Mari S, Scopa T, Varacalli G (2017) Overview of the Prisma space and ground segment and its hyperspectral products. Paper presented at the proceedings of 2017 ieee international geoscience and remote sensing symposium, July 23–28, 2017, Fort Worth, Texas

Harlan SL, Brazel AJ, Prashad L, Stefanov WL, Larsen L (2006) Neighborhood microclimates and vulnerability to heat stress. Soc Sci Med 63:2847–2863. https://doi.org/10.1016/j.socscimed.2006.07.030

Hatt BE, Fletcher TD, Walsh CJ, Taylor SL (2004) The influence of urban density and drainage infrastructure on the concentrations and loads of pollutants in small streams. Environ Manag 34:112–124

Heiden U, Segl K, Roessner S, Kaufmann H (2007) Determination of robust spectral features for identification of urban surface materials in hyperspectral remote sensing data. Remote Sens Environ 111:537–552

Heiden U, Heldens W, Roessner S, Segl K, Esch T, Mueller A (2012) Urban structure type characterization using hyperspectral remote sensing and height information. Landsc Urban Plan 105:361–375. https://doi.org/10.1016/j.landurbplan.2012.01.001

Heldens W, Heiden U, Esch T, Mueller A, Dech S (2017) Integration of remote sensing based surface information into a three-dimensional microclimate model. ISPRS J Photogramm Remote Sens 125:106–124. https://doi.org/10.1016/j.isprsjprs.2017.01.009

Herold M, Roberts D (2005) Spectral characteristics of asphalt road aging and deterioration: implications for remote-sensing applications. Appl Opt 44:4327–4334

Herold M, Gardner ME, Roberts DA (2003) Spectral resolution requirements for mapping urban areas. IEEE Trans Geosci Remote Sens 41:1907–1919

Herold M, Roberts DA, Gardner ME, Dennison PE (2004) Spectrometry for urban area remote sensing—development and analysis of a spectral library from 350 to 2400 nm. Remote Sens Environ 91:304–319

Herold M, Schiefer S, Hostert P, Roberts DA (2007) Applying imaging spectrometry in urban areas. In: Quattrochi DA, Weng QH (eds) Urban remote sensing. CRC Press Inc., Boca Raton, pp 137–161

Hill J, Buddenbaum H, Townsend PA Imaging spectroscopy of forest ecosystems. Surv Geophys (under review)

Hochberg EJ, Roberts DA, Dennison PE, Hulley GC (2015) Special issue on the hyperspectral infrared imager (HyspIRI): emerging science in terrestrial and aquatic ecology, radiation balance and hazards. Remote Sens Environ 167:1–5. https://doi.org/10.1016/j.rse.2015.06.011

Huang Y, Yu B, Zhou J, Hu C, Tan W, Hu Z, Wu J (2013) Toward automatic estimation of urban green volume using airborne LiDAR data and high resolution remote sensing images. Front Earth Sci 7:43–54. https://doi.org/10.1007/s11707-012-0339-6

Imhoff ML, Zhang P, Wolfe RE, Bounoua L (2010) Remote sensing of the urban heat island effect across biomes in the continental USA. Remote Sens Environ 114:504–513. https://doi.org/10.1016/j.rse.2009.10.008

Jilge M, Heiden U, Habermeyer M, Mende A, Juergens C (2017) Detecting unknown artificial urban surface materials based on spectral dissimilarity analysis. Sensors 17:1826

Kennedy C et al (2009) Greenhouse gas emissions from global cities. Environ Sci Technol 43:7297–7302. https://doi.org/10.1021/es900213p

Kruse FA, Boardman JW, Huntington JF (2003) Comparison of airborne hyperspectral data and EO-1 Hyperion for mineral mapping. IEEE Trans Geosci Remote Sens 41:1388–1400. https://doi.org/10.1109/tgrs.2003.812908

Lambin EF et al (2001) The causes of land-use and land-cover change: moving beyond the myths. Glob Environ Change 11:261–269. https://doi.org/10.1016/S0959-3780(01)00007-3

Launeau P et al (2017) Airborne hyperspectral mapping of trees in an urban area. Int J Remote Sens 38:1277–1311. https://doi.org/10.1080/01431161.2017.1285080

Lawler JJ et al (2014) Projected land-use change impacts on ecosystem services in the United States. Proc Natl Acad Sci 111:7492–7497. https://doi.org/10.1073/pnas.1405557111

Leitão P, Schwieder M, Suess S, Okujeni A, Galvão L, van der Linden S, Hostert P (2015) Monitoring natural ecosystem and ecological gradients: perspectives with EnMAP. Remote Sens 7:13098

Licciardi GA, Del Frate F (2011) Pixel unmixing in hyperspectral data by means of neural networks. IEEE Trans Geosci Remote Sens 49:4163–4172. https://doi.org/10.1109/tgrs.2011.2160950

Matsunaga T et al (2017) Current status of hyperspectral imager suite (HISUI) onboard International Space Station (ISS). Paper presented at the proceedings of 2017 IEEE international geoscience and remote sensing symposium, July 23–28, 2017, Fort Worth, Texas

Ngie A, Abutaleb K, Ahmed F, Darwish A, Ahmed M (2014) Assessment of urban heat island using satellite remotely sensed imagery: a review. S Afr Geogr J 96:198–214. https://doi.org/10.1080/03736245.2014.924864

Oke TR (1982) The energetic basis of the urban heat island. Q J R Meteorol Soc 108:1–24. https://doi.org/10.1002/qj.49710845502

Okujeni A, van der Linden S, Tits L, Somers B, Hostert P (2013) Support vector regression and synthetically mixed training data for quantifying urban land cover. Remote Sens Environ 137:184–197. https://doi.org/10.1016/j.rse.2013.06.007

Okujeni A, van der Linden S, Jakimow B, Rabe A, Verrelst J, Hostert P (2014) A comparison of advanced regression algorithms for quantifying urban land cover. Remote Sens 6:6324–6346

Okujeni A, van der Linden S, Hostert P (2015) Extending the vegetation–impervious–soil model using simulated EnMAP data and machine learning. Remote Sens Environ 158:69–80. https://doi.org/10.1016/j.rse.2014.11.009

Okujeni A, van der Linden S, Suess S, Hostert P (2017) Ensemble learning from synthetically mixed training data for quantifying urban land cover with support vector regression. IEEE J Sel Top Appl Earth Obs Remote Sens 10:1640–1650. https://doi.org/10.1109/JSTARS.2016.2634859

Pauleit S, Ennos R, Golding Y (2005) Modeling the environmental impacts of urban land use and land cover change—a study in Merseyside. UK Landsc Urban Plan 71:295–310

Pickett STA et al (2011) Urban ecological systems: scientific foundations and a decade of progress. J Environ Manag 92:331–362. https://doi.org/10.1016/j.jenvman.2010.08.022

Pontius J, Hanavan RP, Hallett RA, Cook BD, Corp LA (2017) High spatial resolution spectral unmixing for mapping ash species across a complex urban environment. Remote Sens Environ 199:360–369. https://doi.org/10.1016/j.rse.2017.07.027

Price JC (1995) Examples of high resolution visible to near-infrared reflectance spectra and a standardized collection for remote sensing studies. Int J Remote Sens 16:993–1000. https://doi.org/10.1080/01431169508954459

Priem F, Canters F (2016) Synergistic use of LiDAR and APEX hyperspectral data for high-resolution urban land cover mapping. Remote Sens 8:787

Priem F, Okujeni A, van der Linden S, Canters F (2016) Use of multispectral satellite imagery and hyperspectral endmember libraries for urban land cover mapping at the metropolitan scale. In: SPIE remote sensing. SPIE, p 100080K. https://doi.org/10.1117/12.2240929

Priem F, Canters F, Okujeni A, van der Linden S (2017) Optimizing mixed spectra generation for regression-based unmixing of land cover in urban areas. In: 2017 joint urban remote sensing event (JURSE), 6–8 March 2017, pp 1–4. https://doi.org/10.1109/jurse.2017.7924554

Ridd MK (1995) Exploring a V–I–S (vegetation–impervious surface–soil) model for urban ecosystem analysis through remote-sensing—comparative anatomy for cities. Int J Remote Sens 16:2165–2185

Roberts DA, Quattrochi DA, Hulley GC, Hook SJ, Green RO (2012) Synergies between VSWIR and TIR data for the urban environment: an evaluation of the potential for the hyperspectral infrared imager (HyspIRI) decadal survey mission. Remote Sens Environ 117:83–101. https://doi.org/10.1016/j.rse.2011.07.021

Roessner S, Segl K, Heiden U, Kaufmann H (2001) Automated differentiation of urban surfaces based on airborne hyperspectral imagery. IEEE Trans Geosci Remote Sens 39:1525–1532

Rosentreter J, Hagensieker R, Okujeni A, Roscher R, Wagner PD, Waske B (2017) subpixel mapping of urban areas using EnMAP data and multioutput support vector regression. IEEE J Sel Top Appl Earth Obs Remote Sens 10:1938–1948. https://doi.org/10.1109/JSTARS.2017.2652726

Schiefer S, Hostert P, Damm A (2006) Correcting brightness gradients in hyperspectral data from urban areas. Remote Sens Environ 101:25–37

Segl K et al (2012) EeteS-The EnMAP End-to-End simulation tool. IEEE J Sel Top Appl Earth Obs Remote Sens 5:522–530. https://doi.org/10.1109/JSTARS.2012.2188994

Seto KC, Sanchez-Rodriguez R, Fragkias M (2010) The new geography of contemporary urbanization and the environment. Ann Rev Environ Resour 35:167–194. https://doi.org/10.1146/annurev-environ-100809-125336

Singh A, Serbin SP, McNeil BE, Kingdon CC, Townsend PA (2015) Imaging spectroscopy algorithms for mapping canopy foliar chemical and morphological traits and their uncertainties. Ecol Appl 25:2180–2197

Small C (2001) Estimation of urban vegetation abundance by spectral mixture analysis. Int J Remote Sens 22:1305–1334

Small C (2003) High spatial resolution spectral mixture analysis of urban reflectance. Remote Sens Environ 88:170–186

Small C (2005) A global analysis of urban reflectance. Int J Remote Sens 26:661–681

Small C, Okujeni A, van der Linden S, Waske B (2018) 6.07—remote sensing of urban environments A2—Liang, Shunlin. In: Comprehensive remote sensing. Elsevier, Oxford, pp 96–127. https://doi.org/10.1016/B978-0-12-409548-9.10380-X

Stewart ID, Oke TR (2012) Local climate zones for urban temperature studies. Bull Am Meteorol Soc 93:1879–1900. https://doi.org/10.1175/bams-d-11-00019.1

Tan J et al (2010) The urban heat island and its impact on heat waves and human health in Shanghai. Int J Biometeorol 54:75–84. https://doi.org/10.1007/s00484-009-0256-x

Tigges J, Lakes T, Hostert P (2013) Urban vegetation classification: benefits of multitemporal RapidEye satellite data. Remote Sens Environ 136:66–75. https://doi.org/10.1016/j.rse.2013.05.001

UN (2018) World urbanization prospects. The 2018 revision. https://esa.un.org/unpd/wup/Publications/Files/WUP2018-KeyFacts.pdf. Accessed 7 June 2018
UN-Habitat (2010) State of the world's cities 2010/2011 - cities for all: bridging the urban divide state of the world's cities reports. UN-Habitat, 224 p
van der Linden S, Hostert P (2009) The influence of urban structures on impervious surface maps from airborne hyperspectral data. Remote Sens Environ 113:2298–2305. https://doi.org/10.1016/j.rse.2009.06.004
van der Linden S, Janz A, Waske B, Eiden M, Hostert P (2007) Classifying segmented hyperspectral data from a heterogeneous urban environment using support vector machines. J Appl Remote Sens 1:013543
van der Meer FD et al (2012) Multi- and hyperspectral geologic remote sensing: A review. Int J Appl Earth Obs Geoinf 14:112–128. https://doi.org/10.1016/j.jag.2011.08.002
Van Wittenberghe S et al (2013) Upward and downward solar-induced chlorophyll fluorescence yield indices of four tree species as indicators of traffic pollution in Valencia. Environ Pollut 173:29–37. https://doi.org/10.1016/j.envpol.2012.10.003
Voogt JA, Oke TR (2003) Thermal remote sensing of urban climates. Remote Sens Environ 86:370–384. https://doi.org/10.1016/S0034-4257(03)00079-8
Weng QH (2001) Modeling urban growth effects on surface runoff with the integration of remote sensing and GIS. Environ Manag 28:737–748
Weng Q, Hu X, Lu D (2008) Extracting impervious surfaces from medium spatial resolution multispectral and hyperspectral imagery: a comparison. Int J Remote Sens 29:3209–3232. https://doi.org/10.1080/01431160701469024
Wetherley EB, Roberts DA, McFadden JP (2017) Mapping spectrally similar urban materials at sub-pixel scales. Remote Sens Environ 195:170–183. https://doi.org/10.1016/j.rse.2017.04.013
Wirion C, Bauwens W, Verbeiren B (2017) Location- and time-specific hydrological simulations with multiresolution remote sensing data in urban areas. Remote Sens 9:645
Wu CS, Murray AT (2003) Estimating impervious surface distribution by spectral mixture analysis. Remote Sens Environ 84:493–505
Yokoya N, Chan J, Segl K (2016) Potential of resolution-enhanced hyperspectral data for mineral mapping using simulated EnMAP and Sentinel-2 images. Remote Sens 8:172
Zhou GQ, Chen WR, Kelmelis JA, Zhang DY (2005) A comprehensive study on urban true orthorectification. IEEE Trans Geosci Remote Sens 43:2138–2147
Zhu Z, Woodcock CE, Rogan J, Kellndorfer J (2012) Assessment of spectral, polarimetric, temporal, and spatial dimensions for urban and peri-urban land cover classification using Landsat and SAR data. Remote Sens Environ 117:72–82. https://doi.org/10.1016/j.rse.2011.07.020

Surveys in Geophysics (2019) 40:489–513
https://doi.org/10.1007/s10712-019-09511-5

# Assessing Vegetation Function with Imaging Spectroscopy

J. A. Gamon[1,2] · B. Somers[3] · Z. Malenovský[4,5,6] · E. M. Middleton[6] · U. Rascher[7] · M. E. Schaepman[8,9]

Received: 14 January 2018 / Accepted: 16 January 2019 / Published online: 15 February 2019


## Abstract

Healthy vegetation function supports diverse biological communities and ecosystem processes, and provides crops, forest products, forage, and countless other benefits. Vegetation function can be assessed by examining dynamic processes and by evaluating plant traits, which themselves are dynamic. Using both trait-based and process-based approaches, spectroscopy can assess vegetation function at multiple scales using a variety of sensors and platforms ranging from proximal to airborne and satellite measurements. Since spectroscopic data are defined by the instruments and platforms available, along with their corresponding spatial, temporal and spectral scales, and since these scales may not always match those of the function of interest, consideration of scale is a necessary focus. For a full understanding of vegetation processes, combined (multi-scale) sampling methods using empirical and theoretical approaches are required, along with improved informatics.

**Keywords** Plant traits · Photosynthesis · Gross primary production (GPP) · Spectroscopy · Multi-scale sampling · Solar-induced fluorescence (SIF) · Spectral reflectance · Imaging spectrometry

✉ J. A. Gamon
jgamon@gmail.com

1 Departments of Earth and Atmospheric Sciences & Biological Sciences, University of Alberta, Edmonton, Alberta T6G 2E1, Canada

2 School of Natural Resources, University of Nebraska, Lincoln, NE 68583, USA

3 Department of Earth and Environmental Sciences, Katholieke Universiteit Leuven (KU Leuven), 3001 Leuven, Belgium

4 Surveying and Spatial Sciences Group, School of Technology Environments and Design, University of Tasmania, Private Bag 76, Hobart, TAS 7001, Australia

5 Remote Sensing Department, Global Change Research Institute CAS, Bělidla 986/4a, 60300 Brno, Czech Republic

6 Biospheric Sciences Laboratory, NASA Goddard Space Flight Center, 8800 Greenbelt Rd, Greenbelt, MD 20771, USA

7 Institute of Bio- and Geosciences, IBG-2: Plant Sciences, Forschungszentrum Jülich, Wilhelm-Johnen-Straße, 52425 Jülich, Germany

8 Remote Sensing Laboratories, Department of Geography, University of Zurich, Winterthurerstrasse 190, 8057 Zurich, Switzerland

9 University Research Priority Program on Global Change and Biodiversity, University of Zurich, Winterthurerstrasse 190, 8057 Zurich, Switzerland

## 1 Introduction

Vegetation plays a crucial role in many Earth system processes, including carbon and water cycling, biogeochemistry, and climate regulation. Additionally, plants provide humans with a wide range of important goods and services, ranging from forest products and fodder to food production. Our society and economy depend upon the healthy function of plants as well as the resilience of ecosystems of which they are a part. Remote sensing is playing an increasingly important role in monitoring, understanding, and maintaining these critical functions.

From a remote sensing perspective, vegetation *function* can have several meanings. One definition refers to the detection or quantification of plant *traits*, structural, physiological, biochemical, or phenological attributes that represent plant functional adaptations or reveal underlying plant ecophysiological processes (Kattge et al. 2011). A complementary definition of vegetation function addresses the *dynamics* of essential physiological, or ecological processes themselves. This definition emphasizes *variation* in key processes over time and space and implies that traits themselves can be dynamic.

In this review, we consider both trait- and process-based approaches used to assess plant function, and explore the history, promise and challenge of integrating these approaches in the context of optical remote sensing. The primary focus is on optical measurements, including spectral reflectance and fluorescence sampled with spectrometers, realizing that other types of remote sensing are also useful. Since our ability to detect plant function is scale-dependent, we start with a brief discussion of scale. Next, we review plant functional traits detectable with remote sensing, followed by a consideration of proximal and airborne remote sensing. We then discuss how global satellite remote sensing has been applied to vegetation assessment along with a presentation of new directions in satellite applications. We conclude with current challenges and opportunities, and provide suggestions and directions that are likely to further advance the study of vegetation function.

## 2 The Concept of Scale

Several aspects of scale, including spatial, temporal, spectral, and biological dimensions, are relevant to any discussion of detecting vegetation function with remote sensing. The dimensions of scale are often intertwined. At finer temporal and spatial scales, underlying mechanisms or traits are best revealed, often requiring fine spectral scales. At progressively larger scales, vegetation function represents the operation of many fine-scale properties and processes that may combine in complex ways due to nonlinear responses and intricate feedbacks. Thus, to a large extent, sampling scale determines which processes are detectable and how. Consequently, any remote sensing observation of vegetation function should include a consideration of scale, with attention to underlying mechanisms at finer scales and resulting emergent properties at larger scales.

Using the example of photosynthesis, a wide range of sampling approaches can be applied according to the particular process of interest and observational scale required (Fig. 1). At the leaf level, ontogenetic and regulatory processes, and corresponding traits, are detectable using proximal spectroscopy and chamber gas exchange measurements (Gamon et al. 1997; Gamon and Surfus 1999). At the canopy level, ranging from individual crowns to more complex plant stands, airborne spectroscopy can be applied to monitor

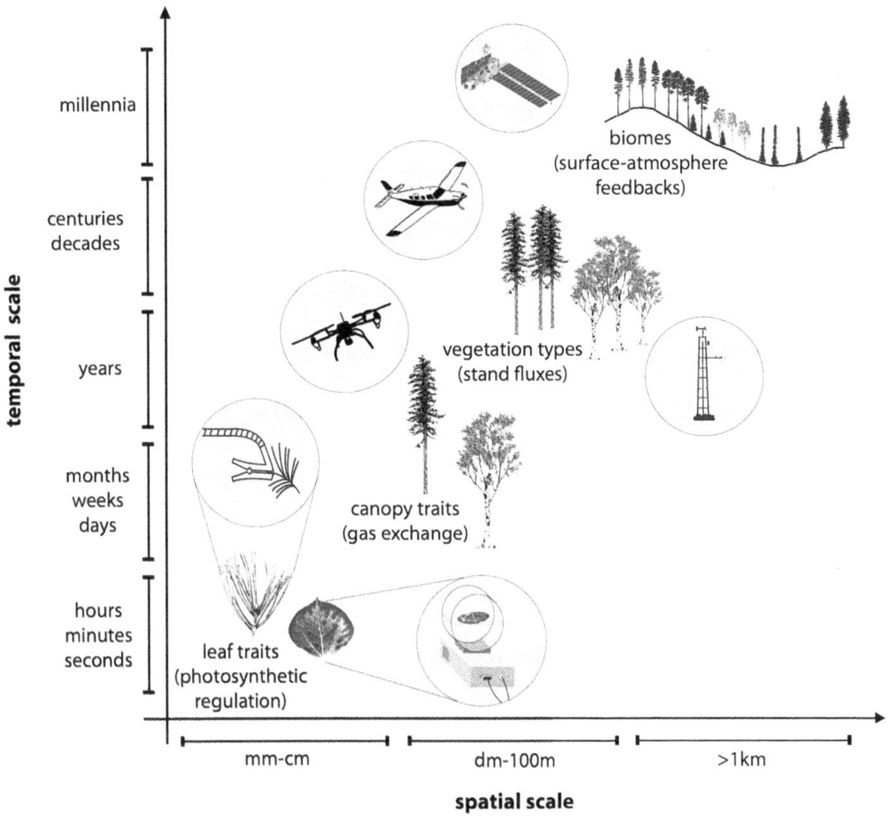


**Fig. 1** Depiction of plant traits (and function) at different spatial and temporal scales. In this example of photosynthetic carbon uptake for evergreen and deciduous trees, relevant traits, corresponding functional processes, and measurement methods and terminology vary with scale (as described in the text) as well as with vegetation type (e.g., evergreen or deciduous) (Table 1). Several measurement approaches are indicated, showing spectroscopic methods in the upper left circles, and corresponding gas exchange methods in the lower right circles. Spatial and temporal scales indicated on the axes are approximate, depicting *relative* rather than absolute values, and refer to the processes and traits, not the measurements, which often do not match the spatial or temporal scales of the relevant processes and traits depicted, leading to scale mismatches

ecosystem photosynthesis and evapotranspiration (Fuentes et al. 2006), often measured independently using eddy covariance. At stand to biome scales, airborne or satellite remote sensing methods can detect changing vegetation productivity using optical or thermal properties, informing us of larger implications, such as altered vegetation productivity resulting from disease (Zarco-Tejada et al. 2018), plant invasion (Asner and Vitousek 2005), or widespread disturbance and climate change (Goetz et al. 2005). At stand to global scales, remote sensing can inform us of vegetation biophysical properties needed to understand emergent processes, such as the influence of surface–atmosphere mass and energy exchange on our atmosphere and climate (Bonan et al. 2002, 2003).

The topic of *how to scale*, or how we understand complex processes across scales, has been well discussed from a variety of perspectives (Levin 1992; Ehleringer and Field 1993; Quattrochi and Goodchild 1997; Enquist et al. 2015), providing useful insights into vegetation functional assessment. A proper study of function should consider processes at multiple scales, typically one level below to one level above the key process being observed (de Wit 1970, as discussed in Jarvis 1993), illustrating the importance of multi-scale sampling strategies (Fig. 1). A sampling strategy should first consider the patch size or length scale (Levin 1992) of the particular process in question. Ideally, remote sensing of vegetation function involves detecting traits (functional attributes) or processes defined at a particular length scale (or patch size) and temporal scale, and matching these scales with the instrument's spatial and temporal resolution and extent. In practice, this is often difficult due to the many sensors, traits, and scales involved and due to the scale mismatch between sensors' technical capabilities and the questions of interest (Fig. 1; Table 1). Due to these technical limitations, the fundamental sampling rules acknowledging effects of scale are often violated in remote sensing. This challenge of scale, often represented as the mismatch between field sampling and remote sensing, has sometimes limited the engagement of terrestrial ecologists or biologists with remote sensing, but is being rectified as the remote sensing capabilities and scaling methodologies evolve.

An important point emerging from this discussion is that there is no single, universally applicable scale for functional studies using remote sensing. Because each remote sensing instrument and platform has a defined sampling scale, often constrained by engineering requirements and physical laws, and because the spatial and temporal resolutions of different remote sensing methods do not always match the operational scale of the vegetation process of interest, multiple platforms and sampling scales should be considered, and additional methods are often needed. These methods may involve "upscaling," the extrapolation from small to large scales through biophysical models or statistical methods (Jung et al. 2011), or "downscaling," the inversion of models or interpretation of large-scale patterns in terms of underlying fine-scale processes (Jacquemoud et al. 2009). Additionally, a nested sampling protocol (Fig. 1), rather than a single remote sensing platform or approach, can improve our understanding of function, as is often used in remote sensing field campaigns (e.g., Sellers et al. 1997). With the proper attention to scale a remote sensing campaign can reveal information regarding underlying mechanisms and emergent properties associated with a particular vegetation function. How and what we detect with remote sensing, including the "best" sampling strategy, can depend on the specific question or hypothesis to be tested, the observer's disciplinary background, the technology at hand, and the realism of the scaling model with respect to the relevant physiological and biophysical processes. Challenges in scaling imaging spectroscopy observations from leaves to vegetation canopies, including topics of natural variability and measurement uncertainty, are further described by Malenovský et al. (2019) (this issue).

## 3 Vegetation Functional Traits

Spectral reflectance offers a powerful alternative to traditional methods of understanding vegetation function because it provides a rapid, consistent, repeatable, nondestructive, and objective sampling method to assess plant traits (functional attributes) from optical properties (Gamon and Surfus 1999). These traits provide information on plant structure, water, and biochemical content (Fig. 2). Consequently, one way to assess vegetation function with remote sensing relies on the concept of plant traits, which can be described at a range of scales (Table 1). Most leaf and plant traits influence vegetation optical properties either directly or indirectly, allowing functional assessment from spectral reflectance and transmittance (Cavender-Bares et al. 2017).

**Table 1** Examples of plant functional attributes and their relevant scales (leaf to landscape) that can be addressed using spectroscopy along with relevant references

| Functional role (scales) | Traits | References |
|---|---|---|
| Gross primary production and energy balance (Canopy—Landscape) | Leaf area index (LAI) | Street et al. (2007) |
| | Leaf angle distribution (LAD) | Wang et al. (2007) |
| | Fraction of PAR absorbed (fPAR) | Myneni et al. (2002) |
| | Albedo | Ollinger et al. (2008), Williamson et al. (2016) |
| | Surface temperature | Schull et al. (2015) |
| | Canopy structure and greenness (NDVI, EVI, NIRv) | Huete et al. (2002), Badgley et al. (2017) |
| Photosynthetic rate, light-use efficiency (Leaf—Stand) | Nitrogen content | Reich et al. (1998) |
| | Specific leaf area (SLA) | Reich et al. (1998), Wright et al. (2004) |
| | Xanthophyll cycle activity (PRI) | Gamon et al. (2001), Gamon et al. (1997) |
| | Chlorophyll and carotenoid pigment levels (CCI) | Gamon et al. (2016) |
| | Solar-induced chlorophyll fluorescence (SIF) | Porcar-Castell et al. (2014), Drusch et al. (2017) |
| | Maximum carboxylation rate (Vcmax) | Serbin et al. (2012), Serbin et al. (2015) |
| Biogeochemical cycling (Leaf) | Nitrogen content | Singh et al. (2015) |
| | Lignin and cellulose content | Singh et al. (2015) |
| | Leaf water content | Serrano et al. (2000), Sims and Gamon (2003) |
| | Specific leaf area (SLA) | Reich et al. (1998), Wright et al. (2004) |
| Stress indication (Leaf—Stand) | Pigment levels (PRI, CCI) | Gamon et al. (2016) |
| | Leaf area index (LAI) | Battaglia et al. (1998) |
| | Leaf angle distribution (LAD) | Ehleringer and Comstock (1987) |
| | Foliar water content | Serrano et al. (2000), Sims and Gamon (2003) |
| | Maximum carboxylation rate (Vcmax) | Serbin et al. (2012), Serbin et al. (2015) |
| | Albedo | Teuling and Seneviratne (2008) |
| | Surface temperature | Schull et al. (2015) |
| | Solar-induced chlorophyll fluorescence (SIF) | Porcar-Castell et al. (2014) |

At the coarser scales (top of table), these attributes are often termed "vegetation biophysical parameters," whereas at the finer scales (moving down the table), many of these attributes are often considered "leaf traits." In reality, there is some overlap across scales, with these attributes providing a continuum of approaches to assessing function. Examples of several vegetation indices useful for functional assessment are also listed, including *NDVI* normalized difference vegetation index, *EVI* enhanced vegetation index, *NIRv* near-infrared reflectance of terrestrial vegetation, *PRI* photochemical reflectance index, *CCI* chlorophyll/carotenoid index, *SIF* solar-induced fluorescence

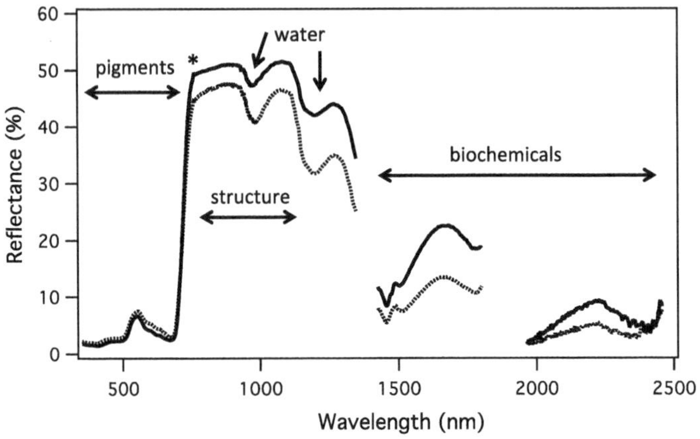


**Fig. 2** Canopy reflectance spectra for deciduous poplar (*Populus balsamifera,* solid line) and evergreen spruce (*Picea glauca, dotted line*) canopies, showing spectral regions with information on pigment content, water content, biochemical content, and leaf or canopy structure. Alternatively, full-spectral information can be used to derive these and other leaf traits, as discussed in the text. Asterisk (*) indicates the location of the $O_2$-A band (fluorescence signal). Data gaps occur in wavelengths of strong atmospheric water absorption. Tree seedlings, sampled June 22, 2016, with a field spectrometer (PSR-3500, Spectral Evolution, Lawrence, MA, USA), in Edmonton, Alberta, Canada

There is a rich literature on plant traits at the leaf scale, much of which is now being related to leaf optical properties via reflectance spectra (e.g., Serbin et al. 2014). Leaf traits are often defined in terms of foliar constituents (e.g., biochemical or water content), or in terms of leaf structure (e.g., surface area, thickness, specific leaf area, or mass per area) (Cornelissen et al. 2003; Kattge et al. 2011; Homolová et al. 2013). Alternatively, leaf traits can be described in more direct functional and dynamic terms, including carboxylation potential, Vcmax (Serbin et al. 2012), photosynthetic rates, or radiation-use efficiency (Gamon et al. 1997). Remotely sensed data do not always map directly onto plant functional traits. Instead, leaf and plant traits can be considered latent variables best represented by transformations or linear combinations of multiple spectral features (Feilhauer et al. 2017). While not all traits are equally retrievable by remote sensing, many hidden traits can be correlated with more accessible traits, facilitating their characterization even when they cannot be directly detected with spectroscopic methods. In these cases, the functional connections between reflectance spectra and functional properties may not yet be entirely clear. Part of the power of spectra for assessing traits is that they provide information from many wavelengths, sometimes into a single metric as represented by a spectral index, or using many wavelengths simultaneously as depicted in a coefficient spectrum (Serbin et al. 2014).

In addition to leaf traits, functionally important vegetation properties can also be defined at larger scales (e.g., canopy traits, vegetation types, and biomes), and these categories can often be viewed as aggregations of underlying leaf or plant traits (van Bodegom et al. 2014). For example, deciduous and evergreen vegetation types tend to have distinct associations of leaf traits that are spectrally discernable (Gamon et al. 1997) (Figs. 1, 2). A clear understanding of canopy and stand structure is needed to understand the "scaleability" of leaf traits, which may depend on the particular trait and context, and can vary with vegetation type. For example, under the right conditions of uniform, closed vegetation stands, xanthophyll cycle pigment activity (an indicator of photosynthetic light-use efficiency) can be directly detected at larger scales (Gamon and Qiu 1999; Stylinski et al. 2002), offering the possibility of stand-scale assessment of photosynthesis using aircraft or satellite remote sensing. However, complex and dynamic canopy structure can readily confound detection of this leaf photoprotective signal (Barton and North 2001; Gitelson et al. 2017). Consequently, accounting for changing Sun angle, view angle, and canopy structure can often improve the interpretation of the underlying physiological signals (Drolet et al. 2008; Hilker et al. 2008; Middleton et al. 2009, 2016; Hilker et al. 2011).

Due to the complexity of canopy structure, some important leaf traits (e.g., nitrogen content) may not always be directly detectable at stand scales (Knyazikhin et al. 2013). However, because foliar nitrogen levels indirectly influence canopy structure, nitrogen levels can often be inferred, even when not directly measured, with remote sensing (Ollinger et al. 2008; Ollinger 2011; Townsend et al. 2013). These examples of leaf traits correlating with higher level traits demonstrate that a proper understanding of the particular context can facilitate the assessment of plant function from remote sensing, even when the underlying leaf traits are not directly detectable at larger scales. Consequently, even at larger scales sampled by aircraft and satellite, the concept of leaf traits can have utility, and can often be expressed in suites of characteristics that collectively define "plant strategies" (Grime 2006) or "optical types" (Ustin and Gamon 2010), both variations on the plant functional type concept. Vegetation functional types are commonly used in many global models of ecosystem function and Earth system science as a way to classify complex vegetation biophysical properties into a few simplified "types," facilitating model simulations (Bonan et al. 2002, 2003; Pavlick et al. 2013; van Bodegom et al. 2014). The concept of vegetation *optical types*, sometimes called *spectral types* or *spectral species* (Feret and Asner 2014), represents an attempt to define functionally significant optical properties (or spectral patterns) related to plant traits. In this way, defining vegetation functional types in terms of optical types (spectral types) can simplify studies of large-scale vegetation processes.

Advances in spectroscopy and statistical techniques have led to reliable methods to assess leaf traits, sometimes rivalling traditional metrics involving structural or wet chemical measurements. Traits can be derived from specific spectral regions using vegetation indices (Table 1; Fig. 2) (e.g. Malenovský et al. 2013). Alternatively, traits can be retrieved from optical measurements using statistical approaches such as partial least squares regression (PLSR, Roelofsen et al. 2014; Serbin et al. 2014). However, these methods have not yet revealed universal relationships. Part of the reason lies in the evolutionary convergence of plant traits; many plant species have evolved similar solutions to the same environmental challenges, and it is often the subtle variation in these features over time or across environmental gradients that can be most informative. Most published studies linking optical properties and plant traits are limited to a single point in time, generally near the center of the growing season (Homolová et al. 2013). Thus, they fail to capture the full variation in conditions or plant traits expressed in the life of an individual organism, canopy, stand, or ecosystem. Not surprisingly, results from one season do not always apply well to datasets collected in another season (Chavana-Bryant et al. 2017). Until we obtain extensive, and uniform, time series of optical data, the full evaluation of plant traits through optical properties will remain a work in progress. Fully understanding the links between spectral space and trait space represents an ongoing challenge, in part due to the large data volume, involving many traits and spectral features along with other independent variables (e.g., time, space, and observation geometry). While progress is being made on the definition and application of optical types (Gamon et al. 1997; Ustin and Gamon 2010; Feret and Asner 2014), we currently lack a comprehensive theory of optical types, or a full understanding of how they relate to plant traits. Opportunities lie in the full exploration of the effects of temporal and spatial scales, a topic more fully explored below (see Sect. 7).

Plant traits can also be related to spectra through forward physical modeling (Schneider et al. 2014) and inversion of leaf or coupled leaf-canopy radiative transfer models at the leaf level (e.g., PROSPECT, Jacquemoud and Baret 1990; Malenovský et al. 2006), or the canopy level using one-dimensional models (e.g., SAIL, Verhoef 1984; or SCOPE, van der Tol et al. 2009) or three-dimensional models (e.g., DART, Gastellu-Etchegorry et al. 2017). For more detailed discussion of such models, see Verrelst et al. (2019) and Malenovský et al. (2019) (both in this volume).

## 4 Proximal Spectrometry

Many of the current advances in spectral reflectance for vegetation function are derived from proximal spectrometers. These include handheld portable spectrometers and ground-based imaging spectrometers (Pinto et al. 2016, 2017), as well as instruments mounted on towers (Hilker et al. 2011), mobile platforms (Gamon et al. 2006; Sims et al. 2006), and unmanned aerial vehicles (UAVs, also known as drones; Burkart et al. 2014; Lucieer et al. 2014; Malenovský et al. 2017). Many of these studies have been coordinated with flux tower observations, facilitating calibrations with independent data and cross-site comparisons (see reviews by Gamon et al. 2010; Balzarolo et al. 2011; Gamon 2015).

A key advantage of proximal sampling has been its ability to provide rich time series of data, which have been difficult to obtain using airborne or satellite platforms. These instruments sample at scales that can be readily related to leaf traits and where process measurements (e.g., carbon or water vapor fluxes, changing pigment content, or dynamic chlorophyll fluorescence) can be independently validated using experimental approaches and coupled measurements at similar scales. They are also particularly well suited to exploring the functional regulatory, ontogenetic and physiological mechanisms that underlie changes in optical properties at larger scales. Hence, proximal instruments provide useful experimental tools to explore new ways to sample vegetation function and to validate optical measurements at larger scales. A disadvantage is their restricted spatial coverage, limited distribution across biomes and vegetation types, and lack of standardization in sensors and protocols, making it difficult to draw general conclusions from individual field studies (Gamon et al. 2010; Gamon 2015). The large number of these studies, combined with improved informatics and statistical or other modeling tools for upscaling, promise to advance our understanding of vegetation function, particularly when combined with aircraft or satellite observations. Proximal sensing can also lead to innovation in sensor technologies and applications that can inform the design and interpretation of future airborne or spaceborne sensors.

## 5 Airborne Remote Sensing

Increasingly, airborne imaging spectrometers are being applied to assess vegetation traits and functional processes. These instruments typically collect measurements in either the "full-range" VSWIR (visible-short-wave infrared) wavelengths or in the more limited VNIR (visible-near-infrared) range. Full-range sensors offer a wider range of options for detecting plant traits, but are more expensive, heavier, and more difficult to maintain and operate. VNIR sensors, while more limited in spectral coverage, are much lighter and less expensive, providing an easier entry point into the world of airborne imaging spectrometry. Depending upon their intended purpose, these instruments range in spectral resolution from 10 nm for typical reflectance applications to ≤ 0.3 nm in the case of airborne spectrometers intended for detection of solar-induced fluorescence (SIF, Rascher et al. 2015; Schaepman et al. 2015; Wieneke et al. 2016). Pixel sizes (IFOV) can approximate 1 $m^2$, a scale fine enough to resolve individual plant canopies in the case of large trees. This combination of fine spatial and spectral resolution is particularly useful for evaluating spatial patterns in plant traits (Asner and Martin 2016).

From the perspective of vegetation dynamics, a primary limitation of airborne platforms has been the difficulty and expense of obtaining regular observations to yield sufficiently large and frequent time series of vegetation function. Typical airborne datasets contain single overpasses, or at best a handful of measurements over a short time span (e.g., several within-day, or "diurnal" acquisitions), which are obtained a few times in a single growing season. Such limited temporal frequency restricts the power for assessing functional processes occurring at different time scales. On the other hand, even single aircraft overpasses can provide useful assessments of spatial patterns in functional traits (Homolová et al. 2014; Asner and Martin 2016) illustrating the benefits of inferring function from trait-based approaches.

Airborne field campaigns with various generations of the Airborne Visible/InfraRed imaging Spectrometer (AVIRIS) sensor, such as the AVIRIS Classic and the AVIRIS-NG, have provided good examples of the power of imaging spectrometry to detect vegetation function. Many of these campaigns have assessed plant function through photosynthesis, water status (Fuentes et al. 2006; Serbin et al. 2015; Somers et al. 2015; Asner et al. 2016a), or nutrient status (Asner and Vitousek 2005). The HyPlant airborne sensor (Rascher et al. 2015), developed in preparation for the forthcoming FLEX satellite mission (Kraft et al. 2013; Drusch et al. 2017), with an expected launch in late 2022, provides very high spectral resolution radiance or reflectance suitable to detect chlorophyll fluorescence and foliar pigment content, both powerful indicators of photosynthetic activity (Rossini et al. 2015; Simmer et al. 2015; Wieneke et al. 2016; Middleton et al. 2017). The Airborne Prism Experiment (APEX), serving as an airborne precursor for a space-based imaging spectrometer mission, demonstrates simultaneous high-resolution spectroscopic retrievals for the atmosphere, the non-vegetated surface, and chlorophyll fluorescence (Schaepman et al. 2015). Numerous airborne campaigns focusing on plant traits have provided promising tests of trait-based approaches to assess vegetation function (Asner et al. 2016b; Schneider et al. 2017). Despite limited revisit time, airborne spectrometry remains a powerful tool for demonstrating the ability of full-spectral information to improve our understanding of vegetation function. Emerging monitoring networks, such as the NEON program (funded by the US National Science Foundation), are beginning to provide airborne imaging spectrometry capability along with proximal spectral measurements across a network of sites in the continental US, offering new opportunities for the systematic exploration of vegetation function across scales for representative ecosystems (Kampe et al. 2010).

# 6 Satellite Remote Sensing

## 6.1 Brief History of Optical Satellites for Vegetation Monitoring

Historically, spatially and temporally extensive series of spectral imagery from space useful for assessing vegetation function have been limited to a few of the over 1000 global orbiting satellites. Most of these have had limited spectral coverage, constraining their utility for vegetation functional studies. Many of the multi-spectral orbiting sensors have been radiometers with a small number (e.g., 2–8) of relatively wide (e.g., ≥ 20–30 nm) spectral bands related to a handful of biophysical traits, some related to green vegetation structure (e.g., LAI and fPAR, Table 1). Long-term repetitive observations have been obtained by the USGS Landsat Thematic Mapper/Enhanced Thematic Mapper (TM/ETM) series, and the NOAA Advanced Very High Resolution Radiometer (AVHRR) series. Despite the fact that the NOAA missions, built primarily for weather observations, were not designed or optimized for vegetation monitoring, they fortuitously have provided important insights into vegetation function over large regions. The discovery that a combination of red and near-infrared wavelengths used together could track seasonality of green vegetation biomass and leaf area index (Rouse et al. 1974; Tucker 1979), has been applied to the interpretation of AVHRR imagery at continental and global scales, providing a revolutionary view of vegetation functional dynamics at continental scales (Tucker et al. 1985), and allowing globally consistent vegetation classifications incorporating seasonal dynamics (DeFries and Townshend 1994).

The MODIS (Moderate Resolution Imaging Spectroradiometer) carried on the Terra (launched in 1999) and Aqua (launched in 2002) satellites have emerged as important sensors for global biospheric monitoring. At its conception in the 1990s, the MODIS data system EOSDIS (the Earth Observing System Data and Information System) was the "largest and most complex data system every constructed" (Marshall 1993). Key to MODIS's success has been the availability of standardized data products, enabling the establishment of a critical terrestrial database spanning many years that has led to scientific advancements in many earth science disciplines. The operational collection and processing of 36 spectral bands across the full shortwave and infrared regions from the Terra and Aqua MODIS have yielded invaluable time series of atmospheric, marine, and terrestrial products. These 36 bands include 20 reflective shortwave and 16 emitted infrared bands. Seven of the reflective bands are used for terrestrial and atmospheric monitoring land discrimination, covering the VSWIR spectral regions, with varying bandwidths (20–50 nm, Fig. 3). Some of these are used to provide a few key land products, including LAI and FPAR that are widely used in studies of vegetation dynamics (Table 1). Additionally, nine narrower (10 nm) "ocean color" bands in the visible spectrum (blue bands, Fig. 3) are also part of the MODIS suite at coarser spatial scales (~ 1 km for nadir views). These ocean bands were not originally intended for vegetation applications, and in earlier MODIS Collections (1–5), they were processed separately from land bands. Consequently, until recently, researchers using the "ocean" bands for terrestrial vegetation studies have had to conduct their own processing and atmospheric corrections, a situation rectified with Collection 6.

Global vegetation products developed from MODIS, including the NDVI, the Enhanced Vegetation Index (EVI; Huete et al. 2002) and related products such as leaf area index (LAI) and the fraction of absorbed photosynthetically active radiation (fPAR, or fAPAR; Myneni et al. 2002), are used to derive gross and net primary production (GPP and NPP) (Table 1). These spectrally based products have been used extensively for assessing regional and global vegetation functional dynamics, including photosynthetic activity and stress or disturbance responses. A key example is the MODIS MOD17 NPP product, based on the light-use efficiency (LUE) model, which provides the first consistent, multi-year depictions of global terrestrial NPP (Running et al. 2004), describing both seasonal phenology and interannual variation in vegetation photosynthetic activity at continental scales, allowing attribution of NPP anomalies to the El Niño–Southern Oscillation (ENSO), drought and other climate factors (Zhao and Running 2010).

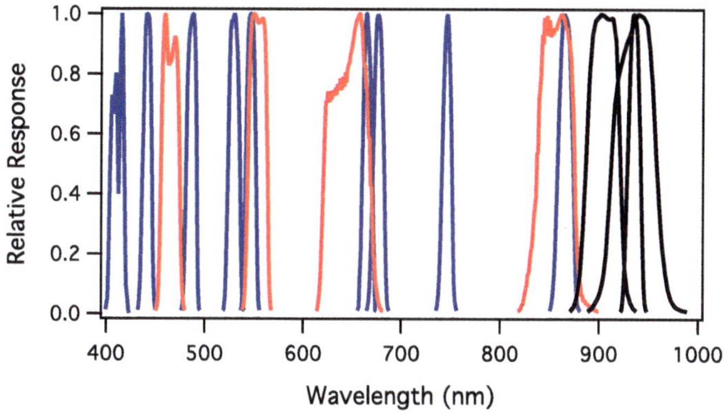


**Fig. 3** Normalized spectral response functions of the MODIS sensors showing ocean bands (blue), land bands (red), and atmospheric bands (black)

New MODIS products for vegetation function continue to emerge. Although MODIS ocean bands were not originally intended for vegetation, the recent MAIAC reprocessing (Lyapustin et al. 2011a, b), as part of MODIS Collection 6, now provides surface reflectance products from both land and ocean bands over terrestrial regions from each MODIS sensor at daily time steps at the equator and multiple times a day in high latitudes due to orbital overlap near the poles. Several studies have demonstrated that combining land and ocean bands over terrestrial regions offers improved assessments of plant traits linked to photosynthetic activity (Rahman et al. 2004; Drolet et al. 2008; Garbulsky et al. 2008; Goerner et al. 2009; Gamon et al. 2016; Middleton et al. 2016). While this new band combination still lacks the spectral detail needed to assess many specific plant traits, it contains sufficient information to detect relative chlorophyll and carotenoid pigment levels, important indicators of plant stress and photosynthetic activity. This new information on plant pigment ratios improves our assessment of photosynthetic phenology, particularly for evergreen vegetation, which has been difficult to assess using greenness indices like NDVI (Gamon et al. 2016). Similarly, a new MODIS product, fAPARchl (the fAPAR fraction attributable to chlorophyll only), has been developed (Zhang et al. 2009, 2013, 2016) and has been proposed as a standard product. These findings are consistent with a history of studies showing that retrievals of plant pigments from remote sensing provide useful information about vegetation photosynthesis, productivity, stress, and overall health (Gitelson et al. 2005; Ustin et al. 2009). In global, cross-biome comparisons, NIRv, a new combination of red and NIR band reflectance from MODIS, yields improved GPP estimates relative to the NDVI (Badgley et al. 2017). These newly emerging indices are a testament to the value of consistent satellite time series and a persistent, accessible data system for advancing our understanding of vegetation function.

Since the historical satellite records contain imagery collected with only a few wide spectral bands, they cannot distinguish many important vegetation functional traits and processes detectable only with narrow-band ($\leq 10$ nm) spectral features. Furthermore, many of the satellite products are temporally aggregated; largely to address problems with varying cloud cover and sensor view angles, past MODIS EVI and NDVI products have been primarily available as composite values at 16-day (and recently at 8-day) time steps, which limits their ability to detect temporal dynamics, particularly during periods of rapid change (e.g., spring green-up or unexpected disturbance events) (Gamon et al. 2013).

Newer MODIS collections (e.g., Collection 6) offer daily time series of MODIS reflectance products, enabling a finer look at vegetation dynamics that offers to better resolve these events. Another limitation of the historical satellite time series has been the requirement to integrate other data (e.g., meteorological data) collected from different instruments on different temporal and spatial scales to facilitate quantitative interpretation of the spectral information (Running et al. 2004; Heinsch et al. 2006). Consequently, while current multi-band satellite products reveal broad patterns of a few key emergent properties of terrestrial ecosystems and biomes at the regional to global scales, they are not able to resolve detailed processes at the scale of individual land management units, or at specific experimental sites within observational networks (e.g., the flux tower network). Consequently, many underlying *mechanisms* remain partly obscured, leaving gaps in our understanding of vegetation function.

Despite their limitations, these global satellite products have demonstrated the power of consistent global surface reflectance products for assessment of terrestrial photosynthetic dynamics. While many underlying mechanisms remain hidden due to their coarse sampling scales, the consistent time series over many decades has provided an invaluable perspective on vegetation change and essential vegetation fields for global Earth system models (Garonna et al. 2016). These models include Dynamic Global Vegetation Models (DGVMs, e.g., Bonan et al. 2003; Pavlick et al. 2013; van Bodegom et al. 2014), soil–vegetation–atmosphere-transfer (SVAT) models depicting surface–atmosphere interactions (Arora 2002), nested physiological-climate models (Sellers et al. 1996a, b), and global models of net primary production (NPP, Running et al. 2004). This power to reveal vegetation function is continually being expanded with new reanalyses, and by adding additional satellite data.

Since 2015, a follow-on satellite to continue the MODIS record is the Joint Polar Satellite System, with a Visible Infrared Imaging Radiometer Suite (VIIRS). Unfortunately, this satellite has fewer bands than MODIS, limiting its utility for vegetation functional assessment. Today, the Copernicus Sentinel program of the European Space Agency (ESA) for the European Commission (EC) is setting new standards for operational multi-spectral Earth Observation with Sentinel-2 and Sentinel-3 (Aschbacher and Milagro-Pérez 2012; Malenovský et al. 2012). The two synchronously flying satellites Sentinel-2A and Sentinel-2B provide 12 optical bands with a spatial resolution of 10–30 meters and a global revisiting time of less than 5 days (Drusch et al. 2012). Besides these operational flagships of ESA, the recently launched (August 2017) VENµS mini-satellite (Israel Space Agency and CNES, France) offers 12 spectral bands (16–40 nm wide) and a unique orbit revisit scheme.

### 6.2 Hyperspectral Satellites

In recent years, several satellite missions demonstrating the capability of imaging spectrometers, also known as hyperspectral scanners, have been placed in orbit. These missions have provided limited geographical coverage, with imagery collected on user request, and include the NASA EO-1/Hyperion and the ESA CHRIS Proba. The Hyperion sensor provided limited image data collection at 30-m spatial resolution upon user request over a 17-year lifespan (2000–2017). All of these datasets are now freely available through the USGS archive. Recent analyses from the Hyperion collection demonstrate the power of imaging spectrometry for evaluating spatial patterns of net ecosystem $CO_2$ exchange (Campbell et al. 2013) and also vegetation functional types based on spectral patterns

(Somers and Asner 2012; Huemmrich et al. 2013). The Compact High Resolution Imaging Spectrometer (CHRIS) sensor has a multi-angular viewing capability (2 forward, a nadir and 2 backward viewing angles) for a better understanding of the bidirectional reflectance distribution function (BRDF) of terrestrial surfaces. It collects up to 150 channels (spectral range of 400–1050 nm) with a ground sampling resolution of 17 or 34 m over preselected sites on demand since 2001. A VIS-NIR spectrometer, the Hyperspectral Imager for the Coastal Ocean (HICO), was flown (2009–2014) on the International Space Station (ISS), and its utility for ecosystem function was demonstrated by Huemmrich et al. (2017).

The hyperspectral or spectroscopic observations acquired from space thus far have provided a strong foundation for the further development of satellite imaging spectrometers with global coverage. Hyperspectral missions of note include the Medium Resolution Imaging Spectrometer (MERIS), which had 15 programmable bands (FWHM ~ 2.5–30 nm) and flew from 2002 to 2012, and the CHRIS-Proba which has 200 narrow bands that can be combined into wider (e.g., 18) bands at varying spatial coverage ≤ 30 m. The two European Sentinel-3 satellites continue the mission concept of MERIS with an increasing number of optical bands (21 vs. 15 bands) and an improved optical performance (Donlon et al. 2012). The DLR Earth Sensing Imaging Spectrometer (DESIS; Eckardt et al. 2015) sponsored by the German Space Agency, was deployed on the International Space Station in June 2018. Additional proposed spaceborne imaging spectrometers promise to further expand our understanding of vegetation function using detectors with much higher spectral, spatial and temporal resolution. Presently, there are several new missions in various stages of planning or deployment. The space agencies of Italy and Israel are preparing the PRISMA mission (Amato et al. 2013) for launch in 2018–2019. The EnMAP mission (Germany) (Guanter et al. 2015a) and NASA's HyspIRI (Hyperspectral Infrared Imager) mission, (Lee et al. 2015), recently evolved into the Surface Biology and Geology (SBG) mission, will follow in the next few years. These emerging satellite instruments offer much richer spectral information than the history of multi-band Earth observation satellites mentioned above, and promise to expand the opportunities for assessing specific plant functional traits and dynamics.

### 6.3 Solar-Induced Fluorescence (SIF)

Recent examples of fortuitous applications of satellite data to assess vegetation function have arisen from several very high spectral resolution but coarse spatial resolution global satellite sensors designed for atmospheric gas sampling, including the Greenhouse gases Observing SATellite (GOSAT), the Global Ozone Monitoring Experiment-2 (GOME-2), and NASA's Orbiting Carbon Observatory (OCO-2). These instruments have extremely high spectral resolution (< 0.5 nm) with the capability of resolving solar Fraunhofer lines and atmospheric (telluric) absorption features that enable detection and quantification of a tiny solar-induced chlorophyll fluorescence (SIF) signal emitted from vegetation. This SIF signal increases with the amount of radiation absorbed by chlorophyll (APAR), and decreases with the amount of absorbed radiation that is dissipated as heat within leaves under stress (Porcar-Castell et al. 2014), providing a direct link to photosynthetic activity (gross primary productivity). These "ultra-spectral" sensors used for SIF detection offer new possibilities for global assessment of terrestrial photosynthesis or gross primary production (GPP), albeit at very coarse temporal and spatial scales (Frankenberg et al. 2011; Joiner et al. 2011; Guanter et al. 2012). A primary drawback of these global satellites has been that retrieval of the weak fluorescence signals requires averaging over large time steps and regions of the planet, obscuring fine-scale dynamics needed to explain underlying mechanism. This limitation is spurring development of new sensors and applications of spectrometry and fluorometry from airborne (Rascher et al. 2015; Schaepman et al. 2015) and proximal platforms (Burkart et al. 2014, 2015; Wyber et al. 2017). These ongoing advances combining airborne and proximal remote sensing offer to further clarify the functional links between passive fluorescence (SIF) and seasonally dynamic pigments levels (Gamon et al. 2016; Springer et al. 2017). Our growing understanding of photosynthetic dynamics will be further improved with the enhanced capabilities of the newly launched TROPOMI sensor (Guanter et al. 2015b), and the planned European Space Agency's FLEX (Fluoresence Explorer) satellite mission, designed specifically to directly address pigments and SIF simultaneously at global scales (Kraft et al. 2013; Drusch et al. 2017).

## 7 Current Challenges and Opportunities

A primary challenge in the remote sensing of vegetation function lies in extracting reliable, quantitative metrics of functional properties and processes that can be universally applied. Part of this challenge derives from the fact that remote sensing methods typically operate at spatial and temporal scales different from the functional processes or traits of interest (Fig. 1). Remarkably, despite this scale mismatch, plant traits and processes can often still be detected, even if the underlying mechanisms of signal transmission and detection remain partly unclear (Townsend et al. 2013; Schweiger et al. 2017). Developing a better understanding of the reasons for this convergence across scales, and understanding why it succeeds in some cases and fails in other cases, remain fundamental tasks. Potential solutions can be found in applying the full range of instrumentation from proximal sensors to global satellites to address scaling methods. Additional solutions lie in integrating physical, biological, and statistical approaches, and further exploring the concept of "optical types," representations of plant traits through patterns in spectral reflectance (Ustin and Gamon 2010), also called "spectral types" or "spectral species" (Feret and Asner 2014). Due to the multiple dimensions and high volume of data involved, greater attention to informatics will be needed. These challenges and opportunities are discussed in more detail below.

### 7.1 Integrating Trait-Based and Process-Based Approaches

The divergent methods to assess vegetation function based on static traits versus dynamic processes often seem to be distinct conceptual and practical approaches. However, these perspectives can be combined for greater insight, and a current opportunity lies in integrating trait- and process-based approaches. Traits, while sometimes applied categorically, can be depicted along a continuum (Wright et al. 2004). This continuum also has a time dimension (Chavana-Bryant et al. 2017). Capturing the full temporal dynamics of plant traits, while rarely attempted, can provide better insights into vegetation processes. The lack of consistent time series of optical properties, except for some proximal ground- or tower-based datasets, limits our current understanding of plant function using remote sensing. The advent of global imaging spectrometers with regular repeat cycles will greatly expand our understanding of vegetation processes at multiple scales, particularly when combined with proximal and airborne spectrometers. Similarly, spectral patterns can reveal information on both states and processes, allowing community properties like spectral diversity to be related to ecosystem processes like productivity (Wang et al. 2016a, b). In this way, integrating traits and processes can help quantify functional diversity (Schneider et al. 2017, Schweiger et al. 2017, 2018) and inform our understanding of ecosystem services (Braun et al. 2017).

## 7.2 Bridging Spatial and Temporal Scales

To what extent leaf traits can be detected at canopy or larger landscape scales observed by global spaceborne missions remains an open question, with good evidence both for and against upscaling of functional traits. The current literature on this topic is not entirely clear, with some authors questioning the detectability of certain leaf traits using optical remote sensing (Knyazikhin et al. 2013) and others noting that they could be detected under the right conditions (Townsend et al. 2013). Not all optical traits scale equally; multiple light scattering and absorption of plant canopies can confound some features, but accentuate other features allowing some leaf traits but not others to be easily detectable at canopy and stand scales. A number of studies report detection of plant functional traits with airborne imaging spectrometry (Asner et al. 2016a, b; Schneider et al. 2017), but the question remains: What exactly is detected as we move to progressively larger scales? A more complete answer to this question requires a clear methodology with well-defined hypotheses and conceptual frameworks, as well as physical models simulating imaging spectroscopy signals.

Scaling limitations are being transcended with new computational methods and sampling approaches. Since remote sensing technology (including observational instruments, computational power, and physical and statistical models) is rapidly changing, the topic of how to assess function across scales needs to be periodically revisited and updated. New statistical methods, including advanced machine learning approaches, are emerging that are greatly facilitating data analysis across scales (Verrelst et al. 2016). Improved comprehensive mathematical approaches to scaling (e.g., physically based radiative transfer models) are also relevant, as described in Verrelst et al. (2019) and Malenovský el al. (2019) (this issue).

While these models capture static vegetation biophysical and structural properties well, most do not adequately describe transient dynamics in plant functional properties mediated by physiological processes, such as chloroplast movement, leaf movement or rapid biochemical and changes associated with light regulation. Although some models are capable of resolving complex photon–vegetation interactions in three-dimensional (3D) space (Gastellu-Etchegorry et al. 2017), they cannot provide the same flexibility and realism that can be obtained from dynamic empirical measurements. Development of new hybrid models is a move in the right direction, where radiative transfer with simulations of dynamic energy budgets at the canopy/landscape scale is combined with photosynthetic biochemical models describing functional leaf optical properties tied to physiological regulation at the leaf scale (e.g., SCOPE, Timmermans et al. 2013). This direction should be considered in models accounting for 3D spatial heterogeneity, which can help us to better understand the impact that less dynamic structural features have on the ability to detect rapidly changing biochemical physiological processes. The development of GUI-driven models, such as the Automated Radiative Transfer Models Operator (ARTMO, http://ipl.uv.es/artmo/) offering wider accessibility in a user-friendly format, represents a recent advance that offers to facilitate the application of models to questions of vegetation function. In all cases, outputs of model simulations need to be tested against empirical measurements for a wide range of vegetation types to help clarify underlying mechanisms.

Clearly, the topic of scaling functional traits needs further study. Because spectra are temporally dynamic, often displaying considerable redundancy (leading to collinear responses), and because plants exhibit functional convergence (Ollinger 2011), it is easy to get the right answer for the wrong reason (or vice versa), i.e., because of non-causal relationships. Without attention to scale, traditional mapping approaches, while they may imply functional differences, may not be sufficient to fully detect functional processes, in part due to complexities of spatial scale, but also due to the temporal dynamics of plant traits (Chavana-Bryant et al. 2017). Unless temporal dynamics are fully considered, there will remain some question as to the accuracy of plant functional characterization.

### 7.3 Conceptual Advances in Integrating Physical and Biological Approaches

An understanding of the scaling questions would benefit from an integration of ecological theory with physical principles of radiative transfer (Kattenborn et al. 2017), but this has yet to be fully accomplished. While a complete radiative transfer modeling capability will have to consider complex physical and physiological processes (e.g., turbulent air transfer and the rapid physiological responses to dynamic light regimes), it can also be better guided by ecological theory that can help explain how canopies, stands, and landscapes are constructed from underlying rules guiding plant evolution and adaptation.

Ecological theory states that form follows function. This principle, which has sometimes been labeled "functional convergence" (Field 1991; Ollinger 2011), allows us to detect functional patterns and differences with imaging spectrometry. An elaboration of this concept states that plant biochemistry and structure have evolved to meet the functional goal of efficient resource capture and organism protection (among other goals), and that detectable differences in these traits with remote sensing reflect different adaptations or evolutionary responses, or "strategies" (Grime 2006). Fundamentally, this is an economic argument, stating that plants tend to follow economic rules when displaying functional properties (Bloom et al. 1985). The scaling of plant traits can be guided by principles of functional convergence, which can involve coordination of plant traits that simplify the challenge of addressing plant function from remote sensing. Different functional solutions can be addressed by a clearer definition of optical types representing optically detectable suites of functionally important plant traits. A better integration of theory with models and empirical measurements will be needed to accomplish these goals.

### 7.4 Informatics Needs for Databases, Archives, Tools, and Protocols

Spectral assessment of vegetation function presents new challenges for informatics in the collection, storage, processing, and archiving of increasingly large data volumes. This challenge exists both for the trait-based approach, where large numbers of plant traits are being assessed with reflectance spectra, and for the process-based approaches, where the additional demands of time-series analyses arise. Currently, unlike the development of databases and related tools that are broadly used by the molecular, genetics, and bioinformatics communities, there is currently no single, widely accepted set of versatile, but standardized data tools for the remote sensing community, greatly hampering advances in this area. While rudimentary online spectral libraries for vegetation are available, most of these cannot easily accommodate large volumes of spectral time series or image data. Most existing spectral libraries focus on a few static *types* (the archetypal mean). However, for functional analysis, we are often more interested in the *variation* over time and space (variance), which is not easily accommodated in current databases. While image databases exist for most satellite sensors, there is no similar community database or archival solution for the variety of current and emerging airborne or field imaging spectrometers, hampering our sharing and understanding of imaging spectroscopy for addressing plant function in a broader context. Both basic research and emerging applications, including phenotyping and precision farming, would greatly benefit from effective data management and processing solutions, which remain sparse to date. Examples include the SPECCHIO database (Hueni et al. 2009, 2011). The terrestrial remote sensing community could learn from other disciplines (e.g., astrophysics), where effective community solutions for handling and sharing large spectral datasets have been developed (Munari et al. 2005; Rayner et al. 2009).

## 8 Societal Benefits and Global Challenges

Further development of remote sensing tools for vegetation functional assessment would provide multiple benefits, both to basic science and to society. Vegetation plays a key role in many global Earth system processes, ranging from biogeochemical cycles to the surface–atmosphere transport of mass and energy that regulates our atmosphere and climate. Vegetation also supports diverse biological communities and provides many essential ecosystem goods and services. Until we can reliably monitor vegetation function at all scales, we cannot fully understand these processes and contributions or how they are changing. For example, longstanding questions about the role of the terrestrial biosphere as a carbon sink, and whether we should expect that sink to start relaxing, thereby accelerating climate change (Schimel 2007) could be better answered with improved remote sensing tools for addressing vegetation function. Similarly, remote sensing offers promising new and integrative ways of assessing biodiversity (Cavender-Bares et al. 2017), species composition (Asner and Vitousek 2005), and emerging diseases (Calderón et al. 2013; Zarco-Tejada et al. 2018) in a changing world. Many promising applications and management practices, including phenotyping, precision agriculture, and early stress and disease detection, could be better addressed with well-designed optical remote sensing methods geared toward accurate, precise, and timely vegetation monitoring and related analytics tools. Development of novel imaging spectroscopy-based remote sensing methods for vegetation function will be essential for monitoring and maintaining healthy and resilient ecosystems in changing world, providing concurrent social and economic benefits.

**Acknowledgements** The authors wish to thank the International Space Science Institute (ISSI) for a stimulating meeting that led to this paper, and three reviewers whose insightful comments and suggestions improved the manuscript. Donnette Thayer assisted with Fig. 1 and Ran Wang with Fig. 2. Data presented in Fig. 2 were collected with support from the National Sciences and Engineering Research Council of Canada (NSERC) and The Canadian Foundation for Innovation (CFI) funding to J. Gamon. Z. Malenovský was supported by the Australian Research Council Future Fellowship: Bridging scales in remote sensing of vegetation stress (FT160100477). The contribution of M.E. Schaepman is supported by the University of Zurich Research Priority Program on Global Change and Biodiversit (URPP GCB). The contribution of B. Somers (Katholieke Universiteit Leuven) is supported by the Belgian Science Policy Office in the framework of the STEREOIII program (projects: SR/00/307, SR/01/349, SR/00/363).

## References

Amato U, Antoniadis A, Carfora MF, Colandrea P, Cuomo V, Franzese M, Pignatti S, Serio C (2013) Statistical classification for assessing PRISMA hyperspectral potential for agricultural land use. IEEE J Sel Top Appl Earth Observ Remote Sens 6:615–625

Arora V (2002) Modeling vegetation as a dynamic component in soil–vegetation–atmosphere transfer schemes and hydrological models. Rev Geophys. https://doi.org/10.1029/2001rg000103

Aschbacher J, Milagro-Pérez MP (2012) The European Earth monitoring (GMES) programme: status and perspectives. Remote Sens Environ 120:3–8

Asner GP, Martin RE (2016) Spectranomics: emerging science and conservation opportunities at the interface of biodiversity and remote sensing. Glob Ecol Conserv 8:212–219. https://doi.org/10.1016/j.gecco.2016.09.010

Asner GP, Vitousek PM (2005) Remote analysis of biological invasion and biogeochemical change. Proc Natl Acad Sci USA 102:4383–4386. https://doi.org/10.1073/pnas.0500823102

Asner GP, Brodrick PG, Anderson CB, Vaughn N, Knapp DE, Martin RE (2016a) Progressive forest canopy water loss during the 2012–2015 California drought. Proc Natl Acad Sci USA 113:E249–E255. https://doi.org/10.1073/pnas.1523397113

Asner GP, Knapp DE, Anderson CB, Martin RE, Vaughn N (2016b) Large-scale climatic and geophysical controls on the leaf economics spectrum. Proc Natl Acad Sci USA 113:E4043–E4051. https://doi.org/10.1073/pnas.1604863113

Badgley G, Field CB, Berry JA (2017) Canopy near-infrared reflectance and terrestrial photosynthesis. Sci Adv 3:5. https://doi.org/10.1126/sciadv.1602244

Balzarolo M et al (2011) Ground-based optical measurements at European flux sites: a review of methods, instruments and current controversies. Sensors 11:7954–7981. https://doi.org/10.3390/s110807954

Barton CVM, North PRJ (2001) Remote sensing of canopy light use efficiency using the photochemical reflectance index—model and sensitivity analysis. Remote Sens Environ 78:264–273. https://doi.org/10.1016/s0034-4257(01)00224-3

Battaglia M, Cherry ML, Beadle CL, Sands PJ, Hingston A (1998) Prediction of leaf area index in eucalypt plantations: effects of water stress and temperature. Tree Physiol 18:521–528

Bloom AJ, Chapin FS, Mooney HA (1985) Resource limitation in plants—an economic analogy. Annu Rev Ecol Syst 16:363–392. https://doi.org/10.1146/annurev.es.16.110185.002051

Bonan GB, Levis S, Kergoat L, Oleson KW (2002) Landscapes as patches of plant functional types: an integrating concept for climate and ecosystem models. Glob Biogeochem Cycles. https://doi.org/10.1029/2000gb001360

Bonan GB, Levis S, Sitch S, Vertenstein M, Oleson KW (2003) A dynamic global vegetation model for use with climate models: concepts and description of simulated vegetation dynamics. Glob Change Biol 9:1543–1566. https://doi.org/10.1046/j.1365-2486.2003.00681.x

Braun D et al (2017) From instantaneous to continuous: using imaging spectroscopy and in situ data to map two productivity-related ecosystem services. Ecol Indic 82:409–419. https://doi.org/10.1016/j.ecolind.2017.06.045

Burkart A, Cogliati S, Schickling A, Rascher U (2014) A novel UAV-based ultra-light weight spectrometer for field spectroscopy. IEEE Sens J 14:62–67. https://doi.org/10.1109/jsen.2013.2279720

Burkart A et al (2015) A method for uncertainty assessment of passive sun-induced chlorophyll fluorescence retrieval using an infrared reference light. IEEE Sens J 15:4603–4611. https://doi.org/10.1109/jsen.2015.2422894

Calderón R, Navas-Cortés JA, Lucena C, Zarco-Tejada PJ (2013) High-resolution airborne hyperspectral and thermal imagery for early detection of *Verticillium* wilt of olive using fluorescence, temperature and narrow-band spectral indices. Remote Sens Environ 139:231–245. https://doi.org/10.1016/j.rse.2013.07.031

Campbell PKE et al (2013) EO-1 hyperion reflectance time series at calibration and validation sites: stability and sensitivity to seasonal dynamics. IEEE J Sel Top Appl Earth Obs Remote Sens 6:276–290. https://doi.org/10.1109/jstars.2013.2246139

Cavender-Bares J, Gamon JA, Hobbie SE, Madritch MD, Meireles JE, Schweiger AK, Townsend PA (2017) Harnessing plant spectra to integrate the biodiversity sciences across biological and spatial scales. Am J Bot 104:966–969. https://doi.org/10.3732/ajb.1700061

Chavana-Bryant C et al (2017) Leaf aging of Amazonian canopy trees as revealed by spectral and physiochemical measurements. New Phytol 214:1049–1063. https://doi.org/10.1111/nph.13853

Cornelissen JHC et al (2003) A handbook of protocols for standardised and easy measurement of plant functional traits worldwide. Aust J Bot 51:335–380. https://doi.org/10.1071/bt02124

de Wit CT (1970) Dynamic concepts in biology. In: Setlik I (ed) Prediction and measurement of photosynthetic productivity. Centre for Agricultural Publishing and Documentation, Wageningen, pp 17–23

DeFries RS, Townshend JRG (1994) NDVI-derived land cover classifications at a global scale. Int J Remote Sens 15:3567–3586

Donlon C et al (2012) The global monitoring for environment and security (GMES) Sentinel-3 mission. Remote Sens Environ 120:37–57. https://doi.org/10.1016/j.rse.2011.07.024

Drolet GG et al (2008) Regional mapping of gross light-use efficiency using MODIS spectral indices. Remote Sens Environ 112:3064–3078. https://doi.org/10.1016/j.rse.2008.03.002

Drusch M et al (2012) Sentinel-2: ESA's optical high-resolution mission for GMES operational services. Remote Sens Environ 120:25–36. https://doi.org/10.1016/j.rse.2011.11.026

Drusch M et al (2017) The FLuorescence EXplorer Mission Concept-ESA's Earth Explorer 8. IEEE Trans Geosci Remote Sens 55:1273–1284. https://doi.org/10.1109/tgrs.2016.2621820

Eckardt A et al (2015) DESIS (DLR Earth Sensing Imaging Spectrometer for the ISS-MUSES platform). In: 2015 IEEE international geoscience and remote sensing symposium (IGARSS), pp 1457–1459

Ehleringer JR, Comstock J (1987) Leaf absorptance and leaf angle: mechanisms for stress avoidance. In: Tenhunen JD, Catarino FM, Lange OL, Oechel WC (eds) Plant response to stress: functional analysis in Mediterranean ecosystems. Springer, Berlin, pp 55–76

Ehleringer JR, Field CB (1993) Scaling physiological processes: leaf to globe, 1st edn. Academic Press, New York

Enquist BJ et al (2015) Scaling from traits to ecosystems: developing a general trait driver theory via integrating trait-based and metabolic scaling theories. In: Pawar S, Woodward G, Dell AI (eds) Advances in ecological research, vol 52. Academic Press, New York, pp 249–318. https://doi.org/10.1016/bs.aecr.2015.02.001

Feilhauer H, Somers B, van der Linden S (2017) Optical trait indicators for remote sensing of plant species composition: predictive power and seasonal variability. Ecol Indic 73:825–833. https://doi.org/10.1016/j.ecolind.2016.11.003

Feret JB, Asner GP (2014) Mapping tropical forest canopy diversity using high-fidelity imaging spectroscopy. Ecol Appl 24:1289–1296

Field CB (1991) Ecological scaling of carbon gain to stress and resource availability. In: Mooney HA, Winner WE, Pell EJ (eds) Response of plants to multiple stresses. Physiological ecology. Academic Press, New York, pp 35–65

Frankenberg C et al (2011) New global observations of the terrestrial carbon cycle from GOSAT: patterns of plant fluorescence with gross primary productivity. Geophys Res Lett 38:6. https://doi.org/10.1029/2011gl048738

Fuentes DA et al (2006) Mapping carbon and water vapor fluxes in a chaparral ecosystem using vegetation indices derived from AVIRIS. Remote Sens Environ 103:312–323. https://doi.org/10.1016/j.rse.2005.10.028

Gamon JA (2015) Reviews and syntheses: optical sampling of the flux tower footprint. Biogeosciences 12:4509–4523. https://doi.org/10.5194/bg-12-4509-2015

Gamon JA, Surfus JS (1999) Assessing leaf pigment content and activity with a reflectometer. New Phytol 143:105–117. https://doi.org/10.1046/j.1469-8137.1999.00424.x

Gamon JA, Qiu HL (1999) Ecological applications of remote sensing at multiple scales. In: Pugnaire F, Valladares F (eds) Handbook of functional plant ecology. Marcel Dekker Inc., New York, pp 805–846

Gamon JA, Serrano L, Surfus JS (1997) The photochemical reflectance index: an optical indicator of photosynthetic radiation use efficiency across species, functional types, and nutrient levels. Oecologia 112:492–501. https://doi.org/10.1007/s004420050337

Gamon JA, Field CB, Fredeen AL, Thayer S (2001) Assessing photosynthetic downregulation in sunflower stands with an optically-based model. Photosynth Res 67:113–125. https://doi.org/10.1023/a:1010677605091

Gamon JA, Cheng YF, Claudio H, MacKinney L, Sims DA (2006) A mobile tram system for systematic sampling of ecosystem optical properties. Remote Sens Environ 103:246–254. https://doi.org/10.1016/j.rse.2006.04.006

Gamon JA et al (2010) SpecNet revisited: bridging flux and remote sensing communities. Can J Remote Sens 36:S376–S390

Gamon JA, Huemmrich KF, Stone RS, Tweedie CE (2013) Spatial and temporal variation in primary productivity (NDVI) of coastal Alaskan tundra: decreased vegetation growth following earlier snowmelt. Remote Sens Environ 129:144–153. https://doi.org/10.1016/j.rse.2012.10.030

Gamon JA et al (2016) A remotely sensed pigment index reveals photosynthetic phenology in evergreen conifers. Proc Natl Acad Sci USA 113:13087–13092. https://doi.org/10.1073/pnas.1606162113

Garbulsky MF, Peñuelas J, Papale D, Filella I (2008) Remote estimation of carbon dioxide uptake by a Mediterranean forest. Glob Change Biol 14:2860–2867. https://doi.org/10.1111/j.1365-2486.2008.01684.x

Garonna I, de Jong R, Schaepman ME (2016) Variability and evolution of global land surface phenology over the past three decades (1982–2012). Glob Change Biol 22:1456–1468. https://doi.org/10.1111/gcb.13168

Gastellu-Etchegorry JP et al (2017) DART: recent advances in remote sensing data modeling with atmosphere, polarization, and chlorophyll fluorescence. IEEE J Sel Top Appl Earth Obs Remote Sens 10:2640–2649. https://doi.org/10.1109/jstars.2017.2685528

Gitelson AA, Viña A, Ciganda V, Rundquist DC, Arkebauer TJ (2005) Remote estimation of canopy chlorophyll content in crops. Geophys Res Lett 32:L08403. https://doi.org/10.1029/2005gl022688

Gitelson AA, Gamon JA, Solovchenko A (2017) Multiple drivers of seasonal change in PRI: Implications for photosynthesis 2. Stand level. Remote Sens Environ 190:198–206. https://doi.org/10.1016/j.rse.2016.12.015

Goerner A, Reichstein M, Rambal S (2009) Tracking seasonal drought effects on ecosystem light use efficiency with satellite-based PRI in a Mediterranean forest. Remote Sens Environ 113:1101–1111. https://doi.org/10.1016/j.rse.2009.02.001

Goetz SJ, Bunn AG, Fiske GJ, Houghton RA (2005) Satellite-observed photosynthetic trends across boreal North America associated with climate and fire disturbance. Proc Natl Acad Sci USA 102:13521–13525. https://doi.org/10.1073/pnas.0506179102

Grime JP (2006) Plant strategies, vegetation processes, and ecosystem properties. Wiley, New York

Guanter L et al (2012) Retrieval and global assessment of terrestrial chlorophyll fluorescence from GOSAT space measurements. Remote Sens Environ 121:236–251. https://doi.org/10.1016/j.rse.2012.02.006

Guanter L et al (2015a) The EnMAP spaceborne imaging spectroscopy mission for earth observation. Remote Sens 7:8830–8857. https://doi.org/10.3390/rs70708830

Guanter L et al (2015b) Potential of the TROPOspheric Monitoring Instrument (TROPOMI) onboard the Sentinel-5 Precursor for the monitoring of terrestrial chlorophyll fluorescence. Atmos Meas Tech 8:1337–1352. https://doi.org/10.5194/amt-8-1337-2015

Heinsch FA et al (2006) Evaluation of remote sensing based terrestrial productivity from MODIS using regional tower eddy flux network observations. IEEE Trans Geosci Remote Sens 44:1908–1925. https://doi.org/10.1109/tgrs.2005.853936

Hilker T, Coops NC, Hall FG, Black TA, Wulder MA, Nesic Z, Krishnan P (2008) Separating physiologically and directionally induced changes in PRI using BRDF models. Remote Sens Environ 112:2777–2788. https://doi.org/10.1016/j.rse.2008.01.011

Hilker T, Gitelson A, Coops NC, Hall FG, Black TA (2011) Tracking plant physiological properties from multi-angular tower-based remote sensing. Oecologia 165:865–876. https://doi.org/10.1007/s00442-010-1901-0

Homolová L, Malenovský Z, Clevers J, Garcia-Santos G, Schaepman ME (2013) Review of optical-based remote sensing for plant trait mapping. Ecol Complex 15:1–16. https://doi.org/10.1016/j.ecocom.2013.06.003

Homolová L, Schaepman ME, Lamarque P, Clevers J, de Bello F, Thuiller W, Lavorel S (2014) Comparison of remote sensing and plant trait-based modelling to predict ecosystem services in subalpine grasslands. Ecosphere. https://doi.org/10.1890/es13-00393.1

Huemmrich KF, Gamon JA, Tweedie CE, Campbell PKE, Landis DR, Middleton EM (2013) Arctic tundra vegetation functional types based on photosynthetic physiology and optical properties. IEEE J Sel Top Appl Earth Obs Remote Sens 6:265–275. https://doi.org/10.1109/jstars.2013.2253446

Huemmrich KF, Campbell PKE, Gao BC, Flanagan LB, Goulden M (2017) ISS as a platform for optical remote sensing of ecosystem carbon fluxes: a case study using HICO. IEEE J Sel Top Appl Earth Obs Remote Sens 10:4360–4375. https://doi.org/10.1109/jstars.2017.2725825

Hueni A, Nieke J, Schopfer J, Kneubühler M, Itten K (2009) The spectral database SPECCHIO for improved long term usability and data sharing. Comput Geosci 35(3):557–565. https://doi.org/10.1016/j.cageo.2008.03.015

Hueni A, Malthus T, Kneubuehler M, Schaepman M (2011) Data exchange between distributed spectral databases. Comput Geosci 37:861–873. https://doi.org/10.1016/j.cageo.2010.12.009
Huete A, Didan K, Miura T, Rodriguez EP, Gao X, Ferreira LG (2002) Overview of the radiometric and biophysical performance of the MODIS vegetation indices. Remote Sens Environ 83:195–213. https://doi.org/10.1016/s0034-4257(02)00096-2
Jacquemoud S, Baret F (1990) PROSPECT: a model of leaf optical properties spectra. Remote Sens Environ 34:75–91. https://doi.org/10.1016/0034-4257(90)90100-z
Jacquemoud S et al (2009) PROSPECT plus SAIL models: a review of use for vegetation characterization. Remote Sens Environ 113:S56–S66. https://doi.org/10.1016/j.rse.2008.01.026
Jarvis P (1993) Prospects for bottom up models. In: Ehleringer JR, Field CB (eds) Scaling physiological processes: leaf to globe, 1st edn. Academic Press, New York, pp 115–126
Joiner J, Yoshida Y, Vasilkov AP, Yoshida Y, Corp LA, Middleton EM (2011) First observations of global and seasonal terrestrial chlorophyll fluorescence from space. Biogeosciences 8:637–651. https://doi.org/10.5194/bg-8-637-2011
Jung M et al (2011) Global patterns of land-atmosphere fluxes of carbon dioxide, latent heat, and sensible heat derived from eddy covariance, satellite, and meteorological observations. J Geophys Res Biogeosci. https://doi.org/10.1029/2010jg001566
Kampe TU, Johnson BR, Kuester M, Keller M (2010) NEON: the first continental-scale ecological observatory with airborne remote sensing of vegetation canopy biochemistry and structure. J Appl Remote Sens. https://doi.org/10.1117/1.3361375
Kattenborn T, Fassnacht FE, Pierce S, Lopatin J, Grime JP, Schmidtlein S (2017) Linking plant strategies and plant traits derived by radiative transfer modelling. J Veg Sci 28:717–727. https://doi.org/10.1111/jvs.12525
Kattge J et al (2011) TRY—a global database of plant traits. Glob Change Biol 17:2905–2935. https://doi.org/10.1111/j.1365-2486.2011.02451.x
Knyazikhin Y et al (2013) Hyperspectral remote sensing of foliar nitrogen content. Proc Natl Acad Sci USA 110:E185–E192. https://doi.org/10.1073/pnas.1210196109
Kraft S et al (2013) FLORIS: phase A status of the fluorescence imaging spectrometer of the Earth Explorer mission candidate FLEX. In: Conference on sensors, systems, and next-generation satellites XVII, Dresden, Germany, 23–26 Sept 2013. Proceedings of SPIE. https://doi.org/10.1117/12.2032060
Lee CM, Cable ML, Hook SJ, Green RO, Ustin SL, Mandl DJ, Middleton EM (2015) An introduction to the NASA Hyperspectral InfraRed Imager (HyspIRI) mission and preparatory activities. Remote Sens Environ 167:6–19. https://doi.org/10.1016/j.rse.2015.06.012
Levin SA (1992) The problem of pattern and scale in ecology. Ecology 73:1943–1967. https://doi.org/10.2307/1941447
Lucieer A, Malenovsky Z, Veness T, Wallace L (2014) HyperUAS-Imaging spectroscopy from a multirotor unmanned aircraft system. J Field Robot 31:571–590. https://doi.org/10.1002/rob.21508
Lyapustin A, Martonchik J, Wang Y, Laszlo I, Korkin S (2011a) Multiangle implementation of atmospheric correction (MAIAC): 1. Radiative transfer basis and look-up tables. J Geophys Res Atmos. https://doi.org/10.1029/2010jd014985
Lyapustin A et al (2011b) Multiangle implementation of atmospheric correction (MAIAC): 2. Aerosol algorithm. J Geophys Res Atmos. https://doi.org/10.1029/2010jd014986
Malenovský Z, Albrechtova J, Lhotakova Z, Zurita-Milla R, Clevers J, Schaepman ME, Cudlin P (2006) Applicability of the PROSPECT model for Norway spruce needles. Int J Remote Sens 27:5315–5340. https://doi.org/10.1080/01431160600762990
Malenovský Z, Rott H, Cihlar J, Schaepman ME, Garcia-Santos G, Fernandes R, Berger M (2012) Sentinels for science: potential of Sentinel-1, -2, and -3 missions for scientific observations of ocean, cryosphere, and land. Remote Sens Environ 120:91–101. https://doi.org/10.1016/j.rse.2011.09.026
Malenovský Z et al (2013) Retrieval of spruce leaf chlorophyll content from airborne image data using continuum removal and radiative transfer. Remote Sens Environ 131:85–102. https://doi.org/10.1016/j.rse.2012.12.015
Malenovský Z, Lucieer A, King DH, Turnbull JD, Robinson SA (2017) Unmanned aircraft system advances health mapping of fragile polar vegetation. Methods Ecol Evol 8:1842–1857. https://doi.org/10.1111/2041-210x.12833
Malenovský Z, Homolová L, Lukeš P, Buddenbaum H, Verrelst J, Alonso L, Schaepman ME, Lauret N, Gastellu-Etchegorry JP (2019) Variability and uncertainty challenges in upscaling imaging spectroscopy observations from leaves to vegetation canopies. Surv Geophys (in review; this issue)
Marshall E (1993) Global change—fitting planet earth into a user-friendly database. Science 261:846. https://doi.org/10.1126/science.261.5123.846

Middleton EM, Cheng Y-B, Hilker T, Black TA, Krishnan P, Coops NC, Huemmrich KF (2009) Linking foliage spectral responses to canopy-level ecosystem photosynthetic light-use efficiency at a Douglas-fir forest in Canada. Can J Remote Sens 35:166–188
Middleton EM, Huemmrich KF, Landis DR, Black TA, Barr AG, McCaughey JH (2016) Photosynthetic efficiency of northern forest ecosystems using a MODIS-derived Photochemical Reflectance Index (PRI). Remote Sens Environ 187:345–366. https://doi.org/10.1016/j.rse.2016.10.021
Middleton EM et al (2017) The 2013 FLEX-US airborne campaign at the Parker Tract loblolly pine plantation in North Carolina, USA. Remote Sens. https://doi.org/10.3390/rs9060612
Munari U, Sordo R, Castelli F, Zwitter T (2005) An extensive library of 2500–10500 Å synthetic spectra. Astron Astrophys 442:1127–1134. https://doi.org/10.1051/0004-6361:20042490
Myneni RB et al (2002) Global products of vegetation leaf area and fraction absorbed PAR from year one of MODIS data. Remote Sens Environ 83:214–231. https://doi.org/10.1016/s0034-4257(02)00074-3
Ollinger SV (2011) Sources of variability in canopy reflectance and the convergent properties of plants. New Phytol 189:375–394. https://doi.org/10.1111/j.1469-8137.2010.03536.x
Ollinger SV et al (2008) Canopy nitrogen, carbon assimilation, and albedo in temperate and boreal forests: functional relations and potential climate feedbacks. Proc Natl Acad Sci USA 105:19336–19341. https://doi.org/10.1073/pnas.0810021105
Pavlick R, Drewry DT, Bohn K, Reu B, Kleidon A (2013) The Jena Diversity-Dynamic Global Vegetation Model (JeDi-DGVM): a diverse approach to representing terrestrial biogeography and biogeochemistry based on plant functional trade-offs. Biogeosciences 10:4137–4177. https://doi.org/10.5194/bg-10-4137-2013
Pinto F et al (2016) Sun-induced chlorophyll fluorescence from high-resolution imaging spectroscopy data to quantify spatio-temporal patterns of photosynthetic function in crop canopies. Plant, Cell Environ 39:1500–1512. https://doi.org/10.1111/pce.12710
Pinto F, Müller-Linow M, Schickling A, Cendrero-Mateo MP, Ballvora A, Rascher U (2017) Multiangular observation of canopy sun-induced chlorophyll fluorescence combining imaging spectroscopy and stereoscopy. Remote Sens 9, article no. 415. https://doi.org/10.3390/rs9050415
Porcar-Castell A et al (2014) Linking chlorophyll a fluorescence to photosynthesis for remote sensing applications: mechanisms and challenges. J Exp Bot 65:4065–4095. https://doi.org/10.1093/jxb/eru191
Quattrochi DA, Goodchild MF (1997) Scale in remote sensing and GIS. CRC Press, Boca Raton
Rahman AF, Cordova VD, Gamon JA, Schmid HP, Sims DA (2004) Potential of MODIS ocean bands for estimating $CO_2$ flux from terrestrial vegetation: a novel approach. Geophys Res Lett. https://doi.org/10.1029/2004gl019778
Rascher U et al (2015) Sun-induced fluorescence—a new probe of photosynthesis: first maps from the imaging spectrometer. HyPlant Glob Change Biol 21:4673–4684. https://doi.org/10.1111/gcb.13017
Rayner JT, Cushing MC, Vacca WD (2009) The infrared telescope facility (IRTF) spectral library: cool stars. Astrophys J Suppl 185:289–432. https://doi.org/10.1088/0067-0049/185/2/289
Reich PB, Ellsworth DS, Walters MB (1998) Leaf structure (specific leaf area) modulates photosynthesis-nitrogen relations: evidence from within and across species and functional groups. Funct Ecol 12:948–958. https://doi.org/10.1046/j.1365-2435.1998.00274.x
Roelofsen HD, van Bodegom PM, Kooistra L, Witte JPM (2014) Predicting leaf traits of herbaceous species from their spectral characteristics. Ecol Evol 4:706–719. https://doi.org/10.1002/ece3.932
Rossini M et al (2015) Red and far red Sun-induced chlorophyll fluorescence as a measure of plant photosynthesis. Geophys Res Lett 42:1632–1639. https://doi.org/10.1002/2014gl062943
Rouse JW, Haas RH, Schell JA, Deering DW (1974) Monitoring vegetation systems in the Great Plains with ERTS. Remote Sensing Center, Texas A&M University, College Station
Running SW, Nemani RR, Heinsch FA, Zhao MS, Reeves M, Hashimoto H (2004) A continuous satellite-derived measure of global terrestrial primary production. Bioscience 54:547–560. https://doi.org/10.1641/0006-3568(2004)054%5b0547:acsmog%5d2.0.co;2
Schaepman ME et al (2015) Advanced radiometry measurements and earth science applications with the Airborne Prism Experiment (APEX). Remote Sens Environ 158:207–219. https://doi.org/10.1016/j.rse.2014.11.014
Schimel D (2007) Carbon cycle conundrums. Proc Natl Acad Sci USA 104:18353–18354. https://doi.org/10.1073/pnas.0709331104
Schneider FD, Letterer R, Morsdorf F, Gastellu-Etchegorry JP, Lauret N, Pfeifer N, Schaepman ME (2014) Simulating imaging spectrometer data: 3D forest modeling based on LiDAR and in situ data. Remote Sens Environ 152:235–250. https://doi.org/10.1016/j.rse.2014.06.015
Schneider FD, Morsdorf F, Schmid B, Petchey OL, Hueni A, Schimel DS, Schaepman ME (2017) Mapping functional diversity from remotely sensed morphological and physiological forest traits. Nat Commun. https://doi.org/10.1038/s41467-017-01530-3

Schull MA, Anderson MC, Houborg R, Gitelson A, Kustas WP (2015) Thermal-based modeling of coupled carbon, water, and energy fluxes using nominal light use efficiencies constrained by leaf chlorophyll observations. Biogeosciences 12:1511–1523. https://doi.org/10.5194/bg-12-1511-2015

Schweiger AK, Schütz M, Risch AC, Kneubühler M, Haller R, Schaepman ME (2017) How to predict plant functional types using imaging spectroscopy: linking vegetation community traits, plant functional types and spectral response. Methods Ecol Evol 8:86–95. https://doi.org/10.1111/2041-210x.12642

Schweiger AK et al (2018) Plant spectral diversity integrates functional and phylogenetic components of biodiversity and predicts ecosystem function. Nat Ecol Evol 2:976. https://doi.org/10.1038/s41559-018-0551-1

Sellers PJ et al (1996a) Comparison of radiative and physiological effects of doubled atmospheric $CO_2$ on climate. Science 271:1402–1406. https://doi.org/10.1126/science.271.5254.1402

Sellers PJ, Los SO, Tucker CJ, Justice CO, Dazlich DA, Collatz GJ, Randall DA (1996b) A revised land surface parameterization (SiB2) for atmospheric GCMs. Part II: the generation of global fields of terrestrial biophysical parameters from satellite data. J Clim 9:706–737. https://doi.org/10.1175/1520-0442(1996)009%3c0706:arlspf%3e

Sellers PJ et al (1997) BOREAS in 1997: experiment overview, scientific results, and future directions. J Geophys Res Atmos 102:28731–28769. https://doi.org/10.1029/97jd03300

Serbin SP, Dillaway DN, Kruger EL, Townsend PA (2012) Leaf optical properties reflect variation in photosynthetic metabolism and its sensitivity to temperature. J Exp Bot 63:489–502. https://doi.org/10.1093/jxb/err294

Serbin SP, Singh A, McNeil BE, Kingdon CC, Townsend PA (2014) Spectroscopic determination of leaf morphological and biochemical traits for northern temperate and boreal tree species. Ecol Appl 24:1651–1669

Serbin SP et al (2015) Remotely estimating photosynthetic capacity, and its response to temperature, in vegetation canopies using imaging spectroscopy. Remote Sens Environ 167:78–87. https://doi.org/10.1016/j.rse.2015.05.024

Serrano L, Ustin SL, Roberts DA, Gamon JA, Peñuelas J (2000) Deriving water content of chaparral vegetation from AVIRIS data. Remote Sens Environ 74:570–581. https://doi.org/10.1016/s0034-4257(00)00147-4

Simmer C et al (2015) Monitoring and modeling the terrestrial system from pores to catchments: the transregional collaborative research center on patterns in the soil–vegetation–atmosphere system. Bull Am Meteorol 96:1765–1787. https://doi.org/10.1175/bams-d-13-00134.1

Sims DA, Gamon JA (2003) Estimation of vegetation water content and photosynthetic tissue area from spectral reflectance: a comparison of indices based on liquid water and chlorophyll absorption features. Remote Sens Environ 84:526–537. https://doi.org/10.1016/s0034-4257(02)00151-7

Sims DA, Luo HY, Hastings S, Oechel WC, Rahman AF, Gamon JA (2006) Parallel adjustments in vegetation greenness and ecosystem $CO_2$ exchange in response to drought in a Southern California chaparral ecosystem. Remote Sens Environ 103:289–303. https://doi.org/10.1016/j.rse.2005.01.020

Singh A, Serbin SP, McNeil BE, Kingdon CC, Townsend PA (2015) Imaging spectroscopy algorithms for mapping canopy foliar chemical and morphological traits and their uncertainties. Ecol Appl 25:2180–2197. https://doi.org/10.1890/14-2098.1.sm

Somers B, Asner GP (2012) Hyperspectral time series analysis of native and invasive species in Hawaiian rainforests. Remote Sens 4:2510–2529. https://doi.org/10.3390/rs4092510

Somers B, Asner GP, Martin RE, Anderson CB, Knapp DE, Wright SJ, Van De Kerchove R (2015) Mesoscale assessment of changes in tropical tree species richness across a bioclimatic gradient in Panama using airborne imaging spectroscopy. Remote Sens Environ 167:111–120. https://doi.org/10.1016/j.rse.2015.04.016

Springer KR, Wang R, Gamon JA (2017) Parallel seasonal patterns of photosynthesis, fluorescence, and reflectance indices in boreal trees. Remote Sens. https://doi.org/10.3390/rs9070691

Street LE, Shaver GR, Williams M, Van Wijk MT (2007) What is the relationship between changes in canopy leaf area and changes in photosynthetic $CO_2$ flux in arctic ecosystems? J Ecol 95:139–150. https://doi.org/10.1111/j.1365-2745.2006.01187.x

Stylinski CD, Gamon JA, Oechel WC (2002) Seasonal patterns of reflectance indices, carotenoid pigments and photosynthesis of evergreen chaparral species. Oecologia 131:366–374. https://doi.org/10.1007/s00442-002-0905-9

Teuling AJ, Seneviratne SI (2008) Contrasting spectral changes limit albedo impact on land-atmosphere coupling during the 2003 European heat wave. Geophys Res Lett. https://doi.org/10.1029/2007gl032778

Timmermans J, Su Z, van der Tol C, Verhoef A, Verhoef W (2013) Quantifying the uncertainty in estimates of surface-atmosphere fluxes through joint evaluation of the SEBS and SCOPE models. Hydrol Earth Syst Sci 17:1561–1573. https://doi.org/10.5194/hess-17-1561-2013

Townsend PA, Serbin SP, Kruger EL, Gamon JA (2013) Disentangling the contribution of biological and physical properties of leaves and canopies in imaging spectroscopy data. Proc Natl Acad Sci USA 110:E1074–E1074. https://doi.org/10.1073/pnas.1300952110

Tucker CJ (1979) Red and photographic infrared linear combinations for monitoring vegetation. Remote Sens Environ 8:127–150. https://doi.org/10.1016/0034-4257(79)90013-0

Tucker CJ, Townshend JRG, Goff TE (1985) African land-cover classification using satellite data. Science 227:369–375. https://doi.org/10.1126/science.227.4685.369

Ustin SL, Gamon JA (2010) Remote sensing of plant functional types. New Phytol 186:795–816. https://doi.org/10.1111/j.1469-8137.2010.03284.x

Ustin SL, Gitelson AA, Jacquemoud S, Schaepman M, Asner GP, Gamon JA, Zarco-Tejada P (2009) Retrieval of foliar information about plant pigment systems from high resolution spectroscopy. Remote Sens Environ 113:S67–S77. https://doi.org/10.1016/j.rse.2008.10.019

van Bodegom PM, Douma JC, Verheijen LM (2014) A fully traits-based approach to modeling global vegetation distribution. Proc Natl Acad Sci USA 111:13733–13738. https://doi.org/10.1073/pnas.1304551110

van der Tol C, Verhoef W, Timmermans J, Verhoef A, Su Z (2009) An integrated model of soil-canopy spectral radiances, photosynthesis, fluorescence, temperature and energy balance. Biogeosciences 6:3109–3129. https://doi.org/10.5194/bg-6-3109-2009

Verhoef W (1984) Light scattering by leaf layers with application to canopy reflectance modeling: the SAIL model. Remote Sens Environ 16:125–141. https://doi.org/10.1016/0034-4257(84)90057-9

Verrelst J et al (2019) Quantifying vegetation biophysical variables from imaging spectroscopy data: a review on retrieval methods. Surv Geophys 1–41. https://doi.org/10.1007/s10712-018-9478-y (this issue)

Verrelst J, Dethier S, Rivera JP, Munoz-Mari J, Camps-Valls G, Moreno J (2016) Active learning methods for efficient hybrid biophysical variable retrieval. IEEE Geosci Remote Sens Lett 13:1012–1016. https://doi.org/10.1109/lgrs.2016.2560799

Wang WM, Li ZL, Su HB (2007) Comparison of leaf angle distribution functions: effects on extinction coefficient and fraction of sunlit foliage. Agric For Meteorol 143:106–122. https://doi.org/10.1016/j.agrformet.2006.12.003

Wang R, Gamon JA, Emmerton CA, Haitao L, Nestola E, Pastorello GZ, Menzer O (2016a) Integrated analysis of productivity and biodiversity in a southern Alberta prairie. Remote Sens. https://doi.org/10.3390/rs8030214

Wang R et al (2016b) Seasonal variation in the NDVI-species richness relationship in a prairie grassland experiment (Cedar Creek). Remote Sens 8:1. https://doi.org/10.3390/rs8020128

Wieneke S, Ahrends H, Damm A, Pinto F, Stadler A, Rossini M, Rascher U (2016) Airborne based spectroscopy of red and far-red sun-induced chlorophyll fluorescence: implications for improved estimates of gross primary productivity. Remote Sens Environ 184:654–667. https://doi.org/10.1016/j.rse.2016.07.025

Williamson SN, Barrio IC, Hik DS, Gamon JA (2016) Phenology and species determine growing-season albedo increase at the altitudinal limit of shrub growth in the sub-Arctic. Glob Change Biol 22:3621–3631. https://doi.org/10.1111/gcb.13297

Wright IJ et al (2004) The worldwide leaf economics spectrum. Nature 428:821–827. https://doi.org/10.1038/nature02403

Wyber R, Malenovský Z, Ashcroft MB, Osmond B, Robinson SA (2017) Do daily and seasonal trends in leaf solar induced fluorescence reflect changes in photosynthesis, growth or light exposure? Remote Sens. https://doi.org/10.3390/rs9060604

Zarco-Tejada PJ et al (2018) Previsual symptoms of *Xylella fastidiosa* infection revealed in spectral plant-trait alterations. Nat Plants 4:432–439. https://doi.org/10.1038/s41477-018-0189-7

Zhang QY, Middleton EM, Margolis HA, Drolet GG, Barr AA, Black TA (2009) Can a satellite-derived estimate of the fraction of PAR absorbed by chlorophyll (FAPAR(chl)) improve predictions of light-use efficiency and ecosystem photosynthesis for a boreal aspen forest? Remote Sens Environ 113:880–888. https://doi.org/10.1016/j.rse.2009.01.002

Zhang QY, Middleton EM, Cheng YB, Landis DR (2013) Variations of foliage chlorophyll fAPAR and foliage non-chlorophyll fAPAR (fAPARchl, fAPARnon-chl) at the Harvard Forest. IEEE J Sel Top Appl Earth Obs Remote Sens 6:2254–2264. https://doi.org/10.1109/jstars.2013.2275176

Zhang Q et al (2016) Integrating chlorophyll fAPAR and nadir photochemical reflectance index from EO-1/Hyperion to predict cornfield daily gross primary production. Remote Sens Environ 186:311–321. https://doi.org/10.1016/j.rse.2016.08.026

Zhao M, Running SW (2010) Drought-induced reduction in global terrestrial net primary production from 2000 through 2009. Science 329:940–943. https://doi.org/10.1126/science.1192666

Surveys in Geophysics (2019) 40:515–551
https://doi.org/10.1007/s10712-018-9492-0

# Spaceborne Imaging Spectroscopy for Sustainable Agriculture: Contributions and Challenges

**Tobias B. Hank[1] · Katja Berger[1] · Heike Bach[2] · Jan G. P. W. Clevers[3] · Anatoly Gitelson[4] · Pablo Zarco-Tejada[5] · Wolfram Mauser[1]**

Received: 16 February 2018 / Accepted: 14 July 2018 / Published online: 26 July 2018


**Abstract**
Agriculture faces the challenge of providing food, fibre and energy from limited land resources to satisfy the changing needs of a growing world population. Global megatrends, e.g., climate change, influence environmental production factors; production and consumption thus must be continuously adjusted to maintain the producer–consumer-equilibrium in the global food system. While, in some parts of the world, smallholder farming still is the dominant form of agricultural production, the use of digital information for the highly efficient cultivation of large areas has become part of agricultural practice in developed countries. Thereby, the use of satellite data to support site-specific management is a major trend. Although the most prominent use of satellite technology in farming still is navigation, Earth Observation is increasingly applied. Some operational services have been established, which provide farmers with decision-supporting spatial information. These services have mostly been boosted by the increased availability of multispectral imagery from NASA and ESA, such as the Landsat or Copernicus programs, respectively. Using multispectral data has arrived in the agricultural commodity chain. Compared to multispectral data, spectrally continuous narrow-band sampling, often referred to as hyperspectral sensing, can potentially provide additional information and/or increased sampling accuracy. However, due to the lack of hyperspectral satellite systems with high spatial resolution, these advantages mostly are not yet used in practical farming. This paper summarizes where hyperspectral data provide additional value and information in an agricultural context. It lists the variables of interest and highlights the contribution of hyperspectral sensing for information-driven agriculture, preparing the application of future operational spaceborne hyperspectral missions.

**Keywords** Agriculture · Hyperspectral · Spaceborne imaging spectroscopy · Food security · Biophysical and biochemical variables

✉ Tobias B. Hank
tobias.hank@lmu.de

Extended author information available on the last page of the article

## 1 Current Challenges of the Global Food System

In pre-industrial times, producers and consumers in the food system were identical. Before the late eighteenth and middle of the nineteenth century, the majority of the European country's population settled in rural areas and was occupied with agricultural activities (Goodwin et al. 2008). Consumption and production of food happened parallel to each other. Nowadays, global large-scale processes like urbanization, which has exponentially increased in the last three decades (Chen et al. 2014), have led to a more complex and spatially divided manifestation of the commodity flows. The global food system, whose most prominent flows are sketched in generalized form in Fig. 1, is supposed to be in a more or less balanced state between the demand by consumers, who increasingly live in cities, and the supply by the producers, who farm the land in a predominantly rural environment. Sites of food production and consumption have become spatially dislocated (Gunasekera and Finnigan 2010) and are now connected via a globally operating food industry, which works within food security frameworks that have been established by regional governance systems (Danbom 1995). The food industry on one side processes agricultural goods and provides food according to the consumer's needs, values and preferences, while it steers the flow of money back to the farmers on the other hand, thus maintaining farm economy.

Profitability is the key of global food production, driving technological and infrastructural investments. In economic terms, agricultural land use, intensification and land expansion take place, under given regulations, where it is profitable to do so (Byerlee et al. 2014). Market forces increase with the number of farmers connected to the global agricultural markets. Thus, economic and regulatory factors, paired with the underlying continued processes of globalization and global environmental change, determine how much food will be produced. Farmers work the land and produce agricultural products with the use of capital, water, energy and technology. The environment, characterized by soil

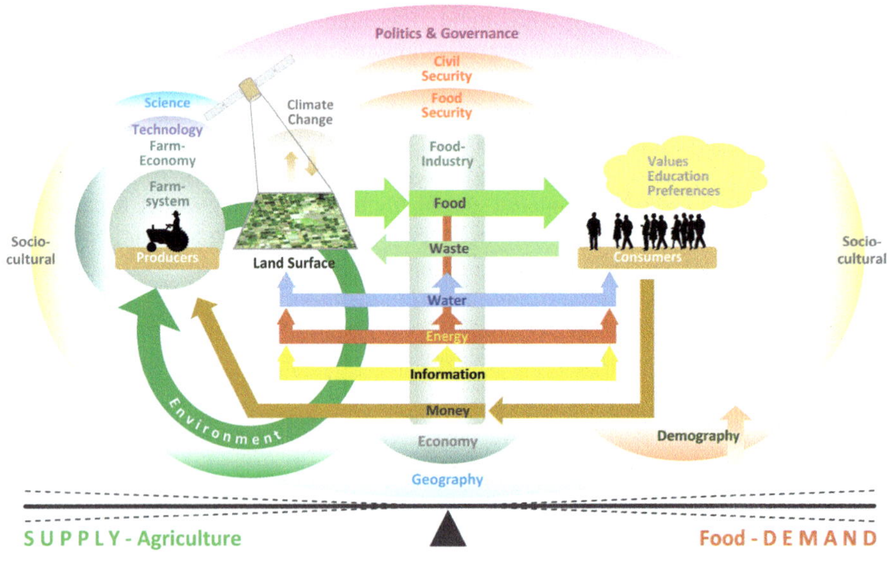


**Fig. 1** Generalization of the most important flows in the Global Food System [modified after www.shiftn.com; figure originally developed for the UK Government Office for Science (GOS 2011)]. It can be observed that one of the important flows is information, which is partly derived from satellite observations

fertility, temperature conditions and available water resources, is their main production factor. Farmers thereby increasingly use science and technology (machinery, fertilizers and plant protection agents, seeds and information, e.g., from remote sensing) to optimize the use of the naturally available production factors and to improve economic benefit (Walter et al. 2017). Environmental production factors, on the other hand, are subject to climate change and environmental degradation within unsustainable farming systems (FAO 2016). At the same time, a growing and increasingly wealthy global population relies on the global food system to satisfy daily food, fibre and energy demands. The global megatrends (climate change, population change, technological change, etc.) gradually cause the supply–demand-balance to shift. The suppliers, i.e. the farmers, are continuously striving to reinstall the equilibrium by adapting and optimizing agricultural production methods. Given the current demographic development, farmers are forced to increase the yields achieved by cultivating a certain acreage of bioproductive land surface, while at the same time protecting their most important production factor, i.e. the environment, from degradation and from emissions (FAO 2016). The threat emanating from emissions thereby is based on two different pathways. The leaching of $NO_3$, which has not been taken up by the crop during growth, pollutes groundwater and thus has a direct effect on human health (Ritter et al. 2002). At the same time, gaseous emissions of $NO_X$ enter the atmosphere and lead to the formation of tropospheric ozone, a greenhouse gas pushing climate change, but also potentially harming crops directly (Robertson et al. 2013). In addition, emissions from excessive use of plant protection agents contribute to environmental pollution (Ritter et al. 2002). One recent trend to mitigate these problems in agricultural production is the application of organic farming methods, which do not rely on the application of mineral fertilizers and plant protection agents (Shams et al. 2017). However, organic farming still is a trade-off between increased environmental protection and reduced overall yields and yield security. Besides organic production, site-specific precision farming is another farming perspective (Blackmore 1994). Thereby, all means of production are spatially optimized to prevent unnecessary expenses and unwanted emission at the same time. Precision farming is an important component of smart farming, which aims at an information-driven optimization of all aspects of a farming system (Bach et al. 2016). Some of the most important applications of smart farming are:

- Using computer-aided information technology, such as Farm Management Information Systems (FMIS), for administration and documentation of work helps the farmer to keep track of expenses on and revenues from individual fields (Zhang et al. 2002).
- Using satellite-guided precision navigation prevents overlap between driving lanes and helps optimizing the distribution of different machines on the farm thus saving fuel and person-hours (Zhang et al. 2002).
- Using long-term Earth Observation helps to identify different soil qualities on a farm and thus helps to optimize soil sampling (Plant et al. 2000).
- Using long-term Earth Observation also helps to identify persistent growth patterns, which indicate the variability of different growth conditions on a farm (e.g., www.talkingfields.de/en/base-map/). This information can be used for the planning of long-term soil enrichment measures, such as increasing soil organic matter content through ploughing green manure. It also can be applied for site-specific planning of seeding density.
- Using real-time Earth Observation, either from satellites, planes, drones or vehicles, can contribute to site-specific planning of four very important management measures (Mulla 2013), as there are (1) fertilization (e.g., Cidad et al. 2001), (2) plant protection

(e.g., Herwitz et al. 2004), (3) irrigation (e.g., Calera et al. 2017), and (4) harvesting (e.g., Yang et al. 2013). Thereby, so-called map-overlay approaches combining different data sources, e.g., satellite sensors and ground-based sensors, seem to represent the most promising method (Auernhammer et al. 1999).

Precision farming thus holds potential for increasing yields on limited land while at the same time-saving resources and preventing environmental pollution (Plant et al. 2000). Current studies show that increased yield levels not necessarily need to be connected to environmental degradation (see review Balafoutis et al. 2017). The so-called green water productivity (kg grain produced per kg water used) has been observed for wheat, rice and maize crops to more or less linearly increase with rising yield levels until an average yield level of 8 tons per hectare is reached (Rockström et al. 2007). Above this point, no further increase in the green water productivity can be observed. Sustainability thus becomes a matter of balancing yield levels in a dimension around this average, because obviously low yield levels, which do not make efficient use of the environmental production factors, are just as unsustainable as too high yields, which drain the natural resources and lead to degradation.

One very important aspect of precision farming is that the method not only is associated with ecologic advantages but also economic profitability can be increased through site-specific farming (Balafoutis et al. 2017), which makes an actual large-scale implementation of precision farming likely. However, the most efficient use of the land surface and thus the most economic and ecologic profit, can only be achieved by farmers who have access to all information concerning the individual needs of their worked land. While in some parts of the world smallholder farming is still the dominant form of agricultural production (Lowder et al. 2016), in developed countries the use of digital information for the highly efficient cultivation of large areas has become part of agricultural practice in recent years (Walter et al. 2017). Thereby, the use of satellite data to support site-specific management is a major trend at least in some developed countries. For instance, in the USA, the market area share of satellite data has tripled during the last 10 years, as can be observed in Fig. 2.

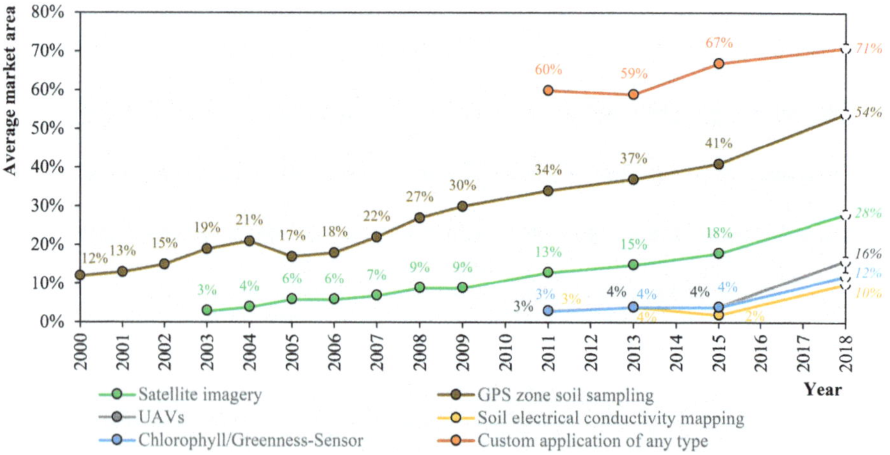


Fig. 2 Development of market area using precision services over time as estimated by dealers of farm equipment in the USA (2018 predicted), showing the strong increase in use of satellite-based information technology in agricultural practice; based on data by Erickson and Widmar (2015)

Although the most prominent use of satellite technology in farming still is navigation via GNSS (Global Navigation Satellite Systems, e.g., for GPS-based soil sampling, see Fig. 2), also Earth Observation (EO) has been increasingly applied (Satellite imagery, Fig. 2). The highest share, however, is allotted to custom applications, which are individually tailored to the information needs of the farmer and which, according to Erickson and Widmar (2015), in many cases also are based on satellite-derived data (e.g., field mapping with GPS) but also on statistical data (e.g., yield monitor analysis). Some operational services have been recently established, which provide farmers with decision-supporting spatial information, e.g., talkingfields (VISTA, Germany; Bach and Angermair 2017; Bach and Mauser 2018) or Irrisat (Ariespace, Italy; Ariespace 2017) etc. The development of these services has mostly been boosted by the increased availability of (cost-free) multispectral imagery from NASA and ESA, such as the Landsat or Copernicus programs, respectively. Using multispectral data has arrived in the agricultural commodity chain. This development comes along with new challenges such as handling big data. Currently, the large amount of data is mostly due to high revisit rates at reasonably high spatial resolution (e.g., Sentinel-2). The big data load so far is not so much due to a detailed spectral sampling of the Earth's surface, although spectrally continuous narrow-band sampling, often referred to as hyperspectral remote sensing (HRS) or imaging spectroscopy (IS; Goetz et al. 1985), can potentially provide additional information and/or increased sampling accuracy compared to multispectral data. Since a few years, HRS-IS technologies have been well accepted by the remote sensing community as an innovative tool for various research applications, not only in agriculture but also in atmospheric sciences, geology, ecology, soil science, limnology, pedology and other areas. However, due to the lack of large-scale hyperspectral EO systems, these advantages mostly are not yet used in practical farming. With some new spaceborne hyperspectral missions being currently developed, e.g., the Italian PRISMA (Labate et al. 2009), US HyspIRI (Roberts et al. 2012), or the German EnMAP (Guanter et al. 2015), the availability of hyperspectral data and its use and application in practical farming will soon be demonstrated and scientifically justified with on-farm research.

This paper summarizes where hyperspectral data can provide additional value and information in an agricultural context. It lists the variables of interest, reviews the retrieval methods and highlights the additional value of hyperspectral sensing for agricultural applications, thus preparing the application of operational spaceborne hyperspectral missions in modern agriculture.

## 2 Earth Observation Opportunities to Support Food Production Efficiency

"Remote sensing data can greatly contribute to the monitoring task by providing timely, synoptic, cost-efficient and repetitive information about the status of the Earth's surface" [Justice et al. (2002) in Atzberger (2013)]. The basis for retrieving information in a spatial way from multispectral data is the assumption of a relation between structural and biochemical traits that describe the state of the canopy and the spectral reflectance that results from the specific variable composition (Mauser et al. 2012). From an agricultural point of view, soil and vegetation properties are of special interest. Observing soils with hyperspectral observation systems for instance allows retrieving information on soil organic carbon and soil iron-oxide content (Stevens et al. 2010), which both can be important indicators for soil fertility. However, the impact of soil analysis through remote sensing for agricultural practice is strongly limited due to three reasons: (1) Retrieval of soil organic carbon content is highly influenced by the current moisture state of the soil, because different moisture states strongly influence soil brightness. Soil moisture, however, is very variable in space and time so that difficulties arise when acquisitions should be compared to in situ measurements (Stevens et al. 2010). (2) Optical sensors operating in the visible (VIS), near-infrared (NIR) and shortwave-infrared (SWIR) are not capable of penetrating the canopy, so soil observations are only feasible for surfaces with very low vegetation coverage (Haubrock et al. 2008). This excludes analysing permanent crops and strongly limits continuous soil monitoring during the growth cycle. (3) Sensors in the VIS, NIR and SWIR domain also are not capable of penetrating the soil surface. The information that might be extracted from optical remote measurements of the soil is restricted to the uppermost soil surface (Haubrock et al. 2008). The true production potential of soils nonetheless lies in the deeper soil horizons that are accessed by the roots of the canopy and which provide water and nutrients to the crop. However, this hidden information on soil column properties becomes secondarily accessible when the plant itself is used as sensor, because the vigour of plants sensitively reacts to soil-related environmental conditions, e.g., to drought and nutrient stress (Basu et al. 2016). In order to make use of the plant response, the relation between crop biochemical and structural traits and the reflectance must precisely be known. It is a commonly promoted limitation of vegetation remote sensing that the spectral signal of actively growing vegetation looks pretty similar across a wide range of different plant types and cultivars (e.g., Nidamanuri and Zbell 2011; Wilson et al. 2014; Zomer et al. 2009). This similarity in the opinion of the authors should not be perceived as drawback, but should rather be interpreted as proof that the underlying physical and chemical processes that contribute to vegetation growth on the Earth's surface (light collection, gas exchange, phenological cycles, carbon assimilation and allocation) more or less apply uniformly for all vegetation types. Due to these uniformly and globally applicable mechanisms, the states of those vegetation variables that leave traces in the spectral reflectance of the canopy can potentially be observed by remote sensing and, consequently, can be analysed through non-crop-specific approaches. Figure 3 gives an overview of spectral domains, which respond to changes in specific vegetation variables, regardless of vegetation type and cultivar.

Figure 3 supports two important findings: (1) Information contained in a vegetation spectrum is rich and diverse. It reaches from biochemical information (pigments and water content) over process information (light use efficiency, LUE) to structural information (leaf area index, LAI, leaf inclination, etc.). (2) The partial disagreement between the hyperspectral and the multispectral reflectance spectrum indicates the limited information content of multispectral sensors for diverse vegetation properties and processes. Very advanced multispectral EO systems, such as ESA's Sentinel-2, already allow deriving valuable information of vegetation properties, e.g., shown by Clevers et al. (2017), Frampton et al. (2013), Herrmann et al. (2011), Vuolo et al. (2016). Nevertheless, hyperspectral sensors, dealing with narrow spectral bands over a continuous spectral range, will be able to detect properties more accurately, especially in the VIS and SWIR domain. Characteristic shapes of individual absorption features caused by diverse biochemical or biophysical plant components may be located spectrally very close to each other. In contrast to hyperspectral data, broadband scanners are not able to resolve the specific wavelength position of these features, which relate for instance to several pigments (anthocyanins, carotenoids) or proteins [see studies from Kokaly et al. (2007), Kokaly and Skidmore (2015), Sahoo et al. (2015), Thenkabail et al. (2000)].

Potentially, all information contained in the spectrum is of interest to farmers. The value of specific variables, i.e. biophysical or biochemical products, for practical precision

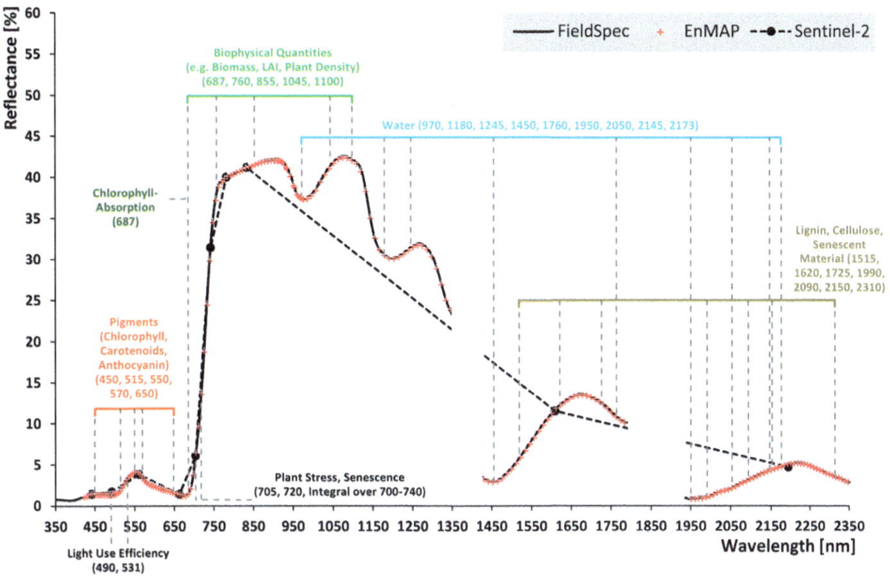


**Fig. 3** Spectral domains that respond to changes of specific variables in vegetation [absorption lines modified after Thenkabail et al. (2013)]. The vital vegetation (Triticum aestivum) spectra in the background are displayed for a quasi-continuous spectral resolution of an ASD FieldSpec4 Standard (black solid line), for a multispectral configuration corresponding to the spectral bands of ESA Sentinel-2 (black dashed line) and for the narrow-band configuration of the future EnMAP Hyperspectral Imager (red crosses). In the case of vital vegetation, specific absorptions of, e.g., cellulose and lignin are obscured by water absorptions in the SWIR and only can be detected when the crop is senescing (see Fig. 4)

farming depends on the scientist's capabilities of retrieving a variable with adequate precision. In order to unlock the information contained in the hyperspectral readings as biophysical and biochemical products for precision farming purposes, several steps are required. This includes the choice of a suitable and capable retrieval method that also accounts for measurement errors of the radiometric data (sensor noise) and field validation data (sampling errors, natural heterogeneity; Baret and Buis 2008). The retrieval process further deals with complex reflectance anisotropies (Walthall 1997), discriminates the signal of overlapping absorption features and finally yet importantly follows the direct relation between cause and effect. For this purpose, physical laws were established inferring biophysical and biochemical variables based on specific knowledge, typically obtained with radiative transfer functions (Verrelst et al. 2015).

The abundance of information that can potentially be derived from vegetation spectra has to be carefully selected for applications in agricultural practice. In short, remotely measured information that is of direct relevance for practical farming should meet four criteria:

1. The derived information must be biophysical/biochemical in the sense that it can be described in physical units: "In particular, the broader availability of air- and space-borne directional imaging spectrometer data supports the estimation of biophysical and biochemical variables with unprecedented accuracy and in calibrated physical units..." (Schaepman et al. 2005). Relative measures, as they are, e.g., derived via vegetation

indices, are of marginal help in agricultural practice, because comparability and transferability are limited (Baret and Buis 2008).
2. The derived information must be directly observable in the spectrum. This means that a direct relation between state variable and the reflectance in a specific spectral domain exists, which can be analysed using different retrieval methods (for a review, see Verrelst et al. 2015).
3. The derived information must be retrievable at a generalized level independent from specific cultivars, because cultivars in most cases are not known to the analysers and thus cannot be taken into account during the retrieval. Generalization capabilities of inversion/retrieval methods are discussed by Durbha et al. (2007) and Kimes et al. (2000).
4. The derived information must be of importance for practical farming (Moran et al. 1997).

The following section aims to identify the biophysical and biochemical variables that are relevant for practical farming according to the criteria mentioned above. The variables are described and their importance for practical precision farming is elaborated. Section 3.1 describes the biochemical leaf (canopy) variables, Sect. 3.2 the canopy biophysical variables and Sect. 3.3 is dedicated to the two variables soil and land use.

# 3 Overview of Biophysical and Biochemical Variables with Direct Relevance for Practical Farming

Providing an overview of the diversity of agriculturally relevant information addressed in the context of this review paper, Table 1 lists the different variables together with their most commonly used physical units, highlights the variable's respective use in practical farming and compares field and remote sampling techniques with special emphasis on the benefits expected from imaging spectroscopy. Note that the variables and retrieval techniques listed in Table 1 primarily apply to homogeneous crop farming. With this, we refer to uniformly cultivated fields at least in the size of several pixels captured by imaging spectrometers. For smallholder farming and geometrically complex crops, other scanning techniques, such as unmanned aerial vehicles (UAV) or low-altitude manned aircrafts, are more appropriate and different types of variables can be derived from such high spatial resolution imagery.

## 3.1 Leaf and Canopy Biochemical Variables

### 3.1.1 Leaf/Canopy Chlorophyll Content

Leaf chlorophyll content (LCC) usually is measured in units of µg chlorophyll per $cm^2$ leaf area, whereas canopy chlorophyll content (CCC) is measured in g chlorophyll per $m^2$ ground area. Both variables indicate the presence of chlorophyll pigments in vegetation. Since chlorophyll is the pigment responsible for the absorption of energy, the abundance of chlorophyll indicates the potential of a plant cell to provide energy for photosynthetic processes and thus forms the basis for vegetation growth on the Earth's surface (Peng et al. 2017). Leaf chlorophyll content and canopy chlorophyll content are one of the most important variables for farming. Because chlorophyll molecules are constructed with the help of four nitrogen atoms in the chlorine-magnesium ligand, a direct relation between the nitrogen supply and the accumulation of chlorophyll exists. CCC was therefore recognized as a

**Table 1** Biochemical and biophysical variables with relevance for agricultural management, which are measured by optical Earth Observation, including their most commonly used physical units, their use in practical farming, common field sampling techniques and the potential of hyperspectral measurements for their retrieval

| Variable | Definition | Physical unit | Use in farming | Field sampling technique | Potential of hyperspectral sensing |
|---|---|---|---|---|---|
| *Canopy/leaf biochemical variables* | | | | | |
| LCC/CCC | Leaf/Canopy Chlorophyll Content | [μg $cm^{-2}$ leaf surface]<br>[g $m^{-2}$ canopy surface] | Nutrient monitoring | Destructive: extraction via organic solvents<br>Non-destructive: SPAD[1]<br>CCC: LCC multiplied with leaf area index (LAI), measured with LI-COR[2] | Continuous sampling of VIS and red edge leads to better discrimination of different pigments |
| $C_{cx}$ | Leaf carotenoid content | [μg $cm^{-2}$ leaf surface] | Nutrient monitoring | Destructive: extraction via organic solvents | Continuous sampling of blue and red edge spectral domains leads to better discrimination of different pigments |
| $C_{ANTH}$ | Leaf anthocyanin content | [μg $cm^{-2}$ leaf surface] | Disease monitoring, air pollution | Liquid chromatography in the laboratory | Pigment content can only be measured by sensors with high spectral resolution in the VIS |
| Cp/N | Leaf protein content/nitrogen content | [μg $cm^{-2}$ leaf surface]/% | Fertilization planning, growth regulation | Dumas combustion/Kjeldahl digestion method (nitrogen) | Canopy chlorophyll content is a proxy of canopy N during vegetative phase. Exploitation of protein absorption features in the SWIR still not fully explored |
| LMA | Leaf Mass Area | [kg $m^{-2}$ leaf surface] | Growth modelling, phenological monitoring | Cutting and drying until constant weight, Planimetry | Can only be assessed via spectral models, which per se are of a hyperspectral nature |
| EWT/CWC | Equivalent Water Thickness, Canopy water content | [mm] | Maturation monitoring, Irrigation monitoring in humid environments | Cutting and drying until constant weight | Hyperspectral observations of the different water absorption features allow discriminating the three phases of water: gas, liquid, ice |

**Table 1** (continued)

| Variable | Definition | Physical unit | Use in farming | Field sampling technique | Potential of hyperspectral sensing |
|---|---|---|---|---|---|
| NPV | Non-photosynthetically active vegetation (lignin + cellulose) | Relative change | Crop residue assessment (Carbon/Nitrogen storage), crop response to seasonal and long-term drought | Cutting, drying and weighing | Spectral resolution ≤ 15 nm in the SWIR required to resolve cellulose and lignin absorption features |
| *Canopy biophysical variables* | | | | | |
| LAI | Leaf Area Index | [$m^2$ leaf surface $m^{-2}$ ground surface] | Biomass monitoring | Destructive: Leaf traps, leaf scanning<br>Non-destructive: Li-Cor | Continuous sampling in red edge and NIR regions provides required retrieval accuracy:<br>Discrimination of structural and chemical variables is improved<br>Discrimination of angular effects is improved |
| ALA | Average Leaf inclination Angle | [°] | Phenological monitoring/drought monitoring | Inclinometer, high-resolution horizontal photograph interpretation | Hyperspectral approaches reduce ambiguity in retrieval |
| Albedo | Average surface reflectivity | [%] | Energy balance modelling | Double pyranometer | Continuous spectrum can be integrated |
| fCOVER | Fractional Cover | [%] | Erosion monitoring | Vertical photographs, planimetry | Hyperspectral approaches reduce ambiguity in spectral unmixing |
| fAPAR | Fraction of absorbed photosynthetically active radiation | [%] | Growth modelling, nutrient monitoring, Phenological monitoring | Line quantum sensors | Hyperspectral observations allow for integrating over the photosynthetically active spectral range. |
| SIF | Sun induced chlorophyll fluorescence | [$mW\ m^{-2}\ sr^{-1}\ nm^{-1}$] | Multiple stress monitoring | Field spectroscopy | Impossible with multispectral sensors. Hyperspectral observation systems with < 3 nm spectral resolution required |

**Table 1** (continued)

| Variable | Definition | Physical unit | Use in farming | Field sampling technique | Potential of hyperspectral sensing |
|---|---|---|---|---|---|
| *Soil and land cover variables* | | | | | |
| LULC | Land use/land cover | [class] | Documentation | Field mapping, statistics | Better discrimination of cultivars through precise spectral measurements |
| SOC | Soil organic carbon content | [%] | Nutrient modelling | Destructive laboratory analysis | Water absorptions must be resolved to discriminate between soil moisture and soil organic carbon content |

[1]SPAD-502Plus handheld non-destructive Chlorophyll-Meter by Konica Minolta Business Solutions Inc.

[2]LAI-2200C handheld non-destructive Plant Canopy Analyzer by LI-COR Biosciences Inc.

valid proxy for canopy nitrogen content (Baret et al. 2007; Clevers and Gitelson 2013) and crop primary production (e.g., Gitelson et al. 2006; Peng et al. 2011). Increasing levels of chlorophyll in the crop indicate the amount of nitrogen that has successfully entered the canopy and thus has contributed to the construction of proteins and light-harvesting pigments. This balance is summarized under the term nitrogen use efficiency (NUE). Rapidly generating high levels of chlorophyll content thus has ecological and economic advantages. On the other hand, over-fertilization can result in unhealthy accumulation of chlorophyll in the leaves, i.e. necrosis, or in weakened structural stability, i.e. lodging, which again inhibits growth. Hyperspectral sensors, due to their continuous sampling of the VIS and red edge spectral domain, are ideally suited to measure leaf chlorophyll content (Gitelson et al. 2003), which is the reason why spectrometers are the tool predominantly used for chlorophyll measurements in laboratory analysis, where the easily soluble chlorophyll can be extracted from the plant tissue using organic solvents (Parry et al. 2014). Satellites, however, observe the canopy as a whole from above. This synoptic view of the canopy is accompanied by ambiguity problems, since for specific spectral domains interaction effects among multiple vegetation traits on measured reflectance exist (Combal et al. 2003). From a practical farming point of view, the leaf chlorophyll content is the most important information for planning fertilization measures during the vegetative phase (Arregui et al. 2006). The challenge remains to discern between structural and chemical traits of the canopy (e.g., Daughtry et al. 2000; Haboudane et al. 2002). This can only be achieved through hierarchical retrieval methods, which take the continuous spectrum into account and where structural canopy properties are determined first and chemical properties are delineated through secondary retrieval iterations, (e.g., Danner et al. 2017).

### 3.1.2 Total Leaf Carotenoid Content

The total carotenoid content of leaves (Car) is mainly composed of the xanthophyll cycle pigments and carotenes. They play an important role in photoprotection, accessory light harvesting and energy transfer (Gitelson et al. 2002; Kong et al. 2017). Carotenoids are present in variable proportions during the differentiation and ageing of leaves, but abiotic stress can inhibit carotenoid production. This includes, for instance, ozone or sulphur dioxide air pollution (Agrawal et al. 1982), heavy metals (Panda et al. 2003), viral attacks (Ibdah et al. 2014) or water deficiency (Mibei et al. 2017). Therefore, carotenoids may serve as indicators for down-regulation of photosynthesis due to environmental stressors.

A review study by Blackburn (2007) confirmed the growing value of hyperspectral remote sensing for plant pigment estimations in ecophysiology, environmental, agricultural and forestry sciences. Extraction of biochemical properties has been accomplished from hyperspectral data using various methods. Thereby the presence of carotenoids is commonly expressed in different units, e.g., as mass per unit surface area (g $m^{-2}$), as mass per unit leaf area (g $cm^{-2}$), or as mass per unit fresh leaf weight (mg $g^{-1}$) e.g., by Yi et al. (2014).

Compared to leaf chlorophyll, there are much fewer studies estimating leaf carotenoid content (Yi et al. 2014) and even less dealing with agricultural crops. Generally, the benefit of hyperspectral data for separating the subtle signals of the different pigments was recognized by several studies (Blackburn 1998; Chappelle et al. 1992; Feret et al. 2008).

### 3.1.3 Total Leaf Anthocyanin Content

Anthocyanins are the most common class of flavonoids. They are responsible for the orange to red, or purple to blue coloration in the tissue depending on the molecule, temperature and pH value, as it can be found, for instance, in blueberry, raspberry, black rice or black soybean (Tanaka et al. 2008). From all the red pigments, including carotenoids and betalains, anthocyanins are the most widespread. Though the full role of anthocyanins still is not completely understood, some essential functions of these pigments have been identified, such as protection against photoinhibition from intense light and mitigation of environmental stresses such as freezing or air pollution. In crops, anthocyanins typically are present when the plants are suffering from drought, freezing or nutrient deficiency (Lee and Gould 2002; Springob et al. 2003).

Leaf anthocyanin content was introduced in the latest version of the leaf optical properties model PROSPECT-D (Féret et al. 2017). The model simulates directional–hemispherical reflectance and transmittance for the spectral range from 400 to 2500 nm and, besides the anthocyanin content, requires leaf chlorophyll content, a leaf structure parameter, carotenoid content, brown pigments, equivalent water thickness and leaf mass per unit area as inputs. Variables can be retrieved from measured leaf spectra by model inversion or from canopy spectra by coupling PROSPECT-D with a radiative transfer model, such as 4SAIL (Jacquemoud et al. 2009; Verhoef and Bach 2003). The use of the model for anthocyanin retrieval is limited to hyperspectral data with high spectral resolution in the VIS spectral domain. The use of narrow-band vegetation indices allows quite accurate retrieval of anthocyanin content against the background of very variable chlorophyll content (Gitelson and Solovchenko 2017; Gitelson et al. 2001, 2009).

### 3.1.4 Leaf Protein Content/Nitrogen Content

With rubisco accounting for 30–50% of nitrogen (N) in green leaves, proteins are the principal N-containing biochemical constituent of plants (Kokaly et al. 2009). Nitrogen, on the other hand, is probably the most important nutrient that plants acquire from the soil. A proper management of nitrogen is a prerequisite for sustainable fertilization in modern agriculture: optimal crop yield from, e.g., high-quality grains, can only be obtained with high uptakes of N (Barraclough et al. 2010). Nitrogen deficiency leads to decreased photosynthetic assimilation and to reduced crop yield in terms of quantity and quality (Jay et al. 2017). Hyperspectral data have been recognized as a promising tool for the non-destructive detection of crop nitrogen for several reasons: the wavelengths important for nitrogen estimation have been found over the whole spectrum due to correlation between nitrogen and other variables (Curran 1989; Homolová et al. 2013). The absorption features of nitrogen-containing biochemical leaf components, such as LCC (6.5% by weight) and proteins may serve as proxies for crop N. The chlorophyll absorption features in the VIS and red edge can be used for nitrogen estimation due to a close relationship of LCC to N during early growth stages (le Maire et al. 2008). During senescence, however, the decomposition of light-harvesting pigments and the translocation of N from leaves, stems and roots lead to gradual changes in the spectral signal of VIS and NIR without being connected to significant changes of total N in the plant. Moving away from the leaf level perspective towards the integration of the full canopy within a specified surface, area acted as a normalization factor, improving the N versus Chl relationship. A very strong linear relationship was established across growing seasons between maize N and Chl contents at the canopy level

(Schlemmer et al. 2013). While monitoring crop nitrogen uptake during the early vegetative growth phases is of major importance for the planning of fertilization measures, nitrogen monitoring during senescent growth stages provides valuable information on yield quality. Continuous information of crop nitrogen during the different growth stages would consequently be of high economic impact. Absorption features of proteins can be found in the SWIR. Recently, protein absorption coefficients also have been integrated into leaf reflectance models and first applications for woody and herbaceous species have been explored (Wang et al. 2015). However, protein absorptions are very shallow and are largely obscured by water absorption features, so that measuring protein content from crop canopies with spectroscopy remains a challenge.

### 3.1.5 Leaf Mass Area

Leaf Mass Area (LMA) denotes the relation of leaf mass to leaf area in a unit of kg dry matter per $m^2$ leaf area. In some publications also its reverse is used, i.e. specific leaf area (SLA, leaf area per unit dry mass; e.g., Ali et al. 2017). This fundamental leaf functional trait plays a key role in ecosystem modelling (Ali et al. 2017; Asner et al. 2011). LMA is a measure of the leaf composition: the first leaves developed by a plant at the beginning of its individual growth cycle usually are rather lightweight, so that the area available for the interception of solar radiation expands rapidly during early growth phases. During later development stages, plants tend to invest more biomass into the structural stability of the leaves, causing the LMA to increase over the course of a growing period and, depending on the crop type, also with increasing LAI of the canopy. However, this process is highly crop-specific and can only be observed through labour-intensive destructive measurements, so that only few data on this variable exist.

LMA is an essential indicator of plant functioning, including photosynthetic and respiratory rates, chemical composition or resistance to herbivory (de la Riva et al. 2016). The importance of LMA for farming compared to the other variables is therefore rather indirect but nonetheless important, in particular regarding the relationship of LMA to photosynthesis–nitrogen relationships (Poorter and Evans 1998).

The spectral recognition of LMA is difficult, because the effect caused by increasing LMA in the leaf reflectance spectrum results in a gradual decrease in reflectance in the NIR shoulder region. A similar effect is achieved by the variations of other structural canopy variables, such as LAI and ALA, and is influenced by illumination/viewing angles. Consequently, a study by Asner et al. (2011) demonstrated that the best-suited wavelengths for accurate LMA determination are found in sections of the spectrum, where the overlap effects with other variables are less pronounced, i.e. in the shortwave-infrared (SWIR) between 1900 and 2500 nm and from 1300 to 1700 nm, as well as in the visible region from 400 to 800 nm. A derivation of LMA from spectral measurements thus is only possible if (1) the full spectrum is taken into account, and (2) all potentially confounding variables are retrieved simultaneously. This can best be achieved through inversion of canopy reflectance models exploiting the full spectral data cube, such as PROSAIL (Jacquemoud et al. 2009) or SLC (Migdall et al. 2009; Verhoef and Bach 2007).

### 3.1.6 Equivalent Water Thickness/Canopy Water Content

The variable Equivalent Water Thickness (EWT) or leaf water content describes the thickness of a theoretical layer of water (in cm), which absorbs radiation according to the Lambert–Beer law (Nobel 2009). Thus, EWT corresponds to the volume of water that is stored within the cells of living vegetation.

A well-known problem is the decoupling of the spectral signal of the three different aggregate states of water, which is necessary for a correct retrieval of EWT. A study showed that the use of hyperspectral data can strongly support the simultaneous estimation of the abundance of gaseous, liquid and frozen water for a defined environment (Green et al. 2006). The authors obtained uncertainties $< 1.5\%$ for all three phases, but there were still problems in retrieving the liquid water in vegetation, making more research in this domain necessary.

For a remote sensor with a defined field of view, it is difficult to decouple the contributions of leaf water content and LAI. Thus, the total canopy water content per unit ground area (CWC, g $m^{-2}$), rather than leaf EWT is usually "observed" or retrieved (Clevers et al. 2010). CWC is a measure for the moisture state of a canopy, which is of interest for practical farming in two aspects:

First, CWC is a variable that describes the maturity state of a crop. Depending on the crop type, crops are harvested at specific thresholds of residual moisture (e.g., Peters 2012). This guarantees optimal storage stability and documents the yield quality in the sense of the true net weight of the harvest. Assessing the residual moisture in a spatially explicit way through remote sensing enables farmers to develop harvesting strategies, e.g., which field or which part of a field to harvest first, and to document the quality of the harvest. Actually, modern harvesting systems already are measuring residual moisture during the harvesting process with infrared sensors installed in combine harvesters (Peters 2012). However, the proximity of the measurement leads to failures, such as clogging of the grain elevators, and the sampling occurs more or less at random and not in a spatially comprehensive way, so that these measurements still are of limited use. Deriving spatial maps, which describe the temporal dynamics of residual moisture based on satellite data, thus would be of high interest for practical farming.

Secondly, CWC can be used to monitor irrigation, which causes significant environmental changes in many parts of the world. Precise information on the extent of irrigation measures is fundamental for global change research, where, e.g., water exchange between the land surface and the atmosphere is explicitly modelled to assess impacts on global food security (Ozdogan et al. 2010). In arid environments, irrigation can be easily detected, because vegetation only prevails in irrigated areas. In humid environments, such as the temperate zone or even within the boundaries of irrigated fields, however, detecting irrigation becomes more difficult (Ozdogan et al. 2010). Only few studies have approached the problem of irrigated area mapping in humid areas (e.g., Thenkabail et al. 2005). Although CWC and EWT are directly connected to the water supply state of crops and the use of CWC or EWT to monitor drought seems rational, the variable may not be directly suited as early drought indicator. This is due to the fact that one of the earliest reactions of plants to water deficit is the maintenance of EWT through stomatal control (Yang and Ling 2004). A decline in EWT, as it can be detected with hyperspectral sensors, thus can be used to document the dehydration of the tissue, but the signal would be detected at a too late point in time to initiate counter management measures. Thermal imaging systems are far better suited to indicate early signs of drought based on the leaf energy balance (Prashar and Jones 2016). For instance, the crop water stress index (CWSI) proved to detect drought stress earlier than EWT estimates (Yang and Ling 2004).

### 3.1.7 Non-photosynthetically Active Vegetation (Lignin + Cellulose)

Non-photosynthetically active vegetation (NPV) elements refer to vegetation compounds that cannot perform photosynthesis, such as plant litter, crop residues, senescing foliage, branches and stems (Zhaoqin and Xulin 2015). NPV is mainly composed of lignin, cellulose and starch, being the most abundant molecules produced by terrestrial photosynthesis. Therefore, the importance of estimating NPV, or vegetation brownness, for understanding terrestrial ecosystem dynamics and changes is evident (Dennison et al. 2016; Okin 2010). For agricultural applications, NPV is particularly interesting in three aspects: (1) Variations in NPV—or the ratio of vital to senescent foliage—can indicate (seasonal) drought events. (2) Non-photosynthetic crop residues on the soil surface significantly reduce soil erosion and enhance soil organic C through improvement of the soil structure. (3) Non-photosynthetic crop residues may contain significant amounts of nitrogen, which enter the soil through ploughing and which must be taken into account when balancing the nitrogen inputs during the following season.

Effective crop residue management systems therefore require rapid site-specific quantification of crop residues over large areas, which can only be obtained with the help of remote sensing technologies (Daughtry et al. 2005). Several studies developed hyperspectral indices for NPV estimation, which are based on the absorption features of cellulose and lignin in the SWIR. For a review of the topic, e.g., see Zhaoqin and Xulin (2015). Comparable to the nitrogen absorption features, the cellulose and lignin absorptions overlap with other variables, e.g., with atmospheric water vapour, and thus must be analysed with high spectral resolution in the SWIR. Figure 4 shows that spectral signatures of soil and NPV, depending on the soil brightness, may appear very similar in VIS and NIR (Fig. 4a, c) or in the SWIR (Fig. 4b, c), respectively. The narrow cellulose absorptions in the SWIR, which cannot be discriminated by broadband sensors (dots in Fig. 4), must be resolved to discriminate between soil and crop residues.

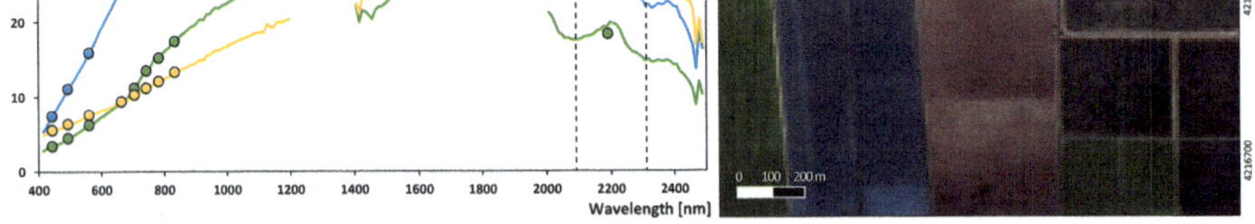


**Fig. 4** Comparison of spectral signatures of senescent cereals (**a**), crop residues (**b**) and bare soil (**c**). The spectra were recorded by the airborne spectrometer AVIRIS (AVIRIS-NG acquisition from 09 October 2016, displayed in true colour, https://avirisng.jpl.nasa.gov/alt_locator/, downloaded 14 February 2018). The dots indicate the reflectance signal, if the AVIRIS spectrum is degraded according to the spectral response function of the Sentinel-2 MSI. It can be observed that high spectral resolution in the SWIR is required to discriminate between non-photosynthetic biomass and soil via the lignin and cellulose absorptions at 2090 and 2310 nm

## 3.2 Canopy Biophysical Variables

### 3.2.1 Leaf Area Index

Leaf Area Index (LAI) denotes the one-sided leaf surface area per square metre ground area, as firstly defined by Watson (1947). LAI has a physical unit of $m^2$ leaf surface per $m^2$ ground surface and therefore is a dimensionless quantity. This variable is of specific importance for practical farming, since LAI defines the leaf surface that is available for exchange processes between the leaf mesophyll and the atmosphere. The entire flux of $CO_2$ and $O_2$ as well as the absorption and emission of radiative energy are scaled via the LAI (Breda 2003). LAI therefore is an important input variable for carbon assimilation models and it helps identifying phenological progress (e.g., Savoy and Mackay 2015), and the accumulation of biomass (e.g., Hank et al. 2015; Vaesen et al. 2001).

For definition, the variable LAI considers the entire leaf surface regardless of the leaf state. LAI therefore must be distinguished from green LAI, which only takes the photosynthetically active leaf surface (i.e. the parts of the canopy that are characterized by the presence of chlorophyll) into account (Haboudane et al. 2004). Additional confusion may arise from using Plant Area Index (PAI; Jonckheere et al. 2004). While the LAI by definition is limited to the leaf fraction of canopies, the PAI takes into account that also stems and fruits contribute to the canopy surface and, at least during some phenological phases, also bear chlorophyll and stomata. Remote sensors, observing the canopy from a large distance, receive mixed signals from all components of the canopy. Moreover, the possible non-random position of all canopy elements, called "clumping effect", influences the spectral signal. Without correcting for clumping, the green LAI or PAI should be considered as effective LAI (or PAI; Chen and Black 1992; Richter et al. 2009). Nevertheless, the leaf surface normally is the most prominently visible physiological element and hence optical sensors are well suited to measure LAI (Jonckheere et al. 2004; Zheng and Moskal 2009). Variations in LAI cause variations in reflectance in visible, red edge and the NIR spectral domains (Berger et al. 2018; Richter et al. 2012; Viña et al. 2011). The more leaf layers are stacked on top of each other, the higher the reflectance in the NIR will become until saturation occurs (Neuwirthová et al. 2017). LAI retrieval methods consequently are based on quantifying the height of the NIR shoulder. Empirical relations between Normalized Difference Vegetation Index (NDVI; Rouse et al. 1973) or similar indices and LAI have been established in the past, which may seem valid for early vegetative growth stages, where LAI and chlorophyll content develop simultaneously. However, a better and more direct estimation of LAI is possible when red edge and NIR (Kira et al. 2016, 2017; Richter et al. 2012; Viña et al. 2011) and the SWIR domain, which provides information on cellulose accumulation (Jacquemoud et al. 1996), are analysed simultaneously.

Hyperspectral systems, together with physically based or hybrid retrieval methods, can help to improve the estimation accuracy of LAI by providing simultaneous sampling of the chlorophyll absorption, the red edge, the NIR shoulder and the SWIR (Lee et al. 2004; Liu et al. 2016). This allows assessing the variable LAI directly and discriminating the structural variable LAI from the biochemical variable leaf chlorophyll content (LCC). It has to be noted that rather precise LAI mapping is also possible with advanced multispectral observation systems, such as ESA's Sentinel-2 Multispectral Imager (MSI), which also covers the spectral domain that is important for measuring LAI. For precise LAI mapping it nonetheless is important that the observation and illumination geometry is well known, because the spectral signals of LAI and observation angle overlap strongly in the NIR shoulder region. Consequently, only retrieval methods that are capable of considering non-linear angular effects are suited for LAI retrieval, such as the inversion of canopy reflectance models. These models, among them the PROSAIL model being the most prominently applied model in the last years for (agricultural) vegetation studies (Berger et al. 2018; Jacquemoud et al. 2009), originally are of a hyperspectral nature, because they have been developed based on laboratory spectroscopy. Using hyperspectral observation systems in combination with hyperspectral canopy reflectance models thus is expected to improve the stability of LAI mapping from time-series observations, which usually are composed of images with different observation and illumination angles.

### 3.2.2 Average Leaf Inclination Angle

The flux of solar radiation per unit leaf area is directly determined by the angle of a leaf's surface to the horizontal, i.e. the leaf angle. Whereas steeper leaf angles enhance the light capture during low solar zenith angles (early morning/late afternoon and winter), they decrease the light absorption during higher solar zenith angles (midday and summer; Falster and Westoby 2003). Therefore, the leaf angle is one of the essential factors for plant thermoregulation, avoiding overheating during midday/summer, while at the same time increasing water use efficiency (King 1997).

The leaf inclination angle distribution (LAD) is commonly categorized into six typical types: spherical, planophile, erectophile, uniform, extremophile or plagiophile (Liang 2003). Several mathematical functions, i.e. leaf inclination distribution functions (LIDF), have been proposed to describe LAD, such as polynomial, ellipsoidal or elliptic distribution, which in turn can be characterized by the average leaf inclination angle (ALA; Jacquemoud et al. 2000). In particular, the ellipsoidal function has been successfully implemented for the SAIL model (Jacquemoud et al. 2000). ALA and LAD also show dynamic behaviour both in diurnal and seasonal cycles, while they may also vary in different canopy layers. In the early morning hours, leaves tend to be more horizontally inclined. Under the influence of sunlight, the transpiration stream in the xylem is activated and the leaves assume a more erect position due to increased cell turgor. Seasonally, growing plants initially penetrate the Earth's surface with vertically inclined leaves. The leaf angle then gradually declines with progressing phenology until the leaves eventually go limp, either temporarily in phases of drought, or finally when the plant matures.

ALA is of prominent interest for farming because measuring leaf angles helps keeping track of the phenological development and potentially allows detecting drought situations (Zhou et al. 2017). Due to its impact on radiative transfer, it also is a very important input parameter for surface energy balance modelling, which is a prerequisite for growth modelling.

Increasing leaf angles lead to reduced reflectance in the NIR shoulder region. This spectral effect is similar to that of decreasing LAI, so that the discrimination of ALA and LAI induced effects on canopy reflectance is difficult (Atzberger 2004). Both structural variables are the dominant drivers of canopy reflectance with the exception of soil reflectance in sparse canopies. However, hyperspectral sensors that equally resolve the NIR shoulder and the far SWIR potentially can help to discriminate these two vegetation traits.

### 3.2.3 Albedo

The albedo of (vegetated) land surfaces is a non-dimensional ratio of the radiation flux reflected from a surface unit into the whole hemisphere and the incoming irradiance from the upper hemisphere (CEOS 2017). Technically, albedo is known as the bi-hemispherical reflectance factor (BHR; Schaepman-Strub et al. 2006), with a value range from 0 to 1. Albedo belongs to the group of Essential Climate Variables (ECV) as defined by the Global Climate Observing System (GCOS) and requires measurement uncertainty of maximal 5% (0.0025) for climate change monitoring purposes.

Moreover, albedo is an essential input parameter for the calculation of evapotranspiration (ET; Su 2002). ET, combining evaporation and transpiration fluxes from the vegetated surface to the atmosphere, plays a key role in the water and energy balance on the Earth's surface and thus is of particular interest for agriculture regarding irrigation management (D'Urso et al. 2010). The advantage of hyperspectral satellite data in estimating surface broadband albedo can be found in the continuous sampling of the electromagnetic spectrum.

### 3.2.4 Fractional Vegetation Cover

The fractional vegetation cover (fCover) corresponds to the complement of the gap fraction in nadir direction (Weiss et al. 2000). FCover is intrinsic to the vegetation canopy implying that it does not depend on illumination geometry. Since fCover belongs to the group of the main biophysical variables and is involved in agriculturally relevant surface processes such as erosion or interception of precipitation, it sometimes can be considered as an indicator of land degradation and thus is important for regional and global climate (change) modelling, and global change monitoring (Jiménez-Muñoz et al. 2009). Fractional cover is a variable that normally becomes visible at scales below the geometric resolution of optical sensors (e.g., Gitelson 2013). Delineation of fractional cover therefore is a prominent example for spectral unmixing techniques, where hyperspectral data provide improved unambiguousness compared to multispectral data, as has been demonstrated for example by Asner and Heidebrecht (2002).

### 3.2.5 Fraction of Absorbed Photosynthetically Active Radiation

The fraction of absorbed photosynthetically active radiation (fAPAR) constitutes a key variable for the energy and carbon balance of ecosystems for various temporal and spatial scales (Gobron and Verstraete 2009). fAPAR depends on canopy structure, optical properties of plant components and on illumination conditions and is very useful as input to a number of primary production models and thus to agricultural information systems.

Today, several operational coarse resolution fAPAR products are available, approximating the daily-integrated value (e.g., the 10:30 solar time instantaneous black-sky fAPAR at time of sensor overpass). NASA, for instance, provides MODIS/TERRA fAPAR Collection 5 products (MOD15A2) continuously since 2000 through the U.S. Geological Survey portal (Martínez et al. 2013).

The use of hyperspectral data for fAPAR delineation seems logical, because only continuous spectral sampling over the VIS spectral domain can measure the amount of radiation that is absorbed by the leaf. Since fAPAR and fCover are secondary variables, which are calculated from primary state variables such as LAI and LCC describing structure and optical properties of leaves and canopy (Weiss et al. 2000), highly accurate estimations of primary state variables from hyperspectral data also contribute to improved information on secondary variables.

### 3.2.6 Solar-Induced Fluorescence

Solar radiative energy absorbed by chlorophyll molecules in leaves will exit the leaf again via three main de-excitation mechanisms: (1) driving photosynthesis, (2) excess energy dissipation as heat or (3) re-emission in the form of chlorophyll fluorescence (ChlF; Maxwell and Johnson 2000). Since these three mechanisms are coupled, ChlF emitted by vegetation can be seen as indicator of the instantaneous plant photosynthetic functioning (e.g., carbon fixation). ChlF carries information on LUE and therefore captures the dynamic behaviour of photosynthesis, or gross primary productivity (GPP), at the relevant scale (Porcar-Castell et al. 2014; Zarco-Tejada et al. 2013a, 2016). Indirectly, ChlF can be used to detect stress in the actual functional status of vegetation before a response in LCC or LAI becomes visible (Meroni et al. 2009). Remote sensing techniques offer the unique ability to assess photosynthesis continuously over large areas by quantifying solar-induced chlorophyll fluorescence (SIF) from satellite or airborne platforms. Unfortunately, the signal is very small compared to land surface reflectance, so that the radiometric sensitivity of the sensors in most cases limits the geometric resolution of SIF that can be achieved. Due to that, the use of unmanned aerial vehicles (UAV) for near-range SIF sensing is promising for some agricultural applications: e.g., key drought-related variables could be obtained via ChlF signals using a micro-hyperspectral imager on board a UAV (Murchie and Lawson 2013; Zarco-Tejada et al. 2012). Moreover, ChlF imaging is recognized as a well-established effective tool for the assessment of bacterial, fungal and viral infections on crop leaves (Bauriegel et al. 2011). For a comprehensive review of the topic see Meroni et al. (2009). Multispectral systems are not able to provide appropriate SIF signals for land surface applications, because the SIF signal is obscured by the bright reflectance of the land surface in the far red. To discriminate the SIF signal, very high spectral resolution is required to measure the fluorescence spectrum within the two oxygen absorption bands ($O_2A$ and $O_2B$), where interference with land surface reflectance is avoided. Data of this kind will for instance be provided by the FLEX (Fluorescence Explorer) mission. The FLORIS (Fluorescence Imaging Spectrometer) instrument on FLEX will provide very high spectral resolution data with 0.3 nm in the spectral domain of the oxygen absorption bands (Coppo et al. 2017). To discriminate the different energy pathways in the leaf, also leaf temperature and the amount of absorbed photosynthetically active radiation (APAR) must be measured. While the temperature measurements can be provided by thermal imagers, e.g., in the case of FLEX this is achieved through a tandem orbit with Sentinel-3, APAR can be measured via spectroscopy. Remote measurements of photosynthetic activity via SIF therefore combine high spectral resolution and spectrally continuous sampling and thus are a true and exclusive hyperspectral application.

Currently, the limited spatial resolution of spaceborne SIF instruments—the future FLEX mission, for instance, will provide 300 m geometric resolution—does not allow small-scale precision farming applications of SIF products. Newest findings, however, reveal that SIF can potentially be used to discriminate between different sources of crop stress, e.g., water-, temperature-, nitrogen-stress (Ač et al. 2015), making SIF measurements one of the most promising future applications in precision farming. To optimally exploit SIF signals from large areas for practical farming, future hyperspectral satellite sensors should aim to simultaneously provide adequate high spectral (~ 0.3 nm) and high spatial resolution (~ 20 m).

## 3.3 Soil and Land Cover Variables

### 3.3.1 Soil Organic Carbon Content

Soil organic carbon (SOC) content is a key soil property. Hyperspectral remote sensing of SOC effectively supports the identification of several environmental concerns, such as soil erosion (e.g., caused by tillage), salinity or soil contamination (Vaudour et al. 2016). Numerous studies have demonstrated the usefulness of imaging spectroscopy to predict SOC compared to time-consuming field-based methods (Castaldi et al. 2018; Forkuor et al. 2017). A good overview of the use of airborne hyperspectral imagery for the monitoring of SOC contents at within-field scale is given by Vaudour et al. (2016).

### 3.3.2 Land Use/Land Cover

The ability of hyperspectral sensors to differentiate terrestrial features by unique spectral signatures is very valuable for classifying distinct land use/cover features, in particular of the vegetative (cropped) surface. Accurate and detailed land use and land cover monitoring is a prerequisite for documenting farming activities such as crop rotation, or intertillage in particular for large agricultural areas. Several studies exploited hyperspectral data capabilities for agricultural land cover classifications and found improvements over multispectral data (Thenkabail et al. 2004). From a practical farming point of view, increased information depth of land cover maps is of particular interest. Hyperspectral sensors providing detailed spectral analysis of pigment composition can be a key for species and/or weed discrimination (Sykas et al. 2013).

# 4 Advantages of Hyperspectral Versus Multispectral Sensing in an Agricultural Context

The rare availability of spaceborne hyperspectral sensors has led to an exchange of arguments between promoters of multispectral and hyperspectral data. Since the construction of hyperspectral sensors compared to multispectral instruments is complicated as well as expensive, users of hyperspectral sensing are encouraged to carefully justify the requirement for high spectral resolution as well as for full spectral coverage for their specific application. While the positive aspects of high spectral resolution can easily be demonstrated for specific applications (e.g., high spectral resolution is needed to discern narrow absorptions of different pigments for vegetation studies, see Sect. 3), the benefits generated from spectrally continuous sampling cannot be explained in such a straightforward way. In some cases, the detection problems in the case of vegetation can indeed be solved by adding one or two very specific bands, while still maintaining a multispectral setup. For instance, the ESA Sentinel-2 MSI was equipped with two extra bands that now can be used to characterize the shape of the chlorophyll absorption. For vegetation science in an agricultural context, the benefits from continuous spectral sampling rather can be found in the fact that the spectral response of different biophysical/

biochemical traits are interacting with each other across different parts of the electromagnetic spectrum. Changes in canopy structure, which can be observed in the NIR/SWIR, will influence the way in which variations in plant chemistry can be observed in the VIS. Constructing observation systems that are capable of tracking only selected parts of the spectrum thus limits the quality of the analysis.

The exhaustive description of variables with relevance for practical farming in Sect. 3 has shown that the number of different traits targeted with remote sensing largely exceeds the number of spectral bands of conventional multispectral sensors. To avoid underdetermined problems, additional spectral measurements are inevitable. Consequently, hyperspectral approaches already have contributed largely to a more effective and more detailed use and exploitation of remotely sensed data in precision agriculture (Nellis et al. 2009). Several studies exploited hyperspectral data capabilities for agricultural land cover classifications. For agricultural crop species discrimination, Thenkabail et al. (2004) found an increase in classification accuracy of up to 43% when using hyperspectral bands compared to broadband Landsat (ETM+). Although there are many examples for improved retrieval accuracy, in some cases hyperspectral data may not provide significant improvement for variable retrieval (Broge and Leblanc 2001; Spanner et al. 1994). Even then, hyperspectral data have large potential to reduce retrieval uncertainties and thus enhance accuracy and stability of the biophysical and biochemical products indirectly. Although some studies confirm this, e.g., Liu et al. (2016), Richter et al. (2012), Verrelst et al. (2016), still more research is needed to properly quantify the gain of hyperspectral measurements compared to multispectral data for specific applications. Generally, advantages of imaging spectroscopy for agriculture can be summarized as follows:

- Higher accuracy of variable retrieval (Lee et al. 2004; Liu et al. 2016);
- Reduced uncertainties in space and time (Ustin et al. 2001);
- Improved automatic interpretation of EO signals by spectrally separating atmosphere, soil, leaves and canopy effects, since ambiguities are reduced;
- Higher accuracy of crop species discrimination (Thenkabail et al. 2004).
- Novel approaches that provide access to variables hidden for multispectral sensors, e.g., SIF.

To make these potentials of hyperspectral data accessible for agricultural management, preprocessing is required to address the problems associated with the high spectral dimensionality provided by hyperspectral sensors: long processing times and the "curse of dimensionality" or Hughes phenomenon. Several approaches exist to solve these issues and to concurrently optimize the retrieval of the variables of interest. In the spectral domain, for instance, feature (band) selection, feature extraction or dimensionality reduction (DR) techniques can be applied (Rivera-Caicedo et al. 2017; van der Maaten et al. 2009). This may include band selection procedures for individual variable retrievals, as for instance performed by Verrelst et al. (2016), who, using two airborne hyperspectral datasets, identified the most sensitive spectral bands for LCC, LAI and CWC measurements. Another example for successful band selection was provided by Thenkabail et al. (2004), who identified a hyperspectral set of optimal bands that best characterize agricultural crop types. The benefits of hyperspectral sensing for crop discrimination can be traced to the fact that different cultivars may show subtle differences in colour (the green colour of the leaves either shifting towards the blue or towards yellowish tones), which cannot be resolved by multispectral sensors. This positive aspect of hyperspectral sensing so far has not been adequately studied and provides a promising opportunity for more research, once hyperspectral sensing from space becomes operational.

Apart from the discipline-specific advantages listed above, bringing hyperspectral observation systems into orbit also enables the use of one sensor for a wide range of different applications. While high spectral resolution may be required for water analysis above all in the VIS, geological applications will require the same spectral resolution in the SWIR. Vegetation studies again, be it for agricultural applications, for natural ecosystems research, or for combined studies investigating the neighbouring effects of managed and natural vegetation, will require the full range of the spectrum. Instead of constructing a number of different multispectral satellites tuned to either application, a hyperspectral sensor can easily serve different applications and the data may be shared across disciplines following the "one key fits all" principle.

Almost no experience so far exists with the use of spaceborne hyperspectral EO, simply because no operational high-quality system has successfully been installed yet. The new kind of data expected from future hyperspectral spaceborne EO missions may enable the development of innovative analysis methods and applications. Hyperspectral algorithm development for vegetation studies so far has been mostly limited to airborne data, which is very heterogeneous in quality, temporal availability and spatial coverage. Hyperspectral algorithms thus are limited in terms of transferability across sensors and space and cannot make full use of the temporal signal, which is of predominant importance for vegetation studies. The temporal characteristics of vegetation development will perfectly be incorporated in hyperspectral time-series, because spaceborne hyperspectral systems will enable repeated measurements of the reflectance of the Earth's surface with homogeneous quality and characteristics all around the globe by applying the same instrument at every location (Houborg et al. 2017).

The installation of hyperspectral sensors thereby will not make the construction of multispectral systems obsolete, but rather will trigger new developments also in the multispectral domain. For instance, the adding of specific spectral bands to multispectral observation systems to assist detecting specific phenomena, such as red edge, can be traced back to hyperspectral studies, which have been used to define the position of these bands (Clevers et al. 2001).

Due to the contradicting effects of increased information content, which comes along with increased noise, the margins of increased accuracy may only be small for some variables. However, most of the variables retrieved from remote sensing are not used directly, but rather are assimilated into complex decision support information systems (e.g., Hank et al. 2015). According to the laws of error propagation, even small margins of increased accuracy may reduce uncertainties in a processing chain significantly. Besides that, some variables can only be uniquely addressed with hyperspectral sensors, because narrow-band high-resolution data are required with the ability to discriminate components that may be grouped by multispectral bands (Adão et al. 2017):

- Hyperspectral data offer the possibility to retrieve biochemicals via specific pigments such as chlorophyll, anthocyanins or carotenoids, and thus give farmers access to primary production, pest and disease monitoring.
- Hyperspectral data allow discriminating the absorptions of atmospheric water vapour from those of cellulose and lignin to quantify non-photosynthetic vegetation and thus give farmers access to improved crop residue management.
- Hyperspectral data support the discrimination of the three phases of water (ice, liquid, gas; Green et al. 2006) and thus give farmers access to plant water status monitoring.

- Hyperspectral data can help to discriminate protein absorptions from the overlaying cellulose, lignin and water absorptions and thus give farmers access to improved fertilization efficiency monitoring.
- Hyperspectral sensing, by combining the strengths of high resolution and wide range continuous spectral sampling, is the only technique that allows tracing the pathways of energy transformation in the leaves of living vegetation via solar-induced fluorescence and thus gives farmers access to investigating impacts of different stress sources (temperature, nitrogen, water).

## 5 Mission Requirements

This paper reviewed the potentials of spaceborne spectroscopy missions for agriculture. It largely discussed aspects of EO, which have not been sufficiently addressed with currently available EO systems. The elements described in this study rather draw a picture of possible improvements and new possibilities using future operational imaging spectroscopy space missions. The following collection of requirements, where no claim is made of completeness, is meant to shape this vision with special emphasis on the challenges, which are encountered when applying hyperspectral data in the context of practical agriculture.

### 5.1 User Requirements for Satellite-Based Technologies

Providing large amounts of satellite data will not per se contribute to the implementation of more efficient farming strategies. Users of satellite-based information in practical agriculture rather require elaborated information products, which must be tailored to the specific needs of farm management. The continuous stream of data, which is generated by EO missions, must be transferred into information products, which assist farmers with decisions on management actions such as irrigation, fertilization or plant protection. A selection of biophysical and biochemical variables, also defined as Level 2B/3 products including fAPAR, LAI, fCover or LCC, is usually generated by special service providers such as the Sentinel toolboxes (Weiss and Baret 2016). To integrate the satellite-based measurements into farming practice, these variables identified as agriculturally relevant in Sect. 3 (Table 1) need to be further refined to directly address the requirements of individual farmers for specific farming decisions. For instance, for fertilization planning, the absolute amount of nitrogen (N) actually stored in the leaves and especially in the grains as well as the remaining N storage in the soil is needed (in kg N $ha^{-1}$). For plant protection and yield management, spatial maps of dry biomass in leaves, stems and grains, provided in units of g dry matter $m^{-2}$ ground surface, are important. The biophysical and biochemical agriculturally relevant variables, which are directly derived from remote sensing, therefore must be processed to higher-level information products, e.g., by assimilating them into models of agricultural production and by embedding them into integrated information systems, which then provide decision-supporting guidance to farmers. To render these EO products valuable for precision agriculture and to increase their acceptance in practical farming, according to the experience of the authors, the following user requirements apply:

- *Data Accuracy* Accuracies must be optimized and remaining uncertainties must be clearly defined. According to the laws of error propagation, even small margins of increased accuracy at the beginning of a processing chain can have large impact on the accuracy of the final result. This applies especially for the variables retrieved from EO, which are used as spatial input parameters for models of agricultural production. As has been elaborated in Sects. 3 and 4, hyperspectral data can help minimizing uncertainties by providing measurements of very specific variables and by contributing to more accurate and stable retrievals. Accuracies with respect to information content must exceed the expert's present capabilities of estimating a variable's value. Uncertainties of satellite-based products normally are defined through comparison with in situ measurements. These measurements again are subject to errors depending on the variable measured and the sampling method applied. The error of the in situ measurement thus defines the maximum detectable accuracy of a satellite measurement. In order to base the decision-making of farmers on remotely sensed products, the retrieval uncertainties therefore should not exceed the uncertainties of in situ measurements. However, one has to realize that in situ measurements mostly are point measurements, whereas farmers need so-called task maps, indicating zones in a field that should be treated as a separate unit with respect to irrigation, fertilization or plant protection. A big advantage of satellite data is the provision of spatial data or maps.

- *Spatial Accuracy* Most information in the context of precision farming refers to spatial datasets, which provide decision-supporting guidance to farmers by being used as map-overlay for the spatial execution of actual management measures. Since spatial applications in practical farming are frequently based on very high precision navigation, such as real-time-kinematic (RTK) in combination with auto-steering, geometric accuracy must be high, i.e. in the metre domain.
- *Time Constraints* Growing crops is a temporally highly dynamic process, which requires different management decisions throughout a growing cycle. For all decisions, up-to-date spatial information on crop and/or soil status is required. Three important time-related aspects must be considered: sensor repetition rate, time for data-processing and transfer of the products to the users. Since farm management measures usually are planned in accordance with the short-term weather forecast, the actual design of the measures occurs on very short notice, i.e. one or two days in advance. Near real-time transfer of the EO-based information products to the users thus is fundamental. Since the satellite data must be processed and ingested into the final information products, the transfer time of the actual satellite data to the value-adding agents must be kept as short as possible ($< 3$ h).
- *Cost Efficiency* Apart from ecological motivation, profit is the major driver in agricultural innovation. The amount of money a farm is going to invest in optimized production depends on the revenue, which is expected from the measure. For site-specific measures, the potential savings in terms of labour, fuel, machinery hours, fertilizers, plant protection agents, water, etc. and the potential increase in revenues in terms of higher yield levels and higher yield quality largely depend on the spatial heterogeneity of the managed sites. The willingness to invest in spatial information therefore depends on the cost efficiency of the solution.

## 5.2 Observational Requirements for Satellite-Based Technologies

As has been described in Sect. 4, hyperspectral satellite missions can potentially be applied to a wide range of different scientific and practical questions. The technical requirements for an imaging spectroscopy mission should not be exclusively driven by a single application. Nonetheless, from the user requirements for agricultural applications listed in Sect. 5.1, some requirements for the design of satellite missions can be derived:

- *Spatial Resolution* The central application of satellite data in agriculture is improved spatial parameterization of models, which are used as decision support systems for site-specific management measures. The requirements for spatial resolution thus are determined by the spatial detail with which farmers are capable of applying measures on their fields. This level of spatial detail currently is limited by the size and type of the available agricultural machinery (fertilizer spreaders, irrigation systems, etc.). Fertilizer spreaders, as they are nowadays applied, usually have working widths of 40 m. A ground sampling distance (GSD) of 20 m thus would optimally resolve ground heterogeneities for the use with these machines, while a GSD of 30 m can be considered to represent the upper limit. However, technology is constantly evolving towards finer resolution. For instance, newest spreader and irrigation systems allow for an independent adjustment of individual sprayer nozzles along the boom. A GSD of 10 metres consequently should be targeted for future missions.
- *Spectral Resolution* Biophysical/biochemical information exhibits various absorption types. For proteins, lignin and cellulose, exhibiting rather shallow but broad absorption features, moderate spectral resolution of about 10 nm in the SWIR seems to be adequate. Other variables instead, such as anthocyanin and carotene, show more narrow absorptions in the VIS and therefore should be addressed at higher spectral resolutions of approx. 6 nm. This level of spectral resolution seems reasonable for spaceborne missions from a technical point of view. However, for the exploitation of SIF signals on the land surface very high spectral resolution $< 3$ nm would be required. The provision of such high spectral resolutions conflicts with the requirements for spatial resolution described above.
- *Spectral Coverage* According to Fig. 3, the spectral features which give access to biophysical/biochemical properties are distributed over a spectral range between 400 and 2500 nm. While some variables can be targeted through specific narrow absorptions, others require analysing contiguous bands. The so-called full range spectral coverage thus should be aimed for.
- *Radiometric Quality* Especially for hyperspectral systems, where the limited amount of available reflected energy is divided into small portions of the electromagnetic spectrum, radiometric quality is of utmost importance for the success of a satellite mission. Since the analysis of vegetation biophysical/biochemical properties is based on different parts of the spectrum, the considered absorptions can either be narrow or shallow or both. Accurate variable retrieval requires both, high signal-to-noise ratio (SNR) as well as high radiometric resolution. Since SNR is highly determined through the characteristics of the detector, it usually decreases with increasing wavelength. For the agriculturally relevant variables addressed by this study, a minimum SNR of approx. 500 in the VNIR and not lower than 150 in the SWIR should be targeted.
- *Radiometric Resolution* Imaging spectroscopy is characterized by high amounts of data. To ensure the success of spaceborne missions, it would be reasonable to keep data rates as low as possible. However, to adequately resolve also subtle differences in reflectance as caused by shallow absorptions, at least 14-bit quantification should be aimed at.
- *Repeat Cycle* The data from an operational mission can be used in nearly all stages of the crop cycle. The phenological development of crops occurs in time-steps of weeks

rather than months. Taking the problems of cloud cover for an optical mission into account, a revisit time of 5 days would be optimal.
- *Platform Agility* Atmospheric correction is a very important but also critical task for the proper use of hyperspectral imagery. Hyperspectral atmospheric retrieval algorithms, being developed based on airborne measurements, are able to deal with reflectance anisotropies. Thus, repeated data takes for the compilation of time-series do not necessarily need to be conducted at nadir but could also be achieved through across-track pointing. Vegetation canopies show strong anisotropy effects through various geometric complexities. Therefore, pointing capabilities of future satellite sensors, both across and along track, could help to improve variable retrieval and thus could trigger the development of new or more sophisticated multiangular retrieval techniques.
- *Spatial Coverage* Since agriculture happens globally and ensuring food security is a global challenge, the data should at least cover the global croplands between 49°S and 66°N, amounting to 1.873 billion hectares of land surface (Thenkabail 2017).
- *Data Policy* The successful transfer of environmental information into practical agriculture largely depends on the acceptance of the guiding information by farmers. Acceptance is generated from high quality and reasonable pricing of the data products. Taking into account that the actual value for practical farming cannot be found in the data itself, but rather enters the product by processing and value adding, data costs should be kept as low as possible. Apart from commercial applications, continuous high-quality data from a hyperspectral mission would also be important for science and education. Following the example of large operational missions such as Landsat or Copernicus, also the data from future hyperspectral missions should be open, free and easily accessible.

The technical requirements summarized in this section were compiled from a scientific point of view, mostly neglecting technical and financial difficulties. Although we took care not to reach too far from technical feasibility, we are aware that not all of the listed requirements will be feasible in the short term and not by one sensor only. Possible trade-offs between the different requirements therefore must be taken into account, which may be judged very differently depending on the application. We believe that some of the valuable work currently done in the context of preparing spectroscopy missions worldwide should be directed to define priorities for technical requirements in a quantitative way.

## 6 Challenges and Future Directions

Currently observed large-scale global developments, such as climate, population and land use changes, challenge the suppliers of food, fibre and energy to a most efficient use of the bioproductive land surface. This paper aimed at summarizing the potential contributions of spaceborne imaging spectroscopy to increased efficiency in practical farming and derived some user and mission requirements from the state of the art as it has been documented in the literature. Although scientific techniques, such as multi-sensor data fusion and 3D modelling, potentially allow for using spaceborne data also for very small-scale applications, we think that the main use of future spaceborne imaging spectroscopy data in operational precision farming will take place on the spatial scale of the satellite data. Accordingly, the findings with respect to the variables and retrieval techniques described here mainly apply to homogeneous crop farming, i.e. to crops that form geometrically homogeneous stands compared to the observation scale of spaceborne sensors and which are cultivated in areas with individual field sizes in the order of 1 hectare and above. Special cultures like vineyards, orchards, plantations may have different requirements, mostly with respect to spatial resolution (e.g., Zarco-Tejada et al. 2013b), which were not reflected in the mission requirements summarized in Sect. 5. Also, smallholder farming, which still is practized on about 12% of global farmland (Lowder et al. 2016), probably will most likely not be able to adapt the new technologies in near future. Efficiency in agriculture is not limited to crop farming; a large part of farming is concerned with livestock husbandry. This implies for instance the monitoring of different grassland biophysical or biochemical variables for the derivation of management characteristics such as degradation or grazing intensity. In summary, for the case of homogeneous crop farming, several variables have been identified and described, which can be measured by hyperspectral sensors and which provide crucial information for decision support tools in the context of farm management. The door for those tools, to enter into practical farming, already is wide open because of several reasons:

- Service providers exist and are growing, who offer EO-based information in an integrative way, so that farmers do not need to specialize in Earth Observation or Geographical Information Systems themselves.
- Most of the farming equipment installed today is capable of performing at least simple variable rate applications, e.g., through variable forward motion speed, which can be used for site-specific management measures. Meanwhile, technology is rapidly evolving, continuously introducing new site-specific management options. The concept of fertigation, where fertilization and irrigation management are coupled through a single spreader system, is one example.
- Finally yet importantly, farmers already are used to trust spectroscopy measurements, because near-infrared spectroscopy (NIRS) is a standard tool in animal feed analysis (e.g., for measuring residual moisture, crude protein, crude oils and fats, starch, sugar, acid/neutral detergent fibre). However, spectroscopy in agricultural practice so far mainly is limited to laboratory spectroscopy and/or near-range measurements, e.g., inside harvesters.

To introduce information derived from spaceborne spectroscopy into agricultural management practice, farmers need to be convinced. They will base their decisions on satellite spectroscopy, if high quality and high stability of the spatial measurements can be demonstrated and confirmed. This can only be achieved through global satellite missions, which are operated based on a dedicated acquisition plan. If high-quality datasets can be compiled on a regular basis, this would allow for a frequent collection of in situ data through dedicated field campaigns and long-term instrumented outdoor research sites. The high spatial and temporal coverage as well as the fact that the same instruments are globally applied through a satellite mission will allow for testing the transferability of all retrieval approaches that have been developed so far using ground-based or airborne data. Moreover, future hyperspectral spaceborne data sets will trigger the development of new and so far unknown applications. The sensor requirements for measuring relevant agricultural variables are very diverse. An imaging spectrometer designed according to the requirements summarized in this document surely is the key to answering some of the questions. Nevertheless, some applications will require different specifications that cannot be achieved with a single instrument. For example, the exploitation of SIF for agricultural purposes will not only need a high-resolution full-range spectrometer, but will additionally require ultra-high spectral resolution in selected parts of the electromagnetic spectrum and surface temperature measurements at the same time. Installing different instruments along one orbit to follow each other in a so-called train constellation would be a sensible strategy, allowing the integration of new technologies into the chain as they become available. Also, vegetation is a dynamic surface, which shows strong diurnal and seasonal variations. To exploit the full potential of multiangular observations, a tandem concept of two identical instruments in parallel flight could be envisioned. Current ambitious satellite programs, such as the ESA Copernicus program, already have proven that Earth Observation activities with data rates in the domain of 10 Petabyte per year are technically feasible. The even increased data rates expected with high-resolution spectroscopy missions will pose new challenges for efficient data downlinking, data storage and data distribution. Advanced on-board processing strategies should be developed to mitigate the data rate problems.

Although information derived from spaceborne imaging spectroscopy can be applied to increase efficiency in agricultural production, one question cannot yet conclusively be answered: Will increased efficiency in agricultural production also result in sustainability, and if so, to what degree? This issue will have to be investigated by future long-term studies.

**Acknowledgements** The paper arises from the Workshop on "Exploring the Earth's Ecosystems on a Global Scale: Requirements, Capabilities and Directions in Spaceborne Imaging Spectroscopy", held at the International Space Science Institute (ISSI), Bern—Switzerland, on 21–25 November 2016. AVIRIS data are copyright 2016 California Institute of Technology and provided by Jet Propulsion Laboratory, California Institute of Technology, under contract with the National Aeronautics and Space Administration (NASA). Thanks to Martin Danner for the simulation of Sentinel-2 like reflectance from field spectrometer and AVIRIS data and to Lukas Graf for the analyses of AVIRIS data. The help of the reviewers and editors is gratefully acknowledged.



## References

Ač A, Malenovský Z, Olejníčková J, Gallé A, Rascher U, Mohammed G (2015) Meta-analysis assessing potential of steady-state chlorophyll fluorescence for remote sensing detection of plant water, temperature and nitrogen stress. Remote Sens Environ 168:420–436. https://doi.org/10.1016/j.rse.2015.07.022

Adão T, Hruška J, Pádua L, Bessa J, Peres E, Morais R, Sousa J (2017) Hyperspectral imaging: a review on UAV-based sensors. Data Process Appl Agric For Remote Sens 9:1110

Agrawal M, Nandi PK, Rao DN (1982) Effect of ozone and sulphur dioxide pollutants separately and in mixture on chlorophyll and carotenoid pigments of Oryza sativa Water. Air Soil Pollut 18:449–454. https://doi.org/10.1007/bf02426052

Ali AM, Darvishzadeh R, Skidmore AK, van Duren I (2017) Specific leaf area estimation from leaf and canopy reflectance through optimization and validation of vegetation indices. Agric For Meteorol 236:162–174. https://doi.org/10.1016/j.agrformet.2017.01.015

Ariespace (2017) Irrisat. https://www.irrisat.com/en/. Accessed 9/11/2017 2017

Arregui LM, Lasa B, Lafarga A, Irañeta I, Baroja E, Quemada M (2006) Evaluation of chlorophyll meters as tools for N fertilization in winter wheat under humid Mediterranean conditions. Eur J Agron 24:140–148. https://doi.org/10.1016/j.eja.2005.05.005

Asner GP, Heidebrecht KB (2002) Spectral unmixing of vegetation, soil and dry carbon cover in arid regions: comparing multispectral and hyperspectral observations. Int J Remote Sens 23:3939–3958. https://doi.org/10.1080/01431160110115960

Asner GP et al (2011) Taxonomy and remote sensing of leaf mass per area (LMA) in humid tropical forests. Ecol Appl 21:85–98
Atzberger C (2004) Object-based retrieval of biophysical canopy variables using artificial neural nets and radiative transfer models. Remote Sens Environ 93:53–67. https://doi.org/10.1016/j.rse.2004.06.016
Atzberger C (2013) Advances in remote sensing of agriculture: context description, existing operational monitoring systems and major information needs. Remote Sens 5:949
Auernhammer H, Demmel M, Maidl FX, Schmidhalter U, Schneider T, Wagner P (1999) An on-farm communication system for precision farming with nitrogen real-time application. Paper presented at the 1999 ASAE/CSAE-SCGR annual international meeting, Toronto, 18–21 July 1999
Bach H, Angermair W (2017) TalkingFields. http://www.talkingfields.de/. Accessed 9.11.2017 2017
Bach H, Mauser W (2018) Sustainable agriculture and smart farming. In: Mathieu P-P, Aubrecht C (eds) Earth observation open science and innovation. Springer, Cham, pp 261–269. https://doi.org/10.1007/978-3-319-65633-5_12
Bach H, Mauser W, Klepper G (2016) Earth observation for food security and sustainable agriculture. In: ESA special publication SP-740 ESA living planet symposium 2016, Prague
Balafoutis A et al (2017) Precision agriculture technologies positively contributing to GHG emissions mitigation. Farm Prod Econ Sustain 9:1339
Baret F, Buis S (2008) Estimating canopy characteristics from remote sensing observations: review of methods and associated problems. Advances in land remote Sensing. Springer, Dordrecht, pp 173–201
Baret F, Houles V, Guerif M (2007) Quantification of plant stress using remote sensing observations and crop models: the case of nitrogen management. J Exp Bot 58:869–880. https://doi.org/10.1093/jxb/erl231
Barraclough PB, Howarth JR, Jones J, Lopez-Bellido R, Parmar S, Shepherd CE, Hawkesford MJ (2010) Nitrogen efficiency of wheat: genotypic and environmental variation and prospects for improvement. Eur J Agron 33:1–11. https://doi.org/10.1016/j.eja.2010.01.005
Basu S, Ramegowda V, Kumar A, Pereira A (2016) Plant adaptation to drought stress. F1000Research 5:F1000, Faculty Rev 1554. https://doi.org/10.12688/f1000research.7678.1
Bauriegel E, Giebel A, Herppich WB (2011) Hyperspectral and chlorophyll fluorescence imaging to analyse the impact of *Fusarium culmorum* on the photosynthetic integrity of infected wheat ears. Sensors (Basel, Switzerland) 11:3765–3779. https://doi.org/10.3390/s110403765
Berger K, Atzberger C, Danner M, D'Urso G, Mauser W, Vuolo F, Hank T (2018) Evaluation of the PROSAIL model capabilities for future hyperspectral model environments: a review study. Remote Sens 10:85
Blackburn GA (1998) Quantifying chlorophylls and caroteniods at leaf and canopy scales: an evaluation of some hyperspectral approaches. Remote Sens Environ 66:273–285. https://doi.org/10.1016/S0034-4257(98)00059-5
Blackburn GA (2007) Hyperspectral remote sensing of plant pigments. J Exp Bot 58:855–867. https://doi.org/10.1093/jxb/erl123
Blackmore S (1994) Precision farming: an introduction. Outlook Agric 23:275–280. https://doi.org/10.1177/003072709402300407
Breda NJ (2003) Ground-based measurements of leaf area index: a review of methods, instruments and current controversies. J Exp Bot 54:2403–2417. https://doi.org/10.1093/jxb/erg263
Broge NH, Leblanc E (2001) Comparing prediction power and stability of broadband and hyperspectral vegetation indices for estimation of green leaf area index and canopy chlorophyll density. Remote Sens Environ 76:156–172. https://doi.org/10.1016/S0034-4257(00)00197-8
Byerlee D, Stevenson J, Villoria N (2014) Does intensification slow crop land expansion or encourage deforestation? Glob Food Secur 3:92–98. https://doi.org/10.1016/j.gfs.2014.04.001
Calera A, Campos I, Osann A, D'Urso G, Menenti M (2017) Remote sensing for crop water management: from ET modelling to services for the end users. Sensors (Basel). https://doi.org/10.3390/s17051104
Castaldi F, Chabrillat S, Jones A, Vreys K, Bomans B, van Wesemael B (2018) Soil organic carbon estimation in croplands by hyperspectral remote APEX data using the LUCAS topsoil database. Remote Sens 10:153
CEOS E (2018) http://database.eohandbook.com/climate/gcosecv.aspx?gcosipECVID=22. Accessed 18 July 2018
Chappelle EW, Kim MS, McMurtrey JE (1992) Ratio analysis of reflectance spectra (RARS): an algorithm for the remote estimation of the concentrations of chlorophyll A, chlorophyll B, and carotenoids in soybean leaves. Remote Sens Environ 39:239–247. https://doi.org/10.1016/0034-4257(92)90089-3
Chen JM, Black TA (1992) Defining leaf area index for non-flat leaves Plant. Cell Environ 15:421–429. https://doi.org/10.1111/j.1365-3040.1992.tb00992.x

Chen M, Zhang H, Liu W, Zhang W (2014) The global pattern of urbanization and economic growth: evidence from the last three decades. PLoS ONE 9:e103799. https://doi.org/10.1371/journal.pone.0103799

Cidad VG, Vrindts E, Baerdemaeker JD Use of very high resolution satellite images for precision farming: recommendations on nitrogen fertilization. In: Europto remote sensing, 2001. SPIE, p 10

Clevers JGPW, Gitelson AA (2013) Remote estimation of crop and grass chlorophyll and nitrogen content using red-edge bands on sentinel-2 and-3. Int J Appl Earth Obs Geoinf 23:344–351. https://doi.org/10.1016/j.jag.2012.10.008

Clevers JGPW, de Jong SM, Epema GF, van der Meer F, Bakker WH, Skidmore AK, Addink EA (2001) MERIS and the red-edge position. Int J Appl Earth Obs Geoinf 3:313–320. https://doi.org/10.1016/S0303-2434(01)85038-8

Clevers JGPW, Kooistra L, Schaepman ME (2010) Estimating canopy water content using hyperspectral remote sensing data. Int J Appl Earth Obs Geoinf 12:119–125. https://doi.org/10.1016/j.jag.2010.01.007

Clevers JGPW, Kooistra L, Van den Brande MM (2017) Using Sentinel-2 data for retrieving LAI and leaf and canopy chlorophyll content of a potato crop. Remote Sens 9:405. https://doi.org/10.3390/rs9050405

Combal B et al (2003) Retrieval of canopy biophysical variables from bidirectional reflectance: using prior information to solve the ill-posed inverse problem. Remote Sens Environ 84:1–15. https://doi.org/10.1016/S0034-4257(02)00035-4

Coppo P, Taiti A, Pettinato L, Francois M, Taccola M, Drusch M (2017) Fluorescence imaging spectrometer (FLORIS) for ESA FLEX mission. Remote Sens 9:649

Curran PJ (1989) Remote sensing of foliar chemistry. Remote Sens Environ 30:271–278. https://doi.org/10.1016/0034-4257(89)90069-2

Danbom DB (1995) From Columbus to ConAgra: the globalization of agriculture and food. In: Bonanno A, Busch L, Friedland WH, Gouveia L, Mingione E (eds) (Lawrence: University Press of Kansas, 1994. viii, 294 pp. Cloth, $40.00, ISBN 0-70060660-2. Paper, $17.95, ISBN 0-7006-0661-0.) J Am Hist 82:676–677. https://doi.org/10.2307/2082206

Danner M, Berger K, Wocher M, Mauser W, Hank T (2017) Retrieval of biophysical crop variables from multi-angular canopy spectroscopy. Remote Sens 9:726. https://doi.org/10.3390/rs9070726

Daughtry CST, Walthall CL, Kim MS, de Colstoun EB, McMurtrey JE (2000) Estimating Corn leaf chlorophyll concentration from leaf and canopy reflectance. Remote Sens Environ 74:229–239. https://doi.org/10.1016/S0034-4257(00)00113-9

Daughtry CST, Hunt ER, Doraiswamy PC, McMurtrey JE (2005) Remote sensing the spatial distribution of crop residues. Agron J 97:864–871. https://doi.org/10.2134/agronj2003.0291

de la Riva EG, Olmo M, Poorter H, Ubera JL, Villar R (2016) Leaf mass per area (LMA) and its relationship with leaf structure and anatomy in 34 mediterranean woody species along a water availability gradient. PLoS ONE 11:e0148788. https://doi.org/10.1371/journal.pone.0148788

Dennison P et al (2016) Global measurement of non-photosynthetic vegetation. NASA, Washington

Durbha SS, King RL, Younan NH (2007) Support vector machines regression for retrieval of leaf area index from multiangle imaging spectroradiometer. Remote Sens Environ 107:348–361. https://doi.org/10.1016/j.rse.2006.09.031

D'Urso G et al (2010) Earth observation products for operational irrigation management in the context of the PLEIADeS project. Agric Water Manag 98:271–282. https://doi.org/10.1016/j.agwat.2010.08.020

Erickson B, Widmar DA (2015) Precision agricultural services dealership survey results. Purdue University Indiana, West Lafayette

Falster DS, Westoby M (2003) Leaf size and angle vary widely across species: what consequences for light interception? New Phytol 158:509–525. https://doi.org/10.1046/j.1469-8137.2003.00765.x

FAO (2016) The state of food and agriculture 2016. Climate change, agriculture and food security. Food and Agriculture Organization of the United Nations, FAO, Rome

Feret J-B et al (2008) PROSPECT-4 and 5: advances in the leaf optical properties model separating photosynthetic pigments. Remote Sens Environ 112:3030–3043. https://doi.org/10.1016/j.rse.2008.02.012

Féret JB, Gitelson AA, Noble SD, Jacquemoud S (2017) PROSPECT-D: towards modeling leaf optical properties through a complete lifecycle. Remote Sens Environ 193:204–215. https://doi.org/10.1016/j.rse.2017.03.004

Forkuor G, Hounkpatin OKL, Welp G, Thiel M (2017) High resolution mapping of soil properties using remote sensing variables in south-western burkina faso: a comparison of machine learning and multiple linear regression models. PLoS ONE 12:e0170478. https://doi.org/10.1371/journal.pone.0170478

Frampton WJ, Dash J, Watmough G, Milton EJ (2013) Evaluating the capabilities of Sentinel-2 for quantitative estimation of biophysical variables in vegetation. ISPRS J Photogramm Remote Sens 82:83–92. https://doi.org/10.1016/j.isprsjprs.2013.04.007
Gitelson AA (2013) Remote estimation of crop fractional vegetation cover: the use of noise equivalent as an indicator of performance of vegetation indices. Int J Remote Sens 34:6054–6066. https://doi.org/10.1080/01431161.2013.793868
Gitelson A, Solovchenko A (2017) Generic algorithms for estimating foliar pigment content. Geophys Res Lett 44:9293–9298. https://doi.org/10.1002/2017GL074799
Gitelson AA, Merzlyak MN, Chivkunova OB (2001) Optical properties and nondestructive estimation of anthocyanin content in plant leaves. Photochem Photobiol 74:38–45
Gitelson AA, Zur Y, Chivkunova OB, Merzlyak MN (2002) Assessing carotenoid content in plant leaves with reflectance spectroscopy. Photochem Photobiol 75:272–281
Gitelson AA, Gritz Y, Merzlyak MN (2003) Relationships between leaf chlorophyll content and spectral reflectance and algorithms for non-destructive chlorophyll assessment in higher plant leaves. J Plant Physiol 160:271–282. https://doi.org/10.1078/0176-1617-00887
Gitelson AA et al (2006) Relationship between gross primary production and chlorophyll content in crops: Implications for the synoptic monitoring of vegetation productivity. J Geophys Res Atmos 111:n/a–n/a. https://doi.org/10.1029/2005jd006017
Gitelson AA, Chivkunova OB, Merzlyak MN (2009) Nondestructive estimation of anthocyanins and chlorophylls in anthocyanic leaves. Am J Bot 96:1861–1868. https://doi.org/10.3732/ajb.0800395
Gobron N, Verstraete MM (2009) ECV T10: fraction of absorbed photosynthetically active radiation (FAPAR) essential climate variables. Global Terrestrial Observing System, Rome
Goetz AF, Vane G, Solomon JE, Rock BN (1985) Imaging spectrometry for Earth remote sensing. Science 228:1147–1153. https://doi.org/10.1126/science.228.4704.1147
Goodwin N, Nelson JA, Ackermann F, Weisskopf T (2008) Consumption and the consumer society. Global Development And Environment Institute, Tufts University, Medford
GOS (2011) Foresight. The future of food and farming—challenges and choices for global sustainability. The Government Office for Science, London
Green RO, Painter TH, Roberts DA, Dozier J (2006) Measuring the expressed abundance of the three phases of water with an imaging spectrometer over melting snow. Water Resour Res. https://doi.org/10.1029/2005WR004509
Guanter L et al (2015) The EnMAP spaceborne imaging spectroscopy mission for earth observation. Remote Sens 7:8830
Gunasekera D, Finnigan J (2010) Track social and economic impacts of food production. Nature 466:920. https://doi.org/10.1038/466920e
Haboudane D, Miller JR, Tremblay N, Zarco-Tejada PJ, Dextraze L (2002) Integrated narrow-band vegetation indices for prediction of crop chlorophyll content for application to precision agriculture. Remote Sens Environ 81:416–426. https://doi.org/10.1016/S0034-4257(02)00018-4
Haboudane D, Miller JR, Pattey E, Zarco-Tejada PJ, Strachan IB (2004) Hyperspectral vegetation indices and novel algorithms for predicting green LAI of crop canopies: modeling and validation in the context of precision agriculture. Remote Sens Environ 90:337–352. https://doi.org/10.1016/j.rse.2003.12.013
Hank T, Bach H, Mauser W (2015) Using a remote sensing-supported hydro-agroecological model for field-scale simulation of heterogeneous crop growth and yield: application for wheat in Central Europe. Remote Sens 7:3934
Haubrock S, Chabrillat S, Kuhnert M, Hostert P, Kaufmann HJ (2008) Surface soil moisture quantification and validation based on hyperspectral data and field measurements. J Appl Remote Sens 2:26
Herrmann I, Pimstein A, Karnieli A, Cohen Y, Alchanatis V, Bonfil DJ (2011) LAI assessment of wheat and potato crops by VENμS and Sentinel-2 bands. Remote Sens Environ 115:2141–2151. https://doi.org/10.1016/j.rse.2011.04.018
Herwitz SR et al (2004) Imaging from an unmanned aerial vehicle: agricultural surveillance and decision support. Comput Electron Agric 44:49–61. https://doi.org/10.1016/j.compag.2004.02.006
Homolová L, Malenovský Z, Clevers JGPW, García-Santos G, Schaepman ME (2013) Review of optical-based remote sensing for plant trait mapping. Ecol Complex 15:1–16. https://doi.org/10.1016/j.ecocom.2013.06.003
Houborg R, McCabe MF, Angel Y, Middleton EM (2017) Time series from hyperion to track productivity in pivot agriculture in saudi arabia. In: 2017 IEEE international geoscience and remote sensing symposium (IGARSS), 23–28 July 2017. pp 3047–3050. https://doi.org/10.1109/igarss.2017.8127641

Ibdah M, Dubey NK, Eizenberg H, Dabour Z, Abu-Nassar J, Gal-On A, Aly R (2014) Cucumber Mosaic Virus as a carotenoid inhibitor reducing *Phelipanche aegyptiaca* infection in tobacco plants. Plant Signal Behav 9:e972146. https://doi.org/10.4161/psb.32096

Jacquemoud S, Ustin SL, Verdebout J, Schmuck G, Andreoli G, Hosgood B (1996) Estimating leaf biochemistry using the PROSPECT leaf optical properties model. Remote Sens Environ 56:194–202. https://doi.org/10.1016/0034-4257(95)00238-3

Jacquemoud S, Bacour C, Poilvé H, Frangi JP (2000) Comparison of four radiative transfer models to simulate plant canopies reflectance: direct and inverse mode. Remote Sens Environ 74:471–481. https://doi.org/10.1016/S0034-4257(00)00139-5

Jacquemoud S et al (2009) PROSPECT + SAIL models: a review of use for vegetation characterization. Remote Sens Environ 113(Supplement 1):S56–S66. https://doi.org/10.1016/j.rse.2008.01.026

Jay S, Maupas F, Bendoula R, Gorretta N (2017) Retrieving LAI, chlorophyll and nitrogen contents in sugar beet crops from multi-angular optical remote sensing: comparison of vegetation indices and PROSAIL inversion for field phenotyping. Field Crops Res 210:33–46. https://doi.org/10.1016/j.fcr.2017.05.005

Jiménez-Muñoz J, Sobrino J, Plaza A, Guanter L, Moreno J, Martinez P (2009) Comparison between fractional vegetation cover retrievals from vegetation indices and spectral mixture analysis: case study of PROBA/CHRIS data over an agricultural area. Sensors 9:768

Jonckheere I, Fleck S, Nackaerts K, Muys B, Coppin P, Weiss M, Baret F (2004) Review of methods for in situ leaf area index determination: part I. Theories, sensors and hemispherical photography. Agric For Meteorol 121:19–35. https://doi.org/10.1016/j.agrformet.2003.08.027

Justice CO et al (2002) An overview of MODIS Land data processing and product status. Remote Sens Environ 83:3–15. https://doi.org/10.1016/S0034-4257(02)00084-6

Kimes DS, Knyazikhin Y, Privette JL, Abuelgasim AA, Gao F (2000) Inversion methods for physically-based models. Remote Sens Rev 18:381–439. https://doi.org/10.1080/02757250009532396

King DA (1997) The functional significance of leaf angle in eucalyptus. Aust J Bot 45:619–639. https://doi.org/10.1071/BT96063

Kira O, Nguy-Robertson AL, Arkebauer TJ, Linker R, Gitelson AA (2016) Informative spectral bands for remote green LAI estimation in C3 and C4 crops. Agric For Meteorol 218–219:243–249. https://doi.org/10.1016/j.agrformet.2015.12.064

Kira O, Nguy-Robertson AL, Arkebauer TJ, Linker R, Gitelson AA (2017) Toward generic models for green LAI estimation in maize and soybean: satellite observations. Remote Sens 9:318

Kokaly RF, Skidmore AK (2015) Plant phenolics and absorption features in vegetation reflectance spectra near 1.66μm. Int J Appl Earth Obs Geoinf 43:55–83. https://doi.org/10.1016/j.jag.2015.01.010

Kokaly RF, Despain DG, Clark RN, Livo KE (2007) Spectral analysis of absorption features N for mapping vegetation cover and microbial communities in yellowstone national park using AVIRIS data. In: Morgan LA (ed) Integrated geoscience studies in the greater yellowstone area—volcanic, tectonic, and hydrothermal processes in the yellowstone geoecosystem. U.S. Geological Survey Professional Paper 1717, 532 p., vol Paper 1717. p 532

Kokaly RF, Asner GP, Ollinger SV, Martin ME, Wessman CA (2009) Characterizing canopy biochemistry from imaging spectroscopy and its application to ecosystem studies. Remote Sens Environ 113:S78–S91. https://doi.org/10.1016/j.rse.2008.10.018

Kong W et al (2017) Estimation of canopy carotenoid content of winter wheat using multi-angle hyperspectral data. Adv Space Res 60:1988–2000. https://doi.org/10.1016/j.asr.2017.07.015

Labate D et al (2009) The PRISMA payload optomechanical design, a high performance instrument for a new hyperspectral mission. Acta Astronaut 65:1429–1436. https://doi.org/10.1016/j.actaastro.2009.03.077

le Maire G et al (2008) Calibration and validation of hyperspectral indices for the estimation of broadleaved forest leaf chlorophyll content, leaf mass per area, leaf area index and leaf canopy biomass. Remote Sens Environ 112:3846–3864. https://doi.org/10.1016/j.rse.2008.06.005

Lee D, Gould K (2002) Why leaves turn red: pigments called anthocyanins probably protect leaves from light damage by direct shielding and by scavenging free radicals. Am Sci 90(6):524–531

Lee K-S, Cohen WB, Kennedy RE, Maiersperger TK, Gower ST (2004) Hyperspectral versus multispectral data for estimating leaf area index in four different biomes. Remote Sens Environ 91:508–520. https://doi.org/10.1016/j.rse.2004.04.010

Liang S (2003) Quantitative remote sensing of land surfaces. Wiley, Hoboken

Liu K, Zhou Q-B, Wu W-B, Xia T, Tang H-J (2016) Estimating the crop leaf area index using hyperspectral remote sensing. J Integr Agric 15:475–491. https://doi.org/10.1016/S2095-3119(15)61073-5

Lowder SK, Skoet J, Raney T (2016) The number, size, and distribution of farms, smallholder farms, and family farms worldwide. World Dev 87:16–29. https://doi.org/10.1016/j.worlddev.2015.10.041

Martínez B, Camacho F, Verger A, García-Haro FJ, Gilabert MA (2013) Intercomparison and quality assessment of MERIS, MODIS and SEVIRI FAPAR products over the Iberian Peninsula. Int J Appl Earth Obs Geoinf 21:463–476. https://doi.org/10.1016/j.jag.2012.06.010
Mauser W, Bach H, Hank T, Zabel F, Putzenlechner B How spectroscopy from space will support world agriculture. In: 2012 IEEE international geoscience and remote sensing symposium, 22–27 July 2012. pp 7321–7324. https://doi.org/10.1109/igarss.2012.6351970
Maxwell K, Johnson GN (2000) Chlorophyll fluorescence—a practical guide. J Exp Bot 51:659–668. https://doi.org/10.1093/jexbot/51.345.659
Meroni M, Rossini M, Guanter L, Alonso L, Rascher U, Colombo R, Moreno J (2009) Remote sensing of solar-induced chlorophyll fluorescence: review of methods and applications. Remote Sens Environ 113:2037–2051. https://doi.org/10.1016/j.rse.2009.05.003
Mibei EK, Ambuko J, Giovannoni JJ, Onyango AN, Owino WO (2017) Carotenoid profiling of the leaves of selected African eggplant accessions subjected to drought stress. Food Sci Nutr 5:113–122. https://doi.org/10.1002/fsn3.370
Migdall S, Bach H, Bobert J, Wehrhan M, Mauser W (2009) Inversion of a canopy reflectance model using hyperspectral imagery for monitoring wheat growth and estimating yield. Precis Agric 10:508–524
Moran MS, Inoue Y, Barnes EM (1997) Opportunities and limitations for image-based remote sensing in precision crop management. Remote Sens Environ 61:319–346. https://doi.org/10.1016/S0034-4257(97)00045-X
Mulla DJ (2013) Twenty five years of remote sensing in precision agriculture: key advances and remaining knowledge gaps. Biosyst Eng 114:358–371. https://doi.org/10.1016/j.biosystemseng.2012.08.009
Murchie EH, Lawson T (2013) Chlorophyll fluorescence analysis: a guide to good practice and understanding some new applications. J Exp Bot 64:3983–3998. https://doi.org/10.1093/jxb/ert208
Nellis DM, Price KP, Rundquist D (2009) Remote sensing of cropland agriculture. In: Warner TA, Nellis DM, Foody GM (eds) The SAGE handbook of remote sensing. SAGE Publications. https://doi.org/10.4135/978-1-8570-2105-9.n26
Neuwirthová E, Lhotáková Z, Albrechtová J (2017) The effect of leaf stacking on leaf reflectance and vegetation indices measured by contact probe during the season. Sensors (Basel, Switzerland) 17:1202. https://doi.org/10.3390/s17061202
Nidamanuri RR, Zbell B (2011) Use of field reflectance data for crop mapping using airborne hyperspectral image. ISPRS J Photogramm Remote Sens 66:683–691. https://doi.org/10.1016/j.isprsjprs.2011.05.001
Nobel PS (2009) Physicochemical and environmental plant physiology, 4th edn. Academic Press, San Diego. https://doi.org/10.1016/B978-0-12-374143-1.00001-6
Okin GS (2010) The contribution of brown vegetation to vegetation dynamics. Ecology 91:743–755
Ozdogan M, Yang Y, Allez G, Cervantes C (2010) Remote sensing of irrigated agriculture: opportunities and challenges. Remote Sens 2:2274
Panda SK, Chaudhury I, Khan MH (2003) Heavy metals induce lipid peroxidation and affect antioxidants in wheat leaves. Biol Plant 46:289–294. https://doi.org/10.1023/a:1022871131698
Parry C, Blonquist JM, Bugbee B (2014) In situ measurement of leaf chlorophyll concentration: analysis of the optical/absolute relationship Plant. Cell Environ 37:2508–2520. https://doi.org/10.1111/pce.12324
Peng Y, Gitelson AA, Keydan G, Rundquist DC, Moses W (2011) Remote estimation of gross primary production in maize and support for a new paradigm based on total crop chlorophyll content. Remote Sens Environ 115:978–989. https://doi.org/10.1016/j.rse.2010.12.001
Peng Y, Nguy-Robertson A, Arkebauer T, Gitelson A (2017) Assessment of canopy chlorophyll content retrieval in maize and soybean: implications of hysteresis on the development of generic algorithms. Remote Sens 9:226
Peters J (2012) On-farm moisture testing of corn silage. Focus Forage 2(6):1–3
Plant RE, Pettygrove GS, Reinert WR (2000) Precision agriculture can increase profits and limit environmental impacts. Calif Agric 54:66–71. https://doi.org/10.3733/ca.v054n04p66
Poorter H, Evans JR (1998) Photosynthetic nitrogen-use efficiency of species that differ inherently in specific leaf area. Oecologia 116:26–37. https://doi.org/10.1007/s004420050560
Porcar-Castell A et al (2014) Linking chlorophyll a fluorescence to photosynthesis for remote sensing applications: mechanisms and challenges. J Exp Bot 65:4065–4095. https://doi.org/10.1093/jxb/eru191
Prashar A, Jones HG (2016) Assessing drought responses using thermal infrared imaging. Methods Mol Biol 1398:209–219. https://doi.org/10.1007/978-1-4939-3356-3_17
Richter K, Atzberger C, Vuolo F, Weihs P, D'Urso G (2009) Experimental assessment of the Sentinel-2 band setting for RTM-based LAI retrieval of sugar beet and maize. Can J Remote Sens 35:230–247
Richter K, Hank TB, Vuolo F, Mauser W, D'Urso G (2012) Optimal exploitation of the sentinel-2 spectral capabilities for crop leaf area index mapping. Remote Sens 4:561

Ritter L, Solomon K, Sibley P, Hall K, Keen P, Mattu G, Linton B (2002) Sources, pathways, and relative risks of contaminants in surface water and groundwater: a perspective prepared for the Walkerton inquiry. J Toxicol Environ Health A 65:1–142

Rivera-Caicedo JP, Verrelst J, Muñoz-Marí J, Camps-Valls G, Moreno J (2017) Hyperspectral dimensionality reduction for biophysical variable statistical retrieval. ISPRS J Photogramm Remote Sens 132:88–101. https://doi.org/10.1016/j.isprsjprs.2017.08.012

Roberts DA, Quattrochi DA, Hulley GC, Hook SJ, Green RO (2012) Synergies between VSWIR and TIR data for the urban environment: an evaluation of the potential for the Hyperspectral Infrared Imager (HyspIRI) Decadal Survey mission. Remote Sens Environ 117:83–101. https://doi.org/10.1016/j.rse.2011.07.021

Robertson GP, Bruulsema TW, Gehl RJ, Kanter D, Mauzerall DL, Rotz CA, Williams CO (2013) Nitrogen–climate interactions in US agriculture. Biogeochemistry 114:41–70. https://doi.org/10.1007/s10533-012-9802-4

Rockström J, Lannerstad M, Falkenmark M (2007) Assessing the water challenge of a new green revolution in developing countries. Proc Natl Acad Sci USA 104:6253–6260. https://doi.org/10.1073/pnas.0605739104

Rouse JW, Haas RH, Schell JA, Deering DW (1973) Monitoring vegetation systems in the Great Plains with ERTS. Paper presented at the Third ERTS Symposium

Sahoo RN, Ray SS, Manjunath KR (2015) Hyperspectral remote sensing of agriculture. Curr Sci 108(5):848–859

Savoy P, Mackay DS (2015) Modeling the seasonal dynamics of leaf area index based on environmental constraints to canopy development. Agric For Meteorol 200:46–56. https://doi.org/10.1016/j.agrformet.2014.09.019

Schaepman ME, Koetz B, Schaepman-Strub G, Itten KI (2005) Spectrodirectional remote sensing for the improved estimation of biophysical and-chemical variables: two case studies. Int J Appl Earth Obs Geoinf 6:271–282. https://doi.org/10.1016/j.jag.2004.10.012

Schaepman-Strub G, Schaepman ME, Painter TH, Dangel S, Martonchik JV (2006) Reflectance quantities in optical remote sensing—definitions and case studies. Remote Sens Environ 103(1):27–42. https://doi.org/10.1016/j.rse.2006.03.002

Schlemmer M, Gitelson A, Schepers J, Ferguson R, Peng Y, Shanahan J, Rundquist D (2013) Remote estimation of nitrogen and chlorophyll contents in maize at leaf and canopy levels. Int J Appl Earth Obs Geoinf 25:47–54. https://doi.org/10.1016/j.jag.2013.04.003

Shams A, Hooshmandan Z, Fard M (2017) Factors affecting wheat farmers' attitudes toward organic farming. Pol J Environ Stud 26:2207–2214. https://doi.org/10.15244/pjoes/69435

Spanner M, Johnson L, Miller J, McCreight R, Freemantle J, Runyon J, Gong P (1994) Remote sensing of seasonal leaf area index across the oregon transect. Ecol Appl 4:258–271. https://doi.org/10.2307/1941932

Springob K, Nakajima J, Yamazaki M, Saito K (2003) Recent advances in the biosynthesis and accumulation of anthocyanins. Nat Prod Rep 20:288–303

Stevens A, Udelhoven T, Denis A, Tychon B, Lioy R, Hoffmann L, van Wesemael B (2010) Measuring soil organic carbon in croplands at regional scale using airborne imaging spectroscopy. Geoderma 158:32–45. https://doi.org/10.1016/j.geoderma.2009.11.032

Su Z (2002) The surface energy balance system (SEBS) for estimation of turbulent heat fluxes. Hydrol Earth Syst Sci 6:85–100. https://doi.org/10.5194/hess-6-85-2002

Sykas D, Karathanassi V, Fountas S A new methodology for the discrimination of plant species and their varieties using hyperspectral data: application on vetch and lentil. In: First international conference on remote sensing and geoinformation of environment, 2013. SPIE, p 8

Tanaka Y, Sasaki N, Ohmiya A (2008) Biosynthesis of plant pigments: anthocyanins, betalains and carotenoids. Plant J 54:733–749. https://doi.org/10.1111/j.1365-313X.2008.03447.x

Thenkabail PS (2017) https://www.usgs.gov/news/new-map-worldwide-croplands-supports-food-and-water-security. Accessed 18 July 2018

Thenkabail PS, Smith RB, De Pauw E (2000) Hyperspectral vegetation indices and their relationships with agricultural crop characteristics. Remote Sens Environ 71:158–182. https://doi.org/10.1016/S0034-4257(99)00067-X

Thenkabail PS, Enclona EA, Ashton MS, Van Der Meer B (2004) Accuracy assessments of hyperspectral waveband performance for vegetation analysis applications. Remote Sens Environ 91:354–376. https://doi.org/10.1016/j.rse.2004.03.013

Thenkabail PS, Schull M, Turral H (2005) Ganges and Indus river basin land use/land cover (LULC) and irrigated area mapping using continuous streams of MODIS data. Remote Sens Environ 95:317–341. https://doi.org/10.1016/j.rse.2004.12.018

Thenkabail PS, Mariotto I, Gumma MK, Middleton EM, Landis DR, Huemmrich KF (2013) Selection of hyperspectral narrowbands (HNBs) and composition of hyperspectral twoband vegetation indices (HVIs) for biophysical characterization and discrimination of crop types using field reflectance and hyperion/EO-1 data. IEEE J Selec Top Appl Earth Obs Remote Sens 6:427–439. https://doi.org/10.1109/JSTARS.2013.2252601

Ustin S, Zarco-Tejada PJ, Asner GP (2001) The role of hyperspectral data in understanding the global carbon cycle. In: Aviris Workshop, Pasadena, California, 27/02–02/03 2001. NASA/JPL

Vaesen K, Gilliams S, Nackaerts K, Coppin P (2001) Ground-measured spectral signatures as indicators of ground cover and leaf area index: the case of paddy rice. Field Crops Res 69:13–25. https://doi.org/10.1016/S0378-4290(00)00129-5

van der Maaten LJP, Postma EO, van den Herik HJ (2009) Dimensionality reduction: a comparative review. Technical Report TiCC TR 2009-005, Maastricht University

Vaudour E, Gilliot JM, Bel L, Lefevre J, Chehdi K (2016) Regional prediction of soil organic carbon content over temperate croplands using visible near-infrared airborne hyperspectral imagery and synchronous field spectra. Int J Appl Earth Obs Geoinf 49:24–38. https://doi.org/10.1016/j.jag.2016.01.005

Verhoef W, Bach H (2003) Simulation of hyperspectral and directional radiance images using coupled biophysical and atmospheric radiative transfer models. Remote Sens Environ 87:23–41

Verhoef W, Bach H (2007) Coupled soil-leaf-canopy and atmosphere radiative transfer modeling to simulate hyperspectral multi-angular surface reflectance and TOA radiance data. Remote Sens Environ 109:166–182

Verrelst J, Camps-Valls G, Muñoz-Marí J, Rivera JP, Veroustraete F, Clevers JGPW, Moreno J (2015) Optical remote sensing and the retrieval of terrestrial vegetation bio-geophysical properties—a review. ISPRS J Photogramm Remote Sens 108:273–290. https://doi.org/10.1016/j.isprsjprs.2015.05.005

Verrelst J, Rivera JP, Gitelson A, Delegido J, Moreno J, Camps-Valls G (2016) Spectral band selection for vegetation properties retrieval using Gaussian processes regression. Int J Appl Earth Obs Geoinf 52:554–567. https://doi.org/10.1016/j.jag.2016.07.016

Viña A, Gitelson AA, Nguy-Robertson AL, Peng Y (2011) Comparison of different vegetation indices for the remote assessment of green leaf area index of crops. Remote Sens Environ 115:3468–3478. https://doi.org/10.1016/j.rse.2011.08.010

Vuolo F et al (2016) Data service platform for sentinel-2 Surface reflectance and value-added products: system use and examples. Remote Sens 8:938

Walter A, Finger R, Huber R, Buchmann N (2017) Opinion: smart farming is key to developing sustainable agriculture. Proc Natl Acad Sci U S A 114:6148–6150. https://doi.org/10.1073/pnas.1707462114

Walthall CL (1997) A study of reflectance anisotropy and canopy structure using a simple empirical model. Remote Sens Environ 61:118–128. https://doi.org/10.1016/S0034-4257(96)00245-3

Wang Z, Skidmore AK, Wang T, Darvishzadeh R, Hearne J (2015) Applicability of the PROSPECT model for estimating protein and cellulose + lignin in fresh leaves. Remote Sens Environ 168:205–218. https://doi.org/10.1016/j.rse.2015.07.007

Watson DJ (1947) Comparative physiological studies in the growth of field crops. I. Variation in net assimilation rate and leaf area between species and varieties, and within and between years. Ann Bot 11:41–76

Weiss M, Baret F (2016) S2ToolBox level 2 products: LAI, FAPAR, FCOVER. Institut National de la Recherche Agronomique (INRA), Avignon

Weiss M, Baret F, Myneni RB, Pragnere A, Knyazikhin Y (2000) Investigation of a model inversion technique to estimate canopy biophysical variables from spectral and directional reflectance data. Agronomie 20:3–22

Wilson J, Zhang C, Kovacs J (2014) Separating crop species in northeastern ontario using hyperspectral data. Remote Sens 6:925

Yang Y, Ling PP (2004) Non-contacting techniques for plant drought stress detection. Am Soc Agric Biol Eng. https://doi.org/10.13031/2013.16168

Yang C, Everitt JH, Du Q, Luo B, Chanussot J (2013) Using high-resolution airborne and satellite imagery to assess crop growth and yield variability for precision agriculture. Proc IEEE 101:582–592. https://doi.org/10.1109/JPROC.2012.2196249

Yi Q, Jiapaer G, Chen J, Bao A, Wang F (2014) Different units of measurement of carotenoids estimation in cotton using hyperspectral indices and partial least square regression. ISPRS J Photogramm Remote Sens 91:72–84. https://doi.org/10.1016/j.isprsjprs.2014.01.004

Zarco-Tejada PJ, González-Dugo V, Berni JAJ (2012) Fluorescence, temperature and narrow-band indices acquired from a UAV platform for water stress detection using a micro-hyperspectral imager and a thermal camera. Remote Sens Environ 117:322–337. https://doi.org/10.1016/j.rse.2011.10.007

Zarco-Tejada PJ, Catalina A, González MR, Martín P (2013a) Relationships between net photosynthesis and steady-state chlorophyll fluorescence retrieved from airborne hyperspectral imagery. Remote Sens Environ 136:247–258. https://doi.org/10.1016/j.rse.2013.05.011

Zarco-Tejada PJ, Guillén-Climent ML, Hernández-Clemente R, Catalina A, González MR, Martín P (2013b) Estimating leaf carotenoid content in vineyards using high resolution hyperspectral imagery acquired from an unmanned aerial vehicle (UAV). Agric For Meteorol 171–172:281–294. https://doi.org/10.1016/j.agrformet.2012.12.013

Zarco-Tejada PJ, González-Dugo MV, Fereres E (2016) Seasonal stability of chlorophyll fluorescence quantified from airborne hyperspectral imagery as an indicator of net photosynthesis in the context of precision agriculture. Remote Sens Environ 179:89–103. https://doi.org/10.1016/j.rse.2016.03.024

Zhang N, Wang M, Wang N (2002) Precision agriculture—a worldwide overview. Comput Electron Agric 36:113–132. https://doi.org/10.1016/S0168-1699(02)00096-0

Zhaoqin L, Xulin G (2015) Remote sensing of terrestrial non-photosynthetic vegetation using hyperspectral, multispectral, SAR, and LiDAR data. Prog Phys Geogr 40:276–304. https://doi.org/10.1177/0309133315582005

Zheng G, Moskal LM (2009) Retrieving leaf area index (LAI) using remote sensing: theories, methods and sensors. Sensors 9:2719

Zhou R et al (2017) Drought stress had a predominant effect over heat stress on three tomato cultivars subjected to combined stress. BMC Plant Biol 17:24. https://doi.org/10.1186/s12870-017-0974-x

Zomer RJ, Trabucco A, Ustin SL (2009) Building spectral libraries for wetlands land cover classification and hyperspectral remote sensing. J Environ Manag 90:2170–2177. https://doi.org/10.1016/j.jenvman.2007.06.028

## Affiliations

**Tobias B. Hank[1] · Katja Berger[1] · Heike Bach[2] · Jan G. P. W. Clevers[3] · Anatoly Gitelson[4] · Pablo Zarco-Tejada[5] · Wolfram Mauser[1]**

[1] Department of Geography, Faculty of Geosciences, Ludwig-Maximilians-Universität München (LMU), Luisenstrasse 37, 80333 Munich, Germany

[2] VISTA – Remote Sensing in Geosciences GmbH, Gabelsbergerstrasse 51, 80333 Munich, Germany

[3] Laboratory of Geo-Information Science and Remote Sensing, Wageningen University and Research, Droevendaalsesteeg 3, 6708 PB Wageningen, The Netherlands

[4] Mapping and Geoinformation Engineering, Faculty of Civil and Environmental Engineering, Israel Institute of Technology (TECHNION), Haifa, Israel

[5] Instituto de Agricultura Sostenible (IAS), Consejo Superior de Investigaciones Cientificas (CSIC), Alameda del Obispo S/n, 14004 Córdoba, Spain

Surveys in Geophysics (2019) 40:553–588
https://doi.org/10.1007/s10712-019-09514-2

# Imaging Spectroscopy of Forest Ecosystems: Perspectives for the Use of Space-borne Hyperspectral Earth Observation Systems

**Joachim Hill**[1] · **Henning Buddenbaum**[1] · **Philip A. Townsend**[2]

Received: 11 June 2018 / Accepted: 23 January 2019 / Published online: 10 February 2019


**Abstract**
The emerging challenges in preserving and managing forest ecosystems are multiscale in terms of space and time, and therefore require spatially and temporally contiguous information sources. Imaging spectroscopy has the potential to contribute information that cannot be raised by other Earth Observation Systems. In particular, the spectral capacity to monitor the distributions of chemical traits, such as canopy foliar nitrogen distribution, and to track changes in water content or the percentage water in plants, has already opened novel pathways toward assessing the global variability of ecosystem functions and services. However, there is an ongoing debate on how to best extract this type of information from the spectral measurements. Empirical approaches have demonstrated their efficiency in a multitude of local studies, but are criticized with respect to poor generalization capacities. Alternative strategies, such as the use of physically based models of leaf or canopy reflectance, or hybrid approaches, have the potential advantage to be more widely applicable. This paper attempts to assess achievements and shortcomings of these strategies and finds that the often-cited disadvantages of using empirical approaches are becoming less pronounced in the light of recent research results. While retrievals based on physically based models on leaf/needle level are close to laboratory quality, results on canopy level available to date still have considerable deficits. Owing to improved instrumental designs, better data calibration, new approaches for compensating canopy effects, and the use of increasingly efficient methods for establishing data-driven models, the scope of empirical approaches has considerably widened and they have been successfully applied to large areas. The future availability of regularly acquired hyperspectral imagery from Earth orbits will substantially contribute to their generalizability.

**Keywords** Imaging spectroscopy · Forest ecosystems · Biochemical traits · Retrieval strategies

✉ Joachim Hill
hillj@uni-trier.de

[1] Environmental Remote Sensing and Geoinformatics, Trier University, 54286 Trier, Germany

[2] Department of Forest and Wildlife Ecology, University of Wisconsin, Madison, WI 53706, USA

## 1 Introduction

Forest ecosystems provide numerous environmental goods and services, such as conservation of biological diversity and climatic control, and, of course, timber and non-timber products (FAO 2010). However, forests and forested ecosystems face numerous risks under increasing pressure due to global warming (Birdsey and Pan 2011; Peng et al. 2011; Swann et al. 2018) and expanding human populations and economies (Hansen et al. 2008). Deforestation associated with conversion of forests to agricultural land and development, legal and illegal logging, drought stress, biotic stress, and increasing wildfire frequency, extent and severity are some of the most important processes which affect forested landscapes (Anderegg et al. 2013a, b; Bond et al. 2010; Ciais et al. 2005).

The European Environmental Agency expects forests in Europe to be among the ecosystems most affected by climate change (EEA 2012). Examples of recent die-offs have been reported from southern and central parts of Europe (e.g., Adams et al. 2017; Bréda 2003). It is expected that forests will become increasingly affected by drought, either during exceptional events or under a long-term drift toward more arid conditions (Ma et al. 2012). Extreme climatic events, such as the heat waves and dry spells experienced during the European summers 2003 and 2018, are expected to occur at increased frequencies (Beniston 2004; Fink et al. 2004). Diffuse tree mortality within forest stands, frequently reported after such extreme events, may be exacerbated by soil properties (mainly local water storage capacity) or decreases in genetic diversity among and within tree species (Lindner et al. 2010; Rennenberg et al. 2006). Due to the direct impact of heat waves and dry spells, forest ecosystems are also becoming more susceptible to pests and diseases. The combined effects bear the risk of a reduction of important ecosystem services, such as carbon sequestration, water retention, soil protection, preserving biodiversity and, last but not least, biomass and timber production (Lindner et al. 2010; Schelhaas et al. 2003; Seidl et al. 2011).

In order to keep track of these effects, monitoring of forest ecosystems is a necessity, both in situ and through remote sensing. While in situ terrestrial forest monitoring usually employs surveys at plot level, logistically it is not feasible to make measurements everywhere that data are needed. Remote sensing offers the potential to sample with reasonable accuracy in inaccessible locations (Asner et al. 2015) while also providing wall-to-wall information at multiple temporal and spatial scales. The emerging challenges in forest management, intricately linked to society's needs to preserve multiple forest values and continue to benefit from all services of forest ecosystems, are multiscale in terms of space and time and therefore need spatially and temporally contiguous information sources. Moreover, since sustainable forest management aims at reconciling economic interests with ecologic concerns, remote sensing must serve both, economically oriented assessments and management requirements, as well as studies of ecological processes and functions (Franklin 2001).

Lausch et al. (2018, 2016, and 2017) summarize this wide scope under the term "forest health," whose proper assessment requires utilitarian as well as ecosystem indicators to be implemented. In their review papers, they provide a detailed summary of indicator methods and strategies to assess forest health. These include various types of Earth Observation systems.

Applications of remote sensing contributing to sustainable forest management are generally presented in four categories that include classification of forest type (i.e., tree species), the assessment of forest structure and available resources (e.g., timber volume, height, age, crown closure), modeling forest ecosystem processes, and forest change detection. For each category, measurable indicators are needed to quantify the effects of management activities and natural phenomena on the sustainability of forest ecosystems on large spatial scales. Some of these indicators can be derived best by imaging spectroscopy data.

We are currently on the brink of a new age of space-borne imaging spectrometers. While the first space-borne imaging spectrometers, CHRIS–Proba and Hyperion, were not able to create a sustainable impact on operational forest remote sensing, the next generation is expected to more reliably deliver a higher data quality. Recently, the imaging spectrometer DESIS (Eckardt et al. 2015) has been installed in the International Space Station. In the near future, several space-borne imaging spectrometers, such as PRISMA (Labate et al. 2009), SHALOM (Feingersh and Ben Dor 2015), HISUI (Matsunaga et al. 2017), EnMAP (Guanter et al. 2015), HYPXIM (Carrere et al. 2013), Surface Biology and Geology (SBG, formerly known as HyspIRI) (Lee et al. 2015), or the recently announced Earth Surface Mineral Dust Source Investigation (EMIT) will be launched. In the Copernicus program, the hyperspectral mission CHIME is in consideration (Nieke and Rast 2018). During the preparation programs to these sensors, especially EnMAP and SBG, large-scale airborne campaigns, have been carried out. So now a few large-scale data sets are available, including some at multiple points in time. Given the unprecedented range of future remote sensing systems in space, an essential question is which of the spectral traits can be better characterized with high spectral, medium spatial resolution instruments. The first instruments will cover only limited areas and are thus expected to be used primarily in scientific contexts. Products developed with scientific instruments may be of general interest for forest practitioners once operational wall-to-wall space-borne imaging spectroscopy becomes available with SBG and CHIME. In the following, we will first highlight which role imaging spectroscopy (IS) data is likely to play in three key application fields of forestry and then outline the scope of this study.

### 1.1 Tree Species Composition

Spatially explicit information on tree species composition of managed and natural forests is relevant for biomass or timber volume and quality estimation, habitat quality assessment, and biodiversity characterization. It thus provides valuable information for nature conservationists as well as for forest managers and is frequently required over large spatial extents. Assuming the same spatial resolution, hyperspectral imaging certainly has the potential to provide higher classification accuracies for forest species mapping than multispectral data (e.g., Asner and Martin 2009; Fassnacht et al. 2014; Foster and Townsend 2004), especially when implemented at local scales. As well, the range of advanced spectral classifiers has been substantially expanded so that it is now possible to make the best use of the available data (e.g., Ghamisi et al. 2017). However, a high degree of species diversity does not necessarily lead to an equally high diversity of traits (Lausch et al. 2016), and large area species mapping is rather boosted by phenological dynamics observed by operational multi- or super-spectral observation systems (Somers and Asner 2013; Stoffels et al. 2015). Considering this and the limited data availability, it is not surprising that only few examples for using hyperspectral data for successful tree species classifications over large and diverse geographic areas have been presented (Fassnacht et al. 2016; Sommer et al. 2016). On the one hand, this might change as soon as space-borne imaging spectrometers will regularly cover large areas. On the other hand, the spatial resolution of proposed sensors may not be sufficient for species identification, especially if species in mixed stands are to be mapped, thus requiring mapping of ecological communities (e.g., Foster and Townsend 2004).

### 1.2 Structural Traits

2D or 3D structural spectral traits depending on leaf arrangement and geometry, tree height or area, density, size, the shape of forest patches as well as fragmentation, complexity, and homogeneity are important characteristics of forests that strongly influence the spectral response of forest canopies. Moreover, forest canopy height, forest extent, and the vertical and horizontal vegetation structure of forests are important proxies for assessing standing timber volume (Nink et al. 2015). There is evidence that multi-resolution multispectral imagery can be used to estimate forest structure based on texture (Wolter et al. 2009) which likely also applies to hyperspectral data, but not with the accuracy and physical realism of active remote sensing data. Given the physical limitations of addressing structure and biomass information over closed canopies with mono-directional, passive optical systems, it is not surprising that hyperspectral data alone will contribute little that goes beyond the information provided by active remote sensing such as airborne laser scanning and synthetic aperture radar (Swatantran et al. 2011).

### 1.3 Detection of Stress and Biochemical Traits

Many studies emphasize the importance of hyperspectral data for characterizing the physiological state of canopies, for example those under stress (e.g., Abdullah et al. 2018; Asner et al. 2004; Dotzler et al. 2015; Lausch et al. 2016; Paz-Kagan et al. 2018). As such, one of the most important application domains of hyperspectral imaging is the observation of biochemical traits. Alterations to biochemical traits such as cellulose, protein, and lignin, changes in the composition and configuration of photosynthetically active pigments in leaves (chlorophyll, xanthophyll, carotenoids) are highly relevant indicators for climate- and site-specific stress conditions, disturbances or resource limitations (Asner et al. 2015; Féret et al. 2017; Singh et al. 2015), although their detection may be hampered by leaf structural changes such as rolling (Chávez et al. 2013). Accurate retrieval of foliar nitrogen, for example, could be used to detect nutrient deficiencies, but also enhance ecosystem process models that describe ecosystem functioning, since nitrogen is an important input parameter of these models. Foliar N is a primary regulator of physiological processes (Wang et al. 2017) and an important indicator of photosynthetic and growth rates (Townsend et al. 2003) since it reflects the investment in photosynthetic enzymes, structure, and light-harvesting complexes of a plant (Singh et al. 2015). Most foliar N (70–80%) is stored in proteins, and less in chlorophyll (Hank et al. 2019; Huber et al. 2008; Schlerf et al. 2010).

The spectral capacity to monitor the distributions of biochemical traits, such as canopy foliar N distribution (Knyazikhin et al. 2013; Schlerf et al. 2010; Serbin et al. 2014; Singh et al. 2015; Townsend et al. 2003), and changes in the water content (Asner et al. 2016; Paz-Kagan et al. 2018), has already revealed the potential to assess indicators for ecosystem functioning on extended spatial dimensions. Pigment-related trait variations also support the observation of shifts in the photosynthesis system (Gamon et al. 1992; Garbulsky et al. 2011; Peñuelas et al. 1995; Suárez et al. 2008), although unveiling the causal relationships behind diurnal, seasonal, and spatial variations of photosynthesis will require instruments with high repetition rates and exceptionally high spectral resolution, capable of tracking changes in the solar-induced chlorophyll fluorescence (Drusch et al. 2017; Rascher et al. 2015).

### 1.4 Scope of the Paper

To make the best use of space-borne imaging spectroscopy data, robust retrieval algorithms that work over large geographic and temporal scales, are needed. Ideally, these should also output uncertainties for all values retrieved. While in the past many imaging spectroscopy measurements were singular studies, in the age of operational space-borne sensors, operational algorithms are required. The way to estimate spectral traits and their variation is intricately linked to the fundamental debate about empirical and physically based retrieval strategies, discussed in the following section.

Since neither tree species classification (with present technologies and data) nor the determination of structural traits is substantially improved by imaging spectroscopy, we will focus on chemical traits. A main scope will be on the comparison of empirical, physically based, and hybrid modeling on the leaf and canopy scale and the question whether empirical approaches can also be generalized to an extent that they become applicable to a wider range of conditions. There are a number of reasons to focus on biochemical traits and the generalizability of retrieval approaches: the availability of radiometrically well-calibrated data and the level-2 processing capacities have constantly improved (Frantz et al. 2016; Masek et al. 2006; Storch et al. 2013; Thompson et al. 2018, 2019; Vermote et al. 2016). In addition, the hyperspectral observation capacities have been increasing during the past years and will further increase with the advent of hyperspectral satellites. Last, but not least, the efficiency of algorithms for building adequate models has been boosted by a range of powerful machine learning algorithms (Fassnacht et al. 2014; Hultquist et al. 2014; Paz-Kagan et al. 2018; Waske et al. 2012).

Here, we provide a review of the literature, with some points illustrated with measurements recently conducted by the authors. The second section provides a comparison of empirical versus physically based modeling; Sect. 3 deals with studies on the estimation of biochemical traits on the leaf and canopy scale, with Sect. 4 providing discussion and conclusions.

## 2 Empirical Versus Physically Based Modeling

### 2.1 Empirical Modeling

In remote sensing, quantitative forest variables like structure parameters and foliar chemical content are estimated either empirically, physically based, or in a hybrid approach (Fig. 1, Féret et al. 2017; Li et al. 2018; Verrelst et al. 2019). Hybrid approaches combine both strategies, e.g., by developing a spectral index based on artificial data created with forward simulations of a radiative transfer model (le Maire et al. 2004, 2008). Verrelst et al. (2019) give a comprehensive overview on retrieval methods.

Empirical models, both parametric and nonparametric, aim at building a statistical relationship between measured traits and an experimental reflectance database. However, these empirical relationships are considered limited in transferability and robustness because of their dependency on the particular conditions under which they were established (field data collection strategy, species diversity, instrumental characteristics,

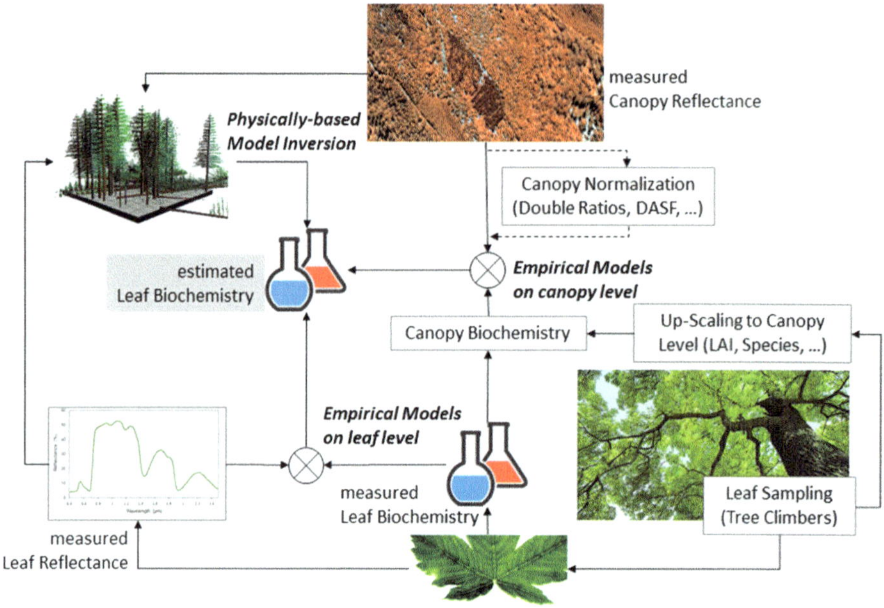


**Fig. 1** Principal workflows of empirical and physically based approaches for estimating biochemical leaf traits on leaf and canopy level

date of acquisition, phenological state of the vegetation, atmospheric, and lighting conditions) (Féret et al. 2017; Li and Wang 2011; Verrelst et al. 2015). Establishing empirical relationship builds on regression algorithms (e.g., multivariate linear regression, principal component regression, partial least-squares regression), machine learning techniques (support vector machines, import vector machines, artificial neural networks, deep learning networks, random forests, or Gaussian processes regression), or unmixing approaches, such as multiple endmember spectral mixture analysis (Roberts et al. 1998) or iterative spectral mixture analysis (Rogge et al. 2006).

In recent years, partial least-squares regression (PLSR) has emerged as the most popular parametric empirical method that can be used to estimate variables of interest even if the explaining variables (spectral reflectance) are redundant and collinear (Atzberger et al. 2010; Wold et al. 2001). The popularity of PLSR is related to its long history (Wold et al. 1984), ease of implementation, relative robustness to new data, and operational computational speed relative to deep learning methods. PLSR generates comparable results to machine learning methods such as Gaussian processes (Wang et al. 2019), but uncertainties need to be assessed through permutation rather than direct calculation (Singh et al. 2015). Asner et al. (2011) and Martin et al. (2018a), for example, used PLSR to measure a wide range of canopy chemicals using field and laser-guided airborne spectroscopy. On the machine learning side (nonparametric empirical models), deep learning networks are increasing in popularity due to improved computational capacities, but other machine learning approaches are also successfully employed. For example, Hultquist et al. (2014) assessed the post-fire burn severity with airborne MASTER data and found that random forest, support vector regression, and Gaussian process regression outperformed conventional multiple regression by 48%, 29%, and 27%, respectively.

Because of the usually high number of inter-correlated spectral channels, it is beneficial for most empirical modeling approaches to reduce the number of spectral features (feature reduction), either by selecting a subset of important bands (feature selection) or by transforming original bands into new features with maximal correlation to the dependent variable of interest (feature extraction) (Fassnacht et al. 2014; Held et al. 2015). Main feature extraction strategies include spectral indices (Féret et al. 2011), wavelets (Cheng et al. 2014; Wang et al. 2017), or linear transformations such as principal components or minimum noise fractions (Green et al. 1998), and singular value decomposition (Rogge et al. 2014). In feature selection approaches, only significant bands are used before establishing a regression model. Popular techniques include stepwise multiple regression (Schlerf et al. 2010), competitive adaptive reweighted sampling (CARS) (Vohland et al. 2016), and band selection based on variable importance in the projection (VIP) score (Abdullah et al. 2018; Heim et al. 2015). Empirical models may also be improved by spectral pre-treatment techniques like first and second derivative, transformation from reflectance to absorption, z-transformation, or Savitzky–Golay smoothing (Buddenbaum and Steffens 2012; Serbin et al. 2014).

## 2.2 Physically Based Modeling

Physically based modeling approaches aim at overcoming the limitations in transferability and robustness by simulating the radiative transfer between light source, canopy, and sensor (Feret et al. 2008). Leaf and canopy reflectance models predict the directional spectral reflectance based on radiation principles and a mathematical description of absorption and scattering processes in leaves, needles, or complete canopies. Foliar or canopy traits are then retrieved from spectral measurements by inverting the models (Combal et al. 2002; Jacquemoud and Baret 1990; Jacquemoud et al. 2009; Koetz et al. 2004; Schlerf and Atzberger 2006; Wang et al. 2018).

The physically based approach involves numerous strategies of varying complexity. There are models for leaf or needle optical properties, 1-D models for homogeneous canopies, and 3-D models for heterogeneous stands. Among the models of leaf optical properties (Miller et al. 2005) the by far the most widely used representative is PROSPECT. While early versions of PROSPECT only considered Chlorophyll a and b (Jacquemoud et al. 2000; Jacquemoud and Baret 1990; Jacquemoud et al. 1996), PROSPECT-4 and -5 added carotenoid content (Feret et al. 2008), and PROSPECT-D added anthocyanins (Féret et al. 2017). The PROSPECT model family has been designed for broadleaf species, but can also be recalibrated for needles (Malenovský et al. 2006a). A popular alternative for needles is LIBERTY (Dawson et al. 1998).

The most simplistic model to up-scale leaf measurements to canopy scale is the Lillesaeter model (Lillesaeter 1982). The model calculates canopy reflectance as $\rho/(1-\tau^2)$, where $\rho$ is leaf reflectance and $\tau$ is leaf transmittance. Leaf-scale reflectance models are often combined with canopy models to account for effects like leaf inclination, multiple scattering of several leaf layers, background effects, and illumination and observation geometry. The SAIL (Scattering by Arbitrarily Inclined Leaves) model (Verhoef 1984) is commonly combined with PROSPECT to simulate reflectance of homogeneous canopies (Berger et al. 2018; Jacquemoud et al. 2009). Since forest canopies tend to be more complicated, further model extensions like GeoSail (Huemmrich 2001), PROFLAIR (Omari et al. 2013), and InFoRM (Atzberger 2000; Schlerf and Atzberger 2006) add realism related to discontinuous canopies, varying tree geometry and density and multiple scattering. More advanced forest reflectance models such as FRT (Kuusk et al. 2014), FLIGHT (Barton and North 2001; North 1996), and 3D ray tracing models like DART (Gastellu-Etchegorry et al. 2004) permit even more differentiated and detailed parametrization of canopy architectures.

The inversion of these models is based either on numerical algorithms, or on searching through lookup tables (LUT) which have been created by using the model in forward mode. Owing to the structural complexity of forest canopies with their diverse architectures, the inversion problem is complex and becomes frequently ill-posed with increasingly complicated models, i.e., different combinations of input parameters may produce almost identical spectra in the forward mode (Combal et al. 2002).

A typical strategy (e.g., Bicheron and Leroy 1999) is to fix some the model parameters to "reasonable" values, and to determine the remaining (free) parameters by minimizing a merit function, i.e., a function of difference between observed and modeled reflectance.

The LUT-based approach identifies the best-fitting spectrum from the collective of precomputed solutions; the corresponding parameters represent the inversion result. The LUT has to be created just once, and only the relatively simple task of finding the best match has to be repeated for each spectrum. This can be much faster, especially for large data sets (e.g., a hyperspectral image) and computationally complex models. But again, if a large number of free parameters need to be determined, LUTs tend to become too large and this may as well lead to ill-posedness. Since the best-fitting spectrum does not always yield the best set of parameters, a range of similarly good fitting spectra can be selected and their mean or median accepted as the inversion result; their standard deviation then provides a measure of uncertainty (Koetz et al. 2007; Locherer et al. 2015).

Figure 1 illustrates the alternative workflows of estimating biochemical traits on leaf and canopy level via empirical or physically based methods. The desired estimations of leaf biochemistry can be obtained directly from the measured canopy reflectance through inversion of a physically based model. If empirical models on canopy levels are to be used, measured leaf biochemistry values are needed that have to be up-scaled to canopy level. It may also be advisable to transform the spectral data using approaches like double ratios or directional area scattering factor (DASF, Knyazikhin et al. 2013, see Sect. 4). A different pathway to estimating leaf biochemistry on canopy level can be taken through measurements of leaf reflectance. Since reflectance measurements are faster and cheaper than chemical analyses they can be used to create a large ensemble of leaf biochemical values that can be up-scaled to canopy level.

## 3 Identifying Biochemical Traits on Leaf and Canopy Level

Spectroscopic studies on the leaf level serve a number of purposes: first, they can supplement or even replace laboratory analyses and help creating reference data sets for studies on canopy level. Second, they provide a basis to understand interactions between plants and radiation without confounding effects of the canopy structure. Generally, identifying biochemical traits is much more straightforward on leaf level than on canopy level (Ustin et al. 2009), although the structural contributions at the canopy level can reinforce empirical relationships at the expense of physical realism (Townsend et al. 2013). Third, results on leaf level can be up-scaled to canopy level (Gara et al. 2018; Malenovský et al. in review).

Reflectance measurements of broadleaves are ideally acquired using an integrating sphere (Asner et al. 2011; Jay et al. 2016; Kokaly et al. 2009), although measurements with a leaf clip are fast and popular. Needle reflectance is a little harder to determine in a reliable way. Malenovský et al. (2006a), Yáñez-Rausell et al. (2014a, b), and Hovi et al. (2017) have presented studies on how to measure needle reflectance with integrating spheres. Alternative ways to determine needle reflectance include measuring needle stacks or mats (Dawson et al. 1998; Schlerf et al. 2010).

A considerable range of biochemical and structural forest variables have been examined using (imaging) spectroscopy (Homolová et al. 2013). They are either expressed as concentrations (per leaf mass) or as content (per leaf area) (Homolová et al. 2013). Since the content of a light-absorbing leaf constituent can be linked to leaf reflectivity via Lambert–Beer's law, this measure is preferred in remote sensing contexts, especially if the value is to be up-scaled to canopy level via leaf area index. Figure 2 shows some of the specific absorption coefficients of pigments and other main optically active leaf ingredients. In the visible part of the spectrum (Fig. 2a), chlorophyll is usually the dominant pigment with strong absorption in the blue and red spectral regions that leads to leaves' green color. Carotenoids (yellow pigments) and anthocyanins (red pigments) dominate in senescent leaves (Féret et al. 2017). The near-infrared region has weak water absorption features around 975 and 1200 nm, strong water absorption features occur around 1450, 1930, and 2600 nm (Asner 1998) (Fig. 2b). Wang et al. (2015) have separated the absorption coefficients of dry matter into cellulose + lignin and nitrogen-containing protein (Fig. 2c), and Curran (1989) has an extensive listing of absorption features associated with specific molecular bonds.

Both empirical models and the inversion of physically based reflectance models for analyzing biochemical traits on the level of leaves and needles have already reached quite a high level of accuracy, in particular for pigments, water content, leaf mass per area, but not for cellulose, lignin, and protein in fresh leaves (Gitelson et al. 2006; Jay et al. 2016; Li et al. 2018; Malenovský et al. 2006a; Wang et al. 2015, 2018).

Asner et al. (2012, 2017) propose seven leaf traits that are functionally relevant to forest diversity and mostly uncorrelated (Asner et al. 2017). Foliar nitrogen (N), water, and LMA, which play important roles in photosynthesis and primary production, have been characterized remotely in various studies (Ustin et al. 2004). Polyphenols act as foliar defense compounds (Kokaly et al. 2009), while lignin is an important leaf structural compound that is recalcitrant to decomposition. Foliar phosphorus (P) and calcium (Ca) are related

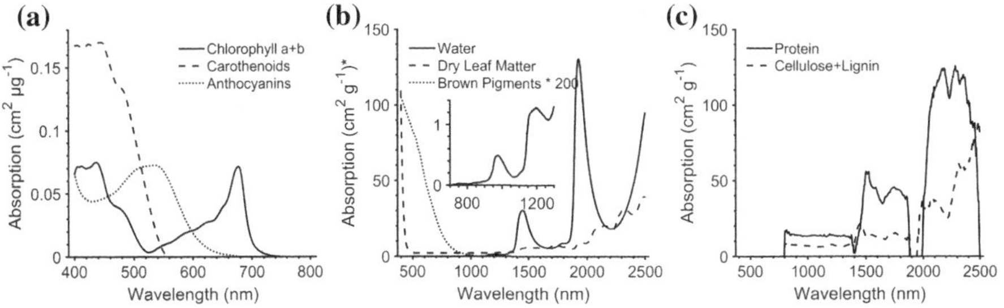


**Fig. 2** **a, b** Specific absorption coefficients of main absorbing leaf constituents according to the PROSPECT-D leaf optical model. Inset in subfigure **b** shows weak water absorption bands near 975 and 1200 nm. Note that the absorption coefficients for brown pigments are given in arbitrary units. **c** Absorption coefficients of protein and lignin + cellulose according to Wang et al. (2015)

to species community turnover in tropical forests and have also been mapped by IS (Asner and Martin 2016; Chadwick and Asner 2016), although relationships with spectra appear correlative and unrelated to currently known absorption features in the VSWIR. As considering the full range of traits is beyond the scope of this paper, we will focus on foliar pigment, water, and nitrogen content.

On the level of tree canopies, either the integrated content of a biochemical trait per ground area can be measured (e.g., canopy water content, CWC, or canopy chlorophyll content, CCC), or the respective content per leaf area (e.g., EWT or leaf chlorophyll content, LCC). These are related by the leaf area index (LAI), i.e., $LCC \cdot LAI = CCC$ and $EWT \cdot LAI = CWC$ (Hank et al. 2019). Where LAI is not estimated, estimates generally refer to traits at the top of the canopy (Singh et al. 2015); note also that some traits vary with canopy position (especially LMA, but not so much N), so that any calibration or validation of an integrated canopy estimate must take into consideration within-canopy variation. The retrieval of biochemical traits of leaves is also substantially compromised through structural properties of the crown layer and the ground reflectance that lead to nonlinear signal mixing through multiple scattering (Ewald et al. 2018). Various approaches for compensating these effects have been presented. These include inversion of physically based reflectance models and approaches like the normalization of spectra through a spectral index correlated with leaf area (double-ratio approach, Colombo et al. 2008) or by physically based normalization factors such as the directional area scattering factor (DASF, Knyazikhin et al. 2013). The selection of illuminated crown pixels is a valid strategy to be applied to either approach, provided the spatial resolution is sufficiently detailed (Asner et al. 2015; Dotzler et al. 2015).

Dotzler et al. (2015) demonstrated the importance of canopy-related effects when investigating the potential of EnMAP and Sentinel-2 to detect early drought stress effects in deciduous forests. Differences between stands on different soil moisture regimes that were statistically significant in moisture stress index (MSI), normalized difference water index (NDWI), and chlorophyll index (CI) disappeared after these indices were normalized to leaf level by using the double-ratio approach of Colombo et al. (2008). The comparison with data from airborne laser scanning suggested that the moisture and pigment differences detected on canopy level were actually due to variations in crown density and volume. The detection of drought-related stress symptoms on leaf level therefore requires that canopy effects be addressed and removed, in particular when spectral information from spectral regions with high transmission and scattering characteristics is used (especially near infrared).

On canopy level, the retrievals of biochemical and structural traits through inversion of physically based models are so far less successful than on leaf level. While most geometric-optical canopy reflectance models [e.g., InFoRM (Atzberger 2000; Schlerf and Atzberger 2006)] are surprisingly good in reproducing measured canopy reflectance as long as important canopy properties (leaf area index, leaf angle distribution, etc.) are known (forward mode), their performance tends to break down in the inversion (where these parameters must be retrieved simultaneously with the biochemical traits).

We summarize accuracies of cited studies in tables. Each study reports the accuracies in a different way and has been conducted under different conditions, so these studies should not be compared with each other directly. Since the coefficient of determination ($R^2$) is the only accuracy measure reported in nearly all studies, we give $R^2$ values of all studies cited although they are quite dependent on the range of values considered. If the root-mean-square error (RMSE) or the RMSE relative to the mean value or the range of values is also reported, that can be found in the respective table. From a statistical standpoint, achievement of a low RMSE and seemingly poor $R^2$ is common when the data range is limited (e.g., often seen with the prediction of carbon concentration, Singh et al. 2015). The resulting maps may exhibit reasonable uncertainty based on RMSE, but the models will be unlikely to be extensible.

### 3.1 Pigment Content Retrieval

Pigment content drives light harvesting to provide the energy used in photosynthesis to drive carbon fixation of plants and is a primary controller of leaf reflectance and transmittance in visible wavelengths, and thus, stand reflectance (Ustin et al. 2009). Independent of using empirical or physically based approaches, most pigment contents on leaf level are retrieved from spectral measurements with moderate to high precision. The accuracies of the studies cited here are summarized in Table 1.

Gitelson and Merzlyak (1997) and Gitelson et al. (2003) showed that simple spectral indices can be used for empirical pigment estimations. Gitelson et al. (2006) further proposed spectral three-band indices for estimating chlorophyll, carotenoids, and anthocyanin contents from leaf reflectance. Fassnacht et al. (2015) introduced a method for merging vegetation indices for carotenoid estimation over a large range of values. Carotenoid contents of the test data set ranged from 16 to 137 mg/m$^2$ and the merged index could explain 88% of the variance. These index-based pigment estimations reach quite high accuracies and have been applied over a wide range of vegetation types (e.g., Beamish et al. 2018; Fawcett et al. 2018; Hernández-Clemente et al. 2012; Sonobe and Wang 2018). Alternatively, Asner et al. (2011) used full-spectrum empirical regression models (PLSR) to estimate numerous foliar traits of many tropical tree species on leaf level using reflectance and transmittance spectra. The highest accuracies for pigments were achieved using transmittance in the full spectral range of 400–2500 nm with $R^2$ values of 0.83, 0.82, and 0.79 for chlorophyll a, chlorophyll b, and carotenoids, respectively. Feilhauer et al. (2015) conducted a multi-method ensemble selection of spectral bands and estimated chlorophyll concentrations of several sets of leaf measurements with large variations in $R^2$ values. The highest accuracies were achieved using an optimized PLSR. Abdullah et al. (2018) used PLSR to estimate chlorophyll content of spruce needles from the reflectance in the 400 to 790 nm range to detect bark beetle attacks. In healthy needles, chlorophyll content was estimated with $R^2=0.64$, in infested needles with $R^2=0.55$.

High to very high leaf-level estimation accuracies are also achieved by inverting the PROSPECT model or by using it for hybrid approaches. Le Maire et al. (2004) used PROSPECT simulated spectra to derive universal leaf chlorophyll content spectral indices. In order to test the applicability of PROSPECT for spruce needles, Malenovský et al. (2006a) applied a constrained inversion of PROSPECT-3.01 and retrieved chlorophyll content with an RMSE of 8.05 μg/cm$^2$. Féret et al. (2011) compared the performance of spectral indices, PLSR, and inversion of PROSPECT-5 for the estimation of chlorophylls and carotenoids on synthetic and real leaf reflectance and transmittance data. They found that the inversion approach is only reliable when both leaf reflectance and transmittance data are available, in which case, the accuracies were higher than using regression. Using PROSPECT-3 inversion, Buddenbaum et al. (2011) explained 0.78 and 0.54 of the variance of SPAD chlorophyll measurements of beech and oak leaves, respectively, that were collected by tree climbers from the crowns of old and young trees. In a greenhouse experiment, Buddenbaum et al. (2012) yielded $R^2=0.73$ for leaf chlorophyll content of potted young beech trees by inverting PROSPECT-5B based on spectra measured with a leaf clip. Dechant

**Table 1** Accuracies and further information of pigment estimation studies

| Study | Pigment | Unit | Range | | | Accuracy | | | Species | Method |
|---|---|---|---|---|---|---|---|---|---|---|
| | | | Min | Max | Mean | $R^2$ | RMSE | RMSE% | | |
| *Leaf level* | | | | | | | | | | |
| Gitelson and Merzlyak (1997) | Chlorophyll a + b | μg/cm² | 0.27 | 62.9 | | ≥ 0.9 | ≤ 5.3[a] | | 8 Different species | Indices |
| Gitelson et al. (2003) | Chlorophyll a + b | μmol/cm² | 1 | 832 | | ≥ 0.95 | ≤ 75 | | 8 Different species | Indices |
| Gitelson et al. (2006) | Chlorophyll a + b | mg/m² | | | 83–408 | ≥ 0.91 | ≤ 51 | < 25 | 6 Different species | Indices |
| Gitelson et al. (2006) | Carotenoids | mg/m² | 16 | 166 | 50–96 | ≥ 0.70 | ≤ 17.2 | ≤ 19 | 3 Different species | Indices |
| Gitelson et al. (2006) | Anthocyanins | mg/m² | | | | > 0.93 | | | 2 Different species | Indices |
| Fassnacht et al. (2015) | Carotenoids | mg/m² | 15.96 | 137.2 | 63.38 | 0.88 | | 8.12 | 3 Tree species | Index fusion |
| Asner et al. (2011) | Chlorophyll a | mg/g | 0.83 | 15.99 | 4.78[b] | 0.83 | | 6.2 | Many tropical tree species | PLSR on transm |
| Asner et al. (2011) | Chlorophyll b | mg/g | 0.29 | 5.83 | 1.78[b] | 0.82 | | 6.2 | Many tropical tree species | PLSR on transm |
| Asner et al. (2011) | Carotenoids | mg/g | 0.35 | 4.32 | 1.41[b] | 0.79 | | 8.1 | Many tropical tree species | PLSR on transm |
| Abdullah et al. (2018) | Chlorophyll a + b | mg/g | 0.4 | 1.1 | | 0.64 | | 0.24[c] | Healthy Spruce Needles | PLSR |
| Abdullah et al. (2018) | Chlorophyll a + b | mg/g | 0.2 | 1 | | 0.55 | | 0.75[c] | Infested Spruce Needles | PLSR |
| Malenovský et al. (2006a) | Chlorophyll a + b | μg/cm² | 25.57 | 85.1 | | | 8.05 | | Norway spruce | Modified PROSPECT-3 inv. |
| Féret et al. (2011) | Chlorophyll a + b | μg/cm² | 0.3 | 106.72 | 32.81 | | 5.38[d] | | Wide range of species | PLSR |
| Féret et al. (2011) | Carotenoids | μg/cm² | 0.04 | 25.3 | 8.51 | | 1.9[d] | | Wide range of species | PLSR |
| Buddenbaum et al. (2011) | Chlorophyll a + b | μg/cm² | 28.73 | 82.35 | 57.75 | 0.7764 | 15.4592 | 26.77 | European Beech | PROSPECT-3 inversion |
| Buddenbaum et al. (2011) | Chlorophyll a + b | μg/cm² | 48.13 | 102.11 | 78.87 | 0.5351 | 8.951 | 11.35 | Oak | PROSPECT-3 inversion |
| Buddenbaum et al. (2012) | Chlorophyll a + b | μg/cm² | 7 | 29 | | 0.73 | | 20.9 | European Beech | PROSPECT-5b inversion |
| Schlerf et al. (2010) | Chlorophyll a + b | mg/g | 1.8 | 4.94 | 3.21 | 0.83 | 0.26 | | Norway spruce | SMR, CM |

**Table 1** (continued)

| Study | Pigment | Unit | Range | | | Accuracy | | | Species | Method |
|---|---|---|---|---|---|---|---|---|---|---|
| | | | Min | Max | Mean | $R^2$ | RMSE | RMSE% | | |
| *Canopy level* | | | | | | | | | | |
| Malenovský et al. (2006b) | Chlorophyll a + b | μg/cm² | 28 | 94 | | 0.72 | 9.53 | | Norway spruce | ANMB605-725 index |
| Schlerf et al. (2010) | Chlorophyll a + b | mg/g | 2.71 | 3.62 | 3.21 | 0.8 | 0.13 | | Norway spruce | SMR, CM |
| Asner et al. (2015) | Chlorophyll a + b | mg/g | 2.65 | 8.53 | | 0.58 ± 0.05 | 1.5 ± 0.71 | 28.14 | Many tropical tree species | PLSR |
| Asner et al. (2015) | Carotenoids | mg/g | 0.69 | 1.84 | | 0.49 ± 0.06 | 0.25 ± 0.08 | 21.2 | Many tropical tree species | PLSR |
| Martin et al. (2018a) | Chlorophyll a + b | mg/g | 2 | 10.3 | 5.6 | 0.21 | 1.68 | 29.8 | Many tropical tree species | PLSR |
| Buddenbaum et al. (2015) | Chlorophyll a + b | μg/cm² | 15 | 40 | | 0.325 | 4.059 | 16.85 | European Beech | PLSR |
| Omari et al. (2013) | Chlorophyll a + b | μg/cm² | 39 | 56 | | 0.2553 | 4.46 | | Trembling aspen, balsam poplar, jack pine | PROFLAIR inversion |

[a]Estimation error

[b]Median

[c]Normalized

[d]PLS-PRESS, LOOCV

et al. (2017) also measured reflectance spectra with a leaf clip and inverted PROSPECT-5B to determine chlorophyll, carotenoid, and water content of leaves instead of chemical analyses, assuming they achieve laboratory level precision.

Figure 3 shows the relationship between chemically determined chlorophyll a + b and carotenoid content of beech (*Fagus sylvatica*) leaves that have been collected by tree climbers in the Donnersberg area in Germany and PROSPECT-D inversion results from spectra recorded using a Spectral Evolution PSR-3500 spectroradiometer with a leaf clip (unpublished data from the authors of this paper). The inversion was done following the approach of Jay et al. (2016) using a numerical minimization in MATLAB (least-squares curve fit with a trust region reflective algorithm). While the chlorophyll inversion results are relatively unbiased, carotenoid content is overestimated by a factor of nearly 2. Whether the reason for this kind of bias lies in the laboratory analysis or in the inversion process is subject of ongoing research.

On canopy level, the results from published studies are inconsistent and in some cases conflicting. Several studies report reasonably high estimation accuracies of foliar pigment content on canopy level (Malenovský et al. 2006b; Schlerf et al. 2010), while others report low accuracies. Malenovský et al. (2006b) reached $R^2 = 0.72$ for canopy chlorophyll content estimation in a Norway spruce stand with a high variability of chlorophyll contents (27.5--93.5 μg/cm$^2$) using their $ANMB_{605-725}$ index and only $R^2 = 0.17$ using the established TCARI/OSAVI index that was originally designed for agricultural crops. Schlerf et al. (2010) used stepwise multiple regression for chlorophyll estimation on stand level in Norway spruce stands with chlorophyll concentrations of 2.7 to 3.6 mg/g dry matter and reached cross-validated $R^2$ values between 0.46 and 0.9, depending on spectral transformation before the regression step. Asner et al. (2015) used airborne high-fidelity visible-to-shortwave infrared spectroscopy to estimate multiple foliar chemicals in tropical forests with PLSR to establish chemical fingerprints of numerous tree species. Independent field validation gave a mean $R^2$ of 0.58 for chlorophyll estimation for several tropical tree species with chlorophyll concentrations of 2.65–8.53 mg/g and an $R^2$ of 0.49 for carotenoid estimation in a range of 0.69–1.84 mg/g. Martin et al. (2018a) evaluated the methods of Asner et al. (2015) on data and field measurements of a wide array of canopies in Bornean forests of Sabah, Malaysia. Despite the strong chlorophyll absorption, the retrieval accuracy of chlorophyll concentration was among the lowest of all variables with $R^2$ values of 0.43, 0.17, and 0.21, respectively, for model calibration, validation, and test for a range

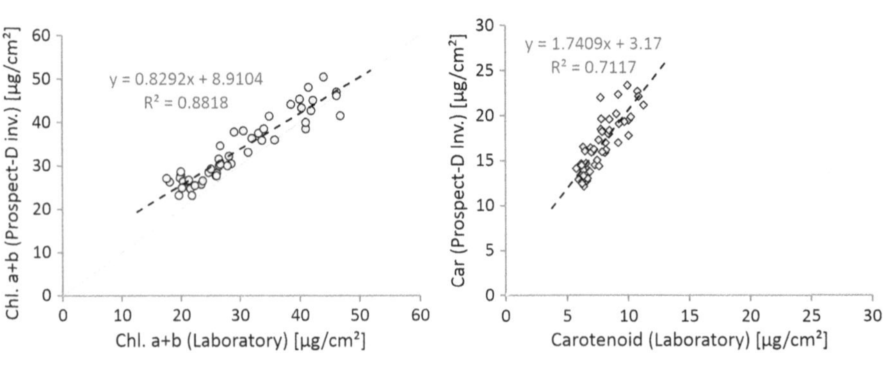


**Fig. 3** PROSPECT-D inversion results versus laboratory measurements of chlorophyll a + b (left) and carotenoids (right)

of 2–10.3 mg/g. Buddenbaum et al. (2015) used field-level IS and PLSR to estimate leaf chlorophyll content of beech seedlings with chlorophyll contents of 15 to 40 µg/cm$^2$. The $R^2$ only reached 0.325, but an interesting finding was that the regression coefficients were significantly different from zero only in the chlorophyll-dominated region of the spectrum, which showed that PLSR was able to determine important bands and ignore the bands outside chlorophyll absorption features.

A physically based model inversion approach for chlorophyll retrieval from Hyperion data is reported in Omari et al. (2013). They inverted the PROFLAIR model (Omari et al. 2009; White et al. 2001) and yielded a low RMSE but only $R^2=0.26$ for LCC in a quite limited range of 40–55 µg/cm$^2$. They ascribed the low $R^2$ value to heterogeneity within the site, uncertainties in preprocessing, the sampling scheme, and the temporal shift between sampling and satellite data acquisition. Due to large LAI variations, CCC also had a larger range and was estimated with $R^2=0.60$ after successful LAI retrieval.

### 3.2 Water Content

Water dominates total vegetation reflectance in the shortwave infrared region of the spectrum. Foliar water can be quantified relative to full turgor, to fresh or dry leaf mass or relative to leaf area. Since water content per area is directly related to light absorption in the leaf, this is the most common way of describing water content in spectroscopy-related studies (Colombo et al. 2008), e.g., use of equivalent water thickness (EWT, g/cm$^2$ or cm) in the PROSPECT model. The fuel moisture content (FMC) is the ratio of EWT and dry matter content and is often used to quantify the susceptibility to fire (Riaño et al. 2005). Further information on the studies cited here is given in Table 2.

Physically based model retrievals of EWT usually yield high accuracies on leaf level. Riaño et al. (2005) estimated EWT and FMC by inversion of radiative transfer models to quantify fire danger. When applied to the LOPEX (Hosgood et al. 1994) leaf data set, EWT and FMC were estimated with $R^2=0.94$ and 0.89, respectively, by inverting PROSPECT. Malenovský et al. (2006a) used unconstrained and constrained inversions of both PROSPECT-3.01 and an adapted version called PROSPECT-3.01.S to retrieve EWT of spruce needles from VNIR spectra. While the unconstrained inversions yielded quite low accuracies, the constrained inversion results were very close to the measured values with RMSE=0.0006 ($R^2$ not reported). Buddenbaum et al. (2011) inverted PROSPECT-3 to estimate EWT of beech and oak leaves with $R^2=0.75$, and Buddenbaum et al. (2012) achieved $R^2=0.87$ with an inversion of PROSPECT-5B for beech leaves. The inversion results were used as a plausibility check to verify the reference measurements of EWT and the leaf spectra.

Similarly, Colombo et al. (2008) tested several spectral indices and regression approaches to estimate EWT on leaf level. Reduced major axis regression exploiting the 1200 nm water absorption feature performed best among the index-based approaches and explained 0.61 of variance, while the inversion of PROSPECT produced a coefficient of determination of 0.65. Cheng et al. (2011) estimate leaf water concentration relative to fresh ($LWC_F$) and dry ($LWC_D$) leaf mass using continuous wavelet analysis (CWA) and several spectral indices. While the spectral indices failed at estimating LWC with $R^2$ values around 0.1, $LWC_F$ and $LWC_D$ were estimated with $R^2=0.68$ and 0.75, respectively, using a combination of six wavelet features. Fang et al. (2017) tested spectral similarity water indices on two freely available leaf data sets (LOPEX93 an PANAMA) and on PROSPECT

**Table 2** Accuracies and further details of water content estimation studies

| Study | Unit | Range | | | Accuracy | | | Species | Method |
|---|---|---|---|---|---|---|---|---|---|
| | | Min | Max | Mean | $R^2$ | RMSE | RMSE% | | |
| *Leaf level* | | | | | | | | | |
| Riaño et al. (2005) | EWT (cm) | 0.001 | 0.05 | | 0.94 | | | LOPEX data set | PROSPECT inversion |
| Malenovský et al. (2006a) | EWT (cm) | 0.0419 | 0.0701 | | | 0.0006 | 55 | Norway spruce | PROSPECT-3.01.S inversion |
| Buddenbaum et al. (2011) | EWT (cm) | 0.00263 | 0.0173 | | 0.75 | | | Beech and Oak | PROSPECT-3 inversion |
| Buddenbaum et al. (2012) | EWT (cm) | 0.00125 | 0.005 | 0.004 | 0.87 | | 11.3 | Beech | PROSPECT-5b inversion |
| Colombo et al. (2008) | EWT (cm) | 0.0091 | 0.0154 | 0.0122 | 0.65 | | | Poplar | PROSPECT inversion |
| Cheng et al. (2011) | $LWC_D$ (%) | 32.31 | 418.2 | 143.6 | 0.71 | | 26.04 | Lianas and trees | Continuous wavelet analysis |
| Cheng et al. (2011) | $LWC_F$ (%) | 24.42 | 80.7 | 57.23 | 0.75 | | 4.34 | Lianas and trees | Continuous wavelet analysis |
| Fang et al. (2017) | EWT (cm) | 0.0037 | 0.0525 | 0.0115 | 0.95 | | 3.63 | LOPEX data set | Spectral index ($SWI_{Double\ SAC}$) |
| Fang et al. (2017) | EWT (cm) | 0.0038 | 0.0255 | 0.0136 | 0.8 | | 7.9 | PANAMA data set | Spectral index ($SWI_{Double\ SAC}$) |
| *Canopy level* | | | | | | | | | |
| Riaño et al. (2005) | FMC (%) | 0 | 200 | | 0.81 | | | Gall oak, rosemary, Rock Rose | PROSPECT–Lillesaeter inversion |
| Colombo et al. (2008) | mean leaf EWT (cm) | 0.0117 | 0.0136 | | 0.57 | | 2.264 | Poplar | Double ratio index MSI/SR |
| Colombo et al. (2008) | $EWT_{canopy}$ (g $m^{-2}$) | 80 | 400 | | 0.818 | | | Poplar | PROSPECT–SAILH inversion |
| Buddenbaum et al. (2015) | EWT (cm) | 0.001 | 0.008 | | 0.7979 | 0.0007 | 9.951 | European Beech | PLSR |
| Buddenbaum et al. (2012) | EWT (cm) | 0.00125 | 0.005 | 0.004 | 0.88 | | | European Beech | PLSR |

simulations and explained up to 0.95 and 0.80 of the variance of LOPEX93 and PANAMA EWT, respectively.

The high compatibility between empirical and physically based model retrievals of EWT is also demonstrated using results from an experiment with oak leaves that were successively dried in an oven (Fig. 4). Panel (a) shows the reflectance spectra of the leaves with colors depending on the EWT. In addition to SWIR, the near-infrared plateau is heavily affected when drying out due to cell plasmolysis, i.e., the leaf structure is changing, while only the shortwave infrared region is affected if only the EWT is changed in a PROSPECT simulation. In panel (b), the result of a PROSPECT-D inversion (numerical inversion as proposed by Jay et al. (2016)) is highly correlated with the laboratory measurements, although the estimated values are systematically underestimated. Panels (c) and (d) show two empirical estimations. For these, the data set was split into a training and a validation subset. All possible combinations of two-band ratio indices were calculated, and their coefficient of determination for estimating EWT was determined with the training subset. The optimal index found was the ratio of wavelengths 1745 and 1455 nm; the result is shown in panel (c). A PLSR was trained on the training subset and applied on both subsets; the result is shown in panel (d). These empirical estimates are unbiased, but may not be transferable to different species.

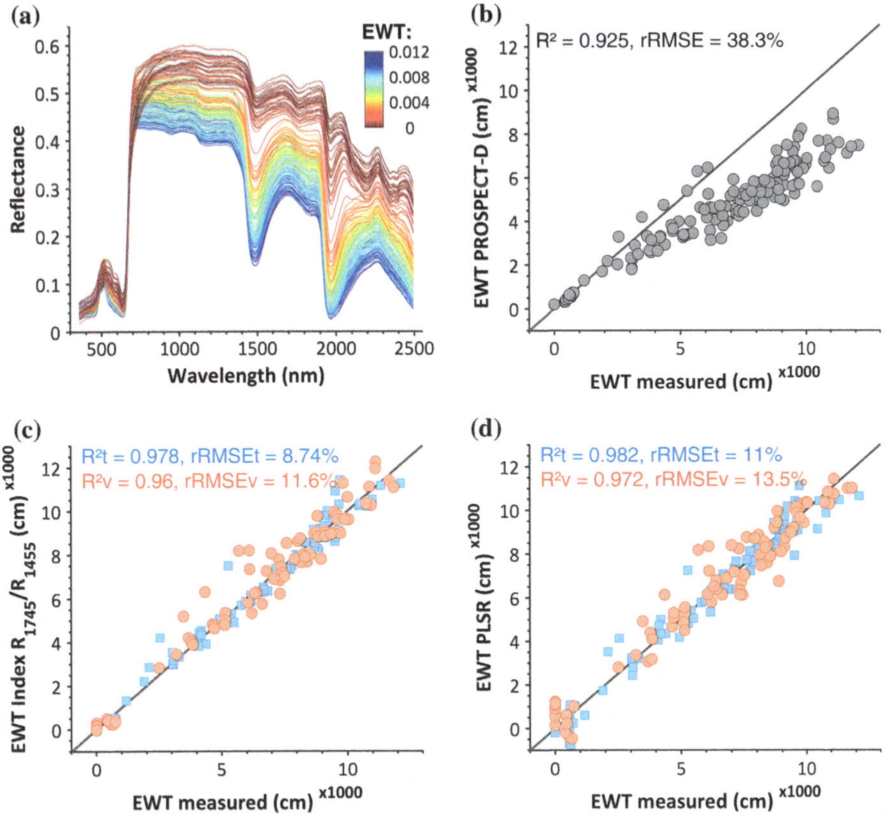


**Fig. 4** Drying experiment of oak leaves. **a** Reflectance spectra colored by EWT, **b** EWT estimation through PROSPECT-D inversion, **c** EWT estimation using the best ratio index, **d** EWT estimation using PLSR, data set split into training and validation sample

In contrast to the other variables discussed here, canopy water content can be derived from IS data during atmospheric correction when absorption bands at 980 and 1160 nm of gaseous, liquid, and solid water are fit to the spectrum (Green et al. 2006; Thompson et al. 2019). Asner et al. (2016) used this approach and characterized the progressive forest canopy water loss during the 2012–2015 drought in California (see also Martin et al. 2018b; Paz-Kagan et al. 2018). Paz-Kagan et al. (2018) mapped the canopy water content of giant Sequoia trees using a random forest machine learning approach. Their results show that IS can contribute to large-scale mapping of environmental influences like drought on forest ecosystems.

Riaño et al. (2005) inverted the PROSPECT–Lillesaeter model to estimate EWT and FMC on canopy level with $R^2=0.75$ for a range of 0–0.05 g/cm$^2$. FMC estimations improved from $R^2=0.62$ to 0.81 when dry matter content from dried leaf samples measured in a laboratory were taken into account. Koetz et al. (2004) used radiative transfer modeling for forest fire fuel estimation. A GeoSail (Huemmrich 2001) inversion yielded low RMSE for EWT retrieval, but also a very low $R^2$. The estimation of CWC showed a better result, but no $R^2$ is reported. Koetz et al. (2007) inverted the GeoSail model on synthetic data and could retrieve EWT values between 0.025 and 0.065 g/cm$^2$ with $R^2=0.92$.

Colombo et al. (2008) used Multispectral Infrared and Visible Imaging Spectrometer (MIVIS) data acquired over poplar plantations in Italy for landscape level estimations of EWT. A double-ratio index (MSI/SR) was used and yielded a cross-validated $R^2$ of 0.57 for the value range of 0.0116–0.0136 g/cm$^2$. Normalizing the moisture stress index (MSI) with the LAI-dependent simple ratio (SR) reduces the influence of LAI on the EWT estimation. An inversion of the PROSPECT + SAILH canopy reflectance model resulted in a relative RMSE of 27% ($R^2=0.818$) for $EWT_{canopy}$, while the mean leaf EWT was only estimated with $R^2=0.048$. This can be explained by a very small range of leaf EWT values, while LAI ranged from 0.5 to 3.5.

White et al. (2007) calculated moisture indices from Hyperion data to detect mountain pine beetle red attack damage. The moisture stress index (MSI) had highest correlation to red attack fraction. They suggest that space-borne hyperspectral imagery may be able to detect pine beetle damages at lower levels of infestation than Landsat imagery. Zhang et al. (2012) used the physically based coupled canopy-leaf radiative transfer model PROSAIL2 to derive the fraction of photosynthetically active radiation absorbed by chlorophyll and the leaf water content of a coniferous forest canopy from Hyperion data. Estimation accuracies are not provided, but the authors stress that their products provide unique information and that using the model inversion strategy, a Markov Chain Monte Carlo (MCMC) method, the globally optimal solution is found.

Buddenbaum et al. (2015, 2012) used field IS to monitor drought-stressed beech seedlings grown in pots in two different years, showing that after water supply was stopped, soil moisture declined first, followed by leaf water content and leaf chlorophyll content (Fig. 6). The drought stress is obvious in the spectra for most plants after some weeks of water deprivation (Fig. 6, red curves). For 2011 data, a PLSR in the visible/near-infrared range yielded $R^2=0.88$ for EWT values from 0.001 to 0.005 g/cm$^2$ (Buddenbaum et al. 2012); for 2012 data, a PLSR in the shortwave infrared range yielded $R^2=0.80$ for a range of 0.005–0.009 g/cm$^2$ (Buddenbaum et al. 2015).

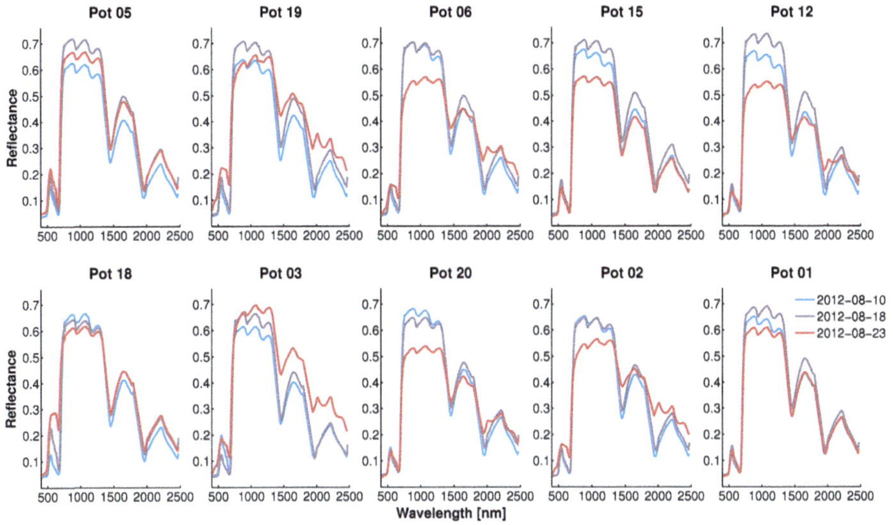


**Fig. 5** Spectra of drought-stressed beech seedlings over the course of 2 weeks (from Buddenbaum et al. 2015)

### 3.3 Leaf Nitrogen Concentration

A variable of special interest is the nitrogen (N) concentration of leaves. Since the protein absorption features in fresh leaves are mostly masked by water absorption, foliar N retrieval is less straightforward than pigment or water content retrieval, but still, some very high accuracies have been reported both on leaf and on canopy level. In some studies (e.g., Serbin et al. 2014), foliar N has been retrieved from dry leaf reflectance, so that the confounding effect of water was removed. Dry spectral approaches based on ground material appear to have good transferability (Fig. 5a), as discussed in more detail below. The accuracies of the studies cited here are listed in Table 3.

Wang et al. (2015) extended the physically based leaf reflectance model PROSPECT-5 by adding protein and cellulose + lignin absorption (see Fig. 2c). They were able to invert the model for fresh and dry leaves with moderate to good accuracies for fresh leaves ($R^2=0.70$ for cellulose + lignin and $R^2=0.47$ for protein) and dry leaves ($R^2=0.79$ for cellulose + lignin and $R^2=0.57$ for protein). A stepwise multiple linear regression approach gave slightly better results ($R^2=0.83$ and 0.66 for cellulose + lignin and protein in fresh leaves, respectively), but may have limited transferability.

Dechant et al. (2017) measured reflectance, leaf mass per area, nitrogen content, and traits linked to photosynthesis, maximum carboxylation capacity ($V_{cmax}$), and maximum electron transport rate ($J_{max}$) of leaves of several tree species. They found clear relationships of the photosynthesis traits with leaf nitrogen content. The nitrogen content per leaf area was estimated via PLSR with $R^2=0.88$. Studying the effects of bark beetle green attack on foliar reflectance and biochemical properties, Abdullah et al. (2018) also employed a PLSR to estimate N concentration and reached $R^2=0.76$ and 0.68 for healthy and bark beetle-infested spruce needles, respectively. Schlerf et al. (2010) compared several spectral preprocessing techniques for estimating N concentration in Norway spruce (*Picea abies*) needles with a stepwise multiple regression approach and could explain 0.59 of the variance (cross-validated) with first derivative spectra, although the range of N in

**Table 3** Accuracies and further details of nitrogen concentration estimation studies

| Study | Unit | Range | | | Accuracy | | | Species | Method |
|---|---|---|---|---|---|---|---|---|---|
| | | Min | Max | Mean | $R^2$ | RMSE | RMSE% | | |
| *Leaf level* | | | | | | | | | |
| Schlerf et al. (2010) | % dry | 0.88 | 1.46 | | 0.59 | 0.073 | 6.1 | Norway Spruce | FD-Stepw |
| Serbin et al. (2014) | % | 0.3 | 6.4 | | 0.97 | 0.13 | 4 | Northern temperate and boreal tree species | PLSR |
| Wang et al. (2015) | g Protein/cm$^2$ | 0.000292 | 0.002 | 0.001 | 0.47 | 0.000275 | 0.17[a] | LOPEX data | PROSPECT-5N inversion (fresh leaves) |
| Wang et al. (2015) | g Protein/cm$^2$ | 0.000292 | 0.002 | 0.001 | 0.66 | 0.000202 | 0.12[a] | LOPEX data | SMLR |
| Dechant et al. (2017) | g/m$^2$ | 0.52 | 3.16 | 1.53 | 0.8808 | 0.1757 | 11.5 | Mixed | PLSR |
| Wang et al. (2018) | g/m$^2$ | 1.43 | 3.68 | 2.78 | 0.51 | 0.851 | 17 | Mixed | Prospect-5/Inform inversion |
| *Canopy level* | | | | | | | | | |
| Townsend et al. (2003) | % | 1.53 | 4.24 | | 0.978 | 0.25 | | Mixed forest | PLSR Hyperion |
| Townsend et al. (2003) | % | 1.53 | 4.24 | | 0.849 | | | Mixed forest | PLSR AVIRIS |
| Coops et al. (2003) | % | 0.77 | 1.58 | 1.22 | 0.62 | | 13 | Eucalypt | PLSR Hyperion |
| Smith et al. (2003) | % | 1 | 2.48 | | 0.79 | 0.19 | | Mixed forest | PLSR AVIRIS |
| Smith et al. (2003) | % | 1 | 2.48 | | 0.6 | 0.25 | | Mixed forest | PLSR Hyperion |
| Huang et al. (2004) | mg/g | 9.8 | 17.8 | 14.35 | 0.8532 | | | Eucalypt | Modified PLSR, HySpex |
| Martin et al. (2008) | % | 0.7 | 2.4 | | 0.83 | 0.19[b] | | Mixed forest | PLSR AVIRIS |
| Martin et al. (2008) | % | 0.7 | 2.43 | | 0.82 | 0.25[b] | | Mixed forest | PLSR Hyperion |
| Ollinger et al. (2008) | % | 0.75 | 2.24 | | 0.79 | | | Mixed | Univariate regression |
| Schlerf et al. (2010) | % dry | 0.88 | 1.46 | | 0.57 | 0.055 | 4.6 | Norway spruce | FD-Stepw |
| Knyazikhin et al. (2013) | % | 0.75 | 2.24 | | 0.607 | | | Mixed | CSC (DASF)-Univariate |
| Singh et al. (2015) | % | 0.96 | 3.31 | 2.2 | 0.84 | 0.23 | 10.5 | Northern temperate and boreal tree species | PLSR |

**Table 3** (continued)

| Study | Unit | Range | | | Accuracy | | | Species | Method |
|---|---|---|---|---|---|---|---|---|---|
| | | Min | Max | Mean | $R^2$ | RMSE | RMSE% | | |
| Asner et al. (2015) | % | 1.28 | 3.33 | | 0.48 | 0.31 | 15.19 | Mixed tropical | PLSR (cal values) |
| Wang et al. (2017) | % | 1.45 | 3.29 | 1.69? | 0.65 | 0.33 | 19 | Mixed | CSC (DASF)-wavelets |
| Wang et al. (2018) | $g/m^2$ | 1.43 | 3.68 | 2.78 | 0.64 | 1.9 | 18 | Mixed | Prospect-5/Inform inversion |
| Martin et al. (2018a, b) | % | 0.6 | 4.46 | 1.79 | 0.54 | 0.43 | 24.4 | Mixed tropical | PLSR (model test) |

[a] nRMSE

[b] Standard error of cross-validation

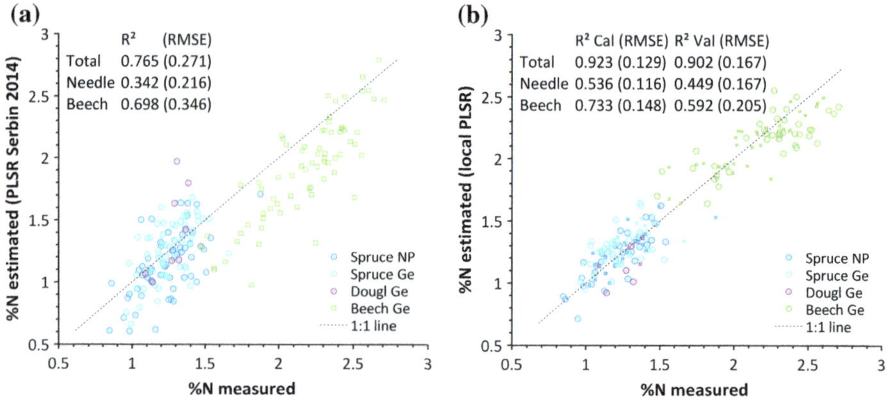


**Fig. 6** Estimation of leaf nitrogen content from dried and powdered samples collected in Germany (NP = Nationalpark Hunsrück-Hochwald, Ge = Gerolstein area). **a** Application of PLSR coefficients from Serbin et al. (2014), **b** local PLSR, small squares represent calibration subsample, large circles are the validation sample

this single-species data set was quite limited (0.88 to 1.46%). Asner et al. (2011) successfully ($R^2 = 0.81$) estimated foliar N concentration of thousands of tropical species from around the world with a PLSR using visible-to-shortwave infrared transmittance spectra, and Serbin et al. (2014) established PLS regressions for estimating seven biochemical traits (N, C, $\delta^{15}$N, SLA, cellulose, ADF, ADL) from ground and dried leaf samples for a large number of North American tree species. The regression models were able to explain between 60% ($\delta^{15}$N) and 97% (N) of the variance in the validation data set. These large samples are an important step toward the development of transferable models. Serbin et al. (2014) have published their PLSR coefficients so that they can be applied to any data set.

In an experiment recently performed by the authors of this paper, leaf samples of Norway spruce, Douglas fir (*Pseudotsuga menziesii*), and European beech (*Fagus sylvatica*) were collected by professional tree climbers at two sites in Germany, the Nationalpark Hunsrück-Hochwald (NP) and near Gerolstein (Ge). The samples were dried, ground, chemically analyzed, and spectrally measured. Figure 6 shows estimations of N content using the PLSR coefficients by Serbin et al. (2014), and a PLSR-based estimation of foliar N content trained on a subsample of our data with $R^2$ values for the total samples and all subsamples. While the locally adapted PLSR yields better results, the use of coefficients from the American data set suggests that transfers are possible.

Large area assessments of ecosystem productivity and health are improved tremendously by reliable estimations of foliar nitrogen content (McNeil et al. 2007a, b; Ollinger et al. 2008; Townsend et al. 2003), just as applications of fertilizers to tree plantations (Albaugh et al. 2003; Allen et al. 2005; Liechty and Fristoe 2013). Large efforts have been undertaken to collect data sets that enable transferable estimations of foliar nitrogen on canopy level.

Townsend et al. (2003) used space-borne EO-1 Hyperion and airborne AVIRIS IS data to map canopy N in Green Ridge State Forest in Maryland, USA, where leaf samples were collected in 27 ground plots. PLSR models explained 97.9% of the variation of N for Hyperion and 84.9% for AVIRIS for N concentrations between 1.2 and 4.1%. Similarly, Smith et al. (2003) explained 79% and 60% of variance of validation data in the range between 1 and 2.5% N using AVIRIS and Hyperion, respectively. Coops et al. (2003) also used Hyperion data. They mapped N concentration of Eucalypt foliage with $R^2=0.62$ for a range of 0.77–1.58%N using PLSR. Huang et al. (2004) also worked in an Eucalypt forest. They estimated N concentrations from continuum-removed HyMap airborne data with stepwise regression, PLSR, and neural networks and reached $R^2=0.85$ for a range of 0.98–1.78% N. Martin et al. (2008) used PLSR to estimate foliar N content on airborne (AVIRIS) and space-borne (Hyperion) hyperspectral data from 137 forested sites in North America, Costa Rica, and Australia. They were able to retrieve N concentrations in the range of 0.7–3.2% with a cross-validated $R^2$ of 0.83 and 0.82 with AVIRIS and Hyperion, respectively, for all sites combined. Singh et al. (2015) built on the leaf-level study by Serbin et al. (2014) and established respective regression models for airborne hyperspectral data. 145 AVIRIS classic data sets from the Northern USA and data from 237 field plots were used to create transferable algorithms for retrieving foliar traits and their uncertainties. They calculated mean stand values of the leaf traits by weighting them by leaf biomass of the trees. Between 0.49 (cellulose content) and 0.88 (SLA) of the variance could be explained by the canopy-level PLSR. N concentration was estimated with $R^2=0.85$ for a range of 0.96–3.31% using the average of 500 prediction models.

Ollinger et al. (2008) developed a regression model for foliar N concentration across several North American mixed forests that exploits a high positive correlation between foliar N and near-infrared (NIR) reflectance. The model explained 79 percent of the variance for a data range of 0.75–2.35% N. The regression model was applied to all forested areas in North America using MODIS satellite data. However, Knyazikhin et al. (2013) pointed out that the positive correlation between reflectance and the content of an absorbing material is counter-intuitive, contending that the observed relationship must be spurious. They showed that the NIR reflectance of closed forest canopies can be explained to a very high degree ($R^2=0.81$) by the broadleaf fraction of a canopy and that the N content is also highly dependent on the broadleaf fraction ($R^2=0.89$). This corresponds to the common knowledge that (a) leaves contain significantly higher amounts of N than needles and (b) that conifers appear darker than broadleaf trees. The lower reflectivity of coniferous trees is mostly due to structural effects: needle shoots have a much higher photon recollision probability due to their higher complexity compared to broadleaves. Knyazikhin et al. (2013) propose a normalizing factor, the DASF, that takes this complexity into account and can be easily computed from remote sensing data. If the reflectance of closed forest stands is divided by the DASF, the structural differences are removed from the signal. The resulting signal is termed canopy scattering coefficient (CSC). By adopting this normalization strategy, Wang et al. (2017) were able to estimate foliar N concentration of mixed forest stands in the Bavarian forest national park from airborne hyperspectral imagery using continuous wavelet analysis with a coefficient of determination of 0.65 for a range of 1.45–3.29% N. Ewald et al. (2018) also stress the influence of forest structure on the prediction of canopy N and P concentrations. They integrated predictor variables derived from airborne laser scanning to improve PLSR models ($R^2_{cv}$ of N estimation was improved from 0.31 to 0.41 for a range of 14–25 g/kg N). The relatively low accuracies are attributed to the structural differences between and within broadleaved and coniferous species and the relatively low range of values.

Foliar N retrievals through physically based model inversions are still rare, partially because there are a range of spectral features associated with N bonds. When coupling the canopy reflectance model InFoRM with their revised leaf-level model PROSPECT-5N (Wang et al. 2015), the achieved accuracy in estimating foliar N concentration was quite limited ($R^2=0.46$), while canopy N content was estimated with $R^2=0.64$ (Wang et al. 2018). The inversion procedure consisted of creating a LUT with 200 000 entries, then iteratively applying three filters to delete improbable parameter combinations from the LUT, and lastly identifying the best-fitting spectra in several different spectral subsets. The mean parameters from the 10, 20, or 100 best-fitting spectra were taken as inversion result. In total, 48 inversion results (4 differently filtered LUTs × 4 spectral subsets × 3 numbers of best fits) were obtained.

Schlerf et al. (2010) reached the highest estimation accuracy for foliar N concentration on Norway spruce (*Picea abies*) canopy level by using a stepwise multiple regression on first derivative spectra from the HyMap sensor with a cross-validated $R^2$ of 0.56. As an extension of the study by Schlerf et al. (2010) that also includes data on beech trees, PLSR models with 2 latent variables for estimating foliar N content were established on reflectance data and on CSC data. Although the PLSR for CSC could not rely on the sharp contrast between coniferous and broadleaf spectra in the NIR, the regression accuracy was even better for CSC data (Fig. 7). Similar to other studies, the high overall coefficient of determination is once more driven by the contrast of N concentrations between spruce and beech. The simple PLSR models were not able to capture as much variation within the species as models with larger numbers of latent variables would. In cases like this, mapping the fraction of coniferous and broadleaved species may suffice to characterize N distribution in forests.

In a further analysis step, the PLSR coefficients for canopy-level N estimation from Singh et al. (2015) were applied to the HyMap data. Since vegetation spectra contain no sharp features, resampling them to a different sensor with the same spectral range should

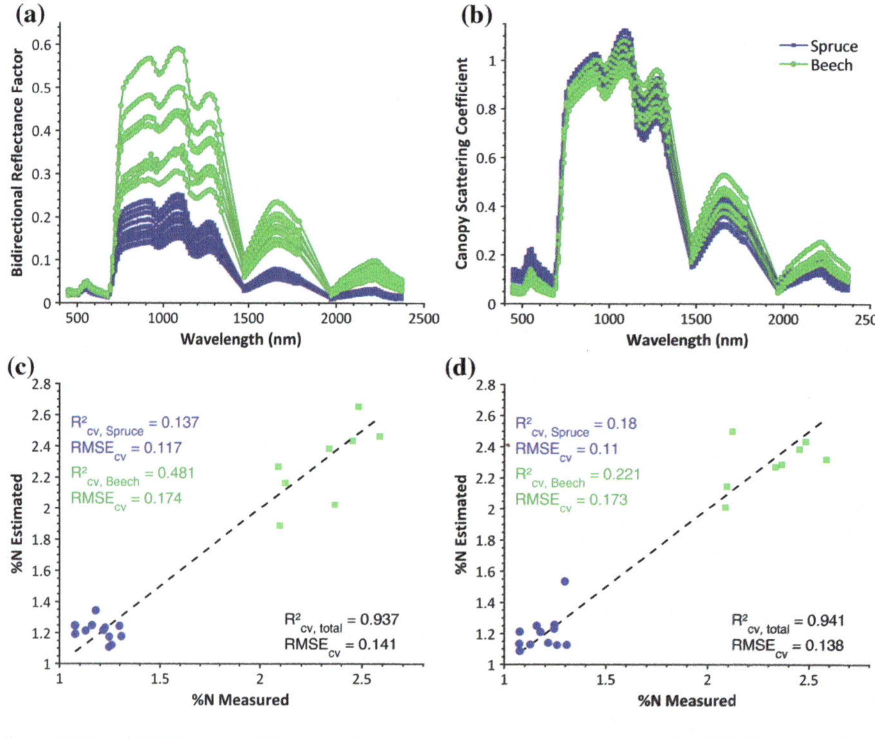


**Fig. 7** BRF and CSC spectra of beech and spruce trees. Lower row: scatter plots of PLSR accuracies for estimating foliar nitrogen content from the respective spectra

theoretically not pose a problem. But although these coefficients had been trained on a very large ensemble of measurements, reliable N estimations were not achieved with them (Fig. 8a). Figure 8b shows the AVIRIS classic spectra used for establishing the PLSR coefficients, and the HyMap spectra these coefficients were applied to. There are clear differences between the atmospherically corrected spectra, especially in the near-infrared region, that may explain the failure of estimating N content with the AVIRIS-trained PLSR coefficients. In addition, the HyMap spruce spectra are significantly darker than the AVIRIS needleleaf tree spectra. Townsend et al. (2017) highlight some of the difficulties of transferring empirical models to different sensors and ecoregions, e.g., different bandwidths, instrument calibration approaches, atmospheric correction schemes, methods of handling variable illumination due to sun-sensor-target geometry (BRDF), and finally variable stand structures.

## 4 Conclusions

Imaging spectroscopy of forest ecosystems has come a long way since the first analyses of foliar biochemistry by Wessman et al. (1989, 1988), and abundant literature on experiments and applications has in principle demonstrated its value for assessing forest ecosystems. It is obvious that mapping biochemical traits, especially those with less dominant absorption features like carotenoids, anthocyanins, or proteins, are one of the specific contributions of IS which cannot be adequately replaced by multi- or super-spectral remote sensing systems, nor with radar systems. The question whether biochemical traits are more efficiently extracted based on carefully established empirical models, through the inversion of physically based reflectance models, or in a hybrid approach, still cannot be answered definitely.

Physically based approaches offer substantial conceptual advantages and most of the available models are capable of accurately reproducing the spectral characteristics of leaves or forest canopies as long as the required parameters are known (forward mode). However, the inversion of reflectance models is more critical. As long as the number of model parameters is limited and their impact on the reflectance signal not overlapping with other parameters, inversion provides reasonably accurate estimates (Omari et al. 2013). Models of optical leaf properties, such as the PROSPECT family, have thus been successfully used

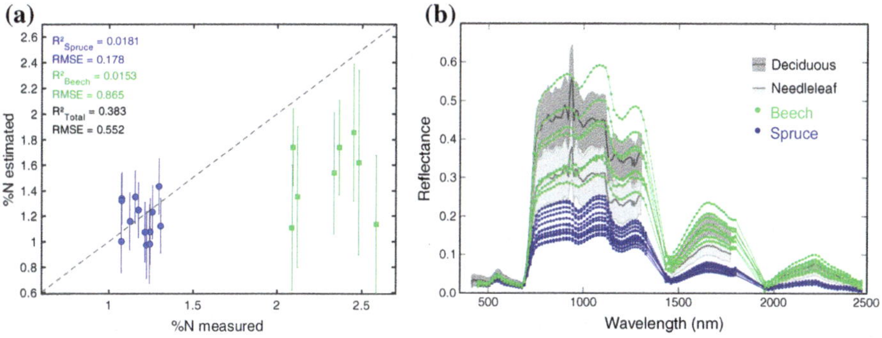


**Fig. 8** **a** Foliar nitrogen content of spruce and beech trees in Germany estimated using coefficients by Singh et al. (2015). **b** AVIRIS Classic spectra used by Singh et al. (2015) (gray), and HyMap spectra of the trees in Germany

in many retrieval studies and are meanwhile accepted as replacement of traditional laboratory analysis (i.e., for measuring leaf water or pigment concentrations) (Dechant et al. 2017). Establishing empirical models for estimating leaf properties requires abundant high-quality calibration data covering the range of trait variation; once calibrated, they can build on a range of powerful regression algorithms able to deal with high dimensionality spectral data and equally reproduce laboratory quality (Martin et al. 2008; Serbin et al. 2014; Singh et al. 2015).

On canopy level, the use of both strategies becomes more difficult. Physically based models require a substantially higher number of parameters to adequately reproduce the scattering and absorption mechanisms within structurally complex forest canopy architectures. The behavior of canopy reflectance models in the forward simulation mode is convincing for a range of models, and increasingly powerful computational resources support the development of even more differentiated simulation concepts. However, difficulties arise when the physically based forest canopy reflectance models are inverted. Although the general principle has been demonstrated, the reported levels of accuracy in retrieving leaf biochemical traits from canopy observations are still limited (e.g., Koetz et al. 2004; Omari et al. 2013; Wang et al. 2018), owing to the ill-posedness of the problem. Strategically, this limitation has been treated by constraining the model inversion, either through including external measurements (e.g., by LAI estimation through LIDAR measurements), fixing some of the model parameters to "reasonable" values, or by imposing constraints (rules) on the inversion algorithm while retrieving the remaining (free) parameters. Wang et al. (2018), for example, filtered their lookup table and restricted the spectral range until acceptable results were found; Riaño et al. (2005) used species-specific dry matter content from measurements of dried leaves to constrain the inversion and get better EWT estimations, and Colombo et al. (2008) manually set different weights for different variables in the cost function of the inversion. The problem with these strategies is that physically based models then may become so much tuned to local conditions that they lose their inherent generalization capacities.

Although powerful retrieval strategies exist, employing empirical approaches for retrieving forest canopy properties becomes demanding, in particular with respect to collecting reference data from elevated tree crowns, and because of the conceptual difficulties in upscaling sampled leaf properties to the level of multi-storey forest canopies. The necessity to collect leaf samples across extended areas and diverse forest types adds further complications. However, a number of studies have demonstrated that this is feasible for a range of canopy traits and within considerably large areas (e.g., Asner et al. 2011; Martin et al. 2008; Ollinger et al. 2008; Singh et al. 2015). Hyperspectral systems or sample collectors mounted on unmanned aerial vehicles may become important tools for collecting reference data to span a bridge between ground-based measurements and space-borne spectroscopy.

Since canopy observations inherently combine leaf biochemical traits with the amount of leaves to estimate a bulk quantity, additional complexity emerges when the objective is to retrieve biochemical traits on leaf level from canopy reflectance data. Strategies for solving this problem that have been proposed (Colombo et al. 2008; Dotzler et al. 2015; Knyazikhin et al. 2013) could demonstrate that spatial differences in pigment and water concentrations identified on canopy scale are not necessarily a consequence of differences at leaf level. However, if the objective of the analysis of imaging spectroscopy data is to generate a map to be used for management or to characterize spatial patterns of important biological quantities, then it may make sense to exploit emergent relationships (e.g., differences between conifer and broadleaf species for estimating canopy N) to produce accurate maps. However, such maps will require good independent validation data and will only be valid within the range of measured conditions. Such work provides a basis for ongoing research and may yield results to enable better understanding of ecological patterns (Asner et al. 2017), but will require more study for associated models to have biophysical realism or to not be ill-posed (Townsend et al. 2013).

Sensu stricto, empirical approaches can only be applied to data sets that are comparable to the training data. First attempts in transferring empirically defined models across large distances to similarly structured forests were contradictory. However, it was demonstrated in this paper that applying PLSR coefficients determined by Serbin et al. (2014) to leaf and needle samples from Western Europe was satisfactory, although outperformed by training a PLSR model on the samples themselves. Transferring PLSR coefficients for canopy estimates (Singh et al. 2015) was clearly less successful, likely due to spectral calibration differences between the AVIRIS and HySpex images and different strategies in scaling leaf traits to canopy level. Free access to collections of spectra and biochemical properties from forests around the world, similar to the Spectranomics database (Carnegie Institution for Science), will be a key element for enhancing satellite-based studies of forest ecosystems in general, and for the development of transferable relationships that go beyond locally restricted studies.

In case the existing technical and methodological incompatibilities can be reduced, e.g., through a more consistent data quality provided by space-borne instruments, it is expected that empirical models can be established which are sufficiently generalizable and transferable between similar forest ecosystems. Since they are computationally fast and therefore suited to be applied in large area monitoring schemes, empirical approaches may establish themselves a valuable alternative to computationally more complex inversions of physically based models.

Hybrid approaches complement pure empirical or physically based approaches. When physically based reflectance models can reproduce canopy reflectance well in the forward mode, these simulated spectra can feed a parametric or nonparametric empirical model to be applied to image data. If too little reference data could be collected in the field to establish a robust statistical relationship, the field data can be used to validate the hybrid model instead. However, a major prerequisite is that globally distributed regular observations with space-borne imaging spectrometers become available to foster the required cross-comparison exercises.

Successful applications of Hyperion data to map canopy chlorophyll content (Omari et al. 2013), canopy water content (White et al. 2007; Zhang et al. 2012), and foliar N (Coops et al. 2003; Martin et al. 2008; Smith et al. 2003; Townsend et al. 2003) prove the great potential of space-borne imaging spectroscopy for forest ecosystem studies. Also the derivation of canopy liquid water content during atmospheric correction, a truly spectroscopic approach, shows the unique value of imaging spectroscopy (Asner et al. 2016; Green et al. 2006; Thompson et al. 2019).

In this perspective, the Italian PRISMA (to be launched in 2019) and the German EnMAP missions (to be launched in 2020) constitute cornerstones in the efforts to regularly map diverse forest ecosystems from space with hyperspectral capacities. Their availability will provide an essential contribution toward optimizing the approaches discussed in this paper. The US-American SBG mission (to be launched in 2025) and the European CHIME mission (under design) will extend the regional coverage of previous satellite systems toward a truly global scope. While their scientific value is not in doubt, the question whether hyperspectral satellite observations will provide added value for many operational forestry applications in comparison with the enormous observation capacities of multispectral sensor systems like Sentinel-2 or Landsat remains open.

**Acknowledgements** The study was supported within the framework of the EnMAP project (Contract No. 50 EE 1530) by the German Aerospace Center (DLR) and the Federal Ministry of Economic Affairs and Energy, and the CalTech Jet Propulsion Laboratory (Contracts 1579654 and 1590148). The authors thank Willy Werner, Dorothee Krieger, Bernhard Backes, Martin Schlerf, Johannes Stoffels, Sandra Dotzler, Barbara Paschmionka, Marion Lusseau, Max Gerhards, and many others who helped gather the data presented here. We also thank the two anonymous reviewers for highly constructive comments that helped improve the manuscript.

## References

Abdullah H, Darvishzadeh R, Skidmore AK, Groen TA, Heurich M (2018) European spruce bark beetle (*Ips typographus*, L.) green attack affects foliar reflectance and biochemical properties. Int J Appl Earth Obs Geoinf 64:199–209. https://doi.org/10.1016/j.jag.2017.09.009

Adams HD et al (2017) A multi-species synthesis of physiological mechanisms in drought-induced tree mortality. Nat Ecol Evol 1:1285–1291. https://doi.org/10.1038/s41559-017-0248-x

Albaugh TJ, Allen HL, Zutter BR, Quicke HE (2003) Vegetation control and fertilization in midrotation *Pinus taeda* stands in the southeastern United States. Ann For Sci 60:619–624. https://doi.org/10.1051/forest:2003054

Allen HL, Fox TR, Campbell RG (2005) What is ahead for intensive pine plantation silviculture in the south? South J Appl For 29:62–69. https://doi.org/10.1093/sjaf/29.2.62

Anderegg LDL, Anderegg WRL, Berry JA (2013a) Not all droughts are created equal: translating meteorological drought into woody plant mortality. Tree Physiol 33:672–683. https://doi.org/10.1093/treephys/tpt044

Anderegg WRL, Kane JM, Anderegg LDL (2013b) Consequences of widespread tree mortality triggered by drought and temperature stress. Nat Clim Change 3:30–36. https://doi.org/10.1038/nclimate1635

Asner GP (1998) Biophysical and biochemical sources of variability in canopy reflectance. Remote Sens Environ 64:234–253. https://doi.org/10.1016/S0034-4257(98)00014-5

Asner GP, Martin RE (2009) Airborne spectranomics: mapping canopy chemical and taxonomic diversity in tropical forests. Front Ecol Environ 7:269–276. https://doi.org/10.1890/070152

Asner GP, Martin RE (2016) Convergent elevation trends in canopy chemical traits of tropical forests. Glob Change Biol 22:2216–2227. https://doi.org/10.1111/gcb.13164

Asner GP, Nepstad D, Cardinot G, Ray D (2004) Drought stress and carbon uptake in an Amazon forest measured with spaceborne imaging spectroscopy. Proc Natl Acad Sci USA 101:6039–6044. https://doi.org/10.1073/pnas.0400168101

Asner GP et al (2011) Spectroscopy of canopy chemicals in humid tropical forests. Remote Sens Environ 115:3587–3598. https://doi.org/10.1016/j.rse.2011.08.020

Asner GP et al (2012) Carnegie Airborne Observatory-2: increasing science data dimensionality via high-fidelity multi-sensor fusion. Remote Sens Environ 124:454–465. https://doi.org/10.1016/j.rse.2012.06.012

Asner GP, Martin RE, Anderson CB, Knapp DE (2015) Quantifying forest canopy traits: imaging spectroscopy versus field survey. Remote Sens Environ 158:15–27. https://doi.org/10.1016/j.rse.2014.11.011

Asner GP, Brodrick PG, Anderson CB, Vaughn N, Knapp DE, Martin RE (2016) Progressive forest canopy water loss during the 2012–2015 California drought. Proc Natl Acad Sci 113:E249–E255. https://doi.org/10.1073/pnas.1523397113

Asner GP et al (2017) Airborne laser-guided imaging spectroscopy to map forest trait diversity and guide conservation. Science 355:385. https://doi.org/10.1126/science.aaj1987

Atzberger C (2000) Development of an invertible forest reflectance model: the INFOR-model. In: Buchroithner M (ed) A decade of trans-European remote sensing cooperation. Proceedings of the 20th EARSeL symposium, Dresden, Germany, 14–16 June 2000, pp 39–44

Atzberger C, Guérif M, Baret F, Werner W (2010) Comparative analysis of three chemometric techniques for the spectroradiometric assessment of canopy chlorophyll content in winter wheat. Comput Electron Agric 73:165–173. https://doi.org/10.1016/j.compag.2010.05.006

Barton CVM, North PRJ (2001) Remote sensing of canopy light use efficiency using the photochemical reflectance index: model and sensitivity analysis. Remote Sens Environ 78:264–273. https://doi.org/10.1016/S0034-4257(01)00224-3

Beamish AL, Coops NC, Hermosilla T, Chabrillat S, Heim B (2018) Monitoring pigment-driven vegetation changes in a low-Arctic tundra ecosystem using digital cameras. Ecosphere 9:e02123. https://doi.org/10.1002/ecs2.2123

Beniston M (2004) The 2003 heat wave in Europe: a shape of things to come? An analysis based on Swiss climatological data and model simulations. Geophys Res Lett. https://doi.org/10.1029/2003GL018857

Berger K, Atzberger C, Danner M, D'Urso G, Mauser W, Vuolo F, Hank T (2018) Evaluation of the PROSAIL model capabilities for future hyperspectral model environments: a review study. Remote Sens 10:85. https://doi.org/10.3390/rs10010085

Bicheron P, Leroy M (1999) A method of biophysical parameter retrieval at global scale by inversion of a vegetation reflectance model. Remote Sens Environ 67:251–266. https://doi.org/10.1016/S0034-4257(98)00083-2

Birdsey R, Pan Y (2011) Ecology: drought and dead trees. Nat Clim Change 1:444–445. https://doi.org/10.1038/nclimate1298

Bond I, Chambwera M, Jones B, Nhantumbo I, Chundama M (2010) REDD+ in dryland forests: issues and prospects for pro-poor REDD in the miombo woodlands of southern Africa, natural resource issues, vol 21. International Institute for Environment and Development, London

Bréda NJJ (2003) Ground-based measurements of leaf area index: a review of methods, instruments and current controversies. J Exp Bot 54:2403–2417. https://doi.org/10.1093/jxb/erg263

Buddenbaum H, Steffens M (2012) The effects of spectral pretreatments on chemometric analyses of soil profiles using laboratory imaging spectroscopy. Appl Environ Soil Sci 2012:12. https://doi.org/10.1155/2012/274903

Buddenbaum H, Pueschel P, Stellmes M, Werner W, Hill J (2011) Measuring water and chlorophyll content on the leaf and canopy scale. EARSeL eProc 10:66–72

Buddenbaum H, Stern O, Stellmes M, Stoffels J, Pueschel P, Hill J, Werner W (2012) Field imaging spectroscopy of beech seedlings under dryness stress. Remote Sens 4:3721–3740. https://doi.org/10.3390/rs4123721

Buddenbaum H et al (2015) Using VNIR and SWIR field imaging spectroscopy for drought stress monitoring of beech seedlings. Int J Remote Sens 36:4590–4605. https://doi.org/10.1080/01431161.2015.1084435

Carrere V et al (2013) HYPXIM: a second generation high spatial resolution hyperspectral satellite for dual applications. In: 5th Workshop on hyperspectral image and signal processing: evolution in remote sensing (WHISPERS), 26–28 June 2013, pp 1–4. https://doi.org/10.1109/whispers.2013.8080685

Chadwick K, Asner G (2016) Organismic-scale remote sensing of canopy foliar traits in lowland tropical forests. Remote Sens 8:87. https://doi.org/10.3390/rs8020087

Chávez R, Clevers J, Herold M, Acevedo E, Ortiz M (2013) Assessing water stress of desert tamarugo trees using in situ data and very high spatial resolution remote sensing. Remote Sens 5:5064. https://doi.org/10.3390/rs5105064

Cheng T, Rivard B, Sánchez-Azofeifa A (2011) Spectroscopic determination of leaf water content using continuous wavelet analysis. Remote Sens Environ 115:659–670. https://doi.org/10.1016/j.rse.2010.11.001

Cheng T, Rivard B, Sánchez-Azofeifa AG, Féret J-B, Jacquemoud S, Ustin SL (2014) Deriving leaf mass per area (LMA) from foliar reflectance across a variety of plant species using continuous wavelet analysis. ISPRS J Photogramm Remote Sens 87:28–38. https://doi.org/10.1016/j.isprsjprs.2013.10.009

Ciais P et al (2005) Europe-wide reduction in primary productivity caused by the heat and drought in 2003. Nature 437:529–533. https://doi.org/10.1038/nature03972

Colombo R, Meroni M, Marchesi A, Busetto L, Rossini M, Giardino C, Panigada C (2008) Estimation of leaf and canopy water content in poplar plantations by means of hyperspectral indices and inverse modeling. Remote Sens Environ 112:1820–1834. https://doi.org/10.1016/j.rse.2007.09.005

Combal B et al (2002) Retrieval of canopy biophysical variables from bidirectional reflectance using prior information to solve the ill-posed inverse problem. Remote Sens Environ 84:1–15. https://doi.org/10.1016/S0034-4257(02)00035-4

Coops NC, Smith M-L, Martin ME, Ollinger SV (2003) Prediction of eucalypt foliage nitrogen content from satellite-derived hyperspectral data. IEEE Trans Geosci Remote Sens 41:1338–1346. https://doi.org/10.1109/TGRS.2003.813135

Curran PJ (1989) Remote sensing of foliar chemistry. Remote Sens Environ 30:271–278. https://doi.org/10.1016/0034-4257(89)90069-2

Dawson TP, Curran PJ, Plummer SE (1998) LIBERTY—modeling the effects of leaf biochemical concentration on reflectance spectra. Remote Sens Environ 65:50–60. https://doi.org/10.1016/S0034-4257(98)00007-8

Dechant B, Cuntz M, Vohland M, Schulz E, Doktor D (2017) Estimation of photosynthesis traits from leaf reflectance spectra: correlation to nitrogen content as the dominant mechanism. Remote Sens Environ 196:279–292. https://doi.org/10.1016/j.rse.2017.05.019

Dotzler S, Hill J, Buddenbaum H, Stoffels J (2015) The potential of EnMAP and Sentinel-2 data for detecting drought stress phenomena in deciduous forest communities. Remote Sens 7:14227–14258. https://doi.org/10.3390/rs71014227

Drusch M et al (2017) The FLuorescence EXplorer Mission Concep—ESA's Earth Explorer 8. IEEE Trans Geosci Remote Sens 55:1273–1284. https://doi.org/10.1109/tgrs.2016.2621820

Eckardt A, Horack J, Lehmann F, Krutz D, Drescher J, Whorton M, Soutullo M (2015) DESIS (DLR earth sensing imaging spectrometer for the ISS-MUSES platform). In: IEEE international geoscience and remote sensing symposium (IGARSS), 26–31 July 2015, pp 1457–1459. https://doi.org/10.1109/igarss.2015.7326053

EEA (2012) Annual environmental indicator report 2012—ecosystem resilience and resource efficiency in a green economy in Europe. European Environment Agency, Copenhagen. https://doi.org/10.2800/487

Ewald M et al (2018) LiDAR derived forest structure data improves predictions of canopy N and P concentrations from imaging spectroscopy. Remote Sens Environ 211:13–25. https://doi.org/10.1016/j.rse.2018.03.038

Fang M, Ju W, Zhan W, Cheng T, Qiu F, Wang J (2017) A new spectral similarity water index for the estimation of leaf water content from hyperspectral data of leaves. Remote Sens Environ 196:13–27. https://doi.org/10.1016/j.rse.2017.04.029

FAO (2010) Guidelines on sustainable forest management in drylands of sub-Saharan Africa. FAO, Rome

Fassnacht F et al (2014) Comparison of feature reduction algorithms for classifying tree species with hyperspectral data on three central European test sites. IEEE J Sel Top Appl Earth Obs Remote Sens 7:2547–2561. https://doi.org/10.1109/jstars.2014.2329390

Fassnacht FE, Stenzel S, Gitelson AA (2015) Non-destructive estimation of foliar carotenoid content of tree species using merged vegetation indices. J Plant Physiol 176:210–217. https://doi.org/10.1016/j.jplph.2014.11.003

Fassnacht FE et al (2016) Review of studies on tree species classification from remotely sensed data. Remote Sens Environ 186:64–87. https://doi.org/10.1016/j.rse.2016.08.013

Fawcett D, Verhoef W, Schläpfer D, Schneider FD, Schaepman ME, Damm A (2018) Advancing retrievals of surface reflectance and vegetation indices over forest ecosystems by combining imaging spectroscopy, digital object models, and 3D canopy modelling. Remote Sens Environ 204:583–595. https://doi.org/10.1016/j.rse.2017.09.040

Feilhauer H, Asner GP, Martin RE (2015) Multi-method ensemble selection of spectral bands related to leaf biochemistry. Remote Sens Environ 164:57–65. https://doi.org/10.1016/j.rse.2015.03.033

Feingersh T, Ben Dor E (2015) SHALOM—a commercial hyperspectral space mission. In: Qian S-E (ed) Optical payloads for space missions. Wiley, Chichester. https://doi.org/10.1002/9781118945179.ch11

Feret J-B et al (2008) PROSPECT-4 and 5: advances in the leaf optical properties model separating photosynthetic pigments. Remote Sens Environ 112:3030–3043. https://doi.org/10.1016/j.rse.2008.02.012

Féret J-B et al (2011) Optimizing spectral indices and chemometric analysis of leaf chemical properties using radiative transfer modeling. Remote Sens Environ 115:2742–2750. https://doi.org/10.1016/j.rse.2011.06.016

Féret JB, Gitelson AA, Noble SD, Jacquemoud S (2017) PROSPECT-D: towards modeling leaf optical properties through a complete lifecycle. Remote Sens Environ 193:204–215. https://doi.org/10.1016/j.rse.2017.03.004

Fink AH, Brücher T, Krüger A, Leckebusch GC, Pinto JG, Ulbrich U (2004) The 2003 European summer heatwaves and drought—synoptic diagnosis and impacts. Weather 59:209–216. https://doi.org/10.1256/wea.73.04

Foster JR, Townsend PA (2004) Linking hyperspectral imagery and forest inventories for forest assessment in the Central Appalachians. In: Yaussy DA, Hix DM, Long RP, Goebel PC (eds) 14th Central hardwood forest conference. U.S. Department of Agriculture, Forest Service, Northeastern Research Station, pp 76–86

Franklin SE (2001) Remote sensing for sustainable forest management. Lewis Publishers, Boca Raton

Frantz D, Röder A, Stellmes M, Hill J (2016) An operational radiometric landsat preprocessing framework for large-area time series applications. IEEE Trans Geosci Remote Sens. https://doi.org/10.1109/tgrs.2016.2530856

Gamon JA, Peñuelas J, Field CB (1992) A narrow-waveband spectral index that tracks diurnal changes in photosynthetic efficiency. Remote Sens Environ 41:35–44. https://doi.org/10.1016/0034-4257(92)90059-s

Gara TW, Skidmore AK, Darvishzadeh R, Wang T (2018) Leaf to canopy upscaling approach affects the estimation of canopy traits. GISci Remote Sens. https://doi.org/10.1080/15481603.2018.1540170

Garbulsky MF, Peñuelas J, Gamon J, Inoue Y, Filella I (2011) The photochemical reflectance index (PRI) and the remote sensing of leaf, canopy and ecosystem radiation use efficiencies: a review and meta-analysis. Remote Sens Environ 115:281–297. https://doi.org/10.1016/j.rse.2010.08.023

Gastellu-Etchegorry JP, Martin E, Gascon F (2004) DART: a 3D model for simulating satellite images and studying surface radiation budget. Int J Remote Sens 25:73–96. https://doi.org/10.1080/0143116031000115166

Ghamisi P, Plaza J, Chen Y, Li J, Plaza AJ (2017) Advanced spectral classifiers for hyperspectral images: a review. IEEE Geosci Remote Sens Mag 5:8–32. https://doi.org/10.1109/mgrs.2016.2616418

Gitelson AA, Merzlyak MN (1997) Remote estimation of chlorophyll content in higher plant leaves. Int J Remote Sens 18:2691–2697. https://doi.org/10.1080/014311697217558

Gitelson AA, Gritz Y, Merzlyak MN (2003) Relationships between leaf chlorophyll content and spectral reflectance and algorithms for non-destructive chlorophyll assessment in higher plant leaves. J Plant Physiol 160:271–282. https://doi.org/10.1078/0176-1617-00887

Gitelson AA, Keydan GP, Merzlyak MN (2006) Three-band model for noninvasive estimation of chlorophyll, carotenoids, and anthocyanin contents in higher plant leaves. Geophys Res Lett. https://doi.org/10.1029/2006gl026457

Green AA, Berman M, Switzer P, Craig MD (1998) A transformation for ordering multispectral data in terms of image quality with implications for noise removal. IEEE Trans Geosci Remote Sens 26:65–74

Green RO, Painter TH, Roberts DA, Dozier J (2006) Measuring the expressed abundance of the three phases of water with an imaging spectrometer over melting snow. Water Resour Res. https://doi.org/10.1029/2005wr004509

Guanter L et al (2015) The EnMAP spaceborne imaging spectroscopy mission for earth observation. Remote Sens 7:8830–8857. https://doi.org/10.3390/rs70708830

Hank TB, Berger K, Bach H, Clevers JGPW, Gitelson A, Zarco-Tejada P, Mauser W (2019) Spaceborne imaging spectroscopy for sustainable agriculture: contributions and challenges. Surv Geophys. https://doi.org/10.1007/s10712-018-9492-0

Hansen MC, Roy DP, Lindquist E, Adusei B, Justice CO, Altstatt A (2008) A method for integrating MODIS and Landsat data for systematic monitoring of forest cover and change in the Congo Basin. Remote Sens Environ 112:2495–2513. https://doi.org/10.1016/j.rse.2007.11.012

Heim R, Jürgens N, Große-Stoltenberg A, Oldeland J (2015) The effect of epidermal structures on leaf spectral signatures of ice plants (Aizoaceae). Remote Sens 7:15862

Held M, Rabe A, Senf C, Svd Linden, Hostert P (2015) Analyzing hyperspectral and hypertemporal data by decoupling feature redundancy and feature relevance. IEEE Geosci Remote Sens Lett 12:983–987. https://doi.org/10.1109/lgrs.2014.2371242

Hernández-Clemente R, Navarro-Cerrillo RM, Zarco-Tejada PJ (2012) Carotenoid content estimation in a heterogeneous conifer forest using narrow-band indices and PROSPECT + DART simulations. Remote Sens Environ 127:298–315. https://doi.org/10.1016/j.rse.2012.09.014

Homolová L, Malenovský Z, Clevers JGPW, García-Santos G, Schaepman ME (2013) Review of optical-based remote sensing for plant trait mapping. Ecol Complex 15:1–16. https://doi.org/10.1016/j.ecocom.2013.06.003

Hosgood B, Jacquemoud S, Andreoli G, Verdebout J, Pedrini G, Schmuck G (1994) Leaf optical properties experiment 93 (LOPEX93). European Commission, Joint Research Center, Ispra

Hovi A, Raitio P, Rautiainen M (2017) A spectral analysis of 25 boreal tree species. Silva Fenn. https://doi.org/10.14214/sf.7753

Huang Z, Turner BJ, Dury SJ, Wallis IR, Foley WJ (2004) Estimating foliage nitrogen concentration from HYMAP data using continuum removal analysis. Remote Sens Environ 93:18–29. https://doi.org/10.1016/j.rse.2004.06.008

Huber S, Kneubühler M, Psomas A, Itten K, Zimmermann NE (2008) Estimating foliar biochemistry from hyperspectral data in mixed forest canopy. For Ecol Manag 256:491–501. https://doi.org/10.1016/j.foreco.2008.05.011

Huemmrich KF (2001) The GeoSail model: a simple addition to the SAIL model to describe discontinuous canopy reflectance. Remote Sens Environ 75:423–431. https://doi.org/10.1016/S0034-4257(00)00184-X
Hultquist C, Chen G, Zhao K (2014) A comparison of Gaussian process regression, random forests and support vector regression for burn severity assessment in diseased forests. Remote Sens Lett 5:723–732. https://doi.org/10.1080/2150704x.2014.963733
Jacquemoud S, Baret F (1990) PROSPECT: a model of leaf optical properties spectra. Remote Sens Environ 34:75–91. https://doi.org/10.1016/0034-4257(90)90100-Z
Jacquemoud S, Ustin SL, Verdebout J, Schmuck G, Andreoli G, Hosgood B (1996) Estimating leaf biochemistry using the PROSPECT leaf optical properties model. Remote Sens Environ 56:194–202. https://doi.org/10.1016/0034-4257(95)00238-3
Jacquemoud S, Bacour C, Poilvé H, Frangi J-P (2000) Comparison of four radiative transfer models to simulate plant canopies reflectance: direct and inverse mode. Remote Sens Environ 74:471–481. https://doi.org/10.1016/S0034-4257(00)00139-5
Jacquemoud S et al (2009) PROSPECT + SAIL models: a review of use for vegetation characterization. Remote Sens Environ 113:S56–S66. https://doi.org/10.1016/j.rse.2008.01.026
Jay S, Bendoula R, Hadoux X, Féret J-B, Gorretta N (2016) A physically-based model for retrieving foliar biochemistry and leaf orientation using close-range imaging spectroscopy. Remote Sens Environ 177:220–236. https://doi.org/10.1016/j.rse.2016.02.029
Knyazikhin Y et al (2013) Hyperspectral remote sensing of foliar nitrogen content. Proc Natl Acad Sci 110:E185–E192. https://doi.org/10.1073/pnas.1210196109
Koetz B, Schaepman M, Morsdorf F, Bowyer P, Itten K, Allgöwer B (2004) Radiative transfer modeling within a heterogeneous canopy for estimation of forest fire fuel properties. Remote Sens Environ 92:332–344. https://doi.org/10.1016/j.rse.2004.05.015
Koetz B, Sun G, Morsdorf F, Ranson KJ, Kneubühler M, Itten K, Allgöwer B (2007) Fusion of imaging spectrometer and LIDAR data over combined radiative transfer models for forest canopy characterization. Remote Sens Environ 106:449–459. https://doi.org/10.1016/j.rse.2006.09.013
Kokaly RF, Asner GP, Ollinger SV, Martin ME, Wessman CA (2009) Characterizing canopy biochemistry from imaging spectroscopy and its application to ecosystem studies. Remote Sens Environ 113(Supplement 1):S78–S91. https://doi.org/10.1016/j.rse.2008.10.018
Kuusk A, Kuusk J, Lang M (2014) Modeling directional forest reflectance with the hybrid type forest reflectance model FRT. Remote Sens Environ 149:196–204. https://doi.org/10.1016/j.rse.2014.03.035
Labate D et al (2009) The PRISMA payload optomechanical design, a high performance instrument for a new hyperspectral mission. Acta Astronaut 65:1429–1436. https://doi.org/10.1016/j.actaastro.2009.03.077
Lausch A, Erasmi S, King D, Magdon P, Heurich M (2016) Understanding forest health with remote sensing-part I—a review of spectral traits, processes and remote-sensing characteristics. Remote Sens 8:1029. https://doi.org/10.3390/rs8121029
Lausch A, Erasmi S, King D, Magdon P, Heurich M (2017) Understanding forest health with remote sensing-part II—a review of approaches and data models. Remote Sens 9:129. https://doi.org/10.3390/rs9020129
Lausch A et al (2018) Understanding forest health with remote sensing, part III: requirements for a scalable multi-source forest health monitoring network based on data science approaches. Remote Sens 10:1120. https://doi.org/10.3390/rs10071120
le Maire G, François C, Dufrêne E (2004) Towards universal broad leaf chlorophyll indices using PROSPECT simulated database and hyperspectral reflectance measurements. Remote Sens Environ 89:1–28. https://doi.org/10.1016/j.rse.2003.09.004
le Maire G et al (2008) Calibration and validation of hyperspectral indices for the estimation of broadleaved forest leaf chlorophyll content, leaf mass per area, leaf area index and leaf canopy biomass. Remote Sens Environ 112:3846–3864. https://doi.org/10.1016/j.rse.2008.06.005
Lee CM, Cable ML, Hook SJ, Green RO, Ustin SL, Mandl DJ, Middleton EM (2015) An introduction to the NASA Hyperspectral InfraRed Imager (HyspIRI) mission and preparatory activities. Remote Sens Environ 167:6–19. https://doi.org/10.1016/j.rse.2015.06.012
Li P, Wang Q (2011) Retrieval of leaf biochemical parameters using PROSPECT inversion: a new approach for alleviating ill-posed problems. IEEE Trans Geosci Remote Sens 49:2499–2506. https://doi.org/10.1109/tgrs.2011.2109390
Li D et al (2018) PROCWT: coupling PROSPECT with continuous wavelet transform to improve the retrieval of foliar chemistry from leaf bidirectional reflectance spectra. Remote Sens Environ 206:1–14. https://doi.org/10.1016/j.rse.2017.12.013

Liechty HO, Fristoe C (2013) Response of midrotation pine stands to fertilizer and herbicide application in the western Gulf coastal plain. South J Appl For 37:69–74
Lillesaeter O (1982) Spectral reflectance of partly transmitting leaves: laboratory measurements and mathematical modeling. Remote Sens Environ 12:247–254. https://doi.org/10.1016/0034-4257(82)90057-8
Lindner M et al (2010) Climate change impacts, adaptive capacity, and vulnerability of European forest ecosystems. For Ecol Manag 259:698–709. https://doi.org/10.1016/j.foreco.2009.09.023
Locherer M, Hank T, Danner M, Mauser W (2015) Retrieval of seasonal leaf area index from simulated EnMAP data through optimized LUT-based inversion of the PROSAIL model. Remote Sens 7:10321
Ma Z et al (2012) Regional drought-induced reduction in the biomass carbon sink of Canada's boreal forests. Proc Natl Acad Sci 109:2423–2427. https://doi.org/10.1073/pnas.1111576109
Malenovský Z et al (in review) Variability and uncertainty challenges in upscaling imaging spectroscopy observations from leaves to vegetation canopies. Surv Geophys
Malenovský Z, Albrechtová J, Lhotáková Z, Zurita-Milla R, Clevers JGPW, Schaepman ME, Cudlín P (2006a) Applicability of the PROSPECT model for Norway spruce needles. Int J Remote Sens 27:5315–5340. https://doi.org/10.1080/01431160600762990
Malenovský Z, Ufer C, Lhotáková Z, Clevers JGPW, Schaepman ME, Albrechtová J, Cudlín P (2006b) A new hyperspectral index for chlorophyll estimation of a forest canopy: area under curve normalised to maximal band depth between 650–725 nm. EARSeL eProc 5:161–172
Martin ME, Plourde LC, Ollinger SV, Smith ML, McNeil BE (2008) A generalizable method for remote sensing of canopy nitrogen across a wide range of forest ecosystems. Remote Sens Environ 112:3511–3519. https://doi.org/10.1016/j.rse.2008.04.008
Martin R, Chadwick K, Brodrick P, Carranza-Jimenez L, Vaughn N, Asner G (2018a) An approach for foliar trait retrieval from airborne imaging spectroscopy of tropical forests. Remote Sens 10:199. https://doi.org/10.3390/rs10020199
Martin RE et al (2018b) Remote measurement of canopy water content in giant sequoias (*Sequoiadendron giganteum*) during drought. For Ecol Manag 419–420:279–290. https://doi.org/10.1016/j.foreco.2017.12.002
Masek JG et al (2006) A Landsat surface reflectance dataset for North America, 1990–2000. IEEE Geosci Remote Sens Lett 3:68–72. https://doi.org/10.1109/lgrs.2005.857030
Matsunaga T et al (2017) Current status of Hyperspectral Imager Suite (HISUI) onboard International Space Station (ISS). In: IEEE international geoscience and remote sensing symposium (IGARSS), 23–28 July 2017, pp 443–446. https://doi.org/10.1109/igarss.2017.8126989
McNeil BE, de Beurs KM, Eshleman KN, Foster JR, Townsend PA (2007a) Maintenance of ecosystem nitrogen limitation by ephemeral forest disturbance: an assessment using MODIS, Hyperion, and Landsat ETM+. Geophys Res Lett. https://doi.org/10.1029/2007gl031387
McNeil BE, Read JM, Driscoll CT (2007b) Foliar nitrogen responses to elevated atmospheric nitrogen deposition in nine temperate forest canopy species. Environ Sci Technol 41:5191–5197. https://doi.org/10.1021/es062901z
Miller J et al (2005) Development of a vegetation fluorescence canopy model—ESTEC contract no. 16365/02/NL/FF—final report
Nieke J, Rast M (2018) Towards the Copernicus Hyperspectral Imaging Mission for the Environment (CHIME). In: IGARSS 2018—IEEE international geoscience and remote sensing symposium, 22–27 July 2018, pp 157–159. https://doi.org/10.1109/igarss.2018.8518384
Nink S, Hill J, Buddenbaum H, Stoffels J, Sachtleber T, Langshausen J (2015) Assessing the suitability of future multi- and hyperspectral satellite systems for mapping the spatial distribution of Norway spruce timber volume. Remote Sens 7:12009–12040. https://doi.org/10.3390/rs70912009
North PRJ (1996) Three-dimensional forest light interaction model using a Monte Carlo method. IEEE Trans Geosci Remote Sens 34:946–956. https://doi.org/10.1109/36.508411
Ollinger SV et al (2008) Canopy nitrogen, carbon assimilation, and albedo in temperate and boreal forests: functional relations and potential climate feedbacks. Proc Natl Acad Sci 105:19336–19341. https://doi.org/10.1073/pnas.0810021105
Omari K, White HP, Staenz K (2009) Multiple scattering within the FLAIR model incorporating the photon recollision probability approach. IEEE Trans Geosci Remote Sens 47:2931–2941. https://doi.org/10.1109/tgrs.2009.2014466
Omari K, White HP, Staenz K, King DJ (2013) Retrieval of forest canopy parameters by inversion of the PROFLAIR leaf-canopy reflectance model using the LUT approach. IEEE J Sel Top Appl Earth Obs Remote Sens 6:715–723. https://doi.org/10.1109/jstars.2013.2240264
Paz-Kagan T et al (2018) Landscape-scale variation in canopy water content of giant sequoias during drought. For Ecol Manag 419–420:291–304. https://doi.org/10.1016/j.foreco.2017.11.018

Peng C et al (2011) A drought-induced pervasive increase in tree mortality across Canada's boreal forests. Nat Clim Change 1:467–471. https://doi.org/10.1038/nclimate1293

Peñuelas J, Baret F, Filella I (1995) Semi-empirical indices to assess carotenoids/chlorophyll a ratio from leaf spectral reflectance. Photosynthetica 31:221–230

Rascher U et al (2015) Sun-induced fluorescence—a new probe of photosynthesis: first maps from the imaging spectrometer HyPlant. Glob Change Biol 21:4673–4684. https://doi.org/10.1111/gcb.13017

Rennenberg H, Loreto F, Polle A, Brilli F, Fares S, Beniwal RS, Gessler A (2006) Physiological responses of forest trees to heat and drought. Plant Biol 8:556–571. https://doi.org/10.1055/s-2006-924084

Riaño D, Vaughan P, Chuvieco E, Zarco-Tejada PJ, Ustin SL (2005) Estimation of fuel moisture content by inversion of radiative transfer models to simulate equivalent water thickness and dry matter content: analysis at leaf and canopy level. IEEE Trans Geosci Remote Sens 43:819–826. https://doi.org/10.1109/tgrs.2005.843316

Roberts DA, Gardner M, Church R, Ustin S, Scheer G, Green RO (1998) Mapping chaparral in the Santa Monica Mountains using multiple endmember spectral mixture models. Remote Sens Environ 65:267–279. https://doi.org/10.1016/S0034-4257(98)00037-6

Rogge DM, Rivard B, Jinkai Z, Jilu F (2006) Iterative spectral unmixing for optimizing per-pixel endmember sets. IEEE Trans Geosci Remote Sens 44:3725–3736. https://doi.org/10.1109/tgrs.2006.881123

Rogge D, Bachmann M, Rivard B, Nielsen AA, Feng J (2014) A spatial–spectral approach for deriving high signal quality eigenvectors for remote sensing image transformations. Int J Appl Earth Obs Geoinf 26:387–398. https://doi.org/10.1016/j.jag.2013.09.007

Schelhaas M-J, Nabuurs G-J, Schuck A (2003) Natural disturbances in the European forests in the 19th and 20th centuries. Glob Change Biol 9:1620–1633. https://doi.org/10.1046/j.1365-2486.2003.00684.x

Schlerf M, Atzberger C (2006) Inversion of a forest reflectance model to estimate structural canopy variables from hyperspectral remote sensing data. Remote Sens Environ 100:281–294. https://doi.org/10.1016/j.rse.2005.10.006

Schlerf M, Atzberger C, Hill J, Buddenbaum H, Werner W, Schüler G (2010) Retrieval of chlorophyll and nitrogen in Norway spruce (*Picea abies* L. Karst.) using imaging spectroscopy. Int J Appl Earth Obs Geoinf 12:17–26. https://doi.org/10.1016/j.jag.2009.08.006

Seidl R, Schelhaas M-J, Lexer MJ (2011) Unraveling the drivers of intensifying forest disturbance regimes in Europe. Glob Change Biol 17:2842–2852. https://doi.org/10.1111/j.1365-2486.2011.02452.x

Serbin SP, Singh A, McNeil BE, Kingdon CC, Townsend PA (2014) Spectroscopic determination of leaf morphological and biochemical traits for northern temperate and boreal tree species. Ecol Appl 24:1651–1669. https://doi.org/10.1890/13-2110.1

Singh A, Serbin SP, McNeil BE, Kingdon CC, Townsend PA (2015) Imaging spectroscopy algorithms for mapping canopy foliar chemical and morphological traits and their uncertainties. Ecol Appl 25:2180–2197. https://doi.org/10.1890/14-2098.1

Smith M-L, Martin ME, Plourde L, Ollinger SV (2003) Analysis of hyperspectral data for estimation of temperate forest canopy nitrogen concentration: comparison between an airborne (AVIRIS) and a spaceborne (Hyperion) sensor. IEEE Trans Geosci Remote Sens 41:1332–1337. https://doi.org/10.1109/tgrs.2003.813128

Somers B, Asner GP (2013) Multi-temporal hyperspectral mixture analysis and feature selection for invasive species mapping in rainforests. Remote Sens Environ 136:14–27. https://doi.org/10.1016/j.rse.2013.04.006

Sommer C, Holzwarth S, Heiden U, Heurich M, Müller J, Mauser W (2016) Feature-based tree species classification using hyperspectral and lidar data in the Bavarian Forest National Park. EARSeL eProc 14:49–70. https://doi.org/10.12760/02-2015-2-05

Sonobe R, Wang Q (2018) Nondestructive assessments of carotenoids content of broadleaved plant species using hyperspectral indices. Comput Electron Agric 145:18–26. https://doi.org/10.1016/j.compag.2017.12.022

Stoffels J et al (2015) Satellite-based derivation of high-resolution forest information layers for operational forest management. Forests 6:1982–2013. https://doi.org/10.3390/f6061982

Storch T, Habermeyer M, Eberle S, Muhle H, Mueller RM (2013) Towards a critical design of an operational ground segment for an Earth observation mission. J Appl Remote Sens 7:075381. https://doi.org/10.1117/1.JRS.7.073581

Suárez L, Zarco-Tejada PJ, Sepulcre-Cantó G, Pérez-Priego O, Miller JR, Jiménez-Muñoz JC, Sobrino J (2008) Assessing canopy PRI for water stress detection with diurnal airborne imagery. Remote Sens Environ 112:560–575. https://doi.org/10.1016/j.rse.2007.05.009
Swann ALS et al (2018) Continental-scale consequences of tree die-offs in North America: identifying where forest loss matters most. Environ Res Lett 13:055014. https://doi.org/10.1088/1748-9326/aaba0f
Swatantran A, Dubayah R, Roberts D, Hofton M, Blair JB (2011) Mapping biomass and stress in the Sierra Nevada using lidar and hyperspectral data fusion. Remote Sens Environ 115:2917–2930. https://doi.org/10.1016/j.rse.2010.08.027
Thompson DR, Natraj V, Green RO, Helmlinger MC, Gao B-C, Eastwood ML (2018) Optimal estimation for imaging spectrometer atmospheric correction. Remote Sens Environ 216:355–373. https://doi.org/10.1016/j.rse.2018.07.003
Thompson DR, Guanter L, Berk A, Gao B-C, Richter R, Schläpfer D, Thome KJ (2019) Retrieval of atmospheric parameters and surface reflectance from visible and shortwave infrared imaging spectroscopy data. Surv Geophys. https://doi.org/10.1007/s10712-018-9488-9
Townsend PA, Foster JR, Chastain RA, Currie WS (2003) Application of imaging spectroscopy to mapping canopy nitrogen in the forests of the central Appalachian Mountains using Hyperion and AVIRIS. IEEE Trans Geosci Remote Sens 41:1347–1354. https://doi.org/10.1109/tgrs.2003.813205
Townsend PA, Serbin SP, Kruger EL, Gamon JA (2013) Disentangling the contribution of biological and physical properties of leaves and canopies in imaging spectroscopy data. Proc Natl Acad Sci 110:E1074. https://doi.org/10.1073/pnas.1300952110
Townsend PA, Wang Z, Singh A (2017) Prospects for universal foliar trait-retrieval algorithms from imaging spectroscopy: cross-site and cross-platform. Paper presented at the 10th EARSeL SIG imaging spectroscopy workshop, Zurich, Switzerland
Ustin SL, Roberts DA, Gamon JA, Asner GP, Green RO (2004) Using imaging spectroscopy to study ecosystem processes and properties. Bioscience 54:523–534. https://doi.org/10.1641/0006-3568(2004)054%5b0523:uistse%5d2.0.co;2
Ustin SL, Gitelson AA, Jacquemoud S, Schaepman M, Asner GP, Gamon JA, Zarco-Tejada P (2009) Retrieval of foliar information about plant pigment systems from high resolution spectroscopy. Remote Sens Environ 113:S67–S77. https://doi.org/10.1016/j.rse.2008.10.019
Verhoef W (1984) Light scattering by leaf layers with application to canopy reflectance modeling: the SAIL model. Remote Sens Environ 16:125–141. https://doi.org/10.1016/0034-4257(84)90057-9
Vermote E, Justice C, Claverie M, Franch B (2016) Preliminary analysis of the performance of the Landsat 8/OLI land surface reflectance product. Remote Sens Environ 185:46–56. https://doi.org/10.1016/j.rse.2016.04.008
Verrelst J, Camps-Valls G, Muñoz-Marí J, Rivera JP, Veroustraete F, Clevers JGPW, Moreno J (2015) Optical remote sensing and the retrieval of terrestrial vegetation bio-geophysical properties—a review. ISPRS J Photogramm Remote Sens 108:273–290. https://doi.org/10.1016/j.isprsjprs.2015.05.005
Verrelst J et al (2019) Quantifying vegetation biophysical variables from imaging spectroscopy data: a review on retrieval methods. Surv Geophys. https://doi.org/10.1007/s10712-018-9478-y
Vohland M, Ludwig M, Harbich M, Emmerling C, Thiele-Bruhn S (2016) Using variable selection and wavelets to exploit the full potential of visible/near infrared spectra for predicting soil properties. J Near Infrared Spectrosc 24:255–269. https://doi.org/10.1255/jnirs.1233
Wang Z, Skidmore AK, Wang T, Darvishzadeh R, Hearne J (2015) Applicability of the PROSPECT model for estimating protein and cellulose + lignin in fresh leaves. Remote Sens Environ 168:205–218. https://doi.org/10.1016/j.rse.2015.07.007
Wang Z et al (2017) Canopy foliar nitrogen retrieved from airborne hyperspectral imagery by correcting for canopy structure effects. Int J Appl Earth Obs Geoinf 54:84–94. https://doi.org/10.1016/j.jag.2016.09.008
Wang Z, Skidmore AK, Darvishzadeh R, Wang T (2018) Mapping forest canopy nitrogen content by inversion of coupled leaf-canopy radiative transfer models from airborne hyperspectral imagery. Agric For Meteorol 253–254:247–260. https://doi.org/10.1016/j.agrformet.2018.02.010
Wang Z, Townsend PA, Schweiger AK, Couture JJ, Singh A, Hobbie SE, Cavender-Bares J (2019) Mapping foliar functional traits and their uncertainties across three years in a grassland experiment. Remote Sens Environ 221:405–416. https://doi.org/10.1016/j.rse.2018.11.016
Waske B, van der Linden S, Oldenburg C, Jakimow B, Rabe A, Hostert P (2012) imageRF—a user-oriented implementation for remote sensing image analysis with Random Forests. Environ Model Softw 35:192–193. https://doi.org/10.1016/j.envsoft.2012.01.014

Wessman CA, Aber JD, Peterson DL, Melillo JM (1988) Remote sensing of canopy chemistry and nitrogen cycling in temperate forest ecosystems. Nature 335:154. https://doi.org/10.1038/335154a0
Wessman CA, Aber JD, Peterson DL (1989) An evaluation of imaging spectrometry for estimating forest canopy chemistry. Int J Remote Sens 10:1293–1316. https://doi.org/10.1080/01431168908903969
White HP, Miller JR, Chen JM (2001) Four-scale linear model for anisotropic reflectance (FLAIR) for plant canopies. I. Model description and partial validation. IEEE Trans Geosci Remote Sens 39:1072–1083. https://doi.org/10.1109/36.921425
White JC, Coops NC, Hilker T, Wulder MA, Carroll AL (2007) Detecting mountain pine beetle red attack damage with EO-1 Hyperion moisture indices. Int J Remote Sens 28:2111–2121. https://doi.org/10.1080/01431160600944028
Wold S, Ruhe A, Wold H, Dunn IW (1984) The collinearity problem in linear regression. The partial least squares (PLS) approach to generalized inverses. SIAM J Sci Stat Comput 5:735–743. https://doi.org/10.1137/0905052
Wold S, Sjöström M, Eriksson L (2001) PLS-regression: a basic tool of chemometrics. Chemometr Intell Lab Syst 58:109–130. https://doi.org/10.1016/S0169-7439(01)00155-1
Wolter PT, Townsend PA, Sturtevant BR (2009) Estimation of forest structural parameters using 5 and 10 meter SPOT-5 satellite data. Remote Sens Environ 113:2019–2036. https://doi.org/10.1016/j.rse.2009.05.009
Yáñez-Rausell L, Malenovský Z, Clevers JGPW, Schaepman ME (2014a) Minimizing measurement uncertainties of coniferous needle-leaf optical properties. Part II: experimental setup and error analysis. IEEE J Sel Top Appl Earth Obs Remote Sens 7:406–420. https://doi.org/10.1109/jstars.2013.2292817
Yáñez-Rausell L, Schaepman ME, Clevers JGPW, Malenovský Z (2014b) Minimizing measurement uncertainties of coniferous needle-leaf optical properties, part I: methodological review. IEEE J Sel Top Appl Earth Obs Remote Sens 7:399–405. https://doi.org/10.1109/jstars.2013.2272890
Zhang Q, Middleton EM, Gao B, Cheng Y (2012) Using EO-1 Hyperion to simulate HyspIRI products for a coniferous forest: the fraction of PAR absorbed by chlorophyll (fAPAR_chl) and leaf water content (LWC). IEEE Trans Geosci Remote Sens 50:1844–1852. https://doi.org/10.1109/tgrs.2011.2169267

Surv Geophys (2019) 40:589–629
https://doi.org/10.1007/s10712-018-9478-y

# Quantifying Vegetation Biophysical Variables from Imaging Spectroscopy Data: A Review on Retrieval Methods

**Jochem Verrelst[1] · Zbyněk Malenovský[2,3,4] · Christiaan Van der Tol[5] · Gustau Camps-Valls[1] · Jean-Philippe Gastellu-Etchegorry[6] · Philip Lewis[7,8] · Peter North[9] · José Moreno[1]**

Received: 2 November 2017 / Accepted: 16 May 2018 / Published online: 1 June 2018


**Abstract** An unprecedented spectroscopic data stream will soon become available with forthcoming Earth-observing satellite missions equipped with imaging spectroradiometers. This data stream will open up a vast array of opportunities to quantify a diversity of biochemical and structural vegetation properties. The processing requirements for such large data streams require reliable retrieval techniques enabling the spatiotemporally explicit quantification of biophysical variables. With the aim of preparing for this new era of Earth observation, this review summarizes the state-of-the-art retrieval methods that have been applied in experimental imaging spectroscopy studies inferring all kinds of vegetation biophysical variables. Identified retrieval methods are categorized into: (1) parametric regression, including vegetation indices, shape indices and spectral transformations; (2)

✉ Jochem Verrelst
jochem.verrelst@uv.es

1 Image Processing Laboratory (IPL), Parc Científic, Universitat de València, Paterna, València 46980, Spain

2 Surveying and Spatial Sciences Group, School of Technology, Environments and Design, University of Tasmania, Private Bag 76, Hobart, TAS 7001, Australia

3 Remote Sensing Department, Global Change Research Institute CAS, Bělidla 986/4a, 60300 Brno, Czech Republic

4 USRA/GESTAR, Biospheric Sciences Laboratory, NASA Goddard Space Flight Center, 8800 Greenbelt Rd, Greenbelt, MD 20771, USA

5 Department of Water Resources, Faculty ITC, University of Twente, P.O. Box 217, 7500 AE Enschede, The Netherlands

6 Centre d'Etudes Spatiales de la Biosphère - UPS, CNES, CNRS, IRD, Université de Toulouse, 31401 Toulouse Cedex 9, France

7 Department of Geography, University College London, Pearson Building, Gower Street, WC1E 6BT London, UK

8 National Centre for Earth Observation, Department of Physics and Astronomy, The University of Leicester, Michael Atiyah Building, LE1 7RH Leicester, UK

9 Department of Geography, Swansea University, Swansea SA2 8PP, UK

nonparametric regression, including linear and nonlinear machine learning regression algorithms; (3) physically based, including inversion of radiative transfer models (RTMs) using numerical optimization and look-up table approaches; and (4) hybrid regression methods, which combine RTM simulations with machine learning regression methods. For each of these categories, an overview of widely applied methods with application to mapping vegetation properties is given. In view of processing imaging spectroscopy data, a critical aspect involves the challenge of dealing with spectral multicollinearity. The ability to provide robust estimates, retrieval uncertainties and acceptable retrieval processing speed are other important aspects in view of operational processing. Recommendations towards new-generation spectroscopy-based processing chains for operational production of biophysical variables are given.

**Keywords** Imaging spectroscopy · Retrieval · Vegetation properties · Parametric and nonparametric regression · Machine learning · Radiative transfer models · Inversion · Uncertainties

## 1 Introduction

Quantitative vegetation variable extraction is fundamental to assess the dynamic response of vegetation to changing environmental conditions. Earth observation sensors in the optical domain enable the spatiotemporally explicit retrieval of plant biophysical variables. This data stream has never been so rich as is foreseen with the new-generation imaging spectrometer missions. The forthcoming EnMAP (Guanter et al. 2015), HyspIRI (Lee et al. 2015), PRISMA (Labate et al. 2009) and FLEX (Drusch et al. 2017) satellite missions will produce large spectroscopic data streams for land monitoring, which will soon become available to a diverse user community. This upcoming vast data stream will not only be standardized (e.g., atmospherically corrected), but will also require reliable and efficient retrieval processing techniques that are accurate, robust and fast.

Since the advent of optical remote sensing science, a variety of retrieval methods for vegetation attribute extraction emerged. Most of these methods have been applied to the data of traditional multispectral sensors (Verrelst et al. 2015), but increasingly they are also applied within imaging spectroscopy studies. This review provides a summary of recently developed methodologies to infer per-pixel biophysical variables from imaging spectroscopy data, covering the visible, near-infrared (NIR) and shortwave infrared spectral regions. Essentially, quantification of surface biophysical variables from spectral data always relies on a model, enabling the interpretation of spectral observations and their translation into a surface biophysical variable. Biophysical variable retrievals, as traditionally described in the terrestrial remote sensing literature, are grouped into two categories: (1) the statistical (or variable-driven) category and (2) the physical (or radiometric data-driven) category (Baret and Buis 2008). Over the last decade, however, both methodological categories expanded into subcategories and combinations thereof. Exemplary is the increasing number of elements of both categories which have been integrated into *hybrid* approaches. This methodological expansion, therefore, demands for a more systematic categorization. From an optical remote sensing point of view, and in line with an earlier, more general review paper (Verrelst et al. 2015), retrieval methods can be classified in the following four methodological categories:

1. *Parametric regression methods* Parametric methods assume an explicit relationship between spectral observations and a specific biophysical variable. Thus, explicit parameterized expressions are built; they are usually based on some physical knowledge of absorption and scattering properties and statistical relationship between the variable and the spectral response. Typically, a band arithmetic formulation is defined (e.g., a spectral index) and then linked to the variable of interest based on a fitting function.
2. *Nonparametric regression methods* Nonparametric methods directly define regression functions according to information from the given spectral data and associated variable, i.e. they are data-driven methods. Hence, in contrast to parametric regression methods, a non-explicit choice is to be made on spectral band relationships, transformation(s) or fitting function. Nonparametric methods can further be split into linear or nonlinear regression methods.
3. *Physically based model inversion methods* Physically based algorithms are applications of physical laws establishing photon interaction cause–effect relationships. Model variables are inferred based on specific knowledge, typically obtained with radiative transfer functions.
4. *Hybrid regression methods* A hybrid-method combines elements of nonparametric statistics and physically based methods. Hybrid models rely on the generic properties of physically based methods combined with the flexibility and computational efficiency of nonparametric nonlinear regression methods.

These categories provide a theoretical framework to organize the myriad of retrieval methods, as well to overview the diversity of published imaging spectroscopy applications based on these methods. However, a few remarks must be considered. One should be aware that the boundaries of these categories are not always clearly defined; for instance, spectral indices are also often used as input into nonparametric methods. Another important aspect is that the majority of the methods reviewed here is not exclusively designed for retrieval of biophysical variables. This especially holds for the statistical methods, whereby a regression model is used to link spectral data with a biophysical variable. In optical remote sensing science, these methods are commonly applied to map any feasible continuous variable, as well in the domains of snow, water or soil properties [see Matthews (2011), Mulder et al. (2011) and Dietz et al. (2012) for reviews]. Nevertheless, to keep this review comprehensive, it is limited to retrieval methods with applications in the domain of vegetation properties mapping. On the other hand, even within these boundaries each of the above methodological categories continues to be expanded with all kinds of spectroscopic data processing applications (e.g., Gewali et al. 2018). The drivers behind this methodological expansion can be found in the: (1) the interminable increase in computational power, (2) the increasing availability and democratizing of spectroscopic data, and (3) the steady progress in imaging spectroscopy sensor technology, which produces each time more sensitive sensors. This progress in imaging spectroscopy technology enables us to infer each time more subtle and highly dynamic vegetation properties from spectral data. For instance, the forthcoming FLEX mission aims to deliver a portfolio of dynamic plant stress and productivity variables based on, among others, the exploitation of sun-induced chlorophyll fluorescence emitted by terrestrial vegetation (Drusch et al. 2017). Hence, this underlines the fact that the list of biophysical variables that can be extracted from imaging spectroscopy is not closed, but instead continues to grow with ongoing progress in spectrometer technology. Consequently, biophysical variables are in this review paper defined as any vegetation property that can be quantified, i.e. any pigments, chemical constituents, structural variables, but also variables related to plant photosynthesis, productivity or diseases. Altogether, the drivers behind methodological expansion are not mutually exclusive, but they strengthen each other, which leads to a rapid progress in the development of advanced retrieval methods that goes hand in hand with improved capabilities to quantify a broad diversity of biophysical variables. As will be demonstrated throughout this review, these trends are resulting in an unprecedented richness of imaging spectroscopy mapping applications.

Regardless of the methodology used or the targeted application, the principal characteristic of spectroscopic data lies in their dense information content embedded in a few hundred spectrally narrow bands. Although such a spectrally dense data source proved to be beneficial for the majority of targeted mapping applications, a key challenge for many retrieval methods is how to deal with spectral multicollinearity, i.e. band redundancy. Special attention, therefore, will be devoted to address common spectroscopic data processing challenges, and solutions will be given how to overcome them. Finally, while imaging spectrometers are so far mostly applied in an experimental context, the developments towards operational systems have manifestly taken off and undoubtedly will lead to new directions and possibilities of Earth observation. In view of getting prepared for these upcoming global spectroscopic data streams, we will close this review with recommendations about the possibilities of integrating promising retrieval approaches into operational schemes.

## 2 Parametric Regression Methods

Parametric regression methods have long been the most popular method to quantify biophysical variables in optical remote sensing, and the field of imaging spectroscopy is no exception to that. This simplest way of developing a regression model explicitly determines parameterized expressions relating a limited number of spectral bands with a biophysical variable of interest. The empirical models rely on a selection of bands with high sensitivity towards the variable of interest, typically in combination with subtle spectral features to reduce undesired effects, related to variations of, for instance, other leaf or canopy properties, background soil reflectance, solar illumination and sensor viewing geometry and atmospheric composition (e.g., Verrelst et al. 2008, 2010). In the following overview, we present common parametric regression methods, which are based on (1) vegetation indices, (2) shape indices and (3) spectral transformations (Fig. 1).

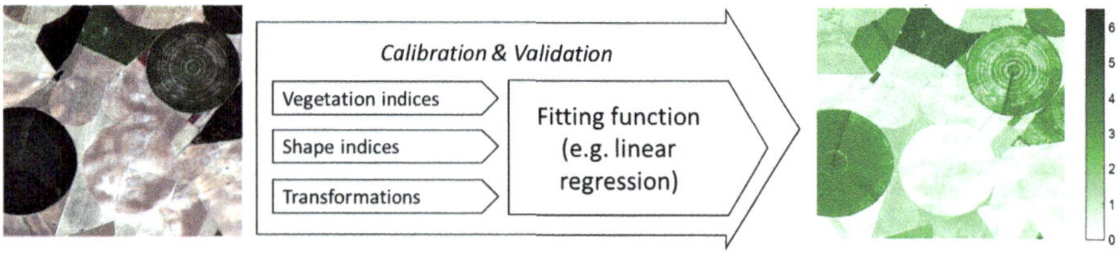


**Fig. 1** Principles of parametric regression. Left: red, green, blue (RGB) subset of a hyperspectral HyMap image (125 bands) over Barrax agricultural site (Spain). Right: illustrative map of a vegetation property (leaf area index (LAI), $m^2/m^2$) as obtained by a two-band normalized difference index and linear regression. The model was validated with a squared correlation coefficient, $R^2$ of 0.89 (RMSE: 0.63; NRMSE 10.1%). It took 0.2 s to produce the map using ARTMO's SI toolbox (Rivera et al. 2014). No uncertainty estimates are provided

## 2.1 Discrete Spectral Band Approaches: Vegetation Indices

Parametric regression models based on vegetation indices (VIs) are by far the oldest and largest group of variable estimation approaches. VIs are defined to enhance spectral features sensitive to a vegetation property, while reducing disturbances by combining some spectral bands into a VI (Clevers 2014; Glenn et al. 2008). The main advantage of VIs is their intrinsic simplicity. VI-based methods found their origin in the first applications of broadband sensor satellites. During the pioneering years of optical remote sensing, only a small set of spectral bands were available and computational power was limited. It led to a long tradition of the development of simple two bands, or at most three to four band indices that continues until today (e.g., Kira et al. 2016). New possibilities have opened with the advent of imaging spectrometers. Optimized narrowband information extraction algorithms were developed based on adaptations of established index formulations, such as simple ratio and normalized difference [see reviews by Clevers (2014), Glenn et al. (2008), Xue and Su (2017)]. On the other hand, the possibilities to develop spectral indices based on a few band combinations grew exponentially, and that demanded more systematic band evaluation methods.

A popular solution involves correlating all possible band combinations according to established index formulations. For two-band index formulations, such as simple ratio or normalized difference, this approach leads to 2D correlation matrices, which enables to visually identify optimal band combinations (e.g., Atzberger et al. 2010; Maire et al. 2004, 2008; Mariotto et al. 2013; Rivera et al. 2014; Thenkabail et al. 2000). Subsequently, given all possible combinations permit to select a 'best-performing index'. Nevertheless, while being mathematically simple, this method is not only tedious—especially when evaluating all possible combinations of more than two bands—but also keeps on being restricted to formulations that make use of a few bands only, with at most using three or four bands. Thus, although the approach is systematic, it continues to underexploit the comprehensive information content hidden in the contiguous spectral data. Moreover, when applying this technique in mapping applications making use of imaging spectroscopy, identical best-performing spectral band combinations for the same biophysical variable have rarely been reported. This suggests that optimized narrowband VIs are strongly case specific and seem to lack generic capacity (Gonsamo 2011; Heiskanen et al. 2013; Mariotto et al. 2013).

More fundamentally, it remains dubious whether relying on transformed data originating from a few discrete bands fully captures the complexity of real-world observation conditions as has been observed by a spectroradiometer. Reducing full-spectrum datasets into simple indices formulations intrinsically leads to remaining spectral information left unexploited. Accordingly, the following two aspects should be considered to ensure optimized use of VIs in a spectroscopic context: (1) *Band selection*. Spectral indices are mathematical functions based on discrete bands, or at best a subset of full spectral information. Thus, the question arises: how do we assess with high enough accuracy whether the most sensitive spectral bands—with respect to biophysical variable retrieval—have been selected? (2) *Formulation*. Enhancing spectral information according to a mathematical transformation should lead to an optimal sensitivity of the spectral signal with respect to the variable of interest. While established formulations such as the simple ratio or normalized difference are commonly used, here the question arises again: how can we be sure whether these linear formulations are the most powerful ones with respect to biophysical variable retrieval? These two questions are almost impossible to resolve considering the unlimited possibilities of band selections together with designing index formulations. Consequently, given their inherent constraints, it can be concluded that VI-based regression models exploit spectroscopic data suboptimally.

## 2.2 Parametric Approaches Based on Spectral Shapes and Spectral Transformations

Because none of the above few band index methods take full advantage of spectroscopic datasets, alternative methods were pursued with the advent of hyperspectral spectroradiometers that allow us to exploit specific absorption regions of the reflectance spectrum. It led to the development of so-called *shape* indices and spectral transformation methods. Shape indices, listed below, extract shape-related information from contiguous spectral signatures for a specific spectral region that is then correlated with a biophysical variable. These types of parametric methods are therefore exclusively applicable to spectroscopic data. The following categories can be identified:

- *Red-edge position (REP) calculations.* Mathematically, the REP inflection point is the position of a wavelength defined as the maximum of the first derivative reflectance between the red and NIR regions, i.e. between 670 and 780 nm (Kanke et al. 2016). The red-edge position is known to be sensitive to multiple biophysical variable variations, both chlorophyll pigments (Delegido et al. 2011) and structural variables, for instance the leaf area index (LAI) (Delegido et al. 2013). Therefore, REP-related methods are typically used to derive canopy chlorophyll content, being the product of LAI and leaf chlorophyll content (Clevers and Kooistra 2012; Li et al. 2017). Many mathematical approaches have been proposed to exploit this region as a sensitive indicator, including: (1) high-order curve fitting (Broge and Leblanc 2001; Clevers et al. 2004; (2) inverted Gaussian models (Cho and Skidmore 2006; Cho et al. 2008; Miller et al. 1990; (3) linear interpolation and extrapolation methods (Cho et al. 2008; Tian et al. 2011; 4) Lagrangian interpolation (Dawson et al. 1998; Pu et al. 2003; (5) rational function application (Baranoski and Rokne 2005); and, more recently, (6) a wavelet-based technique (Li et al. 2017).
- *Derivative-based indices.* Although several of the above-described methods make use of derivatives, e.g., linear extrapolation (Cho and Skidmore 2006) and Lagrangian technique (Dawson et al. 1998), the calculation of a derivative does not have to be restricted to the red edge. The derivative of any spectral region can be calculated and transformed into an index (Elvidge and Chen 1995; Penuelas et al. 1994; Sims and Gamon 2002; Zarco-Tejada et al. 2002). A systematic comparison of first derivative-based indices and conventional indices was performed by Maire et al. (2004) using the leaf optical model PROSPECT. Interestingly, the authors concluded that derivative-based indices are not necessarily better than conventional and properly elaborated indices.
- *Integration-based indices.* Alternatively, some authors proposed to calculate finite integrals of specific spectral regions, typically covering a part of the visible and the red-edge region for LAI or chlorophyll content estimations, into a (normalized) index (Broge and Leblanc 2001; Delegido et al. 2010; Malenovský et al. 2006, 2015; Mutanga et al. 2005; Oppelt and Mauser 2004). Likewise, in a recent study of Pasqualotto et al. (2018) this method exploited the water absorption spectral regions to quantify canopy water content. In these studies, integration-based indices were demonstrated to perform superior to classical vegetation indices, as they exploit more optimally absorption regions embedded in spectroscopic data than indices relying on a reflectance intensity

of few individual bands (Kováč et al. 2013). It can be expected that with the upcoming free availability of imaging spectroscopy data more methods of this kind of that explicitly exploit absorption features related to foliar constituents and pigments will emerge.

- *Continuum removal.* Whereas the above techniques focus on one or more specific spectral regions, continuum removal is a spectral transformation that can be applied over the full spectrum. This technique normalizes reflectance spectra, allowing comparison of individual absorption features with a common baseline (Clark and Roush 1984). The continuum removal transformation enhances and standardizes the specific absorption features related to vegetation properties. Continuum removal can be considered as a standard spectroscopic data processing technique and has found its way into various image processing software packages. Spectroscopic examples of applications include mapping of chlorophyll (Broge and Leblanc 2001; Malenovský et al. 2013, 2017), numerous studies on mapping nitrogen content (Huang et al. 2004; Mitchell et al. 2012; Mutanga and Kumar 2007; Mutanga and Skidmore 2004; Schlerf et al. 2010; Yao et al. 2015), foliar water condition (Stimson et al. 2005), plant stress (Sanches et al. 2014) and grassland biomass (Buchhorn et al. 2013; Cho et al. 2007).
- *Wavelet transform.* Wavelet analysis has been increasingly used to extract information from spectral data, e.g., related to vegetation properties (Rivard et al. 2008). Processing of reflectance spectra with wavelets can be performed as discrete or continuous (CWT) transforms. CWT outputs are directly comparable to the original spectrum and are simple to interpret. In this case, the original spectrum is represented by a set of spectra from small (narrow bandwidth absorption feature and noise) to larger scales (broad features, continuum). By selecting small-scale spectra (i.e. discarding the smallest scale, which contains white noise and high scales related to the continuum), the absorption features of the components are enhanced, preserving the spectral information of the original data (Scafutto et al. 2016). Based on the type of wavelet transform, specific bands sensitive to the targeted variable are then selected (Bao et al. 2017). CWT is often compared in spectroscopic studies against spectral indices and was found to be capable of delivering stronger correlations, e.g., in the detection of wheat aphid pests (Luo et al. 2013), LAI estimation (Huang et al. 2014), nitrogen content and chlorophyll content estimation (He et al. 2015; Kalacska et al. 2015; Luo et al. 2013) and in amplifying spectral separability of alpine wetland grass species (Bao et al. 2017).

Altogether, correlations based on shape indices and spectral transformations are undoubtedly more sophisticated normalization approaches than traditional spectral indices for exploiting the spectral information embedded in spectroscopic data. Moreover, their relatively simple mathematical formulation ensures fast processing. It thus seems logical that these spectral transformation methods became standard spectroscopy image processing techniques. However, these methods alone provide nothing more than spectral transformations and enhancements. When aiming to estimate a biophysical variable, a fitting function—typically a linear least squares fitting, but also exponential, power and polynomial—is still required. Yet it remains questionable whether the selected fitting function is the most suitable one. Moreover, since parametric approaches are based on relatively simple mathematical definitions—as opposed to more advanced methods—no associated uncertainty intervals are provided. Although their strengths lie in their straightforward use and fast processing, with the absence of a per-pixel uncertainty estimate, the performance quality of parametric regression methods as a mapping method is hard to judge. Given the surface diversity captured in a single airborne or spaceborne image, and despite a standard validation exercise for a number of pixels, it still remains unknown how the retrieval quality

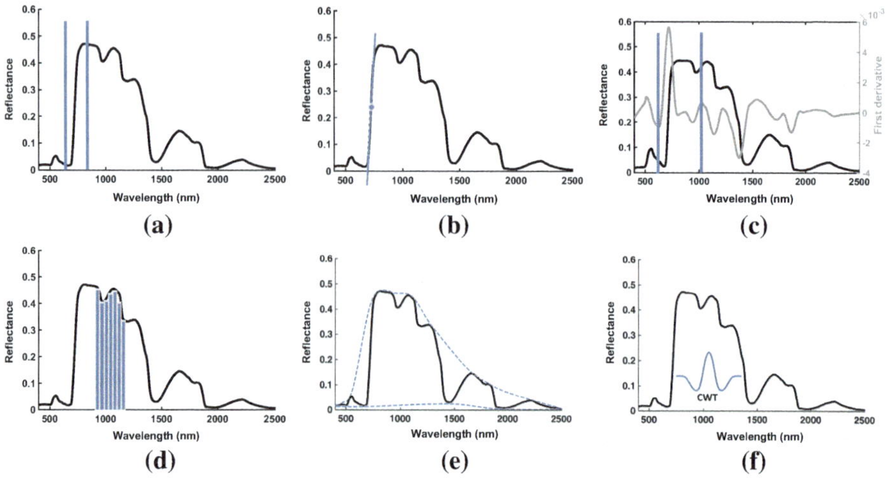


**Fig. 2** Schematic illustrations of parametric regression methods: spectral indices (**a**), red-edge position (REP) calculation (**b**), derivative-based indices (**c**), integral-based indices (**d**), continuum removal (**e**) and wavelet transform (**f**). Note that a fitting function is still required to convert transformations towards a biophysical variable

evolves throughout a complete image. The absence of a quality indicator is, therefore, in our view the main reason why parametric regression methods (Fig. 2) are not recommendable for operational quantification of biophysical variables.

## 3 Nonparametric Regression Methods

Contrary to parametric methods, nonparametric methods optimize the regression algorithm by means of an inherent learning phase based on training data. Essentially, the nonparametric model develops weights (coefficients) adjusted to minimize the estimation error of the variables extracted. This means that no explicit parametrization is required, which practically simplifies the model development, but more expert knowledge to understand and execute these models may be required. Another important advantage of nonparametric methods is the possibility of training with the full-spectrum information. Hence, an explicit selection of spectral bands or transformations is in principle not required. A flexible model is able to combine different data structure features in a nonlinear manner to conform requirements; however, model definition with a too flexible capacity may incur the problem of overfitting the training dataset. To avoid this pitfall, model weights are defined by jointly minimizing the training set approximation error while limiting the model complexity. In view of processing spectroscopic data, a more prevalent problem lies in the so-called *curse of dimensionality* (Hughes phenomenon) (Hughes 1968). Adjacent, contiguous bands carry highly intercorrelated information, which may result in redundant data and possible noise and potentially suboptimal regression performances. As discussed further on, band selection or dimensionality reduction methods that transform the spectral data to lower-dimensional space, while containing the vast majority of the original information, can overcome this problem (Fig. 3).

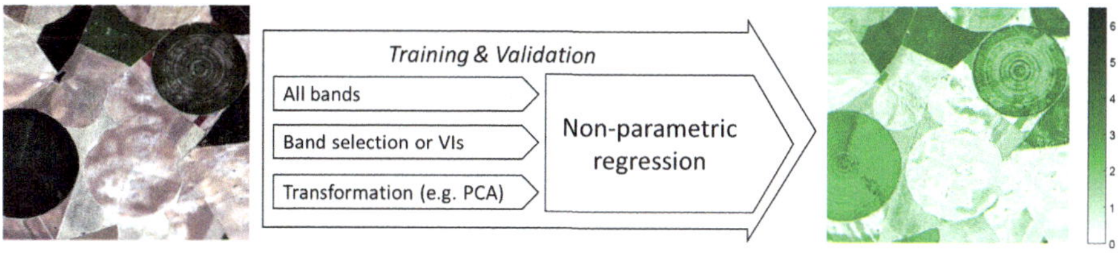


**Fig. 3** Principles of nonparametric regression. Left: red, green, blue (RGB) subset of a hyperspectral HyMap image (125 bands) over Barrax agricultural site (Spain). Right: illustrative map of a vegetation property (leaf area index (LAI), $m^2/m^2$) as obtained by PROSAIL with Gaussian processes regression (GPR). The model was validated with a squared correlation coefficient, $R^2$ of 0.94 (RMSE: 0.39; NRMSE: 6.3%). It took 5.7 s to produce the map using ARTMO's MLRA toolbox (Rivera Caicedo et al. 2014). With GPR also uncertainty estimates are provided (not shown)

## 3.1 Linear Nonparametric Methods

Nonparametric regression algorithms that apply linear transformations are attractive because of their fast performance. These methods became standard methods in chemometric and in image processing software packages. Multivariable linear regression methods can cope with spectroscopic data and typically rely on the estimation of co-variances. When moving towards spectroscopic data, however, this can become problematic when input data quantity is limited with respect to the dimensionality of the dataset. To alleviate collinearity, often linear nonparametric methods are applied in combination with a dimensionality reduction step. Some methods are even intrinsically based on this principle, i.e. principal component regression (PCR) (Wold et al. 1987) and partial least squares regression (PLSR) (Geladi and Kowalski 1986). Common linear nonparametric regression approaches are provided in Table 1, and imaging spectroscopy applications are discussed below.

On the application side, stepwise multiple linear regression (SMLR) is a classical multivariable regression algorithm commonly applied in chemometrics (Atzberger et al. 2010). To evaluate its predictive power, SMLR has been often compared with alternative regression techniques such as PLSR and some studies concluded that PLSR yielded better results when estimating LAI (Darvishzadeh et al. 2008) and canopy chlorophyll content (Atzberger et al. 2010). Also Ramoelo et al. (2011) compared both regression algorithms to estimate foliar nitrogen and phosphorus in combination with continuum removal using field spectrometry. By estimating canopy nitrogen, Miphokasap et al. (2012) demonstrated that the model developed by SMLR led to a higher correlation coefficient and lower errors than model applications based on narrowband VIs. This suggests that nonparametric (full-spectrum) models tend to be more powerful than parametric models. Likewise, Yi et al. (2014) compared SMLR with PLSR and spectral indices for carotenoid estimation in cotton and concluded that best estimations were obtained with PLSR. Likewise, SMLR was compared with PLSR and (nonlinear) machine learning regression algorithms for estimating leaf nitrogen content (Yao et al. 2015). Because of their enhanced flexibility, it may not be a surprise that the nonlinear methods outperformed SMLR and PLSR. This was also observed by various similar studies, as will be addressed in Sect. 3.2.

PCR seems to be more effective in the conversion of spectroscopic data into the estimation of vegetation properties, because the PCA-based dimensionality reduction method is embedded in the method in combination with a linear regression function. Hence, by

**Table 1** Linear nonparametric regression methods applicable to spectroscopic data

| Method | Description | References |
|---|---|---|
| Stepwise multiple linear regression (SMLR) | SMLR recursively applies multiple regression a number of times. Each step removes a variable eliciting the weakest correlation. At the end of the recursive process, a variable set is obtained that is optimally explaining the spectral data distribution | Draper and Smith (2014) |
| Principal components regression (PCR) | PCR is a regression analysis method based on principal components analysis (PCA) estimating regression coefficients. Solutions from PCR are generated performing linear regression of the most relevant components (called scores) obtained after applying PCA | Wold et al. (1987) |
| Partial least squares regression (PLSR) | PLSR is similar to PSR but tackles the co-linearity problem differently than PCR. Applying PCR, regression is performed using PCA scores. These projections are obtained using only input patterns, not outputs. In contrast, PLSR builds the regression model on projections obtained using the partial least squares (PLS) approach. It elicits the directions of maximum input–output cross-covariance. Therefore, PLSR takes both input patterns and output variables into account | Geladi and Kowalski (1986) |
| Ridge (regulated) regression (RR) | RR is the most commonly used method of regularization for ill-posed problems, which are problems that do not have a unique solution. RR deals with co-linearity by allowing a degree of bias in the estimates. Therefore, RR adds a small positive value $\lambda$ to the diagonal elements of the input data covariance matrix. Hence, RR requires finding an optimal value for $\lambda$. Typically, cross-validation is used to reach near optimal values. An important fact about RR is that it enforces the regression coefficients to be lower, but it does not enforce them to be zero. That is, it will not get rid of irrelevant features (bands) but rather minimize their impact on the trained model | Geladi and Kowalski (1986) |
| Least absolute shrinkage and selection operator (LASSO) | Lasso is a extension built on RR, but with a small twist. It also penalizes the regression coefficients absolute size. By this penalization some of the variable estimates may be exactly zero. The larger the penalty, the more the estimates will tend towards zero. This is a convenient approach to automatically perform feature selection, or to deal with correlated predictors | Tibshirani (1996) |

converting the spectral data to a lower-dimensional space automatically overcomes the band redundancy problem. This method has been improved with PLSR, where the projections are optimized in view of the regression. It is, therefore, not a surprise that only few spectroscopic studies examined the predictive power of PCR. Those studies compared PCR against PLSR or against VIs (Atzberger et al. 2010; Fu et al. 2012; Marshall and Thenkabail 2014; Rivera Caicedo et al. 2014; Wang et al. 2017b). Although PCR generally outperformed VIs in explaining variability of a vegetation attribute, in all cases PLSR or any other nonparametric method overran PCR.

PLSR found its way into a broad diversity of imaging spectroscopy applications, especially in the mapping of biochemicals, pigments and vegetation density properties. For instance, PLSR was used in several spectroscopic studies applied to estimate foliage nitrogen content (Coops et al. 2003; Hansen and Schjoerring 2003; Huang et al. 2004). Also Gianelle and Fb (2007) used PLSR to derive grassland phytomass and its total (percentage) nitrogen content from spectroscopic data. Similarly, Cho et al. (2007) and Im et al. (2009) applied PLSR to estimate a diversity of grass and crop biophysical variables (LAI, stem biomass and leaf nutrient concentrations), and Ye et al. (2007) applied PLSR for yield prediction purposes. Beyond individual vegetation attributes, PLSR was recently used to predict landscape-scale fluxes of net ecosystem exchange (NEE) and gross primary productivity (GPP) across multiple timescales (Matthes et al. 2015), and also for the estimation of floristic composition of grassland ecosystems (Harris et al. 2015; Neumann et al. 2016; Roth et al. 2015). At the same time, thanks to its PLS-vectors, PLSR is also increasingly applied for band sensitivity analysis of spectroscopic datasets in view of the targeted application (e.g., Feilhauer et al. 2015; Kiala et al. 2016; Kira et al. 2016; Li et al. 2014a; Neumann et al. 2016). Various experimental studies demonstrated the superior predictive power of PLSR as opposed to VIs for the prediction of multiple vegetation properties, including above-ground biomass, LAI, leaf pigments (chlorophyll, carotenoids), GPP and NEE fluxes, leaf rust disease detection and nutrients concentration (nitrogen and phosphorus concentrations) (Capolupo et al. 2015; Dreccer et al. 2014; Foster et al. 2017; Hansen and Schjoerring 2003; Matthes et al. 2015; Wang et al. 2017a; Yue et al. 2017). However, when compared against machine learning methods, then PLSR no longer appeared to be top performing (Ashourloo et al. 2016; Kiala et al. 2016; Wang et al. 2015; Yao et al. 2015). As will be addressed in Sect. 3.2, this is due to the nonlinear transformation conducted in machine learning methods.

Other linear nonparametric regression methods, such as ridge regression (RR) and LASSO, hardly made it into applications for vegetation properties mapping. Yet a few spectroscopic examples are worth mentioning. For instance, Addink et al. (2007) used RR to map LAI and biomass, and more recently Bratsch et al. (2017) applied LASSO to estimate above-ground biomass quantities among different plant tissue type categories in Alaska. In another biomass estimation study, both RR and LASSO were compared against PLSR (Lazaridis et al. 2010) and also random forests (Zandler et al. 2015). Interestingly, RR and LASSO appeared to be top performing. One may, therefore, wonder why these techniques have not been applied more often. On the other hand, these linear methods are increasingly replaced by their nonlinear counterparts. For instance, RR has been replaced by kernel ridge regression (KRR) (Suykens and Vandewalle 1999), and also PLSR has been redesigned into a kernel version, i.e. the KPLSR, which proved to be more powerful than PLSR for chlorophyll concentration estimation (Arenas-García and Camps-Valls 2008). The family of kernel methods is addressed in Sect. 3.2. That none of these linear nonparametric methods (Fig. 4) deliver uncertainty estimates is another drawback. Similar as in case of parametric regression, without uncertainty estimates it remains questionable

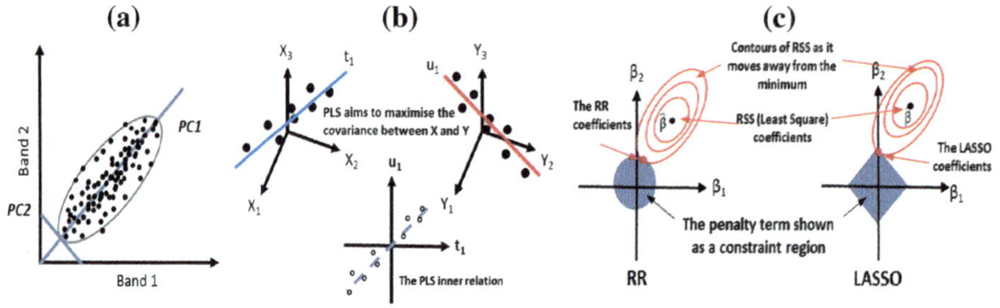


**Fig. 4** Schematic illustrations of principal component (PC) (**a**), partial least squares (PLS) (**b**), ridge regression and LASSO (**c**). PC and PLS are combined with a linear regression model

whether these methods can deliver consistent mapping quality throughout a complete image, or are applicable to other images in space and time.

## 3.2 Nonlinear Nonparametric Methods

When advancing beyond linear transformation techniques, a diversity of nonlinear nonparametric methods has been developed during last few decades. These methods, also referred to as machine learning regression algorithms, apply nonlinear transformations. An important methodological advantage is their capability to capture nonlinear relationships of image features without explicitly knowing the underlying data distribution. Hence, they are developed without assuming a particular probability density distribution, which is the reason why they work well with all kinds of data types. Machine learning methods also offer the possibility to incorporate a prior knowledge and the flexibility to include different data types into the analysis. In principle, they are perfectly suited to process spectroscopic data. In the following sections, examples of the families of (1) decision trees, (2) artificial neural networks and (3) kernel-based regression are explained.

### *3.2.1 Decision Trees*

Decision tree algorithms use a branching method to illustrate every possible outcome of a decision (Table 2). They are more frequently applied in classification than in regression. Only a few decision tree feasibility studies dealing with imaging spectroscopy data are presented in the scientific literature (e.g., Im et al. 2009) most likely because boosted and bagging trees hardly found their way to regression applications. They might be considered as obsolete with the improvements introduced into random forest (RF), which is essentially a specific type of bagging trees. RF builds an ensemble of individual decision trees working with different subsets of features (bands) and eventually different training data points both selected randomly, from which a final prediction is made using particular combination schemes. RF can handle a large number of training samples, does not suffer from overfitting and is robust to outliers and noise (Belgiu and Drăguţ 2016), which makes it an attractive method for spectroscopic mapping applications. RF has recently been made available in various software packages and proved to be a competent regression algorithm. It therefore comes as no surprise that RF gained rapid popularity in imaging spectroscopy mapping of a diverse range of vegetation attributes, including biomass (Adam et al. 2014; Vaglio Laurin et al. 2014), canopy nitrogen (Li et al. 2014) and as indicator of plant species

**Table 2** Decision tree regression methods applicable to spectroscopic data

| Method | Description | References |
|---|---|---|
| Decision trees (DT) | DT learning is based on decision tree predictive modelling. A decision tree is based on a set of hierarchical connected nodes. Each node represents a linear decision based on a specific input feature | Breiman et al. (1984) |
| Boosted trees (BoT) | BoT incrementally builds an ensemble by training each new instance to emphasize the training instances previously mis-modelled | Friedman et al. (2000) |
| Bagging trees (BaT) | BaT is an early ensemble method based on building multiple decision trees by iteratively replacing resampled training data and voting for the decision trees leading to a consensus prediction | Breiman (1996) |
| Random forest (RF) | RF is a specific type of BaT that in constructs a collection of decision trees with controlled variance | Breiman (2001) |

composition (Feilhauer et al. 2017). Some of these studies have compared RF with support vector regression (SVR) or neural networks, but no strong preference towards one or the other method was found, which suggests that all three methods are competitive (Han et al. 2016; Pullanagari et al. 2016). However, just like other machine learning regression methods, RF can face difficulties coping with the collinearity of the spectroscopic data (Rivera-Caicedo et al. 2017). To overcome this problem, RF is often used in combination with sensitive bands or simple transformations in the form of VIs that are known to be sensitive to the targeted vegetation property (Adam et al. 2014; Han et al. 2016; Liang et al. 2016). Alternatively, RF is inherently able to identify sensitive spectral bands, and selection of only those sensitive bands can subsequently improve the regression model (Balzarolo et al. 2015; Feilhauer et al. 2015). Whether applying a band selection method is the most successful strategy, however, remains an open question. Rather than seeking for optimized individual bands, a more elegant solution may lie in firstly applying dimensionality reduction method and then inputting the features of the lower-dimensional space (i.e. components) into the decision tree (Rivera-Caicedo et al. 2017).

### *3.2.2 Artificial Neural Networks*

Artificial neural network (ANN) methods are listed in Table 3. Since the early 1990s, feed-forward and back-propagation ANNs thrived in all kinds of mapping applications, including vegetation properties mapping (Francl and Panigrahi 1997; Kimes et al. 1998; Paruelo and Tomasel 1997). Their strengths lie in their adaptability that can lead to excellent performances. The superiority of ANNs in vegetation properties mapping compared to parametric models (e.g., those based on VIs) has been demonstrated repeatedly in experimental studies (Kalacska et al. 2015; Malenovský et al. 2013; Uno et al. 2005; Wang et al. 2013). Examples of successful spectroscopic applications include the estimation of foliage nitrogen concentrations (Huang et al. 2004) and LAI (Jensen et al. 2012; Neinavaz et al. 2016). In both cited studies, ANN outperformed other linear nonparametric models (e.g., PLSR). Alternative powerful structures involve RBFANNs, BRANNs and RANNs (for explanation, see Table 3). Although these advanced ANNs have been primarily used for classification applications, only recently they were explored to map vegetation properties from spectroscopic data (Chen et al. 2015; Feng et al. 2016; Pôças et al. 2017; Wang et al. 2013). Some of these studies mention the superiority of these advanced ANN designs as compared to standard ANN designs or other machine learning approaches in estimating vegetation properties (Du et al. 2016; Li et al. 2017; Pham et al. 2017).

Applying ANNs to spectroscopic data, nonetheless, can be quite challenging due to the multicollinearity. Feeding many bands into an ANN requires a complex design and consequently a long training time. Just as with decision trees, a popular approach is applying a band selection or the calculation of several sensitive VIs or shape indices such as red-edge position that are then entered either individually or as a combination into the ANN. Various of these band selection studies investigated combinations of VIs that led to the best prediction models (Chen et al. 2015; Feng et al. 2016; Jia et al. 2013; Liang et al. 2015; Mutanga and Kumar 2007; Pôças et al. 2017; Schlerf and Atzberger 2006). As discussed before, it remains questionable whether the selected indices preserve a maximum amount of useful information. On the contrary, when compressing the spectral data using dimensionality reduction methods into a lower-dimensional space, then it is ensured that a maximum amount of spectral information is preserved. This approach was applied, e.g., to assess corn yield (Uno et al. 2005) and phosphorus and nitrogen concentrations (Knox et al. 2011).

**Table 3** Artificial neural network regression methods applicable to spectroscopic data

| Method | Description | References |
|---|---|---|
| Artificial neural networks (ANNs) | ANNs in their basic form are essentially fully connected layered structures of artificial neurons (AN). An AN is basically a pointwise nonlinear function (e.g., a sigmoid or Gaussian function) applied to the output of a linear regression. ANs with different neural layers are interconnected with weighted links. The most common ANN structure is a feedforward ANN, where information flows in a unidirectional forward mode. From the input nodes, data pass hidden nodes (if any) towards the output nodes | Haykin (1999) |
| Back-propagation ANN (BPANN) | The basic type of neural network is multi-layer perceptron, which is feedforward back-propagation ANN. BPANN consists of 2 steps: (1) feedforward the values and (2) calculate the error and propagate it back to the earlier layers. So to be precise, forward-propagation is part of the back-propagation algorithm but comes before back-propagating. This is the most commonly used algorithm when referring to ANN. In many papers using ANN, these standard designs are not explicitly mentioned | Haykin (1999) |
| Radial basis function ANN (RBFANN) | RBFANN is a type of ANN that uses nonlinear radial basis functions (RBFs) as activation functions in the hidden layer. The output of the network is a linear combination of RBFs of the inputs and neuron parameters | Broomhead and Lowe (1988) |
| Recurrent ANN (RANN) | A RANN is a type of ANN that make use of sequential information by introducing loops in the network | Hochreiter and Schmidhuber (1997) |
| Bayesian regularized ANN (BRANN) | BRANNs are more robust than standard BPANNs and can reduce or eliminate the need for lengthy cross-validation. Bayesian regularization is a mathematical process that converts a nonlinear regression into a "well-posed" statistical problem in the manner of a ridge regression | Burden and Winkler (1999) |

It is therefore not surprising that a study comparing PCA vs. indices inputted into ANNs concluded that the PCA-ANN design outperformed VI-ANN designs (Liu and Pan 2017). Moreover, given that only linear transformations are applied in PCA, it may even be that more adaptive dimensionality reduction methods yield superior accuracies when combined with ANN, e.g., partial least squares (PLS), or in the field of nonlinear kernel-based dimensionality reduction methods, e.g., kernel PCA (KPCA) or kernel PLS (KPLS). To ascertain this hypothesis, PCA was compared against 10 alternative dimensionality reduction methods in combination with ANN to carry out LAI estimation. As expected, various alternative dimensionality reduction methods outperformed PCA in developing accurate models (e.g., PLS, KPLS, KPCA) (Rivera-Caicedo et al. 2017).

### *3.2.3 Kernel-Based Machine Learning Regression Methods*

Kernel-based regression methods solve nonlinear regression problems by transferring the data to a higher-dimensional space by a kernel function (Table 4). The flexibility offered by kernel methods allows us to transform almost any linear algorithm that can be expressed in terms of dot products, while still using only linear algebra operations. Kernel methods provide a consistent theoretical framework for developing nonlinear techniques and have useful properties when dealing with a low number of (potentially high-dimensional) training samples, and outliers and noise in the data (Gómez-Chova et al. 2011; Tuia et al. 2018). Given these attractive properties, kernel-based regression methods seem perfectly suited to extract nonlinear information related to vegetation properties from imaging spectroscopy data. Developed in the mid-1990s, among the most popular kernel-based method for classification purposes involves SVM. Its regression version (SVR) gained popularity for the retrieval of continuous vegetation attributes from imaging spectroscopy data in the last decade. Examples include plant height, leaf nitrogen content and leaf chlorophyll content (Karimi et al. 2008; Yang et al. 2011). A multi-output version of SVR was presented by Tuia et al. (2011), with LAI, leaf chlorophyll content and fractional vegetation content being simultaneously estimated. Recently, SVR was used for processing spectroscopic images of sub-decimetre spatial resolution as acquired by low-altitude unmanned aircraft system to infer Antarctic moss vigour (Malenovský et al. 2017). Yet just as with the other advanced regression methods, SVR faces the same difficulties of coping with multicollinearity. Therefore, SVR has been commonly applied in combination with specific spectral subsets or VIs (Lin et al. 2013; Marabel and Alvarez-Taboada 2013), or with wavelet transforms (He et al. 2015). To deal with spectroscopic band redundancy, an advantage of SVR is that it allows band selection (analogous as PLSR and RF), which in principle allows the development of more optimized models (Feilhauer et al. 2015). On the other hand, it is likely that the combination with dimensionality reduction methods will lead to more powerful models (Rivera-Caicedo et al. 2017). To assess its predictive power, various spectroscopic studies compared SVR against similar methods such as SMLR or PLSR, although some band selection method appeared to be essential (Kiala et al. 2016; Wang et al. 2015; Yao et al. 2015). Conversely, when comparing SVR against other machine learning methods such as RF or GPR, then SVR no longer excelled (Pullanagari et al. 2016).

Kernel ridge regression (KRR) emerged as one of the promising upcoming kernel-based regression methods, although only a few spectroscopic studies have used it. For instance, Wang et al. (2011) compared KRR with linear nonparametric methods (multiple linear regression and PLSR) for LAI estimation. The authors concluded that KRR yielded the most accurate estimates. Also Peng et al. (2011) used KRR for the detection of chlorophyll content. Apart

**Table 4** Kernel-based regression methods applicable to spectroscopic data

| Method | Description | References |
|---|---|---|
| Support vector regression (SVR) | The support vector machine (SVM) is a supervised machine learning technique that was invented in the context of the statistical learning theory. It was not until the mid-1990s that an algorithm implementation of the SVM was proposed with the introduction of the kernel trick and the generalization to the non-separable case. SVR is built on the principle of SVM: a nonlinear function is learned by linear learning machine mapping into high-dimensional kernel induced feature space. The capacity of the system is controlled by parameters that do not depend on the dimensionality of feature (bands) space | Vapnik et al. (1997) |
| Kernel ridge regression (KRR) | KRR combines RR with the kernel trick. It thus learns a linear function in the space induced by the respective kernel and the data. For nonlinear kernels, this corresponds to a nonlinear function in the original space. The form of the model learned by KR is identical to SVR. However, different loss functions are used: KRR uses squared error loss, while SVR uses $\epsilon$-insensitive loss combined with RR regularization | Suykens and Vandewalle (1999) |
| Gaussian process regression (GPR) | GPR is based on Gaussian processes (GPs), which generalize Gaussian probability distributions in a function's space. A GP is stochastic since it describes the properties of functions. As in Gaussian distributions, a GP is described by its mean (which for GPs is a function) and covariance (a kernel function). This represents an expected covariance between function values at a given point. Because a GPR model is probabilistic, it is possible to compute the prediction intervals using the trained model | Rasmussen and Williams (2006) |

from these two studies, the spectroscopy vegetation community may not yet be familiar with this method. Solely Rivera Caicedo et al. (2014) had compared KRR against other machine learning algorithms applied to CHRIS (62 bands) and HyMap (125 bands) spectroscopic data for LAI mapping. In that study, KRR not only proved to be a very competitive regression algorithm, but also proved to be extremely fast. This is due to its relatively simple design that requires only one hyperparameter to be tuned. Because of its simplicity, another advantage is that KRR is capable of dealing with collinearity; the method can cope with thousands of contiguous bands. In fact, in the dimensionality reduction comparison study tested with simulated (2100 bands) and HyMap data (Rivera-Caicedo et al. 2017), KRR was the only regression method where dimensionality reduction methods did not lead to improvements as compared to using all bands. In conclusion, KRR emerged as an attractive regression method due to its competitive performance, fast processing and ease of dealing with spectroscopic data.

From all machine learning regression algorithms, probably the most exciting one is Gaussian process regression (GPR). Contrary to other methods, the training phase in GPR takes place in a Bayesian framework, leading to probabilistic outputs (Camps-Valls et al. 2016; Rasmussen and Williams 2006). GPR applied to spectroscopic data started only recently, e.g., for airborne HyMap mapping of leaf chlorophyll content (Verrelst et al. 2013a), and for spaceborne CHRIS mapping of leaf chlorophyll content, LAI and fractional vegetation content (Verrelst et al. 2012a). Of interest is that along with these maps also maps of associated uncertainties (prediction intervals) were provided. Also with an Airborne Hyperspectral Scanner, Roelofsen et al. (2014) applied GPR to map salinity, moisture and nutrient concentrations that in turn were used as inputs for plant association mapping. In the Rivera Caicedo et al. (2014) comparison paper, GPR outperformed the majority of other tested machine learning methods for the prediction of leaf chlorophyll content and LAI from various spectroscopic datasets. Similarly, Ashourloo et al. (2016) concluded that GPR yielded most accurate leaf rust disease detection as compared to VIs, PLSR and SVR. However, GPR is no exception in suffering from radiometric collinearity when many bands are included, and related spectroscopic studies demonstrated that results can be further improved by combining GPR with band selection (Verrelst et al. 2016b) or with dimensionality reduction methods (Rivera-Caicedo et al. 2017). At the same time, alternative GPR versions continue to be developed within the machine learning community. For instance, Lazaro-Gredilla et al. (2014) refined the GPR method by proposing a non-standard variable approximation allowing for accurate inferences in signal-dependent noise scenarios. The so-called variational heteroscedastic GPR (VHGPR) appears to be an excellent alternative for standard GPR, which was demonstrated on a CHRIS dataset where VHGPR outperformed GPR in leaf chlorophyll content estimation. Schematic illustrations of popular nonlinear nonparametric methods are shown in Fig. 5.

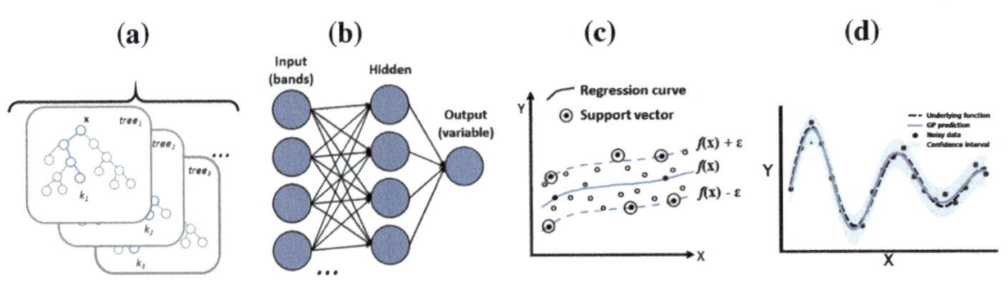


**Fig. 5** Schematic illustrations of random forest (RF) (**a**), neural network (NN) (**b**), support vector regression (SVR) (**c**) and Gaussian processes regression (GPR) (**d**)

## 4 Physically Based Model Inversion Methods

Physically based model inversion is based on physical laws establishing cause–effect relationships. Inferences on model variables are based on generally accepted knowledge embedded in radiative transfer models (RTMs). RTMs are deterministic models that describe absorption and multiple scattering, and some of them even describe the microwave region, thermal emission or sun-induced chlorophyll fluorescence emitted by vegetation (e.g., see Table 5). A diversity of canopy RTMs have been developed over the last three decades with varying degrees of complexity. Gradual increase in RTMs accuracy, yet in complexity too, has diversified RTMs from simple turbid medium RTMs to advanced Monte Carlo RTMs that allow for explicit 3D representations of complex canopy architectures (e.g., see the RAMI exercises (Pinty et al. 2001, 2004; Widlowski et al. 2007, 2011, 2015) for a thorough comparison). This evolution has resulted in an increase in the computational requirements to run the model, which bears implications towards practical applications. From a computational point of view, RTMs can be categorized as either (1) *economically invertible* (or computationally cheap) or (2) *non-economically invertible* models (or computationally expensive). These terms refer to the model complexity and associated run-time constraining the mathematical inversion of such models. Economically invertible models are models with relatively few input parameters and fast processing that enables fast calculations and consequently fast model inversion or rendering of simulated scenes. A well-known example of this category includes the widely used leaf RTM PROSPECT (Feret et al. 2008) coupled with the canopy RTM SAIL (Verhoef 1984a) [combined named as PROSAIL (Jacquemoud et al. 2009a)].

Non-economically invertible RTMs refer to advanced, computationally expensive RTMs, often with a large number of input variables and sophisticated computational and mathematical modelling. These types of RTMs enable the generation of complex or detailed scenes, but at the expense of a significant computational load. In short, the following families of RTMs can be considered as non-economically invertible: (1) Monte Carlo ray-tracing models [e.g., Raytran (Govaerts and Verstraete 1998)], FLIGHT (North 1996) and librat (Lewis 1999)); (2) voxel-based models [e.g., DART (Gastellu-Etchegorry et al. 1996)]; and (3) advanced integrated vegetation and atmospheric transfer models [e.g., SCOPE (Tol et al. 2009) and MODTRAN (Berk et al. 2006)]. Descriptions of advanced canopy RTM models and their latest developments are provided in Table 5. Although these advanced RTMs serve perfectly as virtual laboratories for fundamental research on light–vegetation interactions, they are in general less suitable for retrieval applications, because of either a large number of input variables or a long processing time. Nevertheless, as outlined below, some experimental studies demonstrated that they can as well be applied into inversion schemes, e.g., based on look-up tables and in hybrid strategies.

Regardless of their complexity, they all deliver spectroscopic outputs, typically at 1 nm resolution. Hence, RTM outputs can fit perfectly into inversion schemes of imaging spectroscopy data, while at the same time the simulated data can be resampled to reassemble the band settings of multispectral sensors. Because inversion strategies are usually based on spectral fitting (i.e. only radiometric information is used), the drawback of collinearity complicating regression is not an issue here; however, removal of noisy bands is still a standard and much-needed preprocessing step to enable adequate spectral fitting. Another point to be mentioned is that inversion scheme can only retrieve the RTM input variables. Hence, using this strategy implies that only RTM state variables can be mapped. Yet because the RTM input variables drive the canopy absorbance and scattering mechanisms, the resulting output maps are considered to be physically sound (Knyazikhin et al. 2013; Myneni et al. 1995) (Fig. 6).

Given that in principle only a coupled leaf-canopy RTM and an inversion routine are required for the retrieval of RTM state variables, the approach is generic and generally applicable. Nevertheless, these approaches are not straightforward. First, an RTM has to be selected, whereby a trade-off between the realism and inversion possibility of the RTM has to be made. As discussed above, typically, complex models are more realistic, but they have many variables and consequently challenging to invert, whereas simpler models may be less realistic but easier to invert. Secondly, according to the Hadamard postulates, RTMs are invertible only when an inversion solution is unique and dependent—in a continuous mode—on the variables to be extracted. Unfortunately, this boundary condition is often not met. The inversion of canopy RTMs is frequently underdetermined and ill-posed. The number of unknowns can be much larger than the number of independent observations. This makes physically based retrievals of vegetation properties a challenging task. Several strategies have been proposed to cope with the underdetermined problem of optimizing the inversion process, including (1) *iterative numerical optimization* methods, (2) *look-up table (LUT)-based inversion* (see Fig. 7 for illustrations), or (3) *hybrid* approaches in which LUTs are generated as input for machine learning approaches (see Sect. 5). Below we briefly review some common RTM inversion techniques in view of converting spectroscopic data into maps of RTM leaf and canopy input variables.

*Numerical optimization.* Iterative optimization is a classical technique to invert RTMs in image processing (Botha et al. 2007; Jacquemoud et al. 1995; Zarco-Tejada et al. 2001). The optimization is minimizing a cost function, which estimates the difference between measured and estimated variables by successive input variable iteration. Optimization algorithms are computationally demanding and hence potentially time-consuming depending on the complexity of the RTM and the numbers of image pixels to be processed. However, with the ongoing increase in computational power and open-source availability of optimization libraries, a renaissance of numerical approaches is emerging. Examples of numerical inversion against spectroscopic data include: PROSPECT inversion to retrieve leaf chlorophyll content (Zhang and Wang 2015), retrieval of leaf biochemistry against an improved version of PROSPECT (COSINE) (Jay et al. 2016), and PROSAIL leaf and canopy variables (Bayat et al. 2016; Tol et al. 2016). Despite a gain in computational power, numerical inversion algorithms applied to

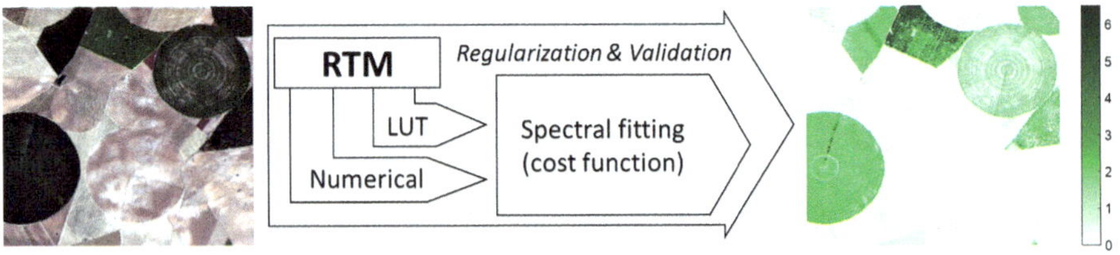


**Fig. 6** Principles of radiative transfer model (RTM) inversion. Left: RGB subset of a hyperspectral HyMap image (125 bands) over Barrax agricultural site (Spain). Right: illustrative map of a vegetation property (LAI, $m^2/m^2$) as obtained by RMSE inversion against a 100,000 PROSAIL LUT (5% noise added, mean of 5% multiple solutions). The model was validated with a $R^2$ of 0.44 (RMSE: 1.85; NRMSE: 31.9%). A systematic underestimation occurred, which in principle implies that the RTM simulated LUT needs to be better parameterized. It took 2315 s to produce the map using ARTMO's LUT-based inversion toolbox (Rivera et al. 2013). Also uncertainty estimates are provided, e.g., in the form of residuals (not shown)

**Table 5** Advanced canopy RTMs commonly used in imaging spectroscopy applications

| RTM | Description |
|---|---|
| SCOPE(Soil–Canopy–Observation of Photosynthesis and Energy fluxes) | SCOPE (Tol et al. 2009) is a soil–vegetation–atmosphere (SVAT) scheme that includes RTMs along with a micrometeorological model for simulating turbulent heat exchange, and a plant physiological model for photosynthesis (Tol et al. 2014). The radiative transfer scheme is based on SAIL (Verhoef 1984b, 1985), extended with a similar radiative transfer for emitted radiation. The emitted radiation includes chlorophyll fluorescence and thermal radiation. Leaf radiative transfer is calculated with Fluspect (Vilfan et al. 2016) which also includes emitted fluorescence radiation. SCOPE is intended as tool to scale processes from leaf to canopy, and to analyse the effects of light scattering. Recent developments include vertical heterogeneity (Yang et al. 2017) and the zeaxanthin–violaxanthin pigment cycles |
| Discrete Anisotropic Radiative Transfer (DART) | DART model is being developed since 1992 as a physically based 3D computer programme (Gastellu-Etchegorry et al. 1996), which simulates radiative budget and remote sensing (airborne and spaceborne) optical image data of natural and urban landscapes for any wavelengths from the ultraviolet to the thermal infrared part of the electromagnetic spectrum (Gastellu-Etchegorry et al. 1999; Guillevic et al. 2003). It computes and provides bottom and top of the atmosphere spectral quantities (i.e. irradiance, exitance and radiance) that are transformed into reflectance or brightness temperature depending on the user DART mode preferences (Gastellu-Etchegorry et al. 2004). Simulated scenes may include the atmosphere, topography and any natural or anthropogenic objects at any geographical location (Grau and Gastellu-Etchegorry 2013). The latest DART optical development includes also the specular reflectance and the light polarization (Gastellu-Etchegorry et al. 2015). Apart of passive remote sensing data, it also simulates active terrestrial and air-/spaceborne light detection and ranging (LiDAR) discrete return, full waveform, multi-pulse and photon counting measurements (Gastellu-Etchegorry et al. 2016; Yin et al. 2016). In case of vegetation, it can also simulate radiative transfer of the solar-induced chlorophyll fluorescence for any virtual 3D Earth scene numerically and as images (Gastellu-Etchegorry et al. 2017) |

**Table 5** (continued)

| RTM | Description |
|---|---|
| Librat | Librat is a 3D Monte Carlo ray-tracing radiative transfer model developed as a library interface to the original ararat (Advanced RAdiometric RAy Tracer) model. The first version of ARARAT was published in 1992 (Lewis and Muller 1993) as part of the Botanical Plant Modelling System (BPMS) (Lewis 1999; Lewis and Muller 1990). Subsequently, the sampling scheme was improved as reported in Saich et al. (2002), and the codes developed into a library in recent years. Librat reads a 3D description of (canopy/soil/topographic) geometry, along with associated information on material scattering properties. The main function in the library then is that a ray is launched from some origin in 3D space, in a specified direction, and the code returns all information about the associated scattering paths and interactions, separated as direct and diffuse components. This core functionality, along with a set of associated sensor models but integrating paths, fired into some volume. It allows for a wide range of radiative transfer calculations, including time-resolved/lidar, splitting of the radiometric information per scattering order as well as straightforward reflectance/transmittance calculations (e.g., Disney et al. 2006; Hancock et al. 2012) |
| FLIGHT | FLIGHT (Barton and North 2001; North 1996) is a Monte Carlo ray-tracing model designed to rapidly simulate light interaction with 3D vegetation canopies at high spectral resolution, and produce reflectance spectra for both forward simulation and for use in inversion (Leonenko et al. 2013). Foliage is represented by structural properties of leaf area, leaf angle distribution, crown dimensions and fractional cover, and the optical properties of leaves, branch, shoot and ground components. The model represents multiple scattering and absorption of light within the canopy and with the ground surface. It has been developed to model 3D canopy photosynthesis (Alton et al. 2007), to simulate waveform and photon counting lidar (Montesano et al. 2015; North et al. 2010) and emitted fluorescence radiation (Hernández-Clemente et al. 2017). Structural data may be specified as a statistical distribution, derived from field measurements (Morton et al. 2014) or by direct inversion from LiDAR data (Bye et al. 2017) |

images are still time-consuming given the many per-pixel iterations and a high number of pixels involved. Hence, in its current form this method stays restricted to computationally fast RTMs in merely experimental settings.

*Look-up table (LUT)* strategies are based on the generation of simulated spectral reflectance scenarios for a high number of plausible combinations of variable value ranges. As such, the inversion problem is reduced to the identification of the modelled reflectance set that resembles most closely the measured one. This process is based on querying the LUT

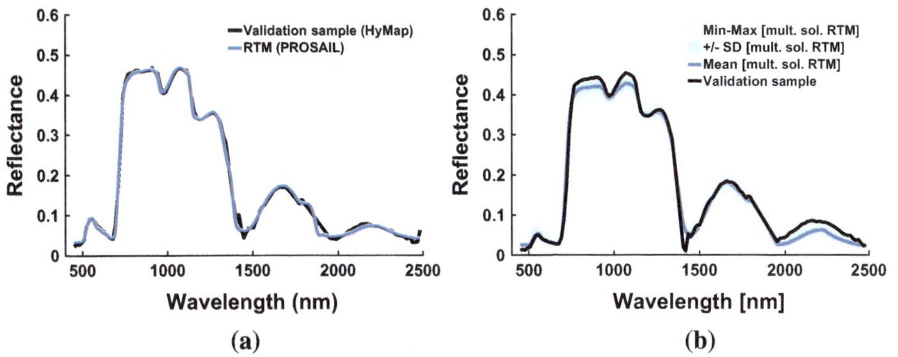


**Fig. 7** Illustrations of numerical inversion (**a**) and LUT-based inversion (**b**). A HyMaP spectrum was inverted against PROSAIL. In the case of LUT-inversion, overview statistics of 5% best multiple solutions are shown

and applying a cost function. A cost function minimizes the summed differences between simulated and measured reflectances for all wavelengths. The main advantage of LUT-based inversion routines over numerical optimization is their computational speed, since the computationally most demanding part of the inversion procedure is completed before the inversion itself (Dorigo et al. 2007). Consequently, LUT-based inversion methods are typically used as a preferred solution in RTM inversion studies. The classical LUT-based inversion approach is based on a RMSE cost function, which continues to be applied until today. This approach proved to be especially successful for chlorophyll (Kempeneers et al. 2008; Omari et al. 2013; Zhang et al. 2008) and LAI mapping. For instance, by using LUT-based inversion routines imaging spectroscopy data have been processed for the mapping of forest LAI (Banskota et al. 2013, 2015), grassland LAI (Atzberger et al. 2015) and LAI over agricultural crops based on UAVs (Duan et al. 2014). To further mitigate the ill-posed problem and optimize the robustness of the LUT-based inversion routines, a diversity of regularization strategies have been explored in inversion applications against spectroscopic data:

- *The use of prior knowledge* to constrain model variables in the development of a LUT (Baret and Buis 2008; Darvishzadeh et al. 2008; Koetz et al. 2005). Prior knowledge typically involves information on the feasible variable ranges for involved vegetation types (Dorigo et al. 2009; Verrelst et al. 2012c). Prior information together with prior distributions is also increasingly applied into a Bayesian context, whereby the inverted values are generated based on likelihoods (Laurent et al. 2013, 2014; Shiklomanov et al. 2016). The advantage of a Bayesian framework is its capability to quantify an inversion uncertainty around an inversion variable.
- *Selection of cost function.* The inverse problem of a nonlinear RTM is based on the minimization of a cost function concurrently measuring the discrepancy between (i) observed and simulated reflectance, and (ii) variables to estimate and the associated prior information (Jacquemoud et al. 2009b). To avoid solutions reaching fixed boundaries, a modified cost function in the LUT search that takes uncertainty of provided prior information into account is sometimes used, e.g., by means of the above-mentioned Bayesian approach. Alternatively, Leonenko et al. (2013) proposed and evaluated over 60 different cost functions dealing with different error distributions. Some more spec-

troscopic studies have evaluated among others the role of cost function (Danner et al. 2017; Locherer et al. 2015) in LUT-based inversion. Although the classical RMSE is a robust cost function, sometimes improvements can be gained with alternative cost functions, e.g., when the LUTs are non-normal distributed.
- The use of *multiple best solutions* in the inversion (mean or median), as opposed to a single best solution (Banskota et al. 2015; Kattenborn et al. 2017; Koetz et al. 2005; Locherer et al. 2015).
- The addition of *artificial noise* (additive or inverse multiplicative white noise) to account for uncertainties linked to measurements and models (Danner et al. 2017; Koetz et al. 2005; Locherer et al. 2015).
- Several spectroscopic studies reported that the relationship between measured and estimated variable perceptibly improves when only specific (sensitive) spectral ranges are selected for model inversion (Darvishzadeh et al. 2012; Meroni et al. 2004; Schlerf et al. 2005). To account for noise in the observations, other spectroscopic studies instead manipulated the observed spectra by applying a smoothing filter (Arellano et al. 2017) or wavelet transforms (Ali et al. 2016; Banskota et al. 2013; Kattenborn et al. 2017). Spectral selection and spectral polishing methods can be applied at the same time in order to enhance the resemblance with the usually more spectrally smooth simulated spectra.

Because of taking sun–target–sensor geometry into account, the use of RTM-based methods has been demonstrated to improve robustness to solar and view angle effects, compared to index-based methods (Kempeneers et al. 2008). Another advantage of RTM inversion routines is that uncertainties are provided as spectral residuals (Rivera et al. 2013) or standard deviations, when mapping multiple solution means (Verrelst et al. 2014). Yet the main drawback lies in its computational burden resulting from too many per-pixel iterations. Although LUT-inversion approaches may speed up the inversion process as opposed to numerical inversion, these inversion routines are still computationally expensive due to the iterative calls of LUT entries on a per-pixel basis. Consequently, despite attempts to optimize inversion algorithms in order to save up computational time for solving inverse radiative transfer problems (Favennec et al. 2016; Gastellu-Etchegorry et al. 2003), in terms of processing speed the RTM inversion routines still run behind statistical methods.

## 5 Hybrid Regression Methods

Having discussed the more fundamental categories of retrieval methods, this section addresses *hybrid* regression methods. Hybrid methods combine the generalization level of physically based methods with the flexibility and computational efficiency of advanced machine learning methods. This approach replaces the ground data needed for training of the parametric or nonparametric models by RTM input variables, which makes it computationally efficient. It is important to note that the hybrid approach does not alleviate the main issues of RTMs, notably that they only include existing knowledge and concepts. Similarly as in the case of LUT-based inversion, RTM simulations build a LUT representing a broad set of canopy realizations and the hybrid approach uses all available data stored in LUT to train a machine learning regression model (Fig. 8).

The awareness in the mid-1990s that ANNs are excellent algorithms to deal with large datasets led to the introduction of hybrid methods based on ANNs trained with generically

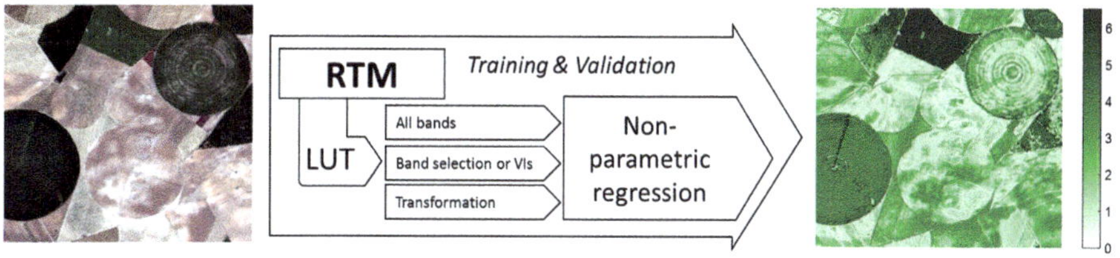


**Fig. 8** Principles of hybrid regression. Left: RGB subset of a hyperspectral HyMap image (125 bands) over Barrax agricultural site (Spain). Right: illustrative map of a vegetation property (LAI, $m^2/m^2$) as obtained by PROSAIL with Gaussian processes regression (GPR) and 15% white noise added. The model was validated with a $R^2$ of 0.88 (RMSE: 0.70; NRMSE: 10.1%). It took 6.3 seconds to produce the map using ARTMO's MLRA toolbox (Rivera Caicedo et al. 2014). With GPR also uncertainty estimates are provided (not shown). Because of not being trained with bare soil spectra, LAI over the non-irrigated parcels is overestimated

applicable RTM-generated data. It led to operational retrieval algorithms for datastreams acquired by multispectral and superspectral sensors [see Verrelst et al. (2015)]. Although this approach is less straightforward in the context of imaging spectroscopy, because of the challenge of collinearity, some recent efforts have been undertaken in exploring this research direction. Noteworthy is the work of Vohland et al. (2010) comparing a numerically optimized ANN with a LUT-based inversion using PROSAIL RTM simulations. Prediction accuracies generally decreased in the following sequence: numerical optimization > LUT > ANN. This would indicate that an ANN may not always be the best choice for inversion applications. However, no dimensionality reduction method was applied, which suggests that the regression model suffered from band collinearity effects. Also Fei et al. (2012) compared a PROSAIL-ANN hybrid approach with a PCA approach. The authors concluded that a PCA transformation into a regression function can mitigate the known reflectance saturation effect of dense canopies to some extent. This PROSAIL-ANN strategy was revisited by Rivera-Caicedo et al. (2017) with alternative dimensionality reduction methods. Although PCA improved accuracies as opposed of using all bands, substantially more improvements were achieved when converting the spectra into components by means canonical correlation analysis (CCA) or orthonormalized PLS (OPLS).

Likewise, inputs from more advanced RTMs were explored to develop specialized hybrid structures. In Malenovský et al. (2013), an ANN was trained based on PROSPECT-DART simulations that explicitly took 3D canopy structures into account to estimate forest leaf chlorophyll content from hyperspectral airborne AISA data. In this approach, the DART simulations went first through a continuum removal transformation. Alternatively, some studies have attempted to move away from ANN models by exploring hybrid structures on the basis of kernel-based machine learning regression algorithms, particularly the popular SVR. For instance, leaf chlorophyll content was estimated based on a PROSAIL-SVR model and applied to imaging spectroscopy (Preidl and Doktor 2011). An analogous concept was applied for a SVR that was trained by PROSPECT-DART simulations in combination with continuum removal transformations, with the purpose of quantifying forest biochemical and structural properties (Homolová and Janoutová 2016). Similarly, Doktor et al. (2014) used a PROSAIL dataset to train a random forest (RF) model to predict LAI and leaf chlorophyll content, and Liang et al. (2016) compared PROSAIL-based hybrid models with SVR and RF for leaf and canopy chlorophyll content estimation from CHRIS data. Finally, Rivera-Caicedo et al. (2017) analysed ensembles of regression algorithms with dimensionality reduction methods to consolidate the most ideal PROSAIL-based (2101 bands) hybrid regression model. This study concluded that compressing PROSAIL data into CCA or OPLS components led to highest accuracies when trained with a GPR model. Altogether, although these studies have only been developed in experimental settings—similar as the operational multispectral hybrid algorithms (e.g., Bacour et al. 2006; Baret et al. 2013)—the hybrid structures can be perfectly implemented into global mapping schemes. When combined with a dimensionality reduction method to suppress collinearity, hybrid methods have a great potential to advance towards operational spectroscopy-based processing schemes.

## 6 Discussion

The mapping of spatially continuous biophysical variables from imaging spectroscopy data is a progressively expanding field of research and development thanks to advances in spectrometer technology and in specialized methods interpreting the acquired spectral data. As a follow-up of an earlier, more general review on retrieval methods applicable to optical remote sensing (Verrelst et al. 2015), here a summary on retrieval methods specifically applied to spectroscopic data has been compiled. Four categories have been summarized: (1) parametric, (2) nonparametric, (3) RTM inversion and (4) hybrid methods. The first two categories are statistical methods commonly used with experimental (field) data, whereas the latter two rely on RTM simulations. A schematic flowchart of the main retrieval methods and their hierarchy is provided in Fig. 9.

While pros and cons of each of these methodological categories have been earlier discussed (Verrelst et al. 2015), here we discuss these categories from the perspective of forthcoming routinely acquired and standardized (e.g., atmospherically corrected) imaging spectroscopy data streams. First of all, the choice of a method bears implications, not only on the retrievability and processing time of mappable vegetation properties, but also on the purpose of the retrieval. Parametric and nonparametric methods rely on ground data for training, which obviously need to be available in order to apply these methods. If they are available, they are the 'shortest' way to the variables of interest, because especially the nonparametric methods do no impose any limitation on the relationship between the spectrum and the variable of interest. In contrast, RTMs describe radiative transfer processes, i.e. they use existing knowledge (as materialized in the models) rather than ground measured data. Retrieval from an RTM through inversion is most useful if one is more interested in the underlying radiative transfer processes (scattering, sun and shade foliage fractions, light distribution within vegetation canopies, relationships between canopy structure and photosynthesis), rather than in merely extracting a specific variable. However, strategies relying on RTM simulations are inherently limited by the input variables of the RTM and, as discussed in Sect. 4, ancillary data and regularization methods may be required to optimize their inversions.

Statistical approaches, on the other hand, possess the flexibility to relate reflectance data with any measured biophysical variable—state variable or not. As demonstrated in Sects. 2 and 3 , this can be any quantifiable attribute, typically in the domains of leaf biochemical constituents (e.g., nitrogen, phosphorus), pigments (e.g., chlorophyll, carotenoids, xanthophylls) or higher-level structural variables (e.g., above-ground biomass, grain yield). The strength of the correlation with validation data typically determines the validity and

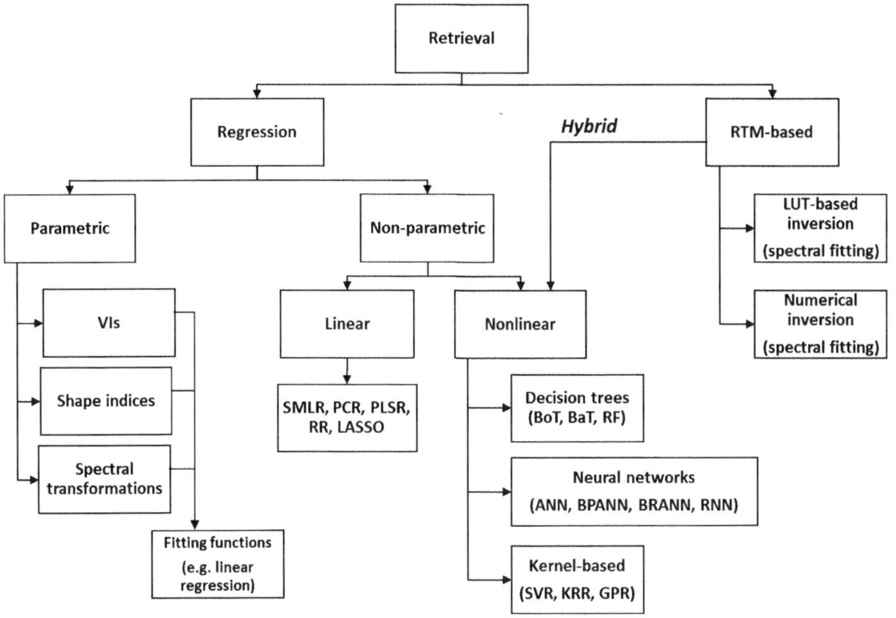


**Fig. 9** Schematic overview of the main retrieval methods

transferability of the statistical model. While this 'seeking for best correlations' can be criticized, because of the absence of a physical basis (Knyazikhin et al. 2013), statistical approaches are becoming increasingly powerful to extract biochemical variables through complex and often indirect relationships. Particularly, machine learning models are powerful in extracting information from subtle variations in spectroscopic data through adaptive, nonlinear relationships. The advantage of these statistical models is that not only variable-specific absorption features can be used for information extraction, but also secondary relationships with variables related to other absorption features that co-vary with the variable of interest can be exploited (Ollinger 2011; Verrelst et al. 2012b). Since high accuracies are often obtained with these methods, they are gaining popularity, not only for quantification of a diversity of vegetation properties, but also in mapping of floristic composition (Feilhauer et al. 2017; Harris et al. 2015; Neumann et al. 2016; Roth et al. 2015).

Regardless of the nature of retrieval method, in view of mapping larger areas, and especially in an operational and global context, what matters is the possibility to provide associated information on the retrieval quality. The characterization of uncertainty is a fundamental requirement for postulating correct scientific conclusions from results and for assimilating results into statistical or mechanistic higher-level models (Cressie et al. 2009). As addressed in Sect. 2, parametric regression methods, i.e. spectral transformation methods in combination with a fitting function, do not provide uncertainty estimates, which undermine their applicability to other images in space and time. Subsequently, while valid when locally calibrated and validated, parametric methods are of little use in an operational context. With regard to inversion routines, uncertainties can be provided as spectral residuals (Rivera et al. 2013) or standard deviations when mapping multiple solution means (Verrelst et al. 2014). Lately, inversion approaches were proposed in a Bayesian framework (Shiklomanov et al. 2016), whereby uncertainties are delivered along with the retrievals. In the case of traditional statistical models, uncertainty estimation has been a complex exercise. Statistical models developed within a Bayesian framework, such as GPR, provide uncertainties together with the predictions (Verrelst et al. 2013b; Camps-Valls et al. 2016), which indicate the probability interval of an estimation relative to the samples used during the training phase. These uncertainties can be used to evaluate GPR model transferability. For example, by mapping the uncertainties Verrelst et al. (2013b) demonstrated that a locally developed regression model can be successfully transported to other images in space and time for the large majority of pixels (i.e. the uncertainty maps were not systematically worse). Similarly, uncertainties can inform about the model performance. It was demonstrated that dimensionality reduction methods applied in GPR models for LAI mapping not only largely speed up the processing, but they also led to lower per-pixel uncertainties as opposed to mapping using all bands (Rivera-Caicedo et al. 2017). In conclusion, in the view of an operational processing need, just as important as the variable retrieval itself is the provision of an associated uncertainty estimate. Uncertainty estimates allow evaluating the method's per-pixel performance and consequently allow evaluating the method's capability to process routinely acquired imaging data. They thus provide a measure of the retrieval fidelity, which can be used to identify and mask out the highly uncertain and non-reliable results.

Another important aspect for operational production of vegetation properties from typically bulky imaging spectroscopy data streams implies computational speed. Generally, the lower the complexity of a model, the faster it will be able to produce maps. This highly favors the application of parametric regression approaches since they consist of only few transformations and equations. Also nonparametric regression algorithms, once trained, can be applied to process an image almost instantaneously. Training of machine learning models is frequently related to the tuning of several free variables with costly cross-validation approaches. These scale poorly with the number of samples (such as in kernel machines) or with the data dimensions (such as in ANNs). Although a trained ANN converts an image into a map quasi-instantly, kernel-based methods require more processing time, because the similarity between each test pixel in the image and those used to train the model has to be estimated. Training can be computationally costly, especially when using a big training dataset, e.g., as in hybrid strategies. A solution to shorten training time could be in size reduction of the training data in a way that maximal relevant information is preserved. This can be achieved by means of dimensionality reduction methods in the spectral domain (Rivera-Caicedo et al. 2017), or by means of intelligent sampling in the sampling domain, e.g., through active learning (Verrelst et al. 2016a).

Considerably longer run-time is expected in the case of inversion routines. Since RTMs take some time to generate simulations, especially for computationally expensive models, and also the evaluation takes place on a per-pixel basis, the iterative inversion routines are computationally expensive leading to relatively slow mapping speeds. In an attempt to accelerate their mapping speed, it has been proposed to approximate the functioning of the original RTM by means of statistical learning called *emulation* (Gómez-Dans et al. 2016; Rivera et al. 2015). Initial experiments to emulate leaf, canopy and atmospheric RTMs demonstrated that emulators can successfully generate spectral output from a limited set of input variable almost instantly, thereby preserving sufficient accuracy as compared to the original RTM (Verrelst et al. 2016c, 2017). Although an emulator reproduces RTM simulations instantly, application of a per-pixel spectral fitting requires many repetitions, which implies that these methods still do not reach the speed of statistical methods.

All in all, having the purpose of advancing towards operational imaging spectroscopy data processing in mind, i.e. reaching globally applicable, accurate and fast estimates, we end up with the following recommendations:

- To enable model transferability to routinely acquired images, retrieval methods must provide associated per-pixel uncertainties as a quality indicator whether the model can perform adequately in another space and time.
- Regarding the computational speed, e.g., in case of repetitive image processing, statistical (i.e. regression) methods are multiple times faster than physically based methods, capable of processing full images in the order of minutes or even seconds.
- In the case of regression methods (experimental or hybrid), multicollinearity of spectroscopic data complicates the development of powerful models. Physically based methods using spectral fitting do not suffer from this problem.
- To mitigate the problem of multicollinearity in regression methods, either band selection or dimensionality reduction methods can be applied before entering the regression. Although band selection is a common practice, more powerful regression models can probably be obtained when using a dimensionality reduction method.

## 7 Conclusions

With forthcoming imaging spectrometer satellite missions, an unprecedented stream of datasets on the terrestrial biosphere will become available. This will require powerful processing techniques enabling quantification of vegetation variables in an operational and global setting. Four categories of retrieval methods have been discussed in this review paper: (1) parametric regression; (2) nonparametric regression; (3) physically based RTM inversion; and, (4) hybrid methods. For each of these categories, various methodological approaches are increasingly applied to imaging spectroscopy data. This literature review synthesized the current state of the art in the field of spectroscopy-based vegetation properties mapping.

Although parametric methods, such as shape indices or spectral transformation, deal well with extracting relevant information embedded in spectroscopic data, their lack of uncertainty estimates makes them unsuitable for operational use. Higher accuracies can be reached with nonlinear nonparametric methods, especially those in the field of machine learning that generate probabilistic outputs, e.g., Gaussian process regression. However, an additional step to mitigate their spectral multicollinearity is deemed necessary. A popular strategy in this respect is selecting a set of vegetation indices or applying spectral transformation before training the machine learning algorithm. It nevertheless remains questionable whether such band selection approaches fully capture all relevant information. Instead, dimensionality reduction methods that enable compressing the large majority of spectral variability into a few components tend to lead to more accurate predictions.

On the other hand, the inversion of physically based RTMs against spectroscopic data is generally applicable and physically sound, but optimizing their inversion strategies is more challenging compared to the regression methods. RTM-based inversion is computationally demanding, and ancillary information is usually required as an input or to regulate the inversion algorithm. Hybrid regression methods, based on the coupling of an RTM with a machine learning regression algorithm, overcome the problem of processing speed. Particularly, Bayesian kernel-based hybrid strategies possess promising features, as they combine speed, flexibility and the provision of uncertainty estimates. Their accuracies and processing speed can be further improved in combination with dimensionality reduction. Altogether, and in the interest of operational spectroscopy-based mapping of vegetation properties, we recommend to further explore the feasibility and implementation of hybrid strategies into the next-generation data processing chains.

**Acknowledgements** Jochem Verrelst was supported by the European Research Council (ERC) under the ERC-2017-STG SENTIFLEX project (grant agreement 755617). Contribution of Zbyněk Malenovský was supported by the Australian Research Council Future Fellowship: Bridging scales in remote sensing of vegetation stress (FT160100477). Gustau Camps-Valls was supported by the ERC under the ERC-CoG-2014 SEDAL project (grant agreement 647423). We thank the two reviewers for their valuable suggestions.

## References

Adam E, Mutanga O, Abdel-Rahman E, Ismail R (2014) Estimating standing biomass in papyrus (Cyperus papyrus L.) swamp: exploratory of in situ hyperspectral indices and random forest regression. Int J Remote Sens 35(2):693–714

Addink E, De Jong S, Pebesma E (2007) The importance of scale in object-based mapping of vegetation parameters with hyperspectral imagery. Photogramm Eng Remote Sens 73(8):905–912

Ali A, Skidmore A, Darvishzadeh R, van Duren I, Holzwarth S, Mueller J (2016) Retrieval of forest leaf functional traits from HySpex imagery using radiative transfer models and continuous wavelet analysis. ISPRS J Photogramm Remote Sens 122:68–80

Alton P, Ellis R, Los S, North P (2007) Improved global simulations of gross primary product based on a separate and explicit treatment of diffuse and direct sunlight. J Geophys Res Atmos 112:D02108

Arellano P, Tansey K, Balzter H, Tellkamp M (2017) Plant family-specific impacts of petroleum pollution on biodiversity and leaf chlorophyll content in the Amazon rainforest of Ecuador. PLoS ONE 12(1):e0169867

Arenas-García J, Camps-Valls G (2008) Efficient kernel orthonormalized PLS for remote sensing applications. IEEE Trans Geosci Remote Sens 46(10):2872–2881

Ashourloo D, Aghighi H, Matkan A, Mobasheri M, Rad A (2016) An investigation into machine learning regression techniques for the leaf rust disease detection using hyperspectral measurement. IEEE J Select Top Appl Earth Observ Remote Sens 9(9):4344–4351

Atzberger C, Guérif M, Baret F, Werner W (2010) Comparative analysis of three chemometric techniques for the spectroradiometric assessment of canopy chlorophyll content in winter wheat. Comput Electron Agric 73(2):165–173

Atzberger C, Darvishzadeh R, Immitzer M, Schlerf M, Skidmore A, le Maire G (2015) Comparative analysis of different retrieval methods for mapping grassland leaf area index using airborne imaging spectroscopy. Int J Appl Earth Observ Geoinf 43:19–31

Bacour C, Baret F, Béal D, Weiss M, Pavageau K (2006) Neural network estimation of LAI, fAPAR, fCover and LAI×Cab, from top of canopy MERIS reflectance data: principles and validation. Remote Sens Environ 105(4):313–325

Balzarolo M, Vescovo L, Hammerle A, Gianelle D, Papale D, Tomelleri E, Wohlfahrt G (2015) On the relationship between ecosystem-scale hyperspectral reflectance and $CO_2$ exchange in European mountain grasslands. Biogeosciences 12(10):3089–3108

Banskota A, Wynne R, Thomas V, Serbin S, Kayastha N, Gastellu-Etchegorry J, Townsend P (2013) Investigating the utility of wavelet transforms for inverting a 3-D radiative transfer model using hyperspectral data to retrieve forest LAI. Remote Sens 5(6):2639–2659

Banskota A, Serbin S, Wynne R, Thomas V, Falkowski M, Kayastha N, Gastellu-Etchegorry JP, Townsend P (2015) An LUT-based inversion of DART model to estimate forest LAI from hyperspectral data. IEEE J Sel Top Appl Earth Observ Remote Sens 8(6):3147–3160

Bao S, Cao C, Chen W, Tian H (2017) Spectral features and separability of alpine wetland grass species. Spectrosc Lett 50(5):245–256

Baranoski G, Rokne J (2005) A practical approach for estimating the red edge position of plant leaf reflectance. Int J Remote Sens 26(3):503–521

Baret F, Buis S (2008) Estimating canopy characteristics from remote sensing observations. Review of methods and associated problems. In: Liang S (ed) Advances in land remote sensing: system, modeling, inversion and application. Springer, New York, pp 171–200

Baret F, Weiss M, Lacaze R, Camacho F, Makhmara H, Pacholcyzk P, Smets B (2013) GEOV1: LAI and FAPAR essential climate variables and FCOVER global time series capitalizing over existing products. Part 1: principles of development and production. Remote Sens Environ 137:299–309
Barton CVM, North P (2001) Remote sensing of canopy light use efficiency using the photochemical reflectance index: model and sensitivity analysis. Remote Sens Environ 78(3):264–273
Bayat B, van der Tol C, Verhoef W (2016) Remote sensing of grass response to drought stress using spectroscopic techniques and canopy reflectance model inversion. Remote Sens 8(7):557
Belgiu M, Drăguţ L (2016) Random forest in remote sensing: a review of applications and future directions. ISPRS J Photogramm Remote Sens 114:24–31
Berk A, Anderson G, Acharya P, Bernstein L, Muratov L, Lee J, Fox M, Adler-Golden S, Chetwynd J, Hoke M, Lockwood R, Gardner J, Cooley T, Borel C, Lewis P, Shettle E (2006) MODTRAN$^{TM}$5: 2006 update. vol 6233 II
Botha E, Leblon B, Zebarth B, Watmough J (2007) Non-destructive estimation of potato leaf chlorophyll from canopy hyperspectral reflectance using the inverted PROSAIL model. Int J Appl Earth Observ Geoinf 9(4):360–374
Bratsch S, Epstein H, Buchhorn M, Walker D, Landes H (2017) Relationships between hyperspectral data and components of vegetation biomass in Low Arctic tundra communities at Ivotuk, Alaska. Environ Res Lett 12(2):025003
Breiman L (1996) Bagging predictors. Mach Learn 24(2):123–140
Breiman L (2001) Random forests. Mach Learn 45(1):5–32
Breiman L, Friedman J, Stone C, Olshen R (1984) Classification and regression trees. The Wadsworth and Brooks-Cole statistics-probability series. Taylor & Francis, London
Broge NH, Leblanc E (2001) Comparing prediction power and stability of broadband and hyperspectral vegetation indices for estimation of green leaf area index and canopy chlorophyll density. Remote Sens Environ 76(2):156–172
Broomhead DS, Lowe D (1988) Radial basis functions, multi-variable functional interpolation and adaptive networks. Tech. rep, Royal Signals and Radar Establishment Malvern (United Kingdom)
Buchhorn M, Walker DA, Heim B, Raynolds MK, Epstein HE, Schwieder M (2013) Ground-based hyperspectral characterization of alaska tundra vegetation along environmental gradients. Remote Sens 5(8):3971–4005
Burden F, Winkler D (1999) Robust QSAR models using bayesian regularized neural networks. J Med Chem 42(16):3183–3187
Bye I, North P, Los S, Kljun N, Rosette J, Hopkinson C, Chasmer L, Mahoney C (2017) Estimating forest canopy parameters from satellite waveform LiDAR by inversion of the FLIGHT three-dimensional radiative transfer model. Remote Sens Environ 188:177–189
Camps-Valls G, Verrelst J, Muñoz-Marí J, Laparra V, Mateo-Jiménez F, Gómez-Dans J (2016) A survey on Gaussian processes for earth observation data analysis. IEEE Geosci Remote Sens Mag 4(2):41–57
Capolupo A, Kooistra L, Berendonk C, Boccia L, Suomalainen J (2015) Estimating plant traits of grasslands from UAV-acquired hyperspectral images: a comparison of statistical approaches. ISPRS Int J Geo-Inf 4(4):2792–2820
Chen B, Wu Z, Wang J, Dong J, Guan L, Chen J, Yang K, Xie G (2015) Spatio-temporal prediction of leaf area index of rubber plantation using HJ-1A/1B CCD images and recurrent neural network. ISPRS J Photogramm Remote Sens 102:148–160
Cho MA, Skidmore AK (2006) A new technique for extracting the red edge position from hyperspectral data: the linear extrapolation method. Remote Sens Environ 101(2):181–193
Cho M, Skidmore A, Corsi F, van Wieren S, Ib Sobhan (2007) Estimation of green grass/herb biomass from airborne hyperspectral imagery using spectral indices and partial least squares regression. Int J Appl Earth Observ Geoinf 9(4):414–424
Cho MA, Skidmore AK, Atzberger C (2008) Towards red-edge positions less sensitive to canopy biophysical parameters for leaf chlorophyll estimation using properties optique spectrales des feuilles (PROSPECT) and scattering by arbitrarily inclined leaves (SAILH) simulated data. Int J Remote Sens 29(8):2241–2255
Clark RN, Roush TL (1984) Reflectance spectroscopy: quantitative analysis techniques for remote sensing applications. J Geophys Res Solid Earth 89(B7):6329–6340
Clevers J (2014) Beyond NDVI: extraction of biophysical variables from remote sensing imagery. Remote Sens Digital Image Process 18:363–381
Clevers J, Kooistra L (2012) Using hyperspectral remote sensing data for retrieving canopy chlorophyll and nitrogen content. IEEE J Sel Top Appl Earth Observ Remote Sens 5(2):574–583

Clevers JGPW, Kooistra L, Salas EAL (2004) Study of heavy metal contamination in river floodplains using the red-edge position in spectroscopic data. Int J Remote Sens 25(19):3883–3895
Coops NC, Smith ML, Martin M, Ollinger SV (2003) Prediction of eucalypt foliage nitrogen content from satellite-derived hyperspectral data. IEEE Trans Geosci Remote Sens 41(6):1338–1346
Cressie N, Calder C, Clark J, Ver Hoef J, Wikle C (2009) Accounting for uncertainty in ecological analysis: the strengths and limitations of hierarchical statistical modeling. Ecol Appl 19(3):553–570
Danner M, Berger K, Wocher M, Mauser W, Hank T (2017) Retrieval of biophysical crop variables from multi-angular canopy spectroscopy. Remote Sens 9(7):726
Darvishzadeh R, Skidmore A, Schlerf M, Atzberger C (2008) Inversion of a radiative transfer model for estimating vegetation LAI and chlorophyll in a heterogeneous grassland. Remote Sens Environ 112(5):2592–2604
Darvishzadeh R, Matkan AA, Dashti Ahangar A (2012) Inversion of a radiative transfer model for estimation of rice canopy chlorophyll content using a lookup-table approach. IEEE J Sel Top Appl Earth Observ Remote Sens 99:1–9
Dawson T, Curran P, Plummer S (1998) LIBERTY—Modeling the effects of leaf biochemical concentration on reflectance spectra. Remote Sens Environ 65(1):50–60
Delegido J, Alonso L, González G, Moreno J (2010) Estimating chlorophyll content of crops from hyperspectral data using a normalized area over reflectance curve (NAOC). Int J Appl Earth Observ Geoinf 12(3):165–174
Delegido J, Verrelst J, Alonso L, Moreno J (2011) Evaluation of sentinel-2 red-edge bands for empirical estimation of green LAI and chlorophyll content. Sensors 11(7):7063–7081
Delegido J, Verrelst J, Meza C, Rivera J, Alonso L, Moreno J (2013) A red-edge spectral index for remote sensing estimation of green LAI over agroecosystems. Eur J Agron 46:42–52
Dietz A, Kuenzer C, Gessner U, Dech S (2012) Remote sensing of snow—a review of available methods. Int J Remote Sens 33(13):4094–4134
Disney M, Lewis P, Saich P (2006) 3D modelling of forest canopy structure for remote sensing simulations in the optical and microwave domains. Remote Sens Environ 100(1):114–132
Doktor D, Lausch A, Spengler D, Thurner M (2014) Extraction of plant physiological status from hyperspectral signatures using machine learning methods. Remote Sens 6(12):12247–12274
Dorigo WA, Zurita-Milla R, de Wit AJW, Brazile J, Singh R, Schaepman ME (2007) A review on reflective remote sensing and data assimilation techniques for enhanced agroecosystem modeling. Int J Appl Earth Observ Geoinf 9(2):165–193
Dorigo W, Richter R, Baret F, Bamler R, Wagner W (2009) Enhanced automated canopy characterization from hyperspectral data by a novel two step radiative transfer model inversion approach. Remote Sens 1(4):1139–1170
Draper NR, Smith H (2014) Applied regression analysis. Wiley, New York
Dreccer M, Barnes L, Meder R (2014) Quantitative dynamics of stem water soluble carbohydrates in wheat can be monitored in the field using hyperspectral reflectance. Field Crops Res 159:70–80
Drusch M, Moreno J, Del Bello U, Franco R, Goulas Y, Huth A, Kraft S, Middleton EM, Miglietta F, Mohammed G et al (2017) The fluorescence explorer mission concept-ESA's earth explorer 8. IEEE Trans Geosci Remote Sens 55(3):1273–1284
Du L, Shi S, Yang J, Sun J, Gong W (2016) Using different regression methods to estimate leaf nitrogen content in rice by fusing hyperspectral LiDAR data and laser-induced chlorophyll fluorescence data. Remote Sens 8(6):526
Duan SB, Li ZL, Wu H, Tang BH, Ma L, Zhao E, Li C (2014) Inversion of the PROSAIL model to estimate leaf area index of maize, potato, and sunflower fields from unmanned aerial vehicle hyperspectral data. Int J Appl Earth Observ Geoinf 26(1):12–20
Elvidge CD, Chen Z (1995) Comparison of broad-band and narrow-band red and near-infrared vegetation indices. Remote Sens Environ 54(1):38–48
Favennec Y, Le Hardy D, Dubot F, Rousseau B, Rousse D (2016) Some speed-up strategies for solving inverse radiative transfer problems. J Phys Conf Ser 676(1):012006
Fei Y, Jiulin S, Hongliang F, Zuofang Y, Jiahua Z, Yunqiang Z, Kaishan S, Zongming W, Maogui H (2012) Comparison of different methods for corn LAI estimation over northeastern China. Int J Appl Earth Observ Geoinf 18:462–471
Feilhauer H, Asner GP, Martin RE (2015) Multi-method ensemble selection of spectral bands related to leaf biochemistry. Remote Sens Environ 164:57–65
Feilhauer H, Somers B, van der Linden S (2017) Optical trait indicators for remote sensing of plant species composition: predictive power and seasonal variability. Ecol Indic 73:825–833

Feng H, Yang F, Li Z, Yang G, Guo J, He P, Wang Y (2016) Hyperspectral estimation of leaf total phosphorus content in apple tree based on optimal weights combination model. Nongye Gongcheng Xuebao/ Trans Chin Soc Agric Eng 32(7):173–180

Feret JB, François C, Asner GP, Gitelson AA, Martin RE, Bidel LPR, Ustin SL, le Maire G, Jacquemoud S (2008) PROSPECT-4 and 5: advances in the leaf optical properties model separating photosynthetic pigments. Remote Sens Environ 112(6):3030–3043

Foster A, Kakani V, Mosali J (2017) Estimation of bioenergy crop yield and N status by hyperspectral canopy reflectance and partial least square regression. Precis Agric 18(2):192–209

Francl L, Panigrahi S (1997) Artificial neural network models of wheat leaf wetness. Agric For Meteorol 88(1–4):57–65

Friedman J, Hastie T, Tibshirani R (2000) Additive logistic regression: a statistical view of boosting. Ann Stat 28(2):337–407

Fu Y, Yang G, Feng H, Xu X, Song X, Wang J (2012) Comparison of winter wheat LAI estimation methods based on hyperspectral dimensionality reduction and vegetation index. Nongye Gongcheng Xuebao/ Trans Chin Soc Agric Eng 28(23):107–113

Gastellu-Etchegorry J, Demarez V, Pinel V, Zagolski F (1996) Modeling radiative transfer in heterogeneous 3-D vegetation canopies. Remote Sens Environ 58(2):131–156

Gastellu-Etchegorry J, Guillevic P, Zagolski F, Demarez V, Trichon V, Deering D, Leroy M (1999) Modeling BRF and radiation regime of boreal and tropical forests: I BRF. Remote Sens Environ 68(3):281–316

Gastellu-Etchegorry J, Gascon F, Esteve P (2003) An interpolation procedure for generalizing a look-up table inversion method. Remote Sens Environ 87(1):55–71

Gastellu-Etchegorry J, Martin E, Gascon F (2004) DART: a 3D model for simulating satellite images and studying surface radiation budget. Int J Remote Sens 25(1):73–96

Gastellu-Etchegorry JP, Yin T, Lauret N, Cajgfinger T, Gregoire T, Grau E, Feret JB, Lopes M, Guilleux J, Dedieu G, Malenovský Z, Cook B, Morton D, Rubio J, Durrieu S, Cazanave G, Martin E, Ristorcelli T (2015) Discrete anisotropic radiative transfer (DART 5) for modeling airborne and satellite spectroradiometer and LIDAR acquisitions of natural and urban landscapes. Remote Sens 7(2):1667–1701

Gastellu-Etchegorry JP, Yin T, Lauret N, Grau E, Rubio J, Cook B, Morton D, Sun G (2016) Simulation of satellite, airborne and terrestrial LiDAR with DART (I): Waveform simulation with quasi-Monte Carlo ray tracing. Remote Sens Environ 184:418–435

Gastellu-Etchegorry J, Lauret N, Yin T, Landier L, Kallel A, Malenovský Z, Al Bitar A, Aval J, Benhmida S, Qi J, Medjdoub G, Guilleux J, Chavanon E, Cook B, Morton D, Nektarios N, Mitraka Z (2017) DART: recent advances in remote sensing data modeling with atmosphere, polarization, and chlorophyll fluorescence. IEEE J Sel Top Appl Earth Observ Remote Sens. Accepted 8 March 2017. (**in press**)

Geladi P, Kowalski B (1986) Partial least-squares regression: a tutorial. Analytica Chimica Acta 185(C):1–17

Gewali U, Monteiro S, Saber E (2018) Machine learning based hyperspectral image analysis: a survey. arXiv:1802.08701

Gianelle D, Guastella F (2007) Nadir and off-nadir hyperspectral field data: strengths and limitations in estimating grassland biophysical characteristics. Int J Remote Sens 28(7):1547–1560

Glenn E, Huete A, Nagler P, Nelson S (2008) Relationship between remotely-sensed vegetation indices, canopy attributes and plant physiological processes: What vegetation indices can and cannot tell us about the landscape. Sensors 8(4):2136–2160

Gómez-Chova L, Muñoz-Marí J, Laparra V, Malo-López J, Camps-Valls G (2011) A review of kernel methods in remote sensing data analysis. In: Optical remote sensing. Springer, Berlin, pp 171–206

Gómez-Dans JL, Lewis PE, Disney M (2016) Efficient emulation of radiative transfer codes using gaussian processes and application to land surface parameter inferences. Remote Sens 8(2):119

Gonsamo A (2011) Normalized sensitivity measures for leaf area index estimation using three-band spectral vegetation indices. Int J Remote Sens 32(7):2069–2080

Govaerts YM, Verstraete MM (1998) Raytran: a Monte Carlo ray-tracing model to compute light scattering in three-dimensional heterogeneous media. IEEE Trans Geosci Remote Sens 36(2):493–505

Grau E, Gastellu-Etchegorry JP (2013) Radiative transfer modeling in the Earth-Atmosphere system with DART model. Remote Sens Environ 139:149–170

Guanter L, Kaufmann H, Segl K, Foerster S, Rogass C, Chabrillat S, Kuester T, Hollstein A, Rossner G, Chlebek C, Straif C, Fischer S, Schrader S, Storch T, Heiden U, Mueller A, Bachmann M, Muhle H, Muller R, Habermeyer M, Ohndorf A, Hill J, Buddenbaum H, Hostert P, van der Linden S, Leitao PJ, Rabe A, Doerffer R, Krasemann H, Xi H, Mauser W, Hank T, Locherer M, Rast M, Staenz K, Sang

B (2015) The EnMAP spaceborne imaging spectroscopy mission for earth observation. Remote Sens 7(7):8830

Guillevic P, Gastellu-Etchegorry J, Demarty J, Prévot L (2003) Thermal infrared radiative transfer within three-dimensional vegetation covers. J Geophys Res Atmos. https://doi.org/10.1029/2002JD002247

Han ZY, Zhu XC, Fang XY, Wang ZY, Wang L, Zhao GX, Jiang YM (2016) Hyperspectral estimation of apple tree canopy LAI based on SVM and RF regression. Guang Pu Xue Yu Guang Pu Fen Xi/Spectrosc Spectr Anal 36(3):800–805

Hancock S, Lewis P, Foster M, Disney M, Muller JP (2012) Measuring forests with dual wavelength lidar: a simulation study over topography. Agric For Meteorol 161:123–133

Hansen PM, Schjoerring JK (2003) Reflectance measurement of canopy biomass and nitrogen status in wheat crops using normalized difference vegetation indices and partial least squares regression. Remote Sens Environ 86(4):542–553

Harris A, Charnock R, Lucas R (2015) Hyperspectral remote sensing of peatland floristic gradients. Remote Sens Environ 162:99–111

Haykin S (1999) Neural networks—a comprehensive foundation, 2nd edn. Prentice Hall, Upper Saddle

He R, Qiao X, Jiang J, Guo H (2015) Retrieving canopy leaf total nitrogen content of winter wheat by continuous wavelet transform. Nongye Gongcheng Xuebao/Trans Chin Soc Agric Eng 31(2):141–146

Heiskanen J, Rautiainen M, Stenberg P, Mõttus M, Vesanto VH (2013) Sensitivity of narrowband vegetation indices to boreal forest LAI, reflectance seasonality and species composition. ISPRS J Photogramm Remote Sens 78:1–14

Hernández-Clemente R, North P, Hornero A, Zarco-Tejada P (2017) Assessing the effects of forest health on sun-induced chlorophyll fluorescence using the FluorFLIGHT 3-D radiative transfer model to account for forest structure. Remote Sens Environ 193:165–179

Hochreiter S, Schmidhuber J (1997) Long short-term memory. Neural Comput 9(8):1735–1780

Homolová L, Janoutová R, Malenovský Z (2016) Evaluation of various spectral inputs for estimation of forest biochemical and structural properties from airborne imaging spectroscopy data. 41:961–966

Huang Z, Turner BJ, Dury SJ, Wallis IR, Foley WJ (2004) Estimating foliage nitrogen concentration from HYMAP data using continuum removal analysis. Remote Sens Environ 93(1):18–29

Huang Y, Tian Q, Wang L, Geng J, Lyu C (2014) Estimating canopy leaf area index in the late stages of wheat growth using continuous wavelet transform. J Appl Remote Sens 8(1):083517

Hughes G (1968) On the mean accuracy of statistical pattern recognizers. IEEE Trans Inf Theory 14(1):55–63

Im J, Jensen JR, Coleman M, Nelson E (2009) Hyperspectral remote sensing analysis of short rotation woody crops grown with controlled nutrient and irrigation treatments. Geocarto Int 24(4):293–312

Jacquemoud S, Baret F, Andrieu B, Danson FM, Jaggard K (1995) Extraction of vegetation biophysical parameters by inversion of the PROSPECT+SAIL models on sugar beet canopy reflectance data. Application to TM and AVIRIS sensors. Remote Sens Environ 52(3):163–172

Jacquemoud S, Verhoef W, Baret F, Bacour C, Zarco-Tejada P, Asner G, François C, Ustin S (2009a) PROSPECT + SAIL models: a review of use for vegetation characterization. Remote Sens Environ 113(SUPPL. 1):S56–S66

Jacquemoud S, Verhoef W, Baret F, Bacour C, Zarco-Tejada P, Asner G, François C, Ustin S (2009b) PROSPECT + SAIL models: a review of use for vegetation characterization. Remote Sens Environ 113(SUPPL. 1):S56–S66

Jay S, Bendoula R, Hadoux X, Féret JB, Gorretta N (2016) A physically-based model for retrieving foliar biochemistry and leaf orientation using close-range imaging spectroscopy. Remote Sens Environ 177:220–236

Jensen R, Hardin P, Hardin A (2012) Estimating urban leaf area index (LAI) of individual trees with hyperspectral data. Photogramm Eng Remote Sens 78(5):495–504

Jia F, Liu G, Liu D, Zhang Y, Fan W, Xing X (2013) Comparison of different methods for estimating nitrogen concentration in flue-cured tobacco leaves based on hyperspectral reflectance. Field Crops Res 150:108–114

Kalacska M, Lalonde M, Moore T (2015) Estimation of foliar chlorophyll and nitrogen content in an ombrotrophic bog from hyperspectral data: Scaling from leaf to image. Remote Sens Environ 169:270–279

Kanke Y, Tubaña B, Dalen M, Harrell D (2016) Evaluation of red and red-edge reflectance-based vegetation indices for rice biomass and grain yield prediction models in paddy fields. Precis Agric 17(5):507–530

Karimi Y, Prasher S, Madani A, Kim S (2008) Application of support vector machine technology for the estimation of crop biophysical parameters using aerial hyperspectral observations. Can Biosyst Eng/Le Genie des biosystems au Canada 50:7.13–7.20

Kattenborn T, Fassnacht F, Pierce S, Lopatin J, Grime J, Schmidtlein S (2017) Linking plant strategies and plant traits derived by radiative transfer modelling. J Veg Sci 28(4):717–727
Kempeneers P, Zarco-Tejada PJ, North PRJ, de Backer S, Delalieux S, Sepulcre-Cantó G, Morales F, van Aardt JAN, Sagardoy R, Coppin P, Scheunders P (2008) Model inversion for chlorophyll estimation in open canopies from hyperspectral imagery. Int J Remote Sens 29(17–18):5093–5111
Kiala Z, Odindi J, Mutanga O, Peerbhay K (2016) Comparison of partial least squares and support vector regressions for predicting leaf area index on a tropical grassland using hyperspectral data. J Appl Remote Sens 10(3):036015
Kimes DS, Nelson RF, Manry MT, Fung AK (1998) Attributes of neural networks for extracting continuous vegetation variables from optical and radar measurements. Int J Remote Sens 19(14):2639–2662
Kira O, Nguy-Robertson A, Arkebauer T, Linker R, Gitelson A (2016) Informative spectral bands for remote green LAI estimation in C3 and C4 crops. Agric For Meteorol 218–219:243–249
Knox N, Skidmore A, Prins H, Asner G, van der Werff H, de Boer W, van der Waal C, de Knegt H, Kohi E, Slotow R, Grant R (2011) Dry season mapping of savanna forage quality, using the hyperspectral Carnegie Airborne Observatory sensor. Remote Sens Environ 115(6):1478–1488
Knyazikhin Y, Schull MA, Stenberg P, Mõttus M, Rautiainen M, Yang Y, Marshak A, Carmona PL, Kaufmann RK, Lewis P et al (2013) Hyperspectral remote sensing of foliar nitrogen content. Proc Natl Acad Sci 110(3):E185–E192
Koetz B, Baret F, Poilvé H, Hill J (2005) Use of coupled canopy structure dynamic and radiative transfer models to estimate biophysical canopy characteristics. Remote Sens Environ 95(1):115–124
Kováč D, Malenovský Z, Urban O, Špunda V, Kalina J, Ač A, Kaplan V, Hanuš J (2013) Response of green reflectance continuum removal index to the xanthophyll de-epoxidation cycle in norway spruce needles. J Exp Bot 64(7):1817–1827
Labate D, Ceccherini M, Cisbani A, De Cosmo V, Galeazzi C, Giunti L, Melozzi M, Pieraccini S, Stagi M (2009) The PRISMA payload optomechanical design: a high performance instrument for a new hyperspectral mission. Acta Astronautica 65(9–10):1429–1436
Laurent V, Verhoef W, Damm A, Schaepman M, Clevers J (2013) A Bayesian object-based approach for estimating vegetation biophysical and biochemical variables from APEX at-sensor radiance data. Remote Sens Environ 139:6–17
Laurent V, Schaepman M, Verhoef W, Weyermann J, Chávez R (2014) Bayesian object-based estimation of LAI and chlorophyll from a simulated Sentinel-2 top-of-atmosphere radiance image. Remote Sens Environ 140:318–329
Lazaridis DC, Verbesselt J, Robinson AP (2010) Penalized regression techniques for prediction: a case study for predicting tree mortality using remotely sensed vegetation indices. Can J For Res 41(1):24–34
Lazaro-Gredilla M, Titsias M, Verrelst J, Camps-Valls G (2014) Retrieval of biophysical parameters with heteroscedastic Gaussian processes. IEEE Geosci Remote Sens Lett 11(4):838–842
le Maire G, François C, Dufrene E (2004) Towards universal broad leaf chlorophyll indices using PROSPECT simulated database and hyperspectral reflectance measurements. Remote Sens Environ 89(1):1–28
le Maire G, François C, Soudani K, Berveiller D, Pontailler JY, Bréda N, Genet H, Davi H, Dufrêne E (2008) Calibration and validation of hyperspectral indices for the estimation of broadleaved forest leaf chlorophyll content, leaf mass per area, leaf area index and leaf canopy biomass. Remote Sens Environ 112(10):3846–3864
Lee CM, Cable ML, Hook SJ, Green RO, Ustin SL, Mandl DJ, Middleton EM (2015) An introduction to the NASA hyperspectral infrared imager (HyspIRI) mission and preparatory activities. Remote Sens Environ 167:6–19. https://doi.org/10.1016/j.rse.2015.06.012 (Special Issue on the Hyperspectral Infrared Imager (HyspIRI))
Leonenko G, North P, Los S (2013) Statistical distances and their applications to biophysical parameter estimation: information measures, M-estimates, and minimum contrast methods. Remote Sens 5(3):1355–1388
Lewis P (1999) Three-dimensional plant modelling for remote sensing simulation studies using the botanical plant modelling system. Agronomie 19(3–4):185–210
Lewis P, Muller J (1993) The advanced radiometric ray tracer: ararat for plant canopy reflectance simulation. Int Arch Photogramm Remote Sens 29:26–26
Lewis P, Muller J (1990) Botanical plant modelling for remote sensing simulation studies. In: 10th Annual International on Geoscience and Remote Sensing Symposium, 1990. IGARSS'90.'Remote Sensing Science for the Nineties', IEEE, pp 1739–1742
Li L, Zhang Q, Huang D (2014a) A review of imaging techniques for plant phenotyping. Sensors (Switzerland) 14(11):20078–20111

Li X, Liu X, Liu M, Wu L (2014b) Random forest algorithm and regional applications of spectral inversion model for estimating canopy nitrogen concentration in rice. Yaogan Xuebao/J Remote Sens 18(4):934–945
Li D, Cheng T, Zhou K, Zheng H, Yao X, Tian Y, Zhu Y, Cao W (2017) WREP: a wavelet-based technique for extracting the red edge position from reflectance spectra for estimating leaf and canopy chlorophyll contents of cereal crops. ISPRS J Photogramm Remote Sens 129:103–117
Liang L, Di L, Zhang L, Deng M, Qin Z, Zhao S, Lin H (2015) Estimation of crop LAI using hyperspectral vegetation indices and a hybrid inversion method. Remote Sens Environ 165:123–134
Liang L, Qin Z, Zhao S, Di L, Zhang C, Deng M, Lin H, Zhang L, Wang L, Liu Z (2016) Estimating crop chlorophyll content with hyperspectral vegetation indices and the hybrid inversion method. Int J Remote Sens 37(13):2923–2949
Lin H, Liang L, Zhang L, Du P (2013) Wheat leaf area index inversion with hyperspectral remote sensing based on support vector regression algorithm. Nongye Gongcheng Xuebao/Trans Chin Soc Agric Eng 29(11):139–146
Liu WY, Pan J (2017) A hyperspectral assessment model for leaf chlorophyll content of Pinus massoniana based on neural network. Chin J Appl Ecol 28(4):1128–1136
Locherer M, Hank T, Danner M, Mauser W (2015) Retrieval of seasonal leaf area index from simulated EnMAP data through optimized LUT-based inversion of the PROSAIL model. Remote Sens 7(8):10321–10346
Luo J, Huang W, Zhao J, Zhang J, Zhao C, Ma R (2013) Detecting aphid density of winter wheat leaf using hyperspectral measurements. IEEE J Sel Top Appl Earth Observ Remote Sens 6(2):690–698
Malenovský Z, Ufer C, Lhotakova Z, Clevers J, Schaepman M, Albrechtova J, Cudlin P (2006) A new hyperspectral index for chlorophyll estimation of a forest canopy: area under curve normalised to maximal band depth between 650 and 725 nm. EARSeL EProc 5(2):161–172
Malenovský Z, Homolová L, Zurita-Milla R, Lukeš P, Kaplan V, Hanuš J, Gastellu-Etchegorry JP, Schaepman ME (2013) Retrieval of spruce leaf chlorophyll content from airborne image data using continuum removal and radiative transfer. Remote Sens Environ 131:85–102
Malenovský Z, Turnbull JD, Lucieer A, Robinson SA (2015) Antarctic moss stress assessment based on chlorophyll content and leaf density retrieved from imaging spectroscopy data. New Phytol 208(2):608–624
Malenovský Z, Lucieer A, King DH, Turnbull JD, Robinson SA (2017) Unmanned aircraft system advances health mapping of fragile polar vegetation. Methods Ecol Evol 8(12):1842–1857
Marabel M, Alvarez-Taboada F (2013) Spectroscopic determination of aboveground biomass in grasslands using spectral transformations, support vector machine and partial least squares regression. Sensors (Switzerland) 13(8):10027–10051
Mariotto I, Thenkabail P, Huete A, Slonecker E, Platonov A (2013) Hyperspectral versus multispectral crop-productivity modeling and type discrimination for the HyspIRI mission. Remote Sens Environ 139:291–305
Marshall M, Thenkabail P (2014) Biomass modeling of four leading world crops using hyperspectral narrowbands in support of HyspIRI mission. Photogramm Eng Remote Sens 80(8):757–772
Matthes J, Knox S, Sturtevant C, Sonnentag O, Verfaillie J, Baldocchi D (2015) Predicting landscape-scale CO2 flux at a pasture and rice paddy with long-term hyperspectral canopy reflectance measurements. Biogeosciences 12(15):4577–4594
Matthews M (2011) A current review of empirical procedures of remote sensing in inland and near-coastal transitional waters. Int J Remote Sens 32(21):6855–6899
Meroni M, Colombo R, Panigada C (2004) Inversion of a radiative transfer model with hyperspectral observations for LAI mapping in poplar plantations. Remote Sens Environ 92(2):195–206
Miller J, Hare E, Wu J (1990) Quantitative characterization of the vegetation red edge reflectance 1. An inverted-Gaussian reflectance model. Remote Sens 11(10):1755–1773
Miphokasap P, Honda K, Vaiphasa C, Souris M, Nagai M (2012) Estimating canopy nitrogen concentration in sugarcane using field imaging spectroscopy. Remote Sens 4(6):1651–1670
Mitchell JJ, Glenn NF, Sankey TT, Derryberry DR, Germino MJ (2012) Remote sensing of sagebrush canopy nitrogen. Remote Sens Environ 124:217–223
Montesano P, Rosette J, Sun G, North P, Nelson R, Dubayah R, Ranson K, Kharuk V (2015) The uncertainty of biomass estimates from modeled ICESat-2 returns across a boreal forest gradient. Remote Sens Environ 158:95–109
Morton DC, Nagol J, Carabajal CC, Rosette J, Palace M, Cook BD, Vermote EF, Harding DJ, North PR (2014) Amazon forests maintain consistent canopy structure and greenness during the dry season. Nature 506(7487):221–224

Mulder V, de Bruin S, Schaepman M, Mayr T (2011) The use of remote sensing in soil and terrain mapping—a review. Geoderma 162(1–2):1–19
Mutanga O, Kumar L (2007) Estimating and mapping grass phosphorus concentration in an African savanna using hyperspectral image data. Int J Remote Sens 28(21):4897–4911
Mutanga O, Skidmore AK (2004) Narrow band vegetation indices overcome the saturation problem in biomass estimation. Int J Remote Sens 25(19):3999–4014
Mutanga O, Skidmore AK, Kumar L, Ferwerda J (2005) Estimating tropical pasture quality at canopy level using band depth analysis with continuum removal in the visible domain. Int J Remote Sens 26(6):1093–1108
Myneni R, Maggion S, Iaquinta J, Privette J, Gobron N, Pinty B, Kimes D, Verstraete M, Williams D (1995) Optical remote sensing of vegetation: modeling, caveats, and algorithms. Remote Sens Environ 51(1):169–188
Neinavaz E, Skidmore A, Darvishzadeh R, Groen T (2016) Retrieval of leaf area index in different plant species using thermal hyperspectral data. ISPRS J Photogramm Remote Sens 119:390–401
Neumann C, Förster M, Kleinschmit B, Itzerott S (2016) Utilizing a PLSR-based band-selection procedure for spectral feature characterization of floristic gradients. IEEE J Sel Top Appl Earth Observ Remote Sens 9(9):3982–3996
North P (1996) Three-dimensional forest light interaction model using a monte carlo method. IEEE Trans Geosci Remote Sens 34(4):946–956
North P, Rosette J, Suárez J, Los S (2010) A Monte Carlo radiative transfer model of satellite waveform LiDAR. Int J Remote Sens 31(5):1343–1358
Ollinger S (2011) Sources of variability in canopy reflectance and the convergent properties of plants. New Phytol 189(2):375–394
Omari K, White H, Staenz K, King D (2013) Retrieval of forest canopy parameters by inversion of the proflair leaf-canopy reflectance model using the lut approach. IEEE J Sel Top Appl Earth Observ Remote Sens 6(2):715–723
Oppelt N, Mauser W (2004) Hyperspectral monitoring of physiological parameters of wheat during a vegetation period using avis data. Int J Remote Sens 25(1):145–159
Paruelo J, Tomasel F (1997) Prediction of functional characteristics of ecosystems: a comparison of artificial neural networks and regression models. Ecol Model 98(2–3):173–186
Pasqualotto N, Delegido J, Wittenberghe SV, Verrelst J, Rivera JP, Moreno J (2018) Retrieval of canopy water content of different crop types with two new hyperspectral indices: water absorption area index and depth water index. Int J Appl Earth Observ Geoinf 67:69–78
Peng Y, Huang H, Wang W, Wu J, Wang X (2011) Rapid detection of chlorophyll content in corn leaves by using least squares-support vector machines and hyperspectral images. Jiangsu Daxue Xuebao (Ziran Kexue Ban)/J Jiangsu Univ (Nat Sci Edn) 32(2):125–128+174
Penuelas J, Gamon JA, Fredeen AL, Merino J, Field CB (1994) Reflectance indices associated with physiological changes in nitrogen- and water-limited sunflower leaves. Remote Sens Environ 48(2):135–146
Pham T, Yoshino K, Bui D (2017) Biomass estimation of Sonneratia caseolaris (l.) Engler at a coastal area of Hai Phong city (Vietnam) using ALOS-2 PALSAR imagery and GIS-based multi-layer perceptron neural networks. GISci Remote Sens 54(3):329–353
Pinty B, Gobron N, Widlowski JL, Gerstl S, Verstraete M, Antunes M, Bacour C, Gascon F, Gastellu JP, Goel N, Jacquemoud S, North P, Qin W, Thompson R (2001) Radiation transfer model intercomparison (RAMI) exercise. J Geophys Res D Atmos 106(D11):11937–11956
Pinty B, Widlowski JL, Taberner M, Gobron N, Verstraete M, Disney M, Gascon F, Gastellu JP, Jiang L, Kuusk A, Lewis P, Li X, Ni-Meister W, Nilson T, North P, Qin W, Su L, Tang S, Thompson R, Verhoef W, Wang H, Wang J, Yan G, Zang H (2004) Radiation Transfer Model Intercomparison (RAMI) exercise: results from the second phase. J Geophys Res D Atmos 109(6):D06210
Pôças I, Gonçalves J, Costa P, Gonçalves I, Pereira L, Cunha M (2017) Hyperspectral-based predictive modelling of grapevine water status in the Portuguese Douro wine region. Int J Appl Earth Observ Geoinf 58:177–190
Preidl S, Doktor D (2011) Comparison of radiative transfer model inversions to estimate vegetation physiological status based on hyperspectral data. 3rd workshop on hyperspectral image and signal processing: evolution in remote sensing (WHISPERS)
Pu R, Gong P, Biging GS, Larrieu MR (2003) Extraction of red edge optical parameters from Hyperion data for estimation of forest leaf area index. IEEE Trans Geosci Remote Sens 41(4):916–921
Pullanagari R, Kereszturi G, Yule I (2016) Mapping of macro and micro nutrients of mixed pastures using airborne AisaFENIX hyperspectral imagery. ISPRS J Photogramm Remote Sens 117:1–10

Ramoelo A, Skidmore AK, Schlerf M, Mathieu R, Heitkönig IM (2011) Water-removed spectra increase the retrieval accuracy when estimating savanna grass nitrogen and phosphorus concentrations. ISPRS J Photogramm Remote Sens 66(4):408–417

Rasmussen CE, Williams CKI (2006) Gaussian processes for machine learning. The MIT Press, New York

Rivard B, Feng J, Gallie A, Sanchez-Azofeifa A (2008) Continuous wavelets for the improved use of spectral libraries and hyperspectral data. Remote Sens Environ 112(6):2850–2862

Rivera Caicedo J, Verrelst J, Muñoz-Marí J, Moreno J, Camps-Valls G (2014) Toward a semiautomatic machine learning retrieval of biophysical parameters. IEEE J Sel Top Appl Earth Observ Remote Sens 7(4):1249–1259

Rivera J, Verrelst J, Leonenko G, Moreno J (2013) Multiple cost functions and regularization options for improved retrieval of leaf chlorophyll content and LAI through inversion of the PROSAIL model. Remote Sens 5(7):3280–3304

Rivera J, Verrelst J, Delegido J, Veroustraete F, Moreno J (2014) On the semi-automatic retrieval of biophysical parameters based on spectral index optimization. Remote Sens 6(6):4924–4951

Rivera JP, Verrelst J, Gómez-Dans J, Muñoz Marí J, Moreno J, Camps-Valls G (2015) An emulator toolbox to approximate radiative transfer models with statistical learning. Remote Sens 7(7):9347

Rivera-Caicedo JP, Verrelst J, Muñoz-Marí J, Camps-Valls G, Moreno J (2017) Hyperspectral dimensionality reduction for biophysical variable statistical retrieval. ISPRS J Photogramm Remote Sens 132:88–101

Roelofsen HD, Kooistra L, van Bodegom PM, Verrelst J, Krol J, Witte JPM (2014) Mapping a priori defined plant associations using remotely sensed vegetation characteristics. Remote Sens Environ 140:639–651

Roth K, Roberts D, Dennison P, Alonzo M, Peterson S, Beland M (2015) Differentiating plant species within and across diverse ecosystems with imaging spectroscopy. Remote Sens Environ 167:135–151

Saich P, Lewis P, Disney M, Thackrah G (2002) Comparison of Hymap/E-SAR data with models for optial reflectance and microwave scattering from vegetation canopies. Retriev Bio Geo Phys Parameters SAR Data Land Appl 475:75–80

Sanches I, Souza Filho C, Kokaly R (2014) Spectroscopic remote sensing of plant stress at leaf and canopy levels using the chlorophyll 680 nm absorption feature with continuum removal. ISPRS J Photogramm Remote Sens 97:111–122

Scafutto R, de Souza Filho C, Rivard B (2016) Characterization of mineral substrates impregnated with crude oils using proximal infrared hyperspectral imaging. Remote Sens Environ 179:116–130

Schlerf M, Atzberger C (2006) Inversion of a forest reflectance model to estimate structural canopy variables from hyperspectral remote sensing data. Remote Sens Environ 100(3):281–294

Schlerf M, Atzberger C, Hill J (2005) Remote sensing of forest biophysical variables using HyMap imaging spectrometer data. Remote Sens Environ 95(2):177–194

Schlerf M, Atzberger C, Hill J, Buddenbaum H, Werner W, Schüler G (2010) Retrieval of chlorophyll and nitrogen in Norway spruce (Picea abies L. Karst.) using imaging spectroscopy. Int J Appl Earth Observ Geoinf 12(1):17–26

Shiklomanov A, Dietze M, Viskari T, Townsend P, Serbin S (2016) Quantifying the influences of spectral resolution on uncertainty in leaf trait estimates through a Bayesian approach to RTM inversion. Remote Sens Environ 183:226–238

Sims DA, Gamon JA (2002) Relationships between leaf pigment content and spectral reflectance across a wide range of species, leaf structures and developmental stages. Remote Sens Environ 81(2–3):337–354

Stimson HC, Breshears DD, Ustin SL, Kefauver SC (2005) Spectral sensing of foliar water conditions in two co-occurring conifer species: Pinus edulis and Juniperus monosperma. Remote Sens Environ 96(1):108–118

Suykens J, Vandewalle J (1999) Least squares support vector machine classifiers. Neural Process Lett 9(3):293–300

Thenkabail P, Smith R, De Pauw E (2000) Hyperspectral vegetation indices and their relationships with agricultural crop characteristics. Remote Sens Environ 71(2):158–182

Tian Y, Yao X, Yang J, Cao W, Zhu Y (2011) Extracting red edge position parameters from ground- and space-based hyperspectral data for estimation of canopy leaf nitrogen concentration in rice. Plant Product Sci 14(3):270–281

Tibshirani R (1996) Regression shrinkage and selection via the lasso. J R Stat Soc Ser B (Methodological) 58:267–288

Tuia D, Verrelst J, Alonso L, Pérez-Cruz F, Camps-Valls G (2011) Multioutput support vector regression for remote sensing biophysical parameter estimation. IEEE Geosci Remote Sens Lett 8(4):804–808

Tuia D, Volpi M, Verrelst J, Camps-Valls G (2018) Advances in kernel machines for image classification and biophysical parameter retrieval. In: Moser G, Zerubia J (eds) Mathematical models for remote sensing image processing. Signals and communication technology. Springer, Cham, pp 399–441
Uno Y, Prasher S, Lacroix R, Goel P, Karimi Y, Viau A, Patel R (2005) Artificial neural networks to predict corn yield from compact airborne spectrographic imager data. Comput Electron Agric 47(2):149–161
Vaglio Laurin G, Cheung-Wai Chan J, Chen Q, Lindsell J, Coomes D, Guerriero L, Del Frate F, Miglietta F, Valentini R (2014) Biodiversity mapping in a tropical West African forest with airborne hyperspectral data. PLoS ONE 9(6):e97910
Van Der Tol C, Verhoef W, Timmermans J, Verhoef A, Su Z (2009) An integrated model of soil-canopy spectral radiances, photosynthesis, fluorescence, temperature and energy balance. Biogeosciences 6(12):3109–3129
Van Der Tol C, Berry J, Campbell P, Rascher U (2014) Models of fluorescence and photosynthesis for interpreting measurements of solar-induced chlorophyll fluorescence. J Geophys Res Biogeosci 119:2312–2327
van der Tol C, Rossini M, Cogliati S, Verhoef W, Colombo R, Rascher U, Mohammed G (2016) A model and measurement comparison of diurnal cycles of sun-induced chlorophyll fluorescence of crops. Remote Sens Environ 186:663–677
Vapnik V, Golowich S, Smola A (1997) Support vector method for function approximation, regression estimation, and signal processing. Adv Neural Inf Process Syst 9:281–287
Verhoef W (1984a) Light scattering by leaf layers with application to canopy reflectance modeling: the SAIL model. Remote Sens Environ 16(2):125–141
Verhoef W (1984b) Light scattering by leaf layers with application to canopy reflectance modeling: the SAIL model. Remote Sens Environ 16(2):125–141
Verhoef W (1985) Earth observation modeling based on layer scattering matrices. Remote Sens Environ 17(2):165–178
Verrelst J, Schaepman M, Koetz B, Kneubuhler M (2008) Angular sensitivity analysis of vegetation indices derived from CHRIS/PROBA data. Remote Sens Environ 112(5):2341–2353
Verrelst J, Schaepman ME, Malenovský Z, Clevers JGPW (2010) Effects of woody elements on simulated canopy reflectance: implications for forest chlorophyll content retrieval. Remote Sens Environ 114(3):647–656
Verrelst J, Alonso L, Camps-Valls G, Delegido J, Moreno J (2012a) Retrieval of vegetation biophysical parameters using Gaussian process techniques. IEEE Trans Geosci Remote Sens 50(5 PART 2):1832–1843
Verrelst J, Alonso L, Camps-Valls G, Delegido J, Moreno J (2012b) Retrieval of vegetation biophysical parameters using Gaussian process techniques. IEEE Trans Geosci Remote Sens 50(5 PART 2):1832–1843
Verrelst J, Romijn E, Kooistra L (2012c) Mapping vegetation density in a heterogeneous river floodplain ecosystem using pointable CHRIS/PROBA data. Remote Sens 4(9):2866–2889
Verrelst J, Alonso L, Rivera Caicedo J, Moreno J, Camps-Valls G (2013a) Gaussian process retrieval of chlorophyll content from imaging spectroscopy data. IEEE J Sel Top Appl Earth Observ Remote Sens 6(2):867–874
Verrelst J, Rivera J, Moreno J, Camps-Valls G (2013b) Gaussian processes uncertainty estimates in experimental Sentinel-2 LAI and leaf chlorophyll content retrieval. ISPRS J Photogramm Remote Sens 86:157–167
Verrelst J, Rivera J, Leonenko G, Alonso L, Moreno J (2014) Optimizing LUT-based RTM inversion for semiautomatic mapping of crop biophysical parameters from sentinel-2 and -3 data: role of cost functions. IEEE Trans Geosci Remote Sens 52(1):257–269
Verrelst J, Camps-Valls G, Muñoz Marí J, Rivera J, Veroustraete F, Clevers J, Moreno J (2015) Optical remote sensing and the retrieval of terrestrial vegetation bio-geophysical properties—a review. ISPRS J Photogramm Remote Sens 108:273–290
Verrelst J, Dethier S, Rivera JP, Munoz-Mari J, Camps-Valls G, Moreno J (2016a) Active learning methods for efficient hybrid biophysical variable retrieval. IEEE Geosci Remote Sens Lett 13(7):1012–1016
Verrelst J, Rivera JP, Gitelson A, Delegido J, Moreno J, Camps-Valls G (2016b) Spectral band selection for vegetation properties retrieval using Gaussian processes regression. Int J Appl Earth Observ Geoinf 52:554–567
Verrelst J, Sabater N, Rivera JP, Muñoz Marí J, Vicent J, Camps-Valls G, Moreno J (2016c) Emulation of leaf, canopy and atmosphere radiative transfer models for fast global sensitivity analysis. Remote Sens 8(8):673

Verrelst J, Rivera Caicedo J, Muñoz Marí J, Camps-Valls G, Moreno J (2017) SCOPE-based emulators for fast generation of synthetic canopy reflectance and sun-induced fluorescence Spectra. Remote Sens 9(9):927

Vilfan N, van der Tol C, Muller O, Rascher U, Verhoef W (2016) Fluspect-B: a model for leaf fluorescence, reflectance and transmittance spectra. Remote Sens Environ 186:596–615

Vohland M, Mader S, Dorigo W (2010) Applying different inversion techniques to retrieve stand variables of summer barley with PROSPECT + SAIL. Int J Appl Earth Observ Geoinf 12(2):71–80

Wang F, Huang J, Lou Z (2011) A comparison of three methods for estimating leaf area index of paddy rice from optimal hyperspectral bands. Precis Agric 12(3):439–447

Wang F, Huang J, Wang Y, Liu Z, Peng D, Cao F (2013) Monitoring nitrogen concentration of oilseed rape from hyperspectral data using radial basis function. Int J Digital Earth 6(6):550–562

Wang J, Wang T, Skidmore A, Shi T, Wu G (2015) Evaluating different methods for grass nutrient estimation from canopy hyperspectral reflectance. Remote Sens 7(5):5901–5917

Wang B, Chen J, Ju W, Qiu F, Zhang Q, Fang M, Chen F (2017a) Limited effects of water absorption on reducing the accuracy of leaf nitrogen estimation. Remote Sens 9(3):291

Wang J, Shen C, Liu N, Jin X, Fan X, Dong C, Xu Y (2017b) Non-destructive evaluation of the leaf nitrogen concentration by in-field visible/near-infrared spectroscopy in pear orchards. Sensors (Switzerland) 17(3):538

Widlowski JL, Pinty B, Clerici M, Dai Y, De Kauwe M, De Ridder K, Kallel A, Kobayashi H, Lavergne T, Ni-Meister W, Olchev A, Quaife T, Wang S, Yang W, Yang Y, Yuan H (2011) RAMI4PILPS: an intercomparison of formulations for the partitioning of solar radiation in land surface models. J Geophys Res G Biogeosci. https://doi.org/10.1029/2010JG001511

Widlowski JL, Taberner M, Pinty B, Bruniquel-Pinel V, Disney M, Fernandes R, Gastellu-Etchegorry JP, Gobron N, Kuusk A, Lavergne T, Leblanc S, Lewis P, Martin E, Mättus M, North P, Qin W, Robustelli M, Rochdi N, Ruiloba R, Soler C, Thompson R, Verhoef W, Verstraete M, Xie D (2007) Third radiation transfer model intercomparison (RAMI) exercise: documenting progress in canopy reflectance models. J Geophys Res D Atmos 112(9):D09111

Widlowski JL, Mio C, Disney M, Adams J, Andredakis I, Atzberger C, Brennan J, Busetto L, Chelle M, Ceccherini G et al (2015) The fourth phase of the radiative transfer model intercomparison (RAMI) exercise: actual canopy scenarios and conformity testing. Remote Sens Environ 169:418–437

Wold S, Esbensen K, Geladi P (1987) Principal component analysis. Chemom Intell Lab Syst 2(1–3):37–52

Xue J, Su B (2017) Significant remote sensing vegetation indices: a review of developments and applications. J Sens. https://doi.org/10.1155/2017/1353691

Yang X, Huang J, Wu Y, Wang J, Wang P, Wang X, Huete A (2011) Estimating biophysical parameters of rice with remote sensing data using support vector machines. Sci China Life Sci 54(3):272–281

Yang P, Verhoef W, van der Tol C (2017) The mSCOPE model: a simple adaptation to the SCOPE model to describe reflectance, fluorescence and photosynthesis of vertically heterogeneous canopies. Remote Sens Environ 201:1–11

Yao X, Huang Y, Shang G, Zhou C, Cheng T, Tian Y, Cao W, Zhu Y (2015) Evaluation of six algorithms to monitor wheat leaf nitrogen concentration. Remote Sens 7(11):14939–14966

Ye X, Sakai K, Manago M, Asada SI, Sasao A (2007) Prediction of citrus yield from airborne hyperspectral imagery. Precis Agric 8(3):111–125

Yi Q, Jiapaer G, Chen J, Bao A, Wang F (2014) Different units of measurement of carotenoids estimation in cotton using hyperspectral indices and partial least square regression. ISPRS J Photogramm Remote Sens 91:72–84

Yin T, Lauret N, Gastellu-Etchegorry JP (2016) Simulation of satellite, airborne and terrestrial LiDAR with DART (II): ALS and TLS multi-pulse acquisitions, photon counting, and solar noise. Remote Sens Environ 184:454–468

Yue J, Yang G, Li C, Li Z, Wang Y, Feng H, Xu B (2017) Estimation of winter wheat above-ground biomass using unmanned aerial vehicle-based snapshot hyperspectral sensor and crop height improved models. Remote Sens 9(7):708

Zandler H, Brenning A, Samimi C (2015) Quantifying dwarf shrub biomass in an arid environment: comparing empirical methods in a high dimensional setting. Remote Sens Environ 158:140–155

Zarco-Tejada P, Miller J, Noland T, Mohammed G, Sampson P (2001) Scaling-up and model inversion methods with narrowband optical indices for chlorophyll content estimation in closed forest canopies with hyperspectral data. IEEE Trans Geosci Remote Sens 39(7):1491–1507

Zarco-Tejada P, Miller J, Mohammed G, Noland T, Sampson P (2002) Vegetation stress detection through chlorophyll+ estimation and fluorescence effects on hyperspectral imagery. J Environ Qual 31(5):1433–1441
Zhang S, Wang Q (2015) Inverse retrieval of chlorophyll from reflected spectra for assimilating branches of drought-tolerant Tamarix ramosissima. IEEE J Sel Top Appl Earth Observ Remote Sens 8(4):1498–1505
Zhang Y, Chen J, Miller J, Noland T (2008) Leaf chlorophyll content retrieval from airborne hyperspectral remote sensing imagery. Remote Sens Environ 112(7):3234–3247

Surveys in Geophysics (2019) 40:631–656
https://doi.org/10.1007/s10712-019-09534-y

# Variability and Uncertainty Challenges in Scaling Imaging Spectroscopy Retrievals and Validations from Leaves Up to Vegetation Canopies

Zbyněk Malenovský[1,2,3] · Lucie Homolová[2] · Petr Lukeš[2] · Henning Buddenbaum[4] · Jochem Verrelst[5] · Luis Alonso[5] · Michael E. Schaepman[6] · Nicolas Lauret[7] · Jean-Philippe Gastellu-Etchegorry[7]

Received: 14 November 2017 / Accepted: 16 April 2019 / Published online: 9 May 2019


## Abstract

Imaging spectroscopy of vegetation requires methods for scaling and generalizing optical signals that are reflected, transmitted and emitted in the solar wavelength domain from single leaves and observed at the level of canopies by proximal sensing, airborne and satellite spectroradiometers. The upscaling embedded in imaging spectroscopy retrievals and validations of plant biochemical and structural traits is challenged by natural variability and measurement uncertainties. Sources of the leaf-to-canopy upscaling variability and uncertainties are reviewed with respect to: (1) implementation of retrieval algorithms and (2) their parameterization and validation of quantitative products through in situ field measurements. The challenges are outlined and discussed for empirical and physical leaf and canopy radiative transfer modelling components, considering both forward and inverse modes. Discussion on optical remote sensing validation schemes includes also description of a multiscale validation concept and its advantages. Impacts of intraspecific and interspecific variability on collected field and laboratory measurements of leaf biochemical traits and optical properties are demonstrated for selected plant species, and field measurement uncertainty sources are listed and discussed specifically for foliar pigments and canopy leaf area index. The review concludes with the main findings and suggestions as how to reduce uncertainties and include variability in scaling vegetation imaging spectroscopy signals and functional traits of single leaves up to observations of whole canopies.

**Keywords** Quantitative remote sensing · Imaging spectroscopy · Retrieval of vegetation traits · Radiative transfer models · Inversion · Variability · Uncertainty · Scaling · Multiscale validation

✉ Zbyněk Malenovský
zbynek.malenovsky@gmail.com

Extended author information available on the last page of the article

## 1 Introduction

Imaging spectroscopy, also known as hyperspectral remote sensing, measures the Earth surface radiance, as a function of the wavelength of the electromagnetic spectrum (from visible to long infrared), the illumination and observation geometry (Sun–object–sensor angularity) and temporal characteristics (acquisition date and revisit frequency) of a particular location represented by geographical coordinates and altitude (Malenovský et al. 2007; Schaepman et al. 2009). Consequently, the scale of an imaging spectroscopy observation is a function of the observation spatial, spectral and temporal extent and resolutions. It depends on spatial characteristics of an observed location, the observational geometry (radiance directions), spectral properties of sensed electromagnetic radiation and temporal specifications of a given data acquisition. Spatial scale of a remote sensing observation, given in the case of space-borne image data by the sensor instantaneous field of view (IFOV), might be due to physical and technical constrains different from the scale required by an observer or data user. For example, in imaging spectroscopy of vegetation one is often interested in biochemical traits of plants at the spatial resolution of single leaves, whereas the remotely sensed data represent whole canopies composed of several plant functional types. Consequently, the scaling techniques changing the size of a measurement unit and allowing us to decipher data acquired at one scale into informational content of another scale were established (Dungan 2001). Two general scaling approaches are being recognized: (1) *upscaling*: the bottom-up scaling of detailed information up to coarser (larger) units and (2) *downscaling*: the top-down scaling, which decomposes signals and information collected at broader scales into elementary constituents of smaller proportions (Marceau and Hay 1999). The general concept of scales in imaging spectroscopy of vegetation is further described by Gamon et al. (2019).

An initial error and uncertainty characterization of an imaging spectrometer was proposed by Schaepman et al. (2015). It has been later extended by Hueni et al. (2017) with an error traceability matrix covering several known error sources. The largest uncertainty in their error propagation scheme was assigned to the creation of vegetation products from a standardized reflectance image using an upscaling scheme. Sources of variability and uncertainty in the process of retrieving plant traits from imaging spectroscopy data of vegetation canopies can be divided into the three broader clusters (Fig. 1): cluster I: sensor design, data acquisition and pre-processing (e.g., sensor calibrations, radiometric, geometrical and atmospheric data corrections; Hueni et al. 2013); cluster II: retrieval algorithm including parameterizations and simplifications imbedded in radiative transfer models (Knyazikhin et al. 2013); and cluster III: field data acquired in situ to constrain retrievals in the form of the prior knowledge and also to validate the retrieved remote sensing products (Mussche et al. 2001; Thimonier et al. 2010).

Assessment of uncertainties related to the acquisition and processing of imaging spectroscopy data requires detailed spectral and radiometric characterization of the sensor performance and operational set-up (including Sun–object–sensor illumination-viewing geometry, acquisition conditions in terms of actual atmospheric parameters including cloud cover) and specifications of the processing chain transforming raw image data to the radiance and reflectance products. Although the magnitudes of these uncertainties are instrument-specific, wavelength-dependent and vary in space and time, their sources are well known and can be quantified (Hueni et al. 2017). For instance, Bachmann et al. (2011) designed a full error propagation concept for airborne imaging spectroscopy data and also analysed the potential influences of spectral and radiometric calibration uncertainties on data products of the upcoming

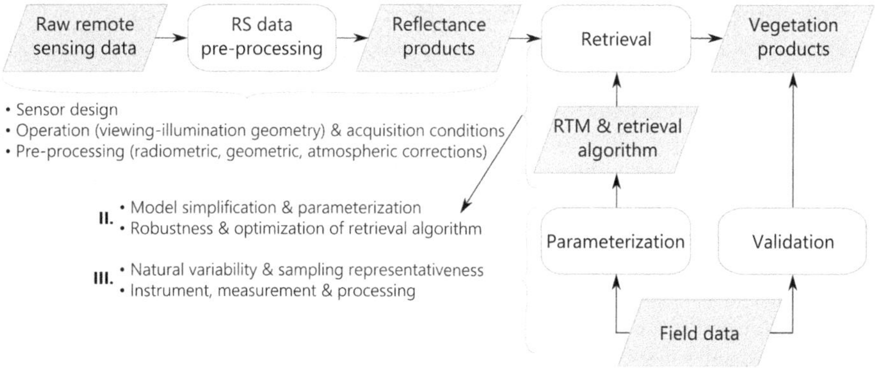


**Fig. 1** Sources of variability and uncertainty encumbering imaging spectroscopy of vegetation originating from: I. sensor design, remote sensing data acquisition and pre-processing, II. design of algorithms retrieving quantitative vegetation traits using empirical and radiative transfer modelling (RTM) and III. parameterization of retrievals and validation of their accuracy using field data

space-borne imaging spectroscopy mission EnMAP (Bachmann et al. 2015). The data quality assessment is part of good practices for many existing operational land-monitoring space-borne multispectral sensors, such as MODIS on Aqua and Terra (Vermote et al. 2002; Lyapustin et al. 2014), ETM+ on Landsat (Feng et al. 2012; Claverie et al. 2015; Vermote et al. 2016) or MSI on Sentinel-2 satellite platforms (Gascon et al. 2017; Gorroño et al. 2017). Due to a high similarity with hyperspectral remote sensing, these practices are adaptable and applicable also on imaging spectroradiometers.

Since cluster I in Fig. 1 (sensor design, data acquisition and pre-processing) does not contain any significant spectral–spatial scaling features, this review focuses on the identification of uncertainty and variability knowledge gaps taking place in the two remaining clusters related to: (1) empirical and physical upscaling retrieval approaches, including radiative transfer modelling (cluster II, Fig. 1) and (2) scaling of field measurements used for parameterization of retrieval algorithms and validation of imaging spectroscopy products (cluster III, Fig. 1). The specific objective is to outline *challenges of natural variability and methodological uncertainty in spectral and spatial (geometrical) scaling of terrestrial vegetation optical properties* (i.e. reflected, transmitted and emitted visible and infrared photon fluxes) from leaves up to canopies. Main attention is paid to retrievals and validation of biochemical and biophysical plant traits (e.g., content of foliar pigments or leaf area index—LAI) obtained from remote sensing imaging spectroscopy data at different spatial scales. Although the review discusses a concept of the multiscale ground air- and space-borne validation scheme, it does not present any quantitative meta-analyses nor error propagation computations, for which actual measurements acquired at several spatial scales simultaneously are required. Also, a temporal scaling of vegetation optical signals is not included, as this topic is extensive and requires a specific publication on its own.

# 2 Upscaling in Retrievals of Vegetation Traits from Imaging Spectroscopy Data

## 2.1 Empirical Approaches Using Statistical Methods

An operationally highly efficient upscaling approach is a direct statistical relationship established between ground measurements of plant traits and imaging spectroscopy observations. This method performs satisfactory for canopies with a vertically homogeneous architecture (e.g., moss and grass turfs, monocultural crops and plantations) that are, however, present less frequently in wild natural landscapes. Upscaling in heterogeneous vegetation canopies that are composed of multiple species, resulting in an irregular distribution of foliage and wood constituents in horizontal and vertical dimensions (e.g., bushlands and forests), requires a more rigorous sampling scheme. Significant investments of time and manpower are needed to collect a sufficient number of representative samples in order to establish a statistically robust regression model. Additionally, the samples must be collected within a short period, close to the time of remote sensing data acquisitions, to ensure the overall temporal consistency. Consequently, the number of samples is in reality often limited by available resources.

To overcome the complexity and high costs of sampling plants in their natural environments, measurements of small-size plant canopies can be for some cases conducted under controlled conditions in the laboratory or in experimental field trials. Such experiments are simpler and time-efficient, but they tend to be specific for a given site, time and plant functional type and consequently difficult to generalize. This approach was applied, for instance, in the study of Buddenbaum et al. (2012, 2015a, b), which investigated the effects of drought stress on young beech (*Fagus sylvatica*) trees under fully controlled irrigation regimes. Proximal hyperspectral images of the beech canopies were recorded once a week during the summer period, when half of the sample trees were cut off from water supply. Leaf reflectance and canopy spectral indices were combined with the partial least square regression (PLSR) (Wold et al. 2001) to upscale leaf chlorophyll and water content of individual leaves to image pixels of beech crowns. In another case study, Kanning et al. (2018) demonstrated the application of drone-based hyperspectral imagery for estimation of LAI and leaf chlorophyll content aiming to assess a potential yield of winter wheat (*Triticum aestivum*). Field measurements of LAI and leaf chlorophyll content were used to calibrate a local PLSR model based on spectral signatures extracted from wheat canopies growing under various *N*-fertilization levels. Direct transfer of an empirical retrieval method established in laboratory to imaging spectroscopy data acquired at different scales and locations has been demonstrated mostly for structurally simple canopies. For example, Malenovský et al. (2015) imposed a gradual water stress on three Antarctic moss species in laboratory-controlled environment. Laboratory measurements of moss traits and complementary spectroscopic measurements were used to train and optimize machine learning algorithms (support vector regressions) estimating chlorophyll content and effective leaf density of moss turfs as the quantitative stress indicators. Subsequent application of the machine learning models on drone-acquired imaging spectroscopy data of two Antarctic moss beds showed an acceptable accuracy of both estimated functional traits (Malenovský et al. 2017). To increase universality and applicability of the empirical upscaling schemes to more complex canopies, various automatized field and laboratory high-throughput plant phenotyping systems combining a large number of multisensor (hyperspectral, chlorophyll fluorescence and thermal) imaging measurements of plants or even small trees have recently been developed (see review by Humplík et al. 2015). A high number of controlled environmental variables in plant phenotyping system allow researchers to build models that are more robust and potentially transferable to other plant types growing at different locations and conditions.

## 2.2 Physically Based Approaches Using Radiative Transfer Models

Mechanistic principals of spectral and spatial scaling effects can be studied with physically based radiative transfer models (RTMs). RTMs are computer algorithms that allow us to scale remotely sensed signals from detailed local to coarser regional and global scales based on our knowledge about light–object physical interactions (Liang 2004). The fundamental role of RTMs is to describe the structural, geometrical and optical effects that modulate remotely sensed signal at different scales. These models are particularly applicable for vegetation canopies, as they are able to simulate the complex nature of vegetation structures leading to specific effects when scaling from leaves to individual plants, fields and stands or whole ecosystems. Reflectance of single leaves is related to canopy hemispherical–directional reflectance function (HDRF) through the combination of structural (e.g., leaf angular distribution, leaf vertical and horizontal clumping and canopy cover causing spatially explicit self-shading effects), optical (reflectance and transmittance of leaves, branches and trunks) and geometrical (solar irradiation and sensor viewing geometries) properties of vegetation remote observations (Knipling 1970). Leaf optical properties simulated with leaf-level RTMs can be physically upscaled in the canopy-level RTMs.

### 2.2.1 Leaf-Level RTMs

The reflection and transmission of light incident on a leaf are determined by biophysical properties of its surface and the inner structures. The incoming photons penetrate deeper into the leaf and interact with its cellular, sub-cellular and intercellular structures, such as mesophyll cells, their organelles and intercellular air spaces. The optical properties of leaves, i.e. amount of absorbed and scattered (reflected and transmitted) light, are driven by the concentrations of light absorbing compounds (e.g., pigment pools, dry matter and water content) and by structural properties of leaves (e.g., thickness and arrangement of inner leaf surfaces). While the specific absorption coefficients of biochemical compounds and refractive index of inner leaf surfaces are usually retrieved from optical measurements, leaf structure can be reconstructed from, for instance, confocal microscopic images of leaf cross sections (Albrechtová et al. 2007). Leaf optical properties can be directly measured (Lukeš et al. 2017; Hovi et al. 2017) as well as modelled using leaf-level RTMs.

One of the most commonly used leaf RTMs is PROSPECT (Jacquemoud and Baret 1990). This model approximates the leaf inner structure by the *plate model* (Allen and Wette 1969, extended for non-compact leaves by Breece and Holmes 1971). The original PROSPECT model requires a relatively small number of input parameters (contents of chlorophylls $a$ and $b$, dry matter and water content, and inner structural parameter $N$), which makes it easy to run in forward and also inverse directions. It was later extended to simulate the influence of leaf carotenoid (Féret et al. 2008) and anthocyanin contents (Féret et al. 2017). It was also incorporated into the FLUSPECT model, which reproduces the chlorophyll fluorescence emission of the photosystems I and II (Vilfan et al. 2016) and spectral manifestation of the photoprotective xanthophyll de-epoxidation cycle (Vilfan et al. 2018). Apart from PROSPECT, Dawson et al. (1998) developed the LIBERTY needle–leaf model to simulate the optical properties of needles, which are anatomically different from plate-like broad leaves. To account for differences in optical properties of the adaxial and abaxial leaf sides, Stuckens et al. (2009) developed the Dorsiventral leaf radiative transfer model (DLM). This model is able to reproduce optical differences between leaf sides based on few additional input parameters related to leaf inner structures.

As demonstrated in numerous studies, the inversion of leaf RTMs allows for non-invasive retrieval of the input model parameters (typically leaf biochemical properties) from measured leaf optical properties (Ustin et al. 2009). For example, Barry et al. (2009) inverted the PROSPECT model to estimate the chlorophyll content of Eucalyptus leaves measured with a spectroradiometer coupled with an integrating sphere. Their chlorophyll retrieval achieved the average root mean square errors (RMSE) of 5 µg $cm^{-2}$ and 3 µg $cm^{-2}$ for juvenile and adult leaves, respectively. Similarly, Malenovský et al. (2006) adjusted and inverted the PROSPECT model for the estimation of its inputs from the optical properties of Norway spruce needles. They minimized the prediction uncertainties in dry matter and water content by implementing the prior information of known inputs' dynamic ranges (i.e. minimal and maximal measured values) and achieved estimations with RMSE of 0.0019 g $cm^{-2}$ and 0.0006 $cm^{-2}$, respectively.

### 2.2.2 Canopy-Level RTMs

A large variety of canopy RTMs have been developed to scale the optical signatures of leaves up to canopy scales (Liang 2004). They can be categorized based on their approach for radiative transfer through vegetation canopies into six groups (Table 1): (1) turbid medium, (2) kernel-driven, (3) spectral invariants, (4) geometrical–optical, (5) discrete geometrical and voxel-based, and (6) Monte Carlo ray-tracing models. Each group has been designed for its specific purpose, which implies that they possess specific conceptual and computational assumptions with different levels of uncertainties. Table 1 summarizes the abilities of the RTM groups to represent canopy spatial and structural heterogeneities. It indicates their computational speed and some examples of the models. It is important to note that Table 1 does not present an exhaustive listing of all existing models. More extensive overview of canopy RTMs is available at, for instance, the Internet portal of the European Council Joint Research Centre (http://rami-benchmark.jrc.ec.europa.eu/HTML/Home.php), which presents a performance comparison of various RTMs carried out within the systematic radiative transfer model intercomparison (RAMI) exercise (Widlowski et al. 2007, 2013, 2015).

Quantitative assessment of forward modelling uncertainty can be conducted through the systematic sensitivity analysis, where the most accurate modelling solution, e.g., photon ray-tracing simulation applied on a detailed vegetation representation, is taken as a reference. The reference results are then compared to outcomes of simulations with a systematically varying input parameterization, more generalized assumptions or increased levels of simplification. Apart from the direct comparison exercise, the global sensitivity analysis (GSA) can be performed to identify the most important inputs that modulate strongly the spectral outputs of RTMs (Verrelst and Rivera-Caicedo 2017). However, GSA is applicable only in case of a simple model with a limited number of input variables (e.g., SAIL; Verhoef 1984). As the theoretical uncertainty of RTMs increases with their increasing complexity, it can be reduced in complex models only by including available prior information. Consequently, the paradigm that the more detailed model is capable of producing more realistic and accurate simulations might not be true, if the uncertainty behind the model inputs (variables and a prior knowledge)

**Table 1** Characteristics of canopy-level radiative transfer models (RTMs) and their spatial error sources

| Type of RTM | Spatial representation of canopy | Ability to capture canopy heterogeneity | Computational speed (Cause) | Examples (Reference) |
|---|---|---|---|---|
| Turbid medium | Horizontally infinite layers of infinitely small leaves with a random spatial distribution and specific statistical angular functions | Cannot describe heterogeneity of forest canopies (contains only an empirical hotspot description) | Very fast (only few input parameters) | SAIL (Verhoef 1984), SCOPE (Van Der Tol et al. 2009) |
| Kernel-driven | Semi-empirical kernels representing isotropic, volumetric and geometrical scattering by 3D objects | The generic geometrical kernel that accounts for shadowing effects between tree crowns (and also for the hot spot effect) | Very fast (reasonable number of input parameters) | Ross-Thick Li-Sparse (Schaaf et al. 2002), Rahman–Pinty–Verstraete (Rahman et al. 1993) |
| Spectral invariants | Spectral invariants—photon recollision probability and directional escape factor, which require a high canopy closure | Semi-empirical approach with physical representation of multiple scatterings, applicable specially in coniferous stands | Very fast (inputs are represented by a small number of spectral invariants) | PARAS (Rautiainen and Stenberg 2005) |
| Geometrical–optical | Geometrically explicit but simple 3D objects with defined shapes and optical properties | Specifically designed to model discontinuous canopies (e.g., forests), but do not have a vertical distribution of biomass | Fast (only geometrical–optical calculations of sunlit and shaded canopy fractions) | FRT (Kuusk et al. 2014), FLIGHT (North 1996) |
| Discrete geometrical (voxel-based flux tracking) | Geometrically explicit 3D objects or voxel representation of landscapes. Each voxel has specific optical (scattering) properties | Exact parameterization of complex vegetation structures, including horizontal and vertical biomass distribution and topography | Computationally intensive (high number of input parameters, the computational time increases with increasing size of simulated scene) | DART (Gastellu-Etchegorry et al. 1996, 2017) |
| Spatially explicit Monte Carlo ray-tracing | Geometrically explicit or voxel 3D representation of objects | Depends on the detail of 3D structures of vegetation canopy created in the model | Computationally highly intensive (high number of input parameters, simulates all photon–object interactions) | Raytran (Govaerts and Verstraete 1998), Librat (Lewis 1999) |

is high. On the other hand, it is important to highlight that novel progressive techniques, such as close-range (ground-, tower- or drone-based) laser scanning (Schneider et al. 2014, 2019; Wallace et al. 2016; Janoutová et al. 2019) and proximal high-resolution imaging spectroscopy (Malenovský et al. 2015, 2017; Wyber et al. 2017), can satisfy the demands of complex RTMs for detailed unbiased inputs and prior information on plant structural, optical and biochemical properties.

A combination of leaf and canopy RTMs in an inverse mode has been frequently applied on imaging spectroscopy data to retrieve various model input parameters, i.e. leaf-level biochemical or canopy structural traits (Schlerf and Atzberger 2006; Homolová et al. 2013; Schneider et al. 2017; Verrelst et al. 2019). The two basic model inversion approaches are: (1) direct mathematical inversion (Laurent et al. 2013, 2014) and (2) indirect approach based on pre-simulated spectral look-up tables (Kimes et al. 2000). In the case of direct inversion, independent simulations are performed for each image pixel separately and model inputs of the most optimal radiative transfer solution are being sought. As the model inversion needs to be run multiple times, the method is not suitable for computationally intensive RTMs, but it allows each retrieval to be performed for different sets of model input parameters of simpler RTMs. This may increase the retrieval accuracy, especially if the site-specific prior information is included. In case of the indirect approach, all model runs are executed and stored in a pre-computed spectral database called *look-up table*. The best fit between the measured and simulated signal is then determined later using optimization or regression algorithms, such as minimization of the merit functions, artificial neural networks, genetic algorithms and many others. Since the inversion algorithms may differ in the retrieval accuracy, their appropriate selection is equally important to the selection of the suitable RTMs (Rivera et al. 2013). Selection of inappropriate model or incorrect set-up of the retrieval algorithm may result in an inversion ambiguity caused by the fact that the different combinations of model inputs result in identical spectral simulations. The ambiguity, also called a mathematically ill-posed problem, can be reduced or even removed by incorporating a prior knowledge on the simulated vegetation canopy (Combal et al. 2002).

Retrievals of vegetation traits via inversion of RTMs have been successfully applied at different spatial scales ranging from individual tree crowns and canopies of forest stands and crops up to global ecosystems. For example, Jay et al. (2016) achieved a high accuracy when retrieving leaf traits from close-range hyperspectral images of five tree species using the PROSPECT leaf model coupled with the COSINE close-range model accounting for specular reflection at leaf surface and local leaf orientation. Malenovský et al. (2013) used adjusted PROSPECT implemented in the discrete anisotropic radiative transfer (DART) model to retrieve leaf chlorophyll content per crown of a Norway spruce (*Picea abies*) forest stand. They generated a RTMs *look-up table* in order to design a new chlorophyll-sensitive and LAI-insensitive optical index, which applied to airborne imaging spectroscopy data showed similar retrieval RMSE of 2.42 μg $cm^{-2}$ as an artificial neural network retrieval. Atzberger and Richter (2012) used the PROSPECT-SAIL model for inversion of crops traits from simulated Sentinel-2 satellite spectral observations. To reduce inversion ambiguity, they introduced object-based retrieval that takes into account also spatial information of several adjacent pixels. This way, they reduced the RMSE of LAI retrieval from 1.46 to 0.54 $m^2$ $m^{-2}$. Strictly speaking, Sentinel-2 MSI as well as MODIS sensors on the Terra and Aqua satellites are not imaging spectroradiometers, but multispectral sensors. Still, MODIS products of LAI and the fraction of absorbed photosynthetically active radiation (Myneni et al. 2002) provide good examples of global vegetation traits produced operationally using canopy radiative transfer solutions that can be retrieved also from upcoming space-borne spectroscopy images.

Validation of vegetation traits retrieved from optical remotely sensed data is an essential part of the estimation process indicating its fidelity. Deviations of remote sensing products from hypothetically *true* values obtained through collection of ground measurements indicate the *physical* uncertainties (see Sect. 3), whereas uncertainties that cannot be assessed through a direct comparison with reference values are called *theoretical* uncertainties. The uncertainties provide the additional complementary information about potentially oversimplified parameterization, inappropriate assumptions or imperfect optimization of RTMs and retrieval algorithms. The *theoretical* uncertainties can be computed in the form of: (1) spectral residuals or standard deviations, when an ensemble of multiple optimal solutions is retrieved (Rivera et al. 2013) or (2) as a likelihood of values estimated in frame of statistical fuzzy approaches such as Bayesian methods (Cooper and Herskovits 1992). Several powerful retrieval methods originating from the field of Bayesian nonparametric statistic, as, for instance, the family of Gaussian processes regression (GPR; Camps-Valls et al. 2016), can deliver high accuracies and at the same time uncertainty intervals of their predictions (Verrelst et al. 2012). The uncertainty intervals indicate how confident the model is in a per-pixel prediction relative to what it has been presented during its training phase. Uncertainty can be used to mask out areas where the actual estimates exceeded the acceptable threshold of 20%, as proposed by the Global Climate Observing System (GCOS 2011). Additionally, it can be used to check transferability of locally trained models through spatial and temporal scales. As an example, Verrelst et al. (2013) trained and successfully validated a GPR model to estimate the total leaf chlorophyll content, which was based on locally selected training pixels of the space-borne image acquired with the CHRIS/PROBA mission of ESA over agricultural fields at Barrax region in Spain. The trained GPR model was then applied to an airborne imagery of a higher spatial resolution, acquired with a CASI imaging spectroradiometer. Per-pixel uncertainties obtained for selected pixels of both datasets indicated a potentially good transferability of the GPR model in case of selected agricultural crops, but they also revealed a failure of the model to retrieve reliably leaf chlorophyll content across the whole airborne image. This validation identified regions where more training data were required to improve robustness and accuracy of the GPR retrieval. More details about the imaging spectroscopy methods for retrieval vegetation traits can be found in Verrelst et al. (2019).

## 3 Upscaling of Ground Measurements for Validation of Imaging Spectroscopy Products

### 3.1 Remote Sensing Validation Schemes

Remote sensing products retrieved empirically or physically from imaging spectroscopy data need to be accompanied with a proper validation indicating of their reliability and fidelity. It is important to note that collections of data for validation purposes as well as for parameterization of the empirical and the physical model-based retrievals are affected by the same sampling and upscaling issues. Validation can be understood as a process of analytical comparison with reference data measured usually in situ, which is presumed to represent the *true* target value (Justice et al. 2000). Validation can be, in a broader sense, considered as a combination of several activities, including comparison with: (1) precise fiducial reference measurements (when a few but very precise measurements are available), (2) other in situ reference measurements (when more but less precise measurements are available), (3) other satellite products, (4) outputs of models and (5) results from monitoring tools assessing spatial and temporal consistency. The term *truth* should be used with care, while keeping in mind that any ground or laboratory measurement is at best an unbiased traceable representation of the true value with an associated uncertainty (Hueni et al. 2017). The uncertainty is a combination of measurement *precision* (i.e. the closeness of two or more measurements represented by numbers of given digits) and *accuracy* (i.e. the closeness of a measured value to a standard of known value). Final ground-*truth* measurement uncertainties are specific to applied measurement protocols, where a sensor ground sampling distance and swath determine the number and spatial extent of field measurements and a vegetation type and traits under investigation determine the most appropriate sampling scheme, instrumentation and data processing. The first step towards a representative number of samples is creation of an appropriately sized elementary sampling unit (ESU). The ESU scheme is a well-established concept for indirect optical measurements of LAI and leaf inclination distribution of agricultural crops and also forest canopies (Weiss et al. 2004; Baret et al. 2005; Morisette et al. 2006). Still, different sampling and measurement strategies are used for LAI measurement than for quantification of leaf pigments, but also for low and homogenous crops than for tall and heterogeneous forest stands (Weiss et al. 2004). Being carried out by tree climbers (Schlerf et al. 2010), shotguns (Féret and Asner 2011), pole pruners (Singh et al. 2015), line launchers (Collis and Harris 1973) or crossbows (Wang et al. 2016), this sampling might be conducted in a less representative way. Singh et al. (2015) measured several leaf traits in mixed forest stands from the lower, middle and upper crown parts and compared different foliage sampling weighting schemes with the remote sensing signal. Based on the results, they used the foliar biomass per species derived from diameter–biomass relationships to scale foliar traits up to the canopy of each test plot. Despite practical difficulties to trace the field and laboratory measurement uncertainties, direct validation using ground *truth* is regarded as a fundamental effort for assessing fidelity of airborne and space-borne vegetation products.

Over the last two decades, several globally produced vegetation products were derived from time series of medium-resolution satellite sensors such as SPOT/Vegetation (Baret et al. 2007), MODIS (Myneni et al. 2002) or MERIS (Bacour et al. 2006) and significant efforts have been spent in their validation and assessment of their spatial and temporal consistency. The field validation ground-*truth* data have been for a variety of vegetation types obtained through dedicated field campaigns coordinated locally or internationally within operational science data networks. One of the largest international networks that operate worldwide is FLUXNET (www.fluxnet.fluxdata.org; Baldocchi et al. 2001). FLUXNET data have been used to validate gross primary production (GPP), LAI, fraction of absorbed photosynthetically active radiation (FPAR) and albedo products of MODIS (Turner et al. 2006). Extensive regional validation campaigns were conducted, for instance, by the BigFoot project for validation of MODIS vegetation products (Cohen and Justice 1999), the VALERI project for validation of the European Space Agency (ESA) instruments such as MERIS (Baret et al. 2005) or through a series of ESA coordinated campaigns such as SEN2FLEX and SEN3EXP for validation of Sentinel-based products. Although these satellite sensors are multispectral, their validation schemes are adjustable and transferable to upcoming hyperspectral space missions, such as the Italian national mission PRISMA (Amato et al. 2013) or the German national mission EnMAP (Guanter et al. 2015).

The importance of data quality control and uncertainty quantification has been addressed in the Quality Assurance Framework for Earth Observation (QA4EO 2009) created by the Committee on Earth Observation Satellites (CEOS), requesting that all datasets and derived products contain a fully traceable indicator of their quality. Consequently, the

Land Product Validation (LPV; http://lpvs.gsfc.nasa.gov) subgroup of CEOS developed a framework for intercomparison and validation of global land products, such as LAI, FPAR and fraction of vegetation cover (fCover). LPV released the community established best practices and guidelines, which adopted a hierarchical approach with four validation stages (Fernandes et al. 2014). In the first stage, product accuracy is assessed from a small (< 30) set of selected locations and time periods by comparison with ground-*truth* or other suitable reference data. In the second stage, the same validation is carried out for a significant set of locations and times, and spatial and temporal consistency is evaluated globally. In the third stage, the product uncertainties and their structure are quantified in a statistically robust way over multiple locations and times globally. The last fourth stage includes systematic and regular updates of the third stage results to match releases of new products, new versions or simply to monitor the performances of the product as long as the satellite time series expands. LPV is searching for supersites with fully characterized land surfaces and vegetation types, which is a prerequisite for their parameterization in 3D radiative transfer models. These sites will serve as test beds for in situ sampling strategies and product algorithm intercomparisons. The LPV activities also resulted in an On Line Interactive Validation Exercise platform (OLIVE; http://calvalportal.ceos.org/web/olive; Weiss et al. 2014) hosted by ESA. OLIVE permits the validation of global LAI, FPAR and fCover products within the two ensembles of sites: (1) 445 BELMANIP2 sites (Baret et al. 2006), containing selected sites of existing sensor networks (FLUXNET, AERONET, VALERI and BigFoot), and (2) 113 DIRECT sites (Garrigues et al. 2008a, b). The OLIVE platform has a potential to be extended towards biophysical products derived from newer medium-spatial resolution satellite data, e.g., from the OLCI sensor onboard of Sentinel-3. However, validation of vegetation products with a higher spatial resolution and inclusion of new products (e.g., leaf chlorophyll content) would require significant changes in the OLIVE validation reference data structures and procedures.

Validation of vegetation products that are not delivered by any satellite platform on the operational basis (e.g., leaf chlorophyll, carotenoid and water content) are less internationally coordinated and their global validation datasets are rare (e.g., Croft et al. in review). ESA funded in 2018 the Fiducial Reference Measurements for Vegetation (FRM4VEG) project (https://frm4veg.org/), aiming to establishing protocols required for traceable in situ measurements of vegetation-related parameters for validation of Copernicus products from multispectral sensors of Sentinel-2, Sentinel-3 and PROBA-V. The fiducial reference measurements should ensure units' traceability in validation schemes that are independent from retrieval procedures and accompanied by an uncertainty budget. The standardized ground *truth* is planned to be measured for remote sensing products as surface reflectance, FPAR and canopy chlorophyll content.

### 3.2 Multiscale Validation Approach

Since the validation data must be collected within a time frame, for which the validated variable remains unchanged, the ground-*truth* datasets are often limited in size due to resource restrictions and time constraints. As it is hard to collect a representative number of ground measurements in few days after the data acquisition, a validation of global remote sensing products with a limited number of sampling points might be unreliable. One possible way to improve the validation process is to apply a multiscale (multilayer) validation approach. The multiscale validation incorporates local-to-regional high-spatial resolution airborne or satellite data, as an upscaling intermediate layer that is placed between field measurements and a medium-to-low-spatial resolution space-borne product (see Fig. 2). This scheme offers an opportunity to split the validation into two consecutive steps. In the first step, pixels of an airborne high-resolution map are compared and validated against the corresponding *truth* of field measurements. In the second step, the airborne vegetation product is enlarged by new clusters of high-resolution pixels and compared to values of corresponding medium-to-low-resolution satellite pixels. The first validation step may take advantage of the very high spatial resolution (< 1 m) allowing to extract from the field sampling locations, e.g., ESU or sampled tree crowns, only those pixels that are not contaminated by unwanted landscape elements (e.g., bare soil or other surrounding vegetation types (Kükenbrink et al. 2019). Within the second step, one may identify and include a higher number of suitable corresponding validation pixel clusters from airborne and space-borne products, which may potentially increase the representativeness of validation. Additionally, spatial aggregation of high-resolution airborne pixels reassembles spectral information of a low-resolution satellite data more precisely than an integration of spatially limited ground measurements. Practical application of the multiscale validation, however, requires fulfilling of several technical conditions. First, to allow the quantitative assessment, all compared variables must be collected simultaneously, i.e.

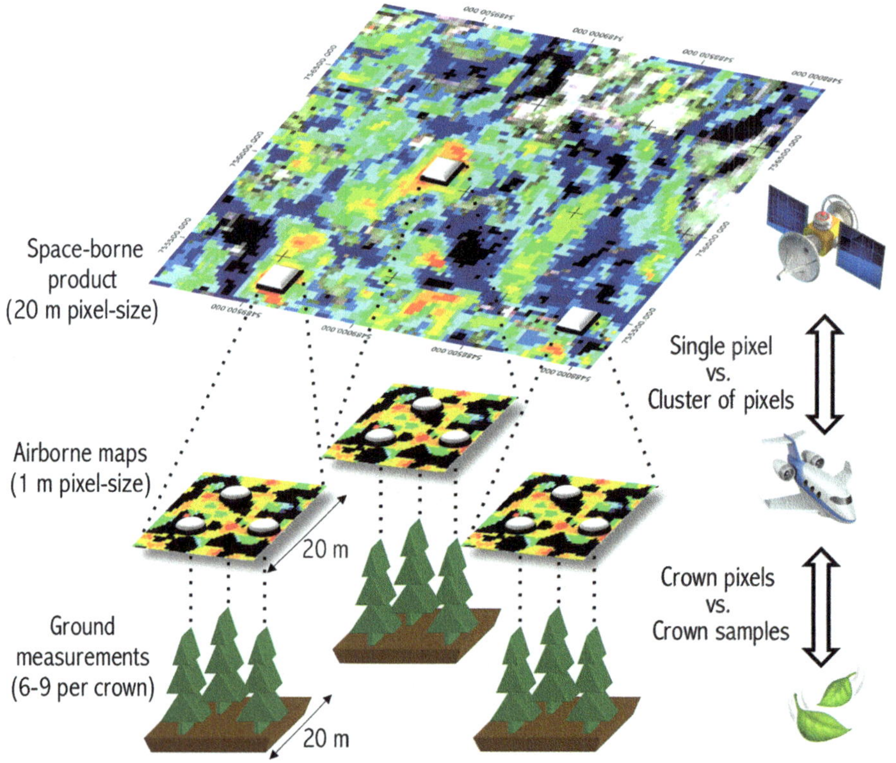


**Fig. 2** Schematic description of a multiscale upscaling strategy for validation of remotely sensed vegetation traits using the three scale levels: (1) ground measurements collected within elementary sampling units (ESU) or from individual trees (i.e. crown samples), (2) clusters of canopy pixels or individual tree crowns (i.e. crown pixels) from a very high-spatial resolution airborne imagery and (3) space-borne pixels of a lower spatial resolution corresponding to pixel clusters of ESU or tree crowns

during the same vegetation growth stage, and have to be expressed in the same physical units, preferably in accordance with the International Systems of Units (SI). Second, all datasets must be spatially co-registered within the same geographical projection system and precisely geocoded with an acceptable positional error (smaller than the data ground sampling distance). Only then can any of the evaluated variables from any location and of any scale be mutually compared. Besides the quantitative accuracy assessment, a spatial consistency of estimated vegetation traits can be assessed from a pattern comparison of the airborne and satellite products. It may reveal the existence of the product spatial inhomogeneity and local anomalies due to insufficiencies in spectral, radiometric, geometrical and atmospheric calibrations and corrections. For instance, a wide field of view of airborne spectroradiometers causes a strong reflectance angular anisotropy, known as the bidirectional reflectance distribution function (BRDF) effect (Weyermann et al. 2014). Since this effect is less prominent in images of orbiting space sensors, one can investigate its impact by comparing both products, preferably accompanied with ground *truth* collected across the whole airborne image swaths.

The potential use and advantages of the multiscale validation scheme has been outlined and demonstrated by Baret et al. (2006) in the case of upscaling the information collected within the standardized ESU to: (1) high-spatial resolution and (2) medium-spatial resolution space-borne observations. Although several multiscale experimental validation campaigns were designed in the similar fashion, e.g., SEN2FLEX (Delegido et al. 2010; Corbari et al. 2013), SEN3EXP (Sobrino et al. 2012) or recently FLEXSense 2018 (Rascher et al. personal communication), the multiscale concept has not been, to our best knowledge, implemented as a standard validation component of any operationally produced global vegetation remote sensing product. Nonetheless, the concept is expected to receive more attention with an increasing affordability and availability of the very high-spatial resolution imaging spectroscopy data produced by manned and recently also unmanned airborne system (UAS) platforms (e.g., Lucieer et al. 2014; Juszak et al. 2017). In particular, UAS platforms might be of a high interest in this sense. They are offering data acquisitions with an unprecedent high detail due to short ground sampling distances and can be deployed multiple times over the same area on demand, which increases representatives and reduces potential errors in their quantitative measurements. Additionally, several studies have demonstrated that UAS is able to provide detailed measurements of vegetation canopy biochemical as well as structural parameters (Zarco-Tejada et al. 2012; Wallace et al. 2016; Malenovský et al. 2017).

### 3.3 Natural Variability and Measurement Uncertainty in Upscaling Validation Schemes

Uncertainty behind in situ measurements of a validation field data is related to the intrinsic natural variability of vegetation parameters in space and time, and to the nature of a measurement method. To illustrate this concept, Fig. 3 shows a hypothetical example of a remote sensing uncertainty assessment demonstrated on five virtual validation sites/scenarios (VS). The measurement and the estimation uncertainties, driven by actual precisions and accuracies, are presented as a purple ellipse. In all cases, the ground-*truth* measurement uncertainty is expected to be smaller than the uncertainty of a corresponding estimate. Hence, the measurement uncertainty is given by an ellipse minor axis and the uncertainty of an estimate is given by its major axis. Horizontal black arrows are showing a real within-site natural variability of a given parameter,

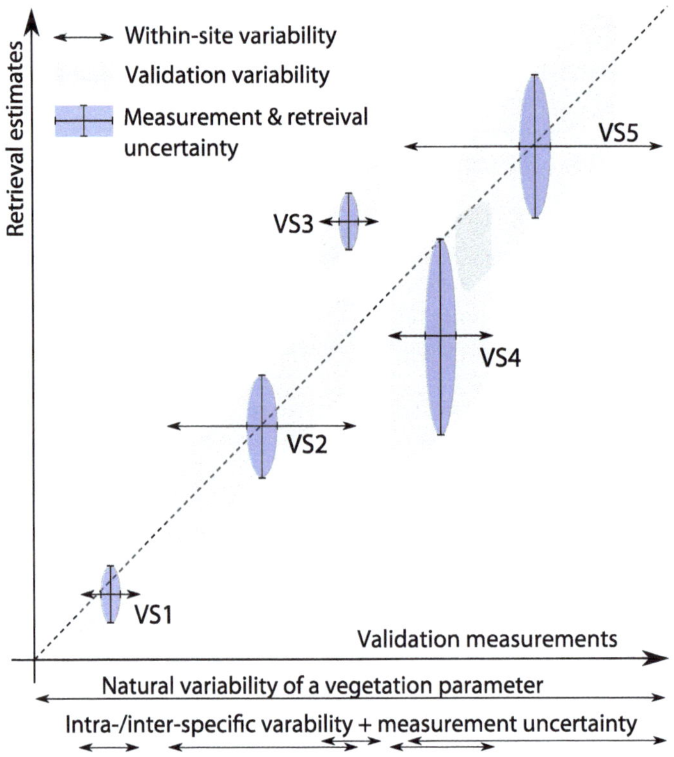


**Fig. 3** Theoretical validation of imaging spectroscopy retrieval estimates for five virtual validation sites (VS; modified according to ESA-ESTEC FLEX Bridge Study Final Report, 2016)

while grey zones illustrate a variability captured in validation measurements collected at given VS. Dashed line indicates the one-to-one line, i.e. the expected best fit between measured and estimated values. When interpreting individual cases shown in Fig. 3, VS1 represents a slightly inaccurate (underestimated) retrieval with a high precision (the retrieval uncertainty interval is small). It is representative for a given site (the within-site variability equals the validation data variability); however, it captures only a small part of the total parameter natural variability. VS2 represents a very accurate but less precise retrieval (the retrieval uncertainty interval is larger than for VS1), which is covering a significant part of a total natural variability; however, it is non-representative for the site. (The within-site variability is larger than the actual variability of validation dataset.) VS3 represents a validation case similar to VS1 (i.e. of a high precision and representative for a given location), but with a low accuracy due to the retrieval overestimation. VS4 is also significantly inaccurate (an underestimated retrieval) and, moreover, most imprecise out of all presented cases (the retrieval uncertainty is the largest); however, it is representative for a given site and covers a significant part of the natural variability. Finally, VS5 represents an ideal scenario, which has a high accuracy without any bias and an acceptable precision. Additionally, VS5 is fully representative for the sampled site and captures a large portion of the existing natural variability. The most accurate, and subsequently successful, validation results should contain cases similar to

VS5 that would be ideally distributed across the entire natural variability of a retrieved vegetation parameter (as shown in Fig. 3).

### 3.3.1 Origins and Impact of Natural Vegetation Variability

Natural variability of vegetation parameters is determined by variations in environmental conditions as light, nutrients and water availability, but also by plant species composition, i.e. species biodiversity. Additionally, an important contribution to the overall variability of leaf-level plant traits, such as specific leaf area, leaf nitrogen, chlorophyll and water content and others, is the intraspecific variability within the single species. While the interspecific variability, i.e. variations between different species, typically explains about 50–90% of the total leaf trait variability (Albert et al. 2010; de Bello et al. 2011), about 20–30% can be explained by the intraspecific variability (Auger and Shipley 2013). Figure 4a shows the interspecific variability in four leaf biochemical traits, serving as inputs into the leaf radiative transfer model PROSPECT (Jacquemoud and Baret 1990), measured for upper-canopy sunlit branches sampled from nine common central European tree species. This graph also illustrates the importance of the sample set representativeness relying on the actual sample size. While the sample collections for spruce and beech species are sufficiently large to capture existing variability of their leaf traits, the size of the remaining leaf sample sets is insufficient ($n \leq 12$) and subsequently diminishing the true natural variability, which is expected to be between 10 and 80 µg of chlorophyll $a+b$ per $cm^2$ of the leaf area. Some tree species exhibit a significant variability in leaf traits within the individuals as an adaptation to canopy light gradients (Niinemets 2010). This is especially prominent in leaves exposed to direct sunlight (sunlit leaves), which contain higher levels of dry matter and photoprotective pigments in comparison with their shaded counterparts (Lichtenthaler et al. 2007). A good example is a shade-tolerant Norway spruce that demonstrates a large variability in leaf traits. Figure 4b shows that except the needle water content, which

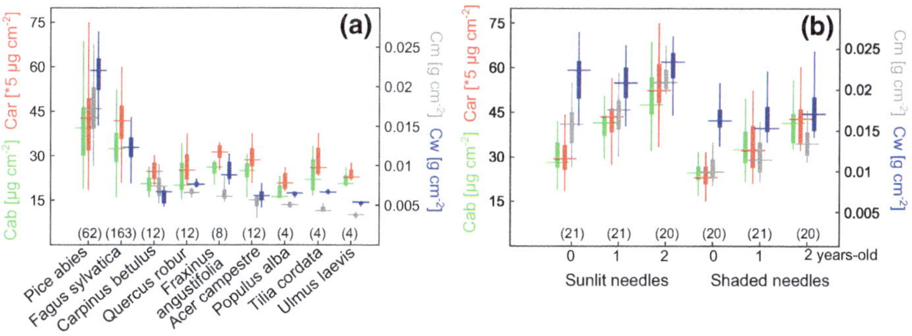


**Fig. 4** Interspecific (**a**) and intraspecific (**b**) species variability of four foliar biochemical traits: chlorophyll $a+b$ (green), total carotenoids (red; ×5), dry matter (grey) and water content (blue). Data were collected at the FLUXNET sites in the Czech Republic (McGloin et al. 2018): Norway spruce site Bílý Kříž (49°30′07.55″N, 18°32′12.74″E; 2016), European beech site Štítná nad Vláří (49°02′09.51″N, 17°58′11.64″E; 2013) and mixed floodplain forest site Lanžhot (48°40′53.57″N, 16°56′46.79″E; 2015). Leaf traits are expressed per hemisurface leaf area (i.e. half of the total leaf area). Horizontal line of each box indicates the median, the bottom and top edges of the box indicate the 25th and 75th percentiles, respectively, and the whiskers extend to 1.5 times of the interquartile range. Numbers in brackets indicate number of collected leaf samples ($n$)

was found to be independent from the age of leaves, all other investigated traits of sunlit and shaded needles respond to the combination of needle age cohorts with a varying availability of direct and diffuse solar radiation.

Differences in leaf biochemistry and inner leaf structure are imprinted in leaf optical properties, i.e. leaf reflectance, transmittance and absorbance (Kokaly et al. 2009), which consequently vary among and within species (Lukeš et al. 2013; Noda et al. 2013; Atherton et al. 2017). Despite identical environmental conditions and similar leaf biochemical properties, the optical properties of the two broadleaf species displayed in Fig. 5, narrow-leafed ash (*Fraxinus angustifolia*) and silver poplar (*Populus alba*), exhibit a different spectral behaviour when comparing the adaxial (upper) and the abaxial (bottom) leaf sides. When considering only the adaxial optical properties, both species are comparable, but the abaxial reflectance of poplar is significantly higher due to the layer of a white scurfy down (i.e. trichomes). This example demonstrates that differences in leaf adaxial and abaxial optical properties, which alternate the canopy reflectance, represent yet another upscaling uncertainty that is seldom considered during ground-*truth* field sampling (Stuckens et al. 2009). This leaf morphological adaptation is of a high importance for plants that are changing the actual leaf angularity in order to protect themselves from adverse environmental conditions of excessive solar irradiation intensity, as, for instance, olive trees (*Olea europaea*).

It is beyond the capabilities of this review to address the full range of variations in plant traits and their spectral properties. Nonetheless, the following suggestions, recommending how to proceed towards better understanding and inclusion of the vegetation-related variability, can be outlined:

- First, combining data from multiple studies and field measurement campaigns into large and open-access databases would help scientists to reveal the *true* size of natural variability in plant traits and spectral properties. It would subsequently become possible to incorporate a large volume of data they contain in remote sensing scaling studies. Several initiatives of such a kind already exist. For instance, the global data-

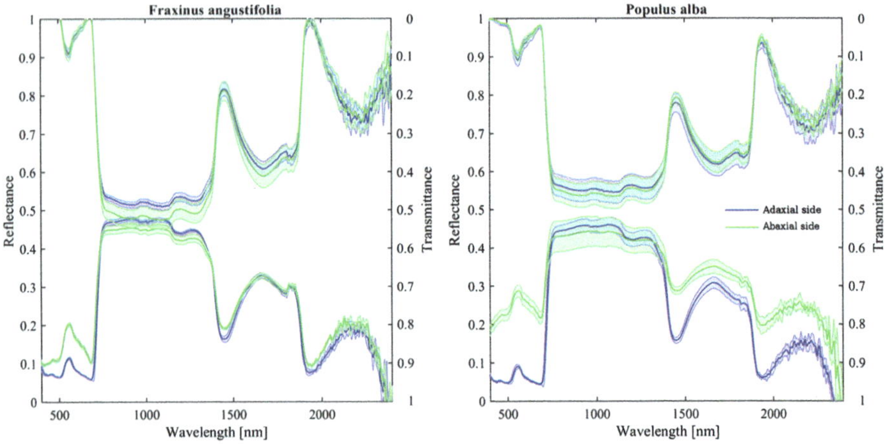


**Fig. 5** Differences in two-sided leaf optical properties, i.e. reflectance (bottom signatures) and transmittance (upper signatures), measured at adaxial side (blue) and abaxial side (green) of two broadleaved species (*Fraxinus angustifolia* and *Populus alba*) sampled at Lanžhot mixed floodplain forest stands (Czech Republic; 48°40′53.57″N, 16°56′46.79″E; 2015)

base of plant traits TRY (https://www.try-db.org, Kattge et al. 2011) has recently been used to derive global maps of specific leaf area, leaf dry matter content, leaf nitrogen, phosphorus concentrations and leaf nitrogen-to-phosphorus ratio (Moreno-Martínez et al. 2018). Field spectroscopy measurements have already been shared for a few years through online spectral databases such as SpecNet (https://specnet.info/, Gamon et al. 2006), EcoSIS (https://ecosis.org/) and Specchio (http://specchio.ch, Hueni et al. 2009). Shared databases, however, require fully standardized protocols for measurements of field traits and optical properties to be compatible.

- Second, a possible way to address large-scale variability in vegetation imaging spectroscopy is to organize multiscale and multidisciplinary measurements campaigns that involve several research teams conducting concurrent field measurements at multiple sites simultaneously with drone-, air- and space-borne imaging spectroscopy acquisitions. Good examples of such joint efforts are multiscale field/flight campaigns supported by ESA in the context of validation of current operational Sentinel missions (e.g., SEN2FLEX in 2005 and SEN3EXP campaign in 2009) and in preparation of the upcoming missions as the FLEX Earth Explorer (e.g., HYFLEX in 2012 or FLEX-Sense campaign in 2018).
- Third, an additional way to better capture plant trait and spectral variability in the physically based retrieval scaling schemes is to improve representation of distinct vegetation functional types in existing RTMs. For example, a genuine architectural and structural representation of heterogenous forest canopies can be efficiently reconstructed from terrestrial laser scans of individual trees or sparse stands (Calders et al. 2015; Schneider et al. 2019). Such a semi-automatic approach allows us to separate non-photosynthetic and photosynthetic tree parts as well as sun- and shade-adapted leaves and adapt accordingly in RTMs vertical distribution of photosynthetic foliar pigment inputs, specifically chlorophylls and carotenoids (e.g., Janoutová et al. 2019).

### 3.3.2 Sources of Field Data Measurement Uncertainty

Most prominent sources of uncertainty in field vegetation measurements are related to sampling methods and to data measurement and post-processing protocols, including technical capabilities of instrumentation. The existing sampling schemes for validation of remote sensing product are outlined in Sects. 3.1 and 3.2. As indicated before, sampling approaches are expected to capture the natural variability of vegetation parameters within an area of interest, but in practice they are limited by available time and resources (manpower and technical equipment). Random sampling methods, which require a minimum prior knowledge on the parameter variability space, assume that collected samples are spatially independent. At the same time, practical constraints and failures in designing the truly random sampling scheme are often enforcing a compromise between a statistically optimal and an experimentally feasible sampling. Therefore, the random sampling methods are less suitable for species rich canopies, where omitting some important species due to the recombination of randomness results in an undersampling (Baraloto et al. 2010). Although no versatile field sampling scheme has been established yet, several attempts were made to standardize sampling schemes and protocols for certain vegetation traits, e.g., the already mentioned VALERI scheme developed for validation of space-borne LAI estimates (Baret et al. 2005).

Several systematic and non-systematic measurement uncertainties of eco-physiologically important vegetation parameters can be reduced only by establishing standardized measurement and data post-processing protocols (Pérez-Harguindeguy et al. 2013) and by using the state-of-the-art, accurate and well-calibrated instrumentation. For example, the uncertainty of optical hand-held devices for quick determination of leaf pigment contents, which are frequently used to validate remote sensing products for crops and broadleaf canopies, is significantly impacted by errors in instrument calibration (Cerovic et al. 2012; Parry et al. 2014). Another example is the determination of canopy or a single-tree crown LAI, which is conducted with several indirect optical methods (Bréda 2003; Jonckheere et al. 2004). LAI measurements by means of hemispherical photography were found to be accurate for short-stature crops (Garrigues et al. 2008a, b), but still challenging in forest canopies, where additional effects of woody elements, foliage clumping and the corresponding appropriate camera exposure settings have to be considered (Zhang et al. 2005; Macfarlane et al. 2007; Thimonier et al. 2010; Liu et al. 2015). Some studies reported disagreement in forest LAI measurements between digital hemispherical photography and other optical indirect instruments, such as the widely used LiCor Plant Canopy Analyser or AcuPAR, especially in dense canopies with a higher LAI (Mussche et al. 2001; Thimonier et al. 2010; Eckrich et al. 2013). The foliage clumping has a pronounced effect on computation of the total LAI (LAIt), especially in coniferous forest stands (Chen et al. 1997). It can enhance the final LAIt uncertainty, if being underrated or, respectively, overrated, especially in forests where trunks are part of its computation. Bao et al. (2018) showed that exclusion of trunks in a spruce (*Picea crassifolia*) forest led to a reduction in the average LAIt by 19.6% and 8.9%, depending on applied clumping estimation method. Hence, the clumping correction coefficient counts for an additional and important error source of LAI measurements that require a standardized measurement protocol.

## 4 Conclusions

Empirical retrievals of vegetation traits from imaging spectroscopy data use direct upscaling approaches, such as statistical regressions established between ground and remotely sensed measurements. Due to their simplicity, they require few inputs and consequently have fewer sources of potential uncertainties. More recent machine learning statistical methods (e.g., Gaussian Processes Regression) can additionally produce assessments of the theoretical uncertainty intervals in the form of the performance likelihood computed during their training phase. However, these methods are tightly linked to the conditions and spatial relations for which they were trained. Hence, their transferability to other locations and applicability on other type of imaging spectroscopy data can potentially result in unacceptable retrieval accuracy. The physically based retrievals that involve radiative transfer simulations of photon interactions with vegetation are in principle more versatile and transferable across scales and space, but inclusion of sophisticated and complex radiative transfer models brings a higher demand on number of input parameters. Field and laboratory measurements, which are collected to satisfy this demand, introduce additional sources of potential uncertainties (i.e. imprecisions and inaccuracies) that have to be quantified and if possible reduced.

The upscaling validation schemes of imaging spectroscopy vegetation products share with the retrieval processes several sources of natural variability and measurement uncertainties, especially those related to field data collections. The magnitude of uncertainty triggered by a number of measured upscaling inputs can be reduced by developing and sharing standardized measurement protocols that use novel and more accurate measurement techniques (e.g., a laser scanning of vegetation structural features). The measurements must be conducted with precise and accurate instrumentation, in which calibrations are refereed to international standards. Additionally, the multiscale ground air- and space-borne validation schemes, carried out within a cooperative international validation campaigns, would increase robustness of vegetation remote sensing products by facilitating direct identification of actual estimation inaccuracies and by collecting sufficient amount of data for a comprehensive computation of imaging spectroscopy error propagation.

Unlike the measurement uncertainty, it is unfeasible and undesirable to diminish the natural variability in plant functional traits. The existing variability must be properly characterized and included in model parameterization and product validation efforts. Regrettably, field measurements often suffer from insufficient sampling designs, resulting in an inadequate cover of the traits' variability. Here, fast technological advancements in high-throughput measurement techniques (as, for instance, rapid repetitive observations from unmanned airborne systems) and an open-access sharing of the plant spectral and functional trait databases can assist scientists with increasing representativeness of ground-*truth* data for upscaling imaging spectroscopy retrievals and validations from leaves to vegetation canopies.

**Acknowledgements** The contribution of ZM was supported by the Australian Research Council Future Fellowship: *Bridging scales in remote sensing of vegetation stress* (FT160100477). The work of LH and PL was supported by the Ministry of Education, Youth and Sports of the Czech Republic by the National Sustainability Program I (NPU I) from Grant Number LO1415. The German Aerospace Center (DLR) and the Federal Ministry of Economics and Technology supported HB within the framework of the EnMAP project (Contract No. 50 EE 1530). JV was supported by the European Research Council (ERC) under the ERC-2017-STG SENTIFLEX project (Grant Agreement 755617). The University of Zurich Research Priority Program on Global Change and Biodiversity (URPP GCB) supported the contribution of MES. The authors acknowledge constructive comments provided by the reviewers that helped to improve the quality of the manuscript.

## Compliance with Ethical Standards

**Conflict of interest** The authors declare that they have no conflict of interest.

## References

Albert CH, Thuiller W, Yoccoz NG, Douzet R, Aubert S, Lavorel S (2010) A multi-trait approach reveals the structure and the relative importance of intra- vs. interspecific variability in plant traits. Funct Ecol 24:1192–1201

Albrechtová J, Janáček J, Lhotáková Z, Radochová B, Kubínová L (2007) Novel efficient methods for measuring mesophyll anatomical characteristics from fresh thick sections using stereology and confocal microscopy: application on acid rain-treated Norway spruce needles. J Exp Bot 58:1451–1461

Allen R, Wette F (1969) Calculation of dynamical surface properties of noble-gas crystals. I. The quasiharmonic approximation. Phys Rev 179:873

Amato U, Antoniadis A, Carfora MF, Colandrea P, Cuomo V, Franzese M, Pignatti S, Serio C (2013) Statistical classification for assessing PRISMA hyperspectral potential for agricultural land use. IEEE J Sel Top Appl Earth Observ Remote Sens 6:615–625

Atherton J, Olascoaga B, Alonso L, Porcar-Castell A (2017) Spatial Variation of leaf optical properties in a boreal forest is influenced by species and light environment. Front Plant Sci 8:309

Atzberger C, Richter K (2012) Spatially constrained inversion of radiative transfer models for improved LAI mapping from future Sentinel-2 imagery. Remote Sens Environ 120:208–218

Auger S, Shipley B (2013) Inter-specific and intra-specific trait variation along short enviromental gradients in an old-growth temperate forest. J Veg Sci 24:419–428

Bachmann M, Adar S, Ben-Dor E, Biesemans J, Briottet X, Grant M, Hanus J, Holzwarth S, Hueni A, Kneubuehler M et al (2011) Towards agreed data quality layers for airborne hyperspectral imagery. In: Proceedings of the 7th EARSeL-SIG-IS, Edinburgh, UK, 11–13 April 2011
Bachmann M, Makarau A, Segl K, Richter R (2015) Estimating the influence of spectral and radiometric calibration uncertainties on EnMAP data products—examples for ground reflectance retrieval and vegetation indices. Remote Sens-Basel 7:10689–10714
Bacour C, Baret F, Béal D, Weiss M, Pavageau K (2006) Neural network estimation of LAI, fAPAR, fCover and LAIxCab, from top of canopy MERIS reflectance data: principles and validation. Remote Sens Environ 105:313–325
Baldocchi DD, Falge E, Gu L, Olson R, Hollinger DY, Running SW, Anthoni P, Bernhofer C, Davis KJ, Evans R, Fuentes J et al (2001) FLUXNET: a new tool to study the temporal and spatial variability of ecosystem-scale carbon dioxide, water vapor, and energy flux densities. Bull Am Meteorol Soc 82:2415–2434
Bao Y, Ni W, Wang D, Yue C, He H, Verbeeck H (2018) Effects of tree trunks on estimation of clumping index and LAI from HemiView and terrestrial LiDAR. Forests 9:144
Baraloto C, Paine TCE, Patiño S, Bonal D, Hérault B, Chave J (2010) Functional trait variation and sampling strategies in species-rich plant communities. Funct Ecol 24:208–216
Baret F, Weiss M, Allard D, Garrigues S, Leroy M, Jeanjean H et al (2005) VALERI: a network of sites and a methodology for the validation of medium spatial resolution land satellite products. Remote Sens Environ 76:36–39
Baret F, Morissette J, Fernandes R, Champeaux JL, Myneni R, Chen J, Plummer S, Weiss M, Bacour C, Garrigue S et al (2006) Evaluation of the representativeness of networks of sites for the global validation and inter-comparison of land biophysical products. Proposition of the CEOS-BELMANIP. IEEE T Geosci Remote 44:1794–1803
Baret F, Hagolle O, Geiger B, Bicheron P, Miras B, Huc M, Berthelot B, Nino F, Weiss M, Samain O et al (2007) LAI, fAPAR and fCover CYCLOPES global products derived from VEGETATION: part 1: principles of the algorithm. Remote Sens Environ 110:275–286
Barry K, Newnham G, Stone C (2009) Estimation of chlorophyll content in Eucalyptus globulus foliage with the leaf reflectance model PROSPECT. Agric For Meteorol 149:1209–1213
Bréda NJJ (2003) Ground-based measurements of leaf area index: a review of methods, instruments and current controversies. J Exp Bot 54:2403–2417
Breece H, Holmes R (1971) Bidirectional scattering characteristics of healthy green soybean and corn leaves in vivo. Appl Opt 10:119–127
Buddenbaum H, Stern O, Stellmes M, Stoffels J, Pueschel P, Hill J, Werner W (2012) Field imaging spectroscopy of beech seedlings under dryness stress. Remote Sens-Basel 4:3721–3740
Buddenbaum H, Rock G, Hill J, Werner W (2015a) Measuring stress reactions of beech seedlings with PRI, fluorescence, temperatures and emissivity from VNIR and thermal field imaging spectroscopy. Eur J Remote Sens 48:263–282
Buddenbaum H, Stern O, Paschmionka B, Hass E, Gattung T, Stoffels J, Hill J, Werner W et al (2015b) Using VNIR and SWIR field imaging spectroscopy for drought stress monitoring of beech seedlings. Int J Remote Sens 36:4590–4605
Calders K, Newnham G, Burt A, Murphy S, Raumonen P, Herold M, Culvenor D, Avitabile V, Disney M, Armston J, Kaasalainen M (2015) Nondestructive estimates of above- ground biomass using terrestrial laser scanning. Methods Ecol Evol 6:198–208
Camps-Valls G, Verrelst J, Muoz-Mar J, Laparra V, Mateo-Jimenez F, Gomez-Dans J (2016) A survey on Gaussian processes for earth observation data analysis. IEEE Geosci Remote Sens Mag 4:58–78
Cerovic ZG, Masdoumier G, Ghozlen NB, Latouche G (2012) A new optical leaf-clip meter for simultaneous non-destructive assessment of leaf chlorophyll and epidermal flavonoids. Physiol Plant 146:251–260
Chen JM, Rich PM, Gower ST, Norman JM, Plummer SE (1997) Leaf area index of boreal forests: theory, techniques and measurements. J Geophys Res 102:29429–429443
Claverie M, Vermote EF, Franch B, Masek JG (2015) Evaluation of the Landsat-5 TM and Landsat-7 ETM+ surface reflectance products. Remote Sens Environ 109:390–403
Cohen W, Justice C (1999) Validating MODIS terrestrial ecology products: linking in situ and satellite measurements. Remote Sens Environ 70:1–4
Collis DG, Harris JWE (1973) Line-throwing gun and cutter for obtaining branches from tree crowns. Can J For Res 3:149–154

Combal B, Baret F, Weiss M, Trubuil A, Macé D, Pragnère A, Myneni R, Knyazikhin Y, Wang L (2002) Retrieval of canopy biophysical variables from bidirectional reflectance. Using prior information to solve the ill-posed inverse problem. Remote Sens Environ 84:1–15

Cooper GF, Herskovits E (1992) A Bayesian method for the induction of probabilistic networks from data. Mach Learn 9:309–347

Corbari C, Sobrino JA, Mancini M, Hidalgo V (2013) Mass and energy flux estimates at different spatial resolutions in a heterogeneous area through a distributed energy-water balance model and remote-sensing data. Int J Remote Sens 34:3208–3230

Croft H, Chen J, Wang R, Mo G, Luo S, Luo X, He L, Gonsamo A, Arabian J, Zhang Y, Simic-Milas A et al The global distribution of leaf chlorophyll content. Remote Sens Environ (in review)

Dawson TP, Curran PJ, Plummer SE (1998) LIBERTY—modeling the effects of leaf biochemical concentration on reflectance spectra. Remote Sens Environ 65:50–60

de Bello F, Lavorel S, Albert CH, Thuiller W, Grigulis K, Dolezal J, Janeček S, Lepš J (2011) Quantifying the relevance of intraspecific trait variability for functional diversity. Methods Ecol Evol 2(2):163–174

Delegido J, Alonso L, González G, Moreno J (2010) Estimating chlorophyll content of crops from hyperspectral data using a normalized area over reflectance curve (NAOC). Int J Appl Earth Obs 12:165–174

Dungan JL (2001) Scaling up and scaling down: the relevance of the support effect on remote sensing of vegetation. In: Tate NJ, Atkinson PM (eds) Modelling scale in geographic information science. Wiley, Chichester, p 277

Eckrich CA, Flaherty EA, Ben-David M (2013) Estimating leaf area index in Southeast Alaska: a comparison of two techniques. PLoS ONE 8:e77642

Feng M, Huang Ch, Channan S, Vermote EF, Masek JG, Townshend JR (2012) Quality assessment of Landsat surface reflectance products using MODIS data. Comput Geosci 38:9–22

Féret J-B, Asner GP (2011) Spectroscopic classification of tropical forest species using radiative transfer modeling. Remote Sens Environ 115:2415–2422

Féret J-B, Francois C, Asner GP, Gitelson AA, Martin RE, Bidel LPR, Ustin SL, le Maire G, Jacquemoud S (2008) PROSPECT-4 and 5: advances in the leaf optical properties model separating photosynthetic pigments. Remote Sens Environ 112:3030–3043

Féret J-B, Gitelson AA, Noble SD, Jacquemoud S (2017) PROSPECT-D: towards modeling leaf optical properties through a complete lifecycle. Remote Sens Environ 193:204–215

Fernandes R, Plummer S, Nightingale J, Baret F, Camacho F, Fang H, Garrigues S, Gobron N, Lang M, Lacaze R, Leblanc S, Meroni M, Martinez B, Nilson T, Pinty B, Pisek J, Sonnentag O, Verger A, Welles J, Weiss M, Widlowski JL (2014) Global leaf area index product validation good practices. Version 2.0. In: Schaepman-Strub G, Román M, Nickeson J (eds) Best practice for satellite-derived land product validation. Land Product Validation Subgroup (WGCV/CEOS), p 76

Gamon JA, Rahman AF, Dungan JL, Schildhauer M, Huemmrich KF (2006) Spectral Network (SpecNet): what is it and why do we need it? Remote Sens Environ 103:227–235

Gamon JA, Somers B, Malenovský Z, Middleton EM, Rascher U, Schaepman ME (2019) Assessing vegetation function with imaging spectroscopy. Surv Geophys. https://doi.org/10.1007/s10712-019-09511-5

Garrigues S, Lacaze R, Baret F, Morisette J, Weiss M, Nickeson J, Fernandes R, Plummer S, Shabanov NV, Myneni R et al (2008a) Validation and intercomparison of global leaf area index products derived from remote sensing data. J Geophys Res 113:G02028

Garrigues S, Shabanov NV, Swanson K, Morisette JT, Baret F, Myneni RB (2008b) Intercomparison and sensitivity analysis of leaf area index retrievals from LAI-2000, AccuPAR, and digital hemispherical photography over croplands. Agric For Meteorol 148:1193–1209

Gascon F, Bouzinac C, Thépaut O, Jung M, Francesconi B, Louis J, Lonjou V, Lafrance B et al (2017) Copernicus Sentinel-2A calibration and products validation status. Remote Sens-Basel 9:584

Gastellu-Etchegorry JP, Demarez V, Pinel V, Zagolski F (1996) Modeling radiative transfer in heterogeneous 3-D vegetation canopies. Remote Sens Environ 58:131–156

Gastellu-Etchegorry JP, Lauret N, Yin T, Landier L, Kallel A, Malenovský Z et al (2017) DART: recent advances in remote sensing data modeling with atmosphere, polarization, and chlorophyll fluorescence. IEEE J Sel Top Appl 10:2640–2649

GCOS (2011) Systematic observation requirements for satellite-based products for climate, 2011 Update, Supplemental Details to the Satellite-Based Component of the Implementation Plan for the Global Observing System for Climate in Support of the UNFCCC. GCOS-154: 138

Gorroño J, Fomferra N, Peters M, Gascon F, Underwood CI, Fox NP, Kirches G, Brockmann C (2017) A radiometric uncertainty tool for the Sentinel 2 mission. Remote Sens-Basel 9:178

Govaerts Y, Verstraete MM (1998) Raytran: a Monte Carlo ray-tracing model to compute light scattering in three-dimensional heterogeneous media. IEEE Trans Geosci Remote 36:493–505
Guanter L, Kaufmann H, Segl K, Foerster S, Rogass C, Chabrillat S, Kuester T, Hollstein A, Rossner G, Chlebek C, Straif C et al (2015) The EnMAP spaceborne imaging spectroscopy mission for earth observation. Remote Sens-Basel 7:8830–8857
Homolová L, Malenovský Z, Clevers JGPW, García-Santos G, Schaepman ME (2013) Review of optical-based remote sensing for plant trait mapping. Ecol Complex 15:1–16
Hovi A, Raitio P, Rautiainen M (2017) A spectral analysis of 25 boreal tree species. Silva Fenn 51:7753
Hueni A, Nieke J, Schopfer J, Kneubühler M, Itten K (2009) The spectral database SPECCHIO for improved long term usability and data sharing. Comput Geosci 37:861–873
Hueni A, Lenhard K, Baumgartner A, Schaepman ME (2013) The APEX (airborne prism experiment-imaging spectrometer) calibration information system. IEEE Trans Geosci Remote Sens 51(11):5169–5180
Hueni A, Damm A, Kneubuehler M, Schläpfer D, Schaepman ME (2017) Field and airborne spectroscopy cross validation—some considerations. IEEE J Sel Top Appl 10:1117–1135
Humplík JF, Lazár D, Husičková A, Spíchal L (2015) Automated phenotyping of plant shoots using imaging methods for analysis of plant stress responses—review. Plant Methods 11:29
Jacquemoud S, Baret F (1990) PROSPECT: a model of leaf optical properties spectra. Remote Sens Environ 34:75–91
Janoutová R, Homolová L, Malenovský Z, Hanuš J, Lauret N, Gastellu-Etchegorry JP (2019) Influence of 3D spruce tree representation on accuracy of airborne and satellite forest reflectance simulated in DART. Forests 10:292
Jay S, Bendoula R, Hadoux X, Féret B, Gorretta N (2016) A physically-based model for retrieving foliar biochemistry and leaf orientation using close-range imaging spectroscopy. Remote Sens Environ 177:220–236
Jonckheere I, Fleck S, Nackaerts K, Muys B, Coppin P, Weiss M, Baret F (2004) Review of methods for in situ leaf area index determination Part I. Theories, sensors and hemispherical photography. Agric For Meteorol 121:19–35
Justice C, Belward A, Morisette J, Lewis P, Privette J, Baret F (2000) Developments in the 'validation' of satellite sensor products for the study of the land surface. Int J Remote Sens 21(17):3383–3390
Juszak I, Iturrate-Garcia M, Gastellu-Etchegorry JP, Schaepman ME, Maximov TC, Schaepman-Strub G (2017) Drivers of shortwave radiation fluxes in Arctic tundra across scales. Remote Sens Environ 193:86–102
Kanning M, Kühling I, Trautz D, Jarmer T (2018) High-resolution UAV-based hyperspectral imagery for LAI and chlorophyll estimations from wheat for yield prediction. Remote Sens 10:2000
Kattge J, Díaz S, Lavorel S, Prentice IC, Leadley P et al (2011) TRY—a global database of plant traits. Glob Change Biol 17:2905–2935
Kimes D, Knyazhikhin Y, Privette J, Abuelgasim A, Gao F (2000) Inversion methods for physically-based models. Remote Sens Rev 18:381–439
Knipling E (1970) Physical and physiological basis for the reflectance of visible and near-infrared radiation from vegetation. Remote Sens Environ 1:155–159
Knyazikhin Y, Schull MA, Stenberg P, Mõttus M, Rautiainen M, Yang Y, Marshak A, Carmona PL, Kaufmann RK, Lewis P, Disney MI et al (2013) Hyperspectral remote sensing of foliar nitrogen content. Proc Natl Acad Sci USA 110:E185–E192
Kokaly RF, Asner GP, Ollinger SV, Martin ME, Wessman CA (2009) Characterizing canopy biochemistry from imaging spectroscopy and its application to ecosystem studies. Remote Sens Environ 113:S78–S91
Kükenbrink D, Hueni A, Schneider FD, Damm A, Gastellu-Etchegorry JP, Schaepman ME, Morsdorf F (2019) Mapping the irradiance field of a single tree: quantifying vegetation induced adjacency effects. IEEE Trans Geosci Remote. https://doi.org/10.1109/tgrs.2019.2895211
Kuusk A, Kuusk J, Lang J (2014) Modeling directional forest reflectance with the hybrid type forest reflectance model FRT. Remote Sens Environ 49:196–204
Laurent VCE, Verhoef W, Damm A, Schaepman ME, Clevers JGPW (2013) A Bayesian object-based approach for estimating vegetation biophysical and biochemical variables from APEX at-sensor radiance data. Remote Sens Environ 139:6–17
Laurent VCE, Schaepman ME, Verhoef W, Weyermann J, Chavez RO (2014) Bayesian object-based estimation of LAI and chlorophyll from a simulated Sentinel-2 top-of-atmosphere radiance image. Remote Sens Environ 140:318–329
Lewis P (1999) Three-dimensional plant modelling for remote sensing simulation studies using the botanical plant modelling system. Agron Agric Environ 19:185–210
Liang S (2004) Quantitative remote sensing of land surfaces. Wiley, Hoboken. ISBN 0-471-28166-2

Lichtenthaler HK, Ač A, Marek MV, Kalina J, Urban O (2007) Differences in pigment composition, photosynthetic rates and chlorophyll fluorescence images of sun and shade leaves of four tree species. Plant Physiol Biochem 45:577–588
Liu Z, Chen JM, Jin G, Qi Y (2015) Estimating seasonal variations of leaf area index using litterfall collection and optical methods in four mixed evergreen-deciduous forests. Agric For Meteorol 209–210:36–48
Lucieer A, Malenovský Z, Veness T, Wallace L (2014) HyperUAS—imaging spectroscopy from a multirotor unmanned aircraft system. J Field Robot 31:571–590
Lukeš P, Stenberg P, Rautiainen M, Mottus M, Vanhatalo K (2013) Optical properties of leaves and needles for boreal tree species in Europe. Remote Sens Lett 4:667–676
Lukeš P, Homolová L, Navrátil M, Hanuš J (2017) Assessing the consistency of optical properties measured in four integrating spheres. Int J Remote Sens 38:3817–3830
Lyapustin A, Wang Y, Xiong X, Meister G, Platnick S, Levy R, Franz B, Korkin S, Hilker T, Tucker J, Hall F, Sellers P, Wu A, Angal A (2014) Scientific impact of MODIS C5 calibration degradation and C6+ improvements. Atmos Meas Tech 7:4353–4365
Macfarlane C, Arndt SK, Livesley SJ, Edgar AC, White DA, Adams MA, Eamus D (2007) Estimation of leaf area index in eucalypt forest with vertical foliage, using cover and fullframe fisheye photography. For Ecol Manag 242:756–763
Malenovský Z, Albrechtová J, Lhotáková Z, Zurita-Milla R, Clevers JGPW, Schaepman ME, Cudlín P (2006) Applicability of the PROSPECT model for Norway spruce needles. Int J Remote Sens 27:5315–5340
Malenovský Z, Bartholomeus HM, Acerbi-Junior FW, Schopfer JT, Painter TH, Epema GF, Bregt AK (2007) Scaling dimensions in spectroscopy of soil and vegetation. Int J Appl Earth Obs 9:137–164
Malenovský Z, Homolová L, Zurita-Milla R, Lukeš P, Kaplan V, Hanuš J, Gastellu-Etchegorry JP, Schaepman ME (2013) Retrieval of spruce leaf chlorophyll content from airborne image data using continuum removal and radiative transfer. Remote Sens Environ 131:85–102
Malenovský Z, Turnbull J, Lucieer A, Robinson SA (2015) Antarctic moss stress assessment based on chlorophyll, water content, and leaf density retrieved from imaging spectroscopy data. New Phytol 208:608–624
Malenovský Z, Lucieer A, King D, Turnbull J, Robinson SA (2017) Unmanned aircraft system advances health mapping of fragile polar vegetation. Methods Ecol Evol 8:1842–1857
Marceau DJ, Hay GJ (1999) Remote sensing contributions to the scale issue. Can J Remote Sens 25:357–366
McGloin R, Šigut L, Havránková K, Dušek J, Pavelka M, Sedlák P (2018) Energy balance closure at a variety of ecosystems in Central Europe with contrasting topographies. Agric For Meteorol 248:418–431
Moreno-Martínez Á, Camps-Valls G, Kattge J, Robinson N, Reichstein M et al (2018) A methodology to derive global maps of leaf traits using remote sensing and climate data. Remote Sens Environ 218:69–88
Morisette JT, Baret F, Privette JL, Myneni RB, Nickeson JE, Garrigues S, Shabanov NV, Weiss M, Fernandes RA, Leblanc SG et al (2006) Validation of global moderate-resolution LAI products: a framework proposed within the CEOS land product validation subgroup. IEEE Trans Geosci Remote 44:1804–1817
Mussche S, Samson R, Nachtergale L, De Schrijver A, Lemeur R, Lust N (2001) A comparison of optical and direct methods for monitoring the seasonal dynamics of leaf area index in deciduous forests. Silva Fenn 35:373–384
Myneni RB, Hoffman S, Knyazikhin Y, Privette JL, Glassy J, Tian Y, Wang Y, Song X, Zhang Y, Smith GR et al (2002) Global products of vegetation leaf area and absorbed PAR from year one of MODIS data. Remote Sens Environ 83:214–231
Niinemets Ü (2010) A review of light interception in plant stands from leaf to canopy in different plant functional types and in species with varying shade tolerance. Ecol Res 25:693–714
Noda HM, Motohka T, Murakamii K, Muraoka H, Nasahara KN (2013) Accurate measurement of optical properties of narrow leaves and conifer needles with a typical integrating sphere and spectroradiometer. Plant Cell Environ 36:1903–1909
North PRJ (1996) Three-dimensional forest light interaction model using a Monte Carlo method. IEEE Trans Geosci Remote 34:946–956
Parry C, Blonquist JM, Bugbee B (2014) In situ measurement of leaf chlorophyll concentration: analysis of the optical/absolute relationship. Plant Cell Environ 37:2508–2520
Pérez-Harguindeguy N, Díaz S, Garnier E, Lavorel S, Poorter H, Jaureguiberry P, Bret-Harte MS et al (2013) New handbook for standardised measurement of plant functional traits worldwide. Aust J Bot 61:167–234

QA4EO Task Team (2009) A quality assurance framework for Earth Observation: Principles. http://qa4eo.org/docs/QA4EO_Principles_v4.0.pdf. Accessed 20 Sept 2017
Rahman H, Verstraete M, Pinty B (1993) Coupled surface-atmosphere reflectance (CSAR) model 1. Model description and inversion on synthetic data. J Geophys Res 98:779–789
Rautiainen M, Stenberg P (2005) Application of photon recollision probability in coniferous canopy reflectance simulations. Remote Sens Environ 96:98–107
Rivera JP, Verrelst J, Leoneko G, Moreno J (2013) Multiple cost functions and regularization options for improved retrieval of leaf chlorophyll content and LAI through inversion of the PROSAIL model. Remote Sens 5:3280–3304
Schaaf C, Gao F, Strahler A, Lucht W, Li X, Tsang T, Strugnell N, Zhang X, Jin Y et al (2002) First operational BRDF, albedo nadir reflectance products from MODIS. Remote Sens Environ 83:135–148
Schaepman ME, Ustin SL, Plaza AJ, Painter TH, Verrelst J, Liang S (2009) Earth system science related imaging spectroscopy—an assessment. Remote Sens Environ 113:S123–S137
Schaepman ME, Jehle M, Hueni A, D'Odorico P, Damm A, Weyerman J, Schneider F, Laurent V, Popp C, Seidel F et al (2015) Advanced radiometry measurements and Earth science applications with the Airborne Prism Experiment (APEX). Remote Sens Environ 158:207–219
Schlerf M, Atzberger C (2006) Inversion of a forest reflectance model to estimate structural canopy variables from hyperspectral remote sensing data. Remote Sens Environ 100:281–294
Schlerf M, Atzberger C, Hill J, Buddenbaum H, Werner W, Schüler G (2010) Retrieval of chlorophyll and nitrogen in Norway spruce (*Picea abies* L. Karst.) using imaging spectroscopy. Int J Appl Earth Obs 12:17–26
Schneider FD, Leiterer R, Morsdorf F, Gastellu-Etchegorry JP, Lauret N, Pfeifer N, Schaepman ME (2014) Simulating imaging spectrometer data: 3D forest modeling based on LiDAR and in situ data. Remote Sens Environ 152:235–250
Schneider FD, Morsdorf F, Schmid B, Petchey OL, Hueni A, Schimel DS, Schaepman ME (2017) Mapping functional diversity from remotely sensed morphological and physiological forest traits. Nat Commun 8:1441
Schneider FD, Kükenbrink D, Schaepman ME, Schimel DS, Morsdorf F (2019) Quantifying 3D structure and occlusion in dense tropical and temperate forests using close-range LiDAR. Agric For Meteorol 268:249–257
Singh A, Serbin SP, McNeil BE, Kingdon CC, Townsend PA (2015) Imaging spectroscopy algorithms for mapping canopy foliar chemical and morphological traits and their uncertainties. Ecol Appl 25:2180–2197
Sobrino JA, Franch B, Mattar C, Jiménez-Muñoz JC, Corbari C (2012) A method to estimate soil moisture from Airborne Hyperspectral Scanner (AHS) and ASTER data: application to SEN2FLEX and SEN3EXP campaigns. Remote Sens Environ 117:415–428
Stuckens J, Verstraeten W, Delalieux S, Swennen R, Coppin P (2009) A dorsiventral leaf radiative transfer model: development, validation and improved model inversion techniques. Remote Sens Environ 113:2560–2573
Thimonier A, Sedivy I, Schleppi P (2010) Estimating leaf area index in different types of mature forest stands in Switzerland: a comparison of methods. Eur J For Res 129:543–562
Turner D, Ritts W, Zhao M, Kurc S, Dunn A, Wofsy S, Small E, Running S (2006) Assessing interannual variation in MODIS-based estimates of gross primary production. IEEE Trans Geosci Remote 44:1899–1907
Ustin SL, Gitelson AA, Jacquemoud S, Schaepman ME, Asner GP, Gamon JA, Zarco-Tejada P (2009) Retrieval of foliar information about plant pigment systems from high resolution spectroscopy. Remote Sens Environ 113:S67–S77
Van Der Tol C, Verhoef W, Timmermans J, Verhoef A, Su Z (2009) An integrated model of soil-canopy spectral radiances, photosynthesis, fluorescence, temperature and energy balance. Biogeosciences 6(12):3109–3129
Verhoef W (1984) Light scattering by leaf layers with application to canopy reflectance modeling: the SAIL model. Remote Sens Environ 16:125–141
Vermote EF, Saleous NZ, Justice CO (2002) Atmospheric correction of MODIS data in the visible to middle infrared: first results. Remote Sens Environ 83:97–111
Vermote E, Justice C, Claverie M, Franch B (2016) Preliminary analysis of the performance of the Landsat 8/OLI land surface reflectance product. Remote Sens Environ 185:46–56
Verrelst J, Rivera-Caicedo J (2017) A global sensitivity analysis toolbox to quantify drivers of vegetation radiative transfer models. In: Petropoulos G, Srivastava PK (eds) sensitivity analysis in earth observation modelling. Elsevier, Amsterdam, pp 319–339

Verrelst J, Alonso L, Camps-Valls G, Delegido J, Moreno J (2012) Retrieval of vegetation biophysical parameters using Gaussian process techniques. IEEE Trans Geosci Remote Sens 50:1832–1843
Verrelst J, Alonso L, Rivera-Caicedo J, Moreno J, Camps-Valls G (2013) Gaussian process retrieval of chlorophyll content from imaging spectroscopy data. IEEE J Sel Top Appl Earth Observ Remote Sens 6:867–874
Verrelst J, Malenovský Z, van der Tol C, Camps-Valls G, Gastellu-Etchegorry JP, Lewis P, North P, Moreno J (2019) Quantifying vegetation biophysical variables from imaging spectroscopy data: a review on retrieval methods. Surv Geophys. https://doi.org/10.1007/s10712-018-9478-y
Vilfan N, van der Tol C, Muller O, Rascher O, Verhoef W (2016) Fluspect-B: a model for leaf fluorescence, reflectance and transmittance spectra. Remote Sens Environ 186:596–615
Vilfan N, van der Tol C, Yang P, Wyber R, Malenovský Z, Robinson SA, Verhoef W (2018) Extending Fluspect to simulate xanthophyll driven leaf reflectance dynamics. Remote Sens Environ 211:345–356
Wallace L, Lucieer A, Malenovský Z, Turner D, Vopěnka P (2016) Assessment of forest structure using two UAV techniques: a comparison of airborne laser scanning and structure from motion (SfM) point clouds. Forests 7:62
Wang Z, Wang T, Darvishzadeh R, Skidmore AK, Jones S, Suarez L, Woodgate W, Heiden U, Heurich M, Hearne J (2016) Vegetation indices for mapping canopy foliar nitrogen in a mixed temperate forest. Remote Sens-Basel 8:491
Weiss M, Baret F, Smith GJ, Jonckheere I, Coppin P (2004) Review of methods for in situ leaf area index (LAI) determination Part II. Estimation of LAI, errors and sampling. Agric For Meteorol 121:37–53
Weiss M, Baret F, Block T, Koetz B, Burini A, Scholze B, Lecharpentier P, Brockmann C, Fernandes R et al (2014) On Line Validation Exercise (OLIVE): a web based service for the validation of medium resolution land products. Application to FAPAR products. Remote Sens-Basel 6:4190–4216
Weyermann J, Damm A, Kneubühler M, Schaepman ME (2014) Correction of reflectance anisotropy effects of vegetation on airborne spectroscopy data and derived products. IEEE Trans Geosci Remote 52:616–627
Widlowski JL, Taberner M, Pinty B, Bruniquel-Pinel V, Disney M, Fernandes R, Gastellu-Etchegorry JP, Gobron N, Kuusk A, Lavergne T, Leblanc S et al (2007) The third RAdiation transfer Model Intercomparison (RAMI) exercise: documenting progress in canopy reflectance modelling. J Geophys Res 112(D09111):28
Widlowski JL, Pinty B, Lopatka M, Atzberger C, Buzica D, Chelle M, Disney M, Gastellu-Etchegorry JP, Gerboles M, Gobron N, Grau E et al (2013) The fourth radiation transfer model intercomparison (RAMI-IV): proficiency testing of canopy reflectance models with ISO-13528. J Geophys Res 118(D09111):13
Widlowski JL, Mio C, Disney M, Adams J, Andredakis I, Atzberger C, Brennan J, Busetto L, Chelle M, Ceccherini G, Colombo R et al (2015) The fourth phase of the radiative transfer model intercomparison (RAMI) exercise: actual canopy scenarios and conformity testing. Remote Sens Environ 169:418–437
Wold S, Sjöström M, Eriksson L (2001) PLS-regression: a basic tool of chemometrics. Chemometr Intell Lab 58:109–130
Wyber R, Malenovský Z, Ashcroft M, Osmond B, Robinson S (2017) Do daily and seasonal trends in leaf solar induced fluorescence reflect changes in photosynthesis, growth or light exposure? Remote Sens-Basel 9:604
Zarco-Tejada PJ, González-Dugo V, Berni JAJ (2012) Fluorescence, temperature and narrow-band indices acquired from a UAV platform for water stress detection using a micro-hyperspectral imager and a thermal camera. Remote Sens Environ 117:322–337
Zhang Y, Chen JM, Miller JR (2005) Determining digital hemispherical photograph exposure for leaf area index estimation. Agric For Meteorol 133(1–4):166–181

**Publisher's Note** Springer Nature remains neutral with regard to jurisdictional claims in published maps and institutional affiliations.

## Affiliations

**Zbyněk Malenovský[1,2,3] · Lucie Homolová[2] · Petr Lukeš[2] · Henning Buddenbaum[4] · Jochem Verrelst[5] · Luis Alonso[5] · Michael E. Schaepman[6] · Nicolas Lauret[7] · Jean-Philippe Gastellu-Etchegorry[7]**

1 Surveying and Spatial Sciences Group, School of Technology, Environments and Design, University of Tasmania, Private Bag 76, Hobart, TAS 7001, Australia

2 Global Change Research Institute CAS, Remote Sensing Department, Bělidla 986/4a, 603 00 Brno, Czech Republic

3 USRA/GESTAR, NASA Goddard Space Flight Center, Biospheric Sciences Laboratory, 8800 Greenbelt Rd, Greenbelt, MD 20771, USA

4 Environmental Remote Sensing and Geoinformatics, Trier University, 54286 Trier, Germany

5 Image Processing Laboratory (IPL), Parc Científic, Universitat de València, 46980 Paterna, Valencia, Spain

6 Remote Sensing Laboratories, Department of Geography, University of Zurich, Winterthurerstrasse 190, 8057 Zurich, Switzerland

7 Centre d'Etudes Spatiales de la Biosphère - UPS, CNES, CNRS, IRD, Université de Toulouse, 31401 Toulouse Cedex 9, France

Surveys in Geophysics (2019) 40:657–687
https://doi.org/10.1007/s10712-018-9485-z

# Synergies of Spaceborne Imaging Spectroscopy with Other Remote Sensing Approaches

Luis Guanter[1] · Maximilian Brell[1] · Jonathan C.-W. Chan[2] · Claudia Giardino[3] · Jose Gomez-Dans[4] · Christian Mielke[1] · Felix Morsdorf[5] · Karl Segl[1] · Naoto Yokoya[6]

Received: 8 December 2017 / Accepted: 2 July 2018 / Published online: 16 July 2018


**Abstract**
Imaging spectroscopy (IS), also commonly known as hyperspectral remote sensing, is a powerful remote sensing technique for the monitoring of the Earth's surface and atmosphere. Pixels in optical hyperspectral images consist of continuous reflectance spectra formed by hundreds of narrow spectral channels, allowing an accurate representation of the surface composition through spectroscopic techniques. However, technical constraints in the definition of imaging spectrometers make spectral coverage and resolution to be usually traded by spatial resolution and swath width, as opposed to optical multispectral (MS) systems typically designed to maximize spatial and/or temporal resolution. This complementarity suggests that a synergistic exploitation of spaceborne IS and MS data would be an optimal way to fulfill those remote sensing applications requiring not only high spatial and temporal resolution data, but also rich spectral information. On the other hand, IS has been shown to yield a strong synergistic potential with non-optical remote sensing methods, such as thermal infrared (TIR) and light detection and ranging (LiDAR). In this contribution we review theoretical and methodological aspects of potential synergies between optical IS and other remote sensing techniques. The focus is put on the evaluation of synergies between spaceborne optical IS and MS systems because of the expected availability of the two types of data in the next years. Short reviews of potential synergies of IS with TIR and LiDAR measurements are also provided.

**Keywords** Imaging spectroscopy · Multispectral remote sensing · Synergy · Data fusion · Spatial enhancement · Thermal infrared · LiDAR

## 1 Introduction

Imaging spectroscopy (IS) in the optical domain, also known as hyperspectral remote sensing, is an Earth observation technique based on spectrally contiguous measurements of the solar light reflected by the Earth's surface and atmosphere (Goetz et al. 1985). Each pixel in the resulting hyperspectral images contains a continuous spectrum

✉ Luis Guanter
guanter@gfz-potsdam.de

Extended author information available on the last page of the article

sampling absorption features which can be linked to the pixel composition. Due to this generic measurement principle, IS provides an accurate representation of geobiophysical parameters, which can be used to infer quantitative information on a wide range of Earth's surface parameters and processes.

The rise and consolidation of IS as a powerful remote sensing technique for land monitoring over the last 3 decades has mostly relied on airborne spectrometers. In particular, the NASA Jet Propulsion Laboratory's Airborne Visible/Infrared Imaging Spectrometer (AVIRIS) (Green et al. 1998), covering the 400–2500 nm spectral range (visible to shortwave infrared, VSWIR) with 10-nm-wide spectral channels, has been used in a large number of campaigns across different continents and ecosystems (e.g., Thompson et al. 2017). From a satellite perspective, the Hyperion spectrometer onboard NASA's Earth Observing One (EO-1) spacecraft was a technology demonstration project which operated between 2001 and 2017 (Ungar et al. 2003). Hyperion acquired hyperspectral images with a 30-m ground sampling distance (GSD), a 7-km horizontal swath and a spectral coverage and sampling similar to that of AVIRIS, although with a much lower radiometric performance and overall data quality. Other satellite IS projects, in this case restricted to the visible near-infrared (VNIR, 400–1000 nm) spectral region, are the Compact High Resolution Imaging Spectrometer (CHRIS) on ESA's Proba-1 microsatellite (Barnsley et al. 2004), which has been operating since 2001, and the Hyperspectral Imager for the Coastal Ocean (HICO) (Lucke et al. 2011), developed by NASA and the US Office of Naval Research and operating onboard the International Space Station (ISS) between 2009 and 2015.

After those technology demonstration projects, several scientific missions expected to deliver accurate spectroscopic measurements are scheduled for the next years. In particular, the Environmental Mapping and Analysis Program (EnMAP) is a German mission which will measure in the VSWIR spectral range with an average spectral sampling of 10 nm, a 30-m GSD, a 30-km swath width and a revisit time under quasi-nadir observation of less than 4 weeks (Guanter et al. 2015). These characteristics are shared by the Italian Space Agency's PRISMA (Hyperspectral Precursor of the Application Mission) (Candela et al. 2016), which in addition presents a panchromatic channel with a 5-m GSD. Other projects, such as NASA's Hyperspectral Infrared Imager (HyspIRI) (Lee et al. 2015), and the Italian-Israeli SHALOM (Spaceborne Hyperspectral Applicative Land and Ocean Mission), currently awaiting the final decision for implementation, could follow EnMAP and PRISMA by mid-2020s.

In general, upcoming space-based VSWIR IS missions such as EnMAP and PRISMA are expected to provide hyperspectral data in a higher data rate and radiometric and spectral quality than their predecessor Hyperion. However, due to trade-offs in spectrometer design between spatial resolution, spectral resolution, swath width and signal-to-noise ratio (SNR), spaceborne IS missions are usually designed to acquire data with a moderate GSD (typically 30 m) as well as with a small across-track swath, which results in a nadir revisit time of up to 4 weeks. It must also be mentioned that EnMAP and PRISMA are "site-oriented" missions, which means that they are tasked on a daily basis to acquire images over selected sites, as opposed to "wall-to-wall" mapping missions with systematic full global coverage. Temporal resolution will be improved by EnMAP and PRISMA through across-track pointing, but this can only happen over a limited number of sites per day due to the high impact of platform pointing maneuvers on mission operations. Those sampling limitations will hamper the use of EnMAP, PRISMA and similar missions for those applications requiring high temporal resolution (e.g., those dealing with water and vegetation) or spatial resolution (e.g., land-cover mapping or mineral exploration).

The EnMAP and PRISMA spaceborne IS missions are expected to coexist with a number of other satellite missions based on different measurement principles, and especially with optical multispectral (MS) missions. An overview of some operating and upcoming satellite missions mentioned in this paper is provided in Table 1. For example, the ESA/Copernicus Sentinel-2 mission (Drusch et al. 2012), which has an optical multispectral imager (MSI) as main payload, is planned for long-term operations and will coexist with EnMAP, PRISMA and other future IS missions. Sentinel-2 MSI has a wide spatial coverage (290 km swath), VSWIR spectral coverage (13 spectral channels between 440 and 2200 nm), high spatial resolution (10 spectral channels at 10- or 20-m ground sampling distance), high temporal resolution (5-day revisit time) and open data policy. Also Landsat-8/9 missions (Roy et al. 2014) offer spatially continuous VSWIR MS data and will coexist with EnMAP and PRISMA. Optical MS missions such as Sentinel-2 and Landsat hold a strong potential for synergistic use with EnMAP-like IS missions, since the poorer spectral information of the multispectral data is compensated by their improved spatial coverage, temporal resolution and (in the case of Sentinel-2) spatial resolution. This is illustrated in Fig. 1. The combination of optical IS and MS missions can thus be used for a temporal and/or spatial enhancement of the rich spectral information provided by the IS data set. This is also the rationale for the inclusion of a panchromatic channel in the PRISMA mission concept (Candela et al. 2016) and for the joint deployment of Hyperion and the Advanced Land Imager (ALI) onboard the EO-1 platform (Fig. 2).

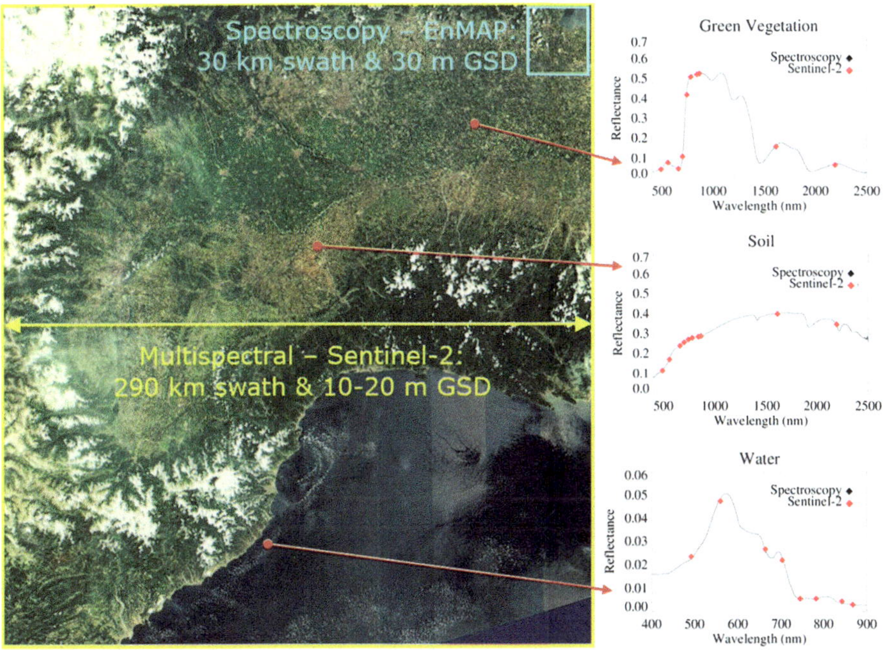


**Fig. 1** First image from Sentinel-2A covering parts of Italy and France in June 2015. The spatial coverage of Sentinel-2 (in yellow) is compared to that of the EnMAP imaging spectroscopy mission (in blue). Panels on the right-hand side compare surface reflectance spectra as acquired by an imaging spectrometer and the 10–20-m channels in the multispectral imager (MSI) on Sentinel-2. GSD stands for ground sampling distance

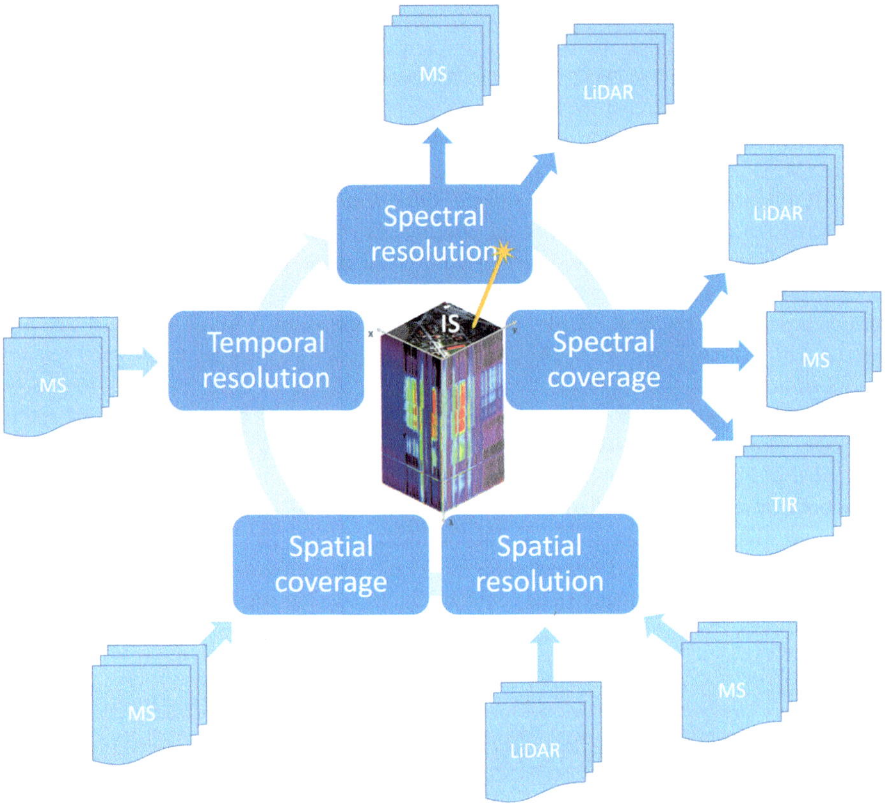


**Fig. 2** Schematic view of how synergies between imaging spectroscopy and other remote sensing techniques can be used to improve spectral, spatial and temporal sampling in the resulting data set. IS stands for imaging spectroscopy, MS for optical multispectral data and TIR for thermal infrared

A different type of synergy with IS data would be the one consisting in the combination of IS measurements with those derived from non-optical instruments carrying fundamentally different information. This would be, for instance, the combination of IS and light detection and ranging (LiDAR) data, which can be very useful for, e.g., the classification of urban objects or the characterization of vegetation covers. The latter is main purpose of the Carnegie Airborne Observatory intended for ecological research, which combines airborne spectroscopy and a dual-laser waveform LiDAR scanner (Asner et al. 2012). Regarding other spectral ranges, the HyspIRI mission concept relies on the combination of spaceborne hyperspectral VSWIR and thermal infrared (TIR) measurements in order to address a number of scientific questions related to the Earth's ecosystems (Lee et al. 2015). Synergies between IS and TIR measurements are also considered for the ESA FLuorescence EXplorer (FLEX) mission (Drusch et al. 2017). FLEX will provide spectroscopic measurements in the 500–800-nm spectral window at high spectral resolution ($< 3$ nm) and low spatial resolution (GSD = 300 m) for the retrieval of chlorophyll fluorescence and other plant biochemical parameters. FLEX will fly in tandem with the Sentinel-3 mission adding TIR measurements necessary for the interpretation of the fluorescence measurements from FLEX. On the other hand, an exceptional wealth of information for ecosystem research will

**Table 1** Observation parameters of some of the satellite missions mentioned in this paper

| | Spectral sampling | Spectral coverage | GSD (optical) (m) | Revisit time | Status | Note |
|---|---|---|---|---|---|---|
| EnMAP | IS | VSWIR | 30 | 27 days nadir, ~ 4 days with 30° pointing | Launch ~ 2020 | Acquisitions on request |
| PRISMA | IS | VSWIR | 30 | 29 days nadir, ~ 7 days with 15° pointing | Launch ~ 2018 | Acquisitions on request; 5 m PAN |
| HyspIRI | IS | VSWIR and TIR | 30 | 5–16 days | Under evaluation | Global Mapper with 185-km swath |
| Hyperion | IS | VSWIR | 30 | ~ 6 days with 22° pointing | End in 2017 | Flight with MS ALI |
| CHRIS-Proba | IS | VNIR | 17–34 | ~ 6 days with 30° pointing | Operating | Multiangular capabilities |
| Landsat | MS | VSWIR and TIR | 30 | 16 days | Operating | Long data record |
| Sentinel-2 | MS | VSWIR | 10–20 | 5 days | Operating | High spatio-temporal resolution |
| Sentinel-3 | MS | VNIR and TIR | 300 | 1 days | Operating | Focus on ocean monitoring |
| ASTER | MS | VSWIR and TIR | 15–30 | 16 days | Operating | Good SWIR sampling (until 2008) |

*IS* imaging spectrometer, *MS* multispectral instrument, *VSWIR* visible to shortwave infrared, *VNIR* visible near-infrared, *SWIR* shortwave infrared, *TIR* thermal infrared, *GSD* ground sampling distance, *PAN* panchromatic channel, *ALI* Advanced Land Imager become available from the combined operation of a series of remote sensing instruments to be deployed at the ISS in 2018. This will include the Global Ecosystem Dynamics Investigation (GEDI) LiDAR, the Ecosystem Spaceborne Thermal Radiometer Experiment on Space Station (ECOSTRESS), and the Orbiting Carbon Observatory 3 (OCO-3), all three from NASA, and the VSWIR Hyperspectral Imager Suite (HISUI) from the Japanese Ministry of Economy, Trade, and Industry (METI). The combination of those four instruments will provide key information on canopy structure (GEDI), evapotranspiration and stress (ECOSTRESS), chlorophyll fluorescence (OCO-3) and ecosystem composition and plant traits (HISUI), which will be used to investigate vegetation functioning at the ecosystem scale during the time period in which the 4 instruments will be operated (Stavros et al. 2017).

This contribution reviews potential synergies between IS data and other remote sensing techniques. The focus is on the discussion of theoretical aspects and methodological issues, rather than on a comprehensive review of single examples in the literature. Special attention is put on the assessment of synergies between space-based IS and optical MS missions because of the open and large-scale data availability expected for the next years, thanks in particular to the coexistence of EnMAP, PRISMA, Sentinel-2 and Landsat missions. Theoretical considerations and some examples of those synergies between IS and optical MS missions are presented in Sect. 2. Potential synergies of IS with TIR and LiDAR measurements will be discussed in Sect. 3. A summary of key points and a discussion of the implications of synergistic approaches for the design and exploitation of future IS missions will be finally provided in Sect. 4.

## 2 Synergies of Imaging Spectroscopy with Optical MS Data

### 2.1 Approaches for Synergistic Use of Optical IS and MS Data

Synergies between spaceborne IS and optical MS measurements, e.g., from EnMAP/PRISMA and Sentinel-2, respectively, could be developed in at least two different directions:

- Enhancement of the spatial resolution of the IS data through fusion with higher-resolution MS data
- Improvement of mapping capabilities through the joint exploitation of MS and IS data sets

The fundamental basis for these two types of synergistic approaches and some examples are discussed hereinafter in this section.

#### 2.1.1 Enhancement of Spatial Resolution of IS Data Through Fusion with MS Data

Recently, considerable attention has been paid to the development of resolution enhancement techniques for IS imagery via IS and MS data fusion (Yokoya et al. 2017). The resolution enhancement of IS imagery can be performed by fusing a low-resolution IS image with a higher-resolution MS image, where both images are acquired over the same Earth's surface in the same season under similar atmospheric and illumination conditions. The resolution-enhanced IS data have the high spatial resolution of the MS sensor and the high spectral resolution of the IS sensor. Such high-order image products, which can be generated by using operational satellites (e.g., EnMAP and Sentinel-2), have the potential to enable a variety of new IS applications on a global scale, including high-resolution mapping of minerals, urban surface materials and plant species.

The final goal for the sharpening of IS data would be the accurate reconstruction at high spatial resolution of not only the broadband spectral shape, but also the single absorption features present in the spectrum. From a theoretical point of view, an absorption feature in the IS data can only be spatially sharpened with higher-resolution MS data if (and only if) the IS spectra represent pure spectra at the MS resolution, and at least one of the following conditions holds:

1. the absorption feature is wide enough to be sampled by the MS instrument (e.g., chlorophyll or iron, see Fig. 3a)
2. the material causing the absorption feature of interest also presents absorption features in other parts of the spectrum which are wide enough to be sampled by the MS instrument (e.g., liquid water presents absorption features with varying depth within the entire 950–2500-nm window, see Fig. 3b)
3. the material causing the absorption feature of interest tends to coexist with other materials presenting absorption features in parts of the spectrum sampled by the MS instrument (e.g., chlorophyll and liquid water contents tend to covary in healthy vegetation)

The increase of information content of the spatially enhanced IS data would depend on which of those three conditions applies. In the case of (1), the absorption feature is already sampled at high spatial resolution in the MS data, but the spatially enhanced IS data set would allow the application of band fitting retrieval algorithms, which can lead to more robust retrievals. Concerning (2), weaker absorption features can be spatially enhanced through leverage with other parts of the spectrum at which the same material presents absorption features, the advantage of this being that the sharpened narrow features may be less affected by confounding factors than the wider ones sampled by the MS spectrum. As for (3), the spatial enhancement of absorption features would only map the statistical coexistence of different materials represented in the spectrum, and the resulting

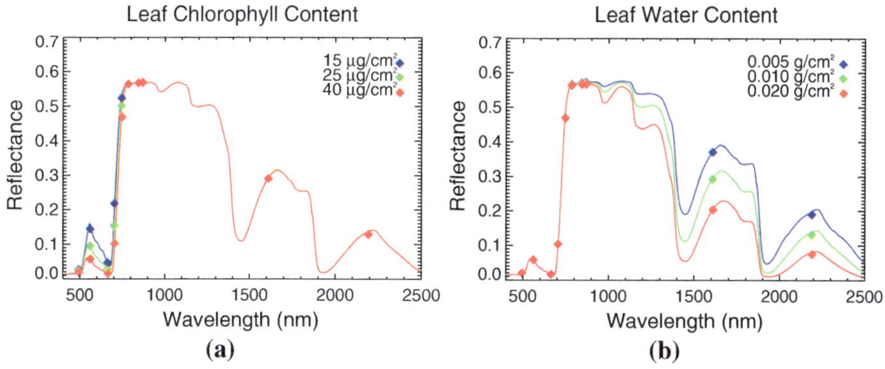


**Fig. 3** Top-of-canopy reflectance spectra at full spectral resolution and resampled to the 10–20-m channels of Sentinel-2 MSI for **a** different values of leaf chlorophyll content, and **b** different values of leaf water content

sharpened features would not represent actual changes in the surface composition or condition for those pixels in which the coexistence between the two materials deviates from the expectation.

To solve the IS and MS data fusion problem, researchers have proposed various methods in the last decade. The existing literature can be categorized into two groups. The first group of methods is based on pan-sharpening. Pan-sharpening is a technique that fuses low-resolution MS and higher-resolution panchromatic images to create a high-resolution MS image. Since pan-sharpening can be regarded as a special case of IS and MS data fusion, significant effort has been devoted to extending and generalizing existing pan-sharpening techniques for IS and MS data fusion. Representative methods include component substitution (Chen et al. 2014), multiresolution analysis (Selva et al. 2015) and patch-wise sparse representation methods (Grohnfeldt et al. 2013).

The second group of methods solves the problem through the analysis of the latent spectral characteristics of the observed scene based on a subspace spanned by a set of basis vectors or spectral signatures of underlying materials (so-called endmembers). This approach includes various methods based on matrix factorization (Yokoya et al. 2012), spectral unmixing (Lanaras et al. 2015) and Bayesian probability (Wei et al. 2015). For instance, unmixing-based methods reconstruct a high-resolution IS image by estimating spectral signatures of endmembers and high-resolution fractional abundances from the input IS and MS images, respectively.

Methods above have been recently compared with extensive experiments in a review paper by Yokoya et al. (2017). Multiresolution analysis-based methods and unmixing-based methods demonstrated good and stable performance with different fusion scenarios. Current research on the development of algorithms for IS and MS data fusion is focused on combining different approaches to further improve the reconstruction performance.

Figure 4 presents an example of IS and MS data fusion using simulated EnMAP and Sentinel-2 images over an urban area of Brussels, Belgium. The fusion procedure is

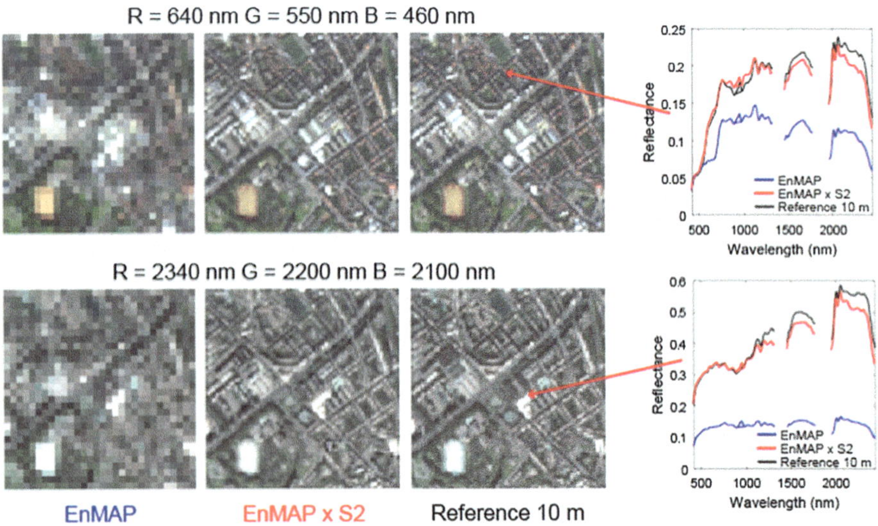


**Fig. 4** Example of spatial enhancement of IS data. Color composite images of simulated EnMAP, EnMAP-Sentinel-2 fused and reference data over an urban area of Brussels (Belgium) are shown together with spectral signatures for two locations

composed of two steps: (1) self-sharpening of Sentinel-2 data that sharpens the 20-m-GSD bands by the 10-m-GSD bands, and (2) the fusion of EnMAP and 10-m-GSD Sentinel-2 data. The multiresolution analysis-based fusion technique presented in Selva et al. (2015) was used for both steps. As shown in the color composite images in Fig. 4, the spatial information content is significantly improved. On the other hand, the spectral profiles in Fig. 4 indicate that the spectral quality is variable particularly in the SWIR range where the spectral coverage of Sentinel-2 is limited with only two bands. The importance of the overlap of spectral response functions between two sensors is discussed in Yokoya et al. (2017).

Some publications have appeared in recent years reporting the impact of resolution enhancement of IS imagery on spectral unmixing (Yokoya et al. 2016) and land-cover classification (Chan and Yokoya 2016). Due to inevitable spectral distortions in the resolution-enhanced data, the use of external spectral libraries does not always work for classification or spectral unmixing. In contrast, it has been shown that good results can be obtained in classification or spectral unmixing by using reference spectra acquired from each fused image (Yokoya et al. 2017, 2016).

Research on quality assessment of resolution-enhanced products is also important from a practical viewpoint; however, very few publications can be found compared to those dealing with algorithm development. Quantitative evaluation of resolution-enhanced data is usually performed within simulation studies because reference data are necessary to quantify reconstruction performance. When resolution-enhanced IS image products are generated from operational satellites (e.g., EnMAP and Sentinel-2), quantitative quality assessment without reference is required to provide spectral quality at each pixel so that users can identify and select reliable pixel spectra. The standard technique for this purpose is to examine consistency between the input images and degraded versions of the fused image using quality measures (Palsson et al. 2016). There is still room for investigation on how to integrate the consistency information obtained for each of the two input images.

### 2.1.2 Improvement of Mapping Capabilities Through the Joint Exploitation of MS and IS Data Sets

The high spectral resolution and coverage of IS instruments can also be used improved spatio-temporal monitoring with MS data. This could be achieved through, at least, (1) analysis of IS results for the interpretation of colocated MS measurements and (2) the extrapolation of IS-based information to the broad spatial and temporal coverage of the MS data set.

*Analysis of IS results for the interpretation of MS measurements* The richer spectral information in the IS data complements the wider-area and higher spatio-temporal mapping capabilities of the MS instrument. For example, Milewski et al. (2017) combined EO-1 Hyperion data with multitemporal and multispectral Landsat acquisitions in order to analyze the spatial distribution of surface evaporite minerals and changes in a Namibian Kalahari salt pan, which is a semiarid depositional environment associated with episodic flooding events. The dynamic of the surface crusts was evaluated through change detection analysis based on a time series of Landsat acquisitions (1984–2015), whereas a hyperspectral image from Hyperion was used to map the spatial distribution of the major crust types (halite, gypsum, calcite/sepiolite and disturbed, dark crust) and their abundances through spectral mixture analysis (SMA). The combined information from the hyperspectral and multispectral data sets could then be exploited to spatially differentiate and map depositional environments over the whole salt pan.

In a different study, Mielke et al. (2014a) assessed the potential of combined IS and MS spaceborne data for mapping the spatial extent of mine waste surfaces in South Africa. For that task, the broadband iron feature depth (IFD) index was proposed as a potential proxy for mine waste. The IFD derived from Landsat was found to be in good agreement with primary and secondary iron-bearing minerals mapped from Hyperion data, which suggests that a combination of IS data for mineral identification with MS data for repetitive area-wide mapping of the IFD as a mine waste proxy is a promising synergistic application of IS and MS data. The use of the IFD index for geological applications based on combined IS and MS data will be further discussed in Sect. 2.2.2.

Also dealing with geological mapping, Bishop et al. (2011) employed a two-step progressive approach, first to locate target areas characterized by hydrothermal mineral alteration using Advanced Spaceborne Thermal Emission and Reflection Radiometer (ASTER) VNIR and SWIR data, and secondly, to attempt detailed mineral mapping using Hyperion's spectral information.

*Extrapolation of IS-based information to the broad spatial and temporal coverage of the MS data* In a different type of synergetic use of IS and MS data, the richer spectroscopic information delivered by the imaging spectrometer over a given area can be used to enhance the mapping potential of MS observations over a wider area than the one sampled by the IS data set.

For instance, Hubbard et al. (2003) combined Hyperion, EO–1 Advanced Land Imager (ALI) and the co-orbiting ASTER data to map hydrothermally altered rocks associated with volcanic systems over the Central Andes. The mineral maps derived from Hyperion data with the Tetracorder expert system (Clark et al. 2003) were used to adjust image display thresholds in the alteration mineral maps derived from ALI and ASTER over a much broader area than the Hyperion coverage alone. Hyperion data were also used for the interpretation of ASTER and ALI mapping results as well as for their radiometric and atmospheric correction. A similar setup was used by Hubbard and Crowley (2005) for mineral mapping over the Chilean–Bolivian Altiplano.

Schmid et al. (2005) used IS data to support the mapping of geophysical parameters in space and time with Landsat data. In particular, they developed an approach to monitor changes in wetlands in Central Spain. For this purpose, an SMA approach was used with a temporal series of Landsat data to detect changes in the wetland over time. The spectral endmembers for the SMA were extracted from hyperspectral data acquired during an airborne campaign, resampled to Landsat's TM and ETM+ spectral responses. It was used for change analysis at different and temporal scales, showing the feasibility of exploiting spectral endmembers derived from hyperspectral information in the analysis of MS data.

Concerning vegetation, optical MS measurements of vegetation reflectance spectra are generally sensitive to canopy water content (CWC), but quantitative estimates of CWC can only be reliably derived from IS radiance data through physical modeling of vapor and liquid water spectral features in the near-infrared. Asner et al. (2016) extrapolated CWC maps derived from airborne IS data over some sites to the entire state of California by means of a deep learning technique establishing empirical relationships between IS-based CWC and a series of parameters retrieved from Landsat and other ancillary data sets. The resulting California-wide CWC maps were analyzed to assess the forest canopy water loss during the 2012–2015 California drought.

*Model inversion and assimilation of multitemporal data sets* Strategies for the combination of IS and MS data can be based on a physical modeling of the geophysical parameters to be inferred from the remote sensing data, which complements the empirical approaches discussed in Sects. 2.1.1 and 2.1.2. Broadly speaking, this physical modeling means that the scene (e.g., land surface vegetation, atmosphere) can be parameterized by a finite set of metrics (e.g., leaf area index, leaf pigment concentrations, soil optical properties). These parameters can all be combined in a so-called *state vector*. Each observation, be it IS or MS, would provide an *inference* of the scene parameters: an uncertainty-quantified estimate of the state vector, or more precisely, the probability density function of the state vector. Parameters that can be inferred well because a particular observation has a strong sensitivity to it will be characterized by low uncertainty, whereas parameters with a large uncertainty will be symptomatic of low sensitivity in the observations. The mapping from observation to state vector is accomplished by inverting the observations using an *observation operator*, typically a physical model based on radiative transfer theory that produces a prediction of the sensor observations as a function of the state vector.

In the case of multitemporal data sets consisting of both MS and IS acquisitions, dynamic physically based models allow us to blend in observations from the two sensors in a physically consistent manner. In its simplest form, we can imagine a scenario where MS and IS sensors fly over the same scene simultaneously. We can then invert the observations from one sensor and use the inferred state vector probability density function as the *a priori* distribution of the second observation, which will result in a physically based combination of both, respecting the characteristics of both sensors. The use of an observation operator allows one to account for different sensor characteristics (different illumination/viewing geometries, spectral sampling, spatial resolution, etc.). The retrieval system is in this case combining two independent inferences of the same scene, acquired at the same time, to come to a solution that is consistent with both. An example will be discussed in Sect. 2.2.3.

If observations do not happen simultaneously, a first approach might be to assume that within a temporal window, the scene changes little, and thus they can be assumed simultaneous. However, this assumption can be broken in a few days for the land surface and in less than an hour for atmospheric parameters. Dynamic models might be used to propagate the state vector at one location from one time step to another. The simplest possible dynamic model is to assume that nothing is changing, but to use this model under the assumption that it is wrong, and thus as model propagates the state over longer and longer time gaps will add uncertainty to the original estimate. This uncertainty inflation approach is the basis of temporal regularization (e.g., Lewis et al. 2012). More sophisticated approaches will use a dynamic model that describes the evolution of the state vector, but the concept of model uncertainty is still important, as even in the hypothetical case that the model was perfect, changes in the scene would render it wrong.

The approach described above has been widely used for dynamical systems, as well as climate studies, where it often is referred to as "data assimilation." A number of standard techniques, such as Kalman and particle filters, variational approaches, have been exploited to this end with Earth observation data. The same approaches can be used for IS and MS data combinations, provided that a reasonable observation operator is used. An example is shown in Fig 5, where we show the EO-LDAS system introduced in Lewis et al. (2012) being used to invert observations of surface directional reflectance from MS sensors, Landsat 8 and Sentinel-2. In this experiment, Landsat 8 has fewer observations than Sentinel-2, so individual inversions result in a very poor description of the dynamics of the parameter evolution over a year. It is also obvious that the single observations have large error bars, a consequence of the limited information content on each observation. The situation improves with Sentinel-2, as the number of observations increases. However, the dynamics are not well described by this experiment: The clustering of observations results in an incomplete retrieval of the temporal evolution of the different parameters. Once we start

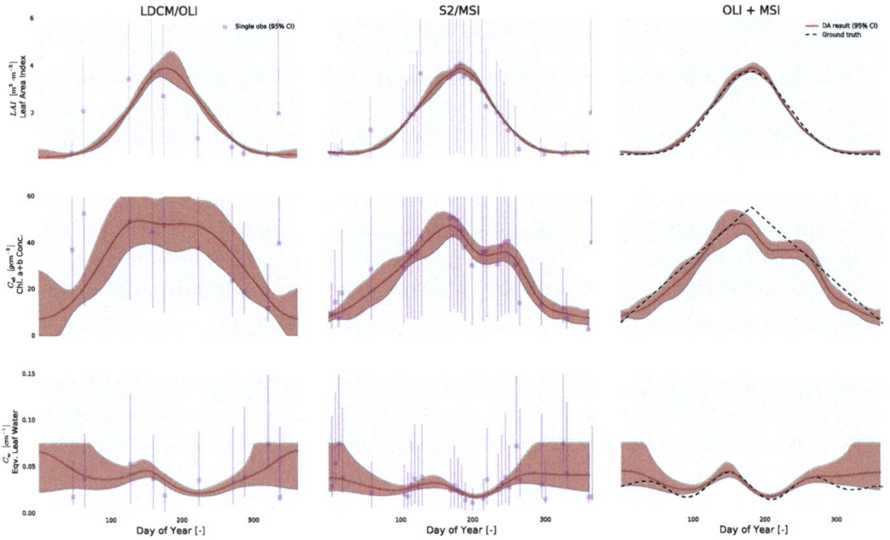


**Fig. 5** Synthetic experiment demonstrating "data assimilation" system to invert land surface parameters from MS observations. Here, we have simulated a scene where leaf chlorophyll $C_{ab}$, leaf area index *LAI* and equivalent leaf water $C_w$ vary over time. The retrieved *LAI*, $C_{ab}$ and $C_w$ are shown in each row (top to bottom, respectively). Each column shows the results of retrieving the parameters from Landsat 8 (left), Sentinel-2 MSI (center) and a combination of both. The dots with error bars refer to single observation inversions (mean and 1.96 times standard deviation), whereas the filled region shows the results of extending the inversions with a dynamic model and a variational data assimilation system

applying a dynamic model, we get interpolation, but also reduced uncertainties, with both Sentinel-2 and Landsat 8 being able to provide a reasonable path of the trajectory of the parameters. Once the dynamic model is established, it is straightforward to combine the observations from both sensors, resulting in a further reduced uncertainty.

## 2.2 Examples of Synergistic Use of IS and MS Data

The theoretical discussion of potential synergies between optical IS and MS data presented in Sect. 2.1 is complemented in this section with a series of examples of how such synergies work for selected study cases including land use and land-cover mapping, mineral exploration, vegetation parameter retrieval and water applications.

### 2.2.1 Land Use and Land-Cover Mapping

Land use and land-cover (LULC) mapping is crucial to many scientific investigations from local to global scales. For decades, land-cover maps are used for urban planning and a multitude of environmental monitoring applications such as urban expansion, forest inventory, biodiversity, land surface modeling. LULC mapping with satellite data is one of the most widely investigated subjects.

In the last 2 decades, extensive efforts have been devoted to understand IS data for LULC mapping. However, airborne IS data are expensive and pose big processing challenges when coverage is very large, whereas the typical 30-m GSD of spaceborne IS missions is in general too coarse for many applications. To tackle this spatial issue, one possibility is to apply superresolution image reconstruction algorithms. Superresolution-enhanced hyperspectral VNIR CHRIS/Proba data (9 m) have been tested for land-cover classification and unmixing (Chan et al. 2011). Demarchi et al. (2012) applied the same methodology for subpixel mapping. The impression is that these superresolution-enhanced data sets have not been satisfactorily evaluated and hence their real potential remains uncertain. A major issue is the difficulty in compiling such data set with reliable groundtruth. Another issue is the algorithm evaluation method: How should the accuracy be evaluated for data sets acquired at different spatial resolution. Traditional accuracy measures for land-cover classification have long been criticized as limited and problematic (Foody 2002).

Recent development in pan-sharpening techniques and image fusion and the highly anticipated new spaceborne IS missions (e.g., EnMAP, HyspIRI) have provided new momentum in the spatial enhancement of satellite IS images. While the issues related to an appropriate evaluation method still exist, one significant obstacle has been overcome—there will be real accessible data sets, such as EnMAP and Sentinel-2. Many methods fusing high-resolution MS data and low-resolution IS data have been proposed targeting such future data sets. However, real examples of applying a fused and spatially enhanced IS image for LULC classification are still comparatively rare.

To illustrate the usefulness of IS-MS fused images for LULC mapping, an APEX data set acquired from the Brussels capital region has been used. It has 288 bands between 400 and 2500 nm and is acquired at a GSD of 2.4 m (Chan and Yokoya 2016). An EnMAP image with 242 bands is simulated with the EnMAP end-to-end simulator tool (Segl et al. 2012) at a 30-m GSD to mimic low-resolution IS data. A Sentinel-2 data set with 10 bands at 10- and 20-m GSDs is also simulated using the S2eteS Sentinel-2 scene simulator (Segl et al. 2015). A generic 13-class land-cover classification scheme is adapted for the study area: larch, pines, ash, maple, oak, beech, grassland (cropland, lawn and parks), buildings, roads, soil (bare soil, fallowed field, construction site), tennis court, football field and water surface (artificial lake and canal). Figure 6 shows the false color composite of the study area.

We compared the classification accuracy of EnMAP (30 m), Sentinel-2 (10 and 20 m) and fused IS image at 10 m. The fusion approach has been described in Sect. 2.1.1. A groundtruth IS image at 10 m with the same spectral configuration as EnMAP is also simulated. All data sets are upscaled at 10 m with the same number of rows and columns for comparison. A total of 1095 pixels are blind-tested for accuracy. Two advanced classification algorithms, Rotation Forest (RF) (Rodriguez et al. 2006) and Canonical Correlation Forest (CCF) (Rainforth and Wood 2015), are investigated. Figure 6b shows the classification map generated from an IS image at 10-m resolution which is used as the benchmark for comparison; a legend with only 8 colors is used for easy visualization. Table 2 shows the overall accuracy, average class accuracy and kappa values of the classification results. In general, CCF performs better than RF. The benchmark 10-m IS data have achieved 70–74% (RF-CCF) overall accuracy. With Sentinel-2, the overall accuracy is 66–72%. With EnMAP at 30 m, the O.A. is 61–67%. For the fused IS data set, accuracies are 68–73%. The performance of the enhanced data is a little lower than the benchmark IS, moderately higher than the Sentinel-2, but significantly (5%) better than the 30-m EnMAP.

Figure 7 clearly shows that the fused IS image reveals important details such as road networks comparable to Sentinel-2 but at a higher accuracy. Extraction and classification of urban objects such as road and buildings are understandably too challenging with EnMAP at 30-m resolution. This explains the high performance of Sentinel-2. Comparison between MS and IS for land-cover classification have been widely studied, and

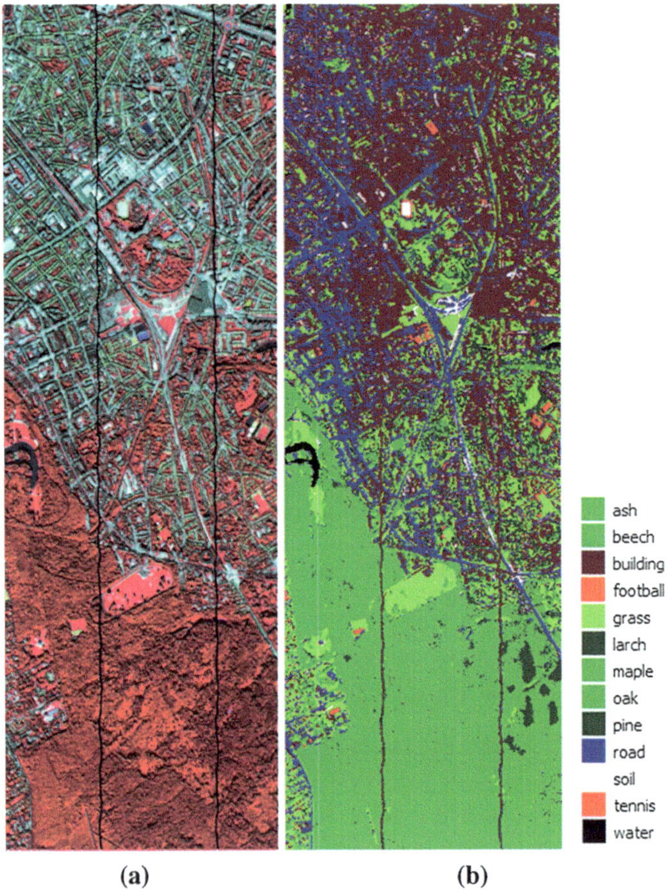


**Fig. 6** Land use and land-cover mapping over Brussels. A false color composite of the study area from airborne APEX IS data is shown in (**a**), and a classification map using Canonical Correlation Forest (CCF) is depicted in (**b**)

depending on the application, IS imagery does not always have superior performance (Xu and Gong 2007). A MS-IS fusion approach is more suitable for challenging problems that require very rich spectral information and are better addressed with IS data. Our example shows that fusion of high-resolution MS and low-resolution IS images can achieve synergies in terms of a significant improvement in classification details as compared with the low-resolution data and higher class accuracies as compared to the MS data. We expect the fusion synergies to have a significantly greater impact on specific LULC applications which require spatial details to characterize range, combination, distribution and clustering of species, for example, urban mapping (Herold et al. 2004), vegetation species mapping (Chan and Paelinckx 2008) and biodiversity information required for environmental assessments (Bush et al. 2017). Given the fact that spaceborne IS data at 30-m GSD will only be available in the next few years, the novelty of a potential fused IS data at high GSD (10 m) with large coverage is almost certain to attract new research momentum with innovative LULC applications.

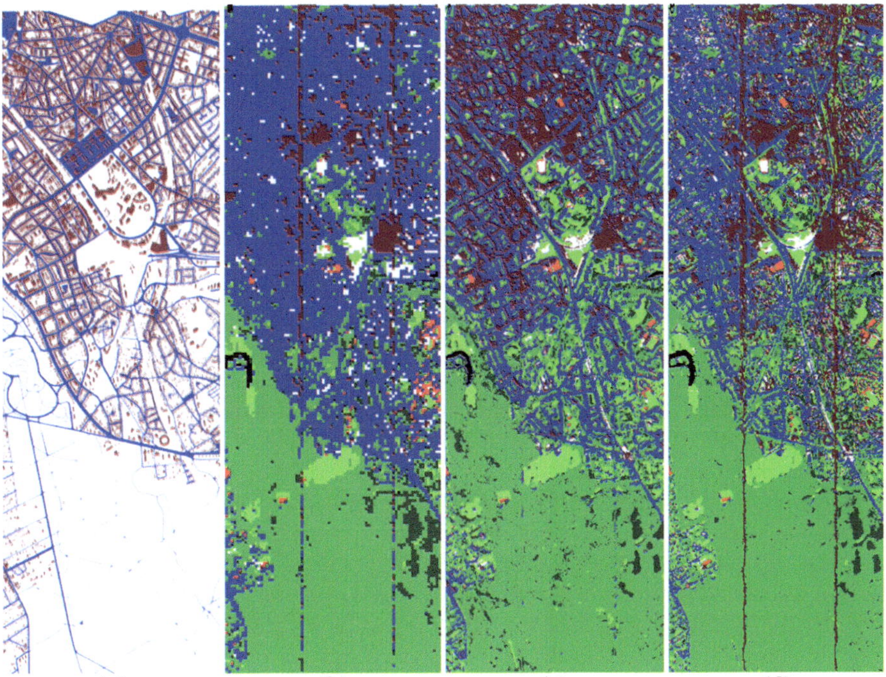


**Fig. 7** Ground-truth map of road and building classes (**a**) and classification results for EnMAP (**b**), Sentinel-2 (**c**) and fused image (**d**)

**Table 2** Comparison of classification accuracies (given as percentages) with 13 classes

| Classifier | Rotation Forests | | | Canonical Correlation Forests | | |
|---|---|---|---|---|---|---|
| Accuracy | OA | AA | Kappa | OA | AA | Kappa |
| Reference | 69.11 | 75.87 | 0.63 | 74.45 | 72.73 | 0.70 |
| EnMAP | 61.98 | 62.80 | 0.56 | 67.63 | 66.11 | 0.62 |
| Sentinel-2 | 66.85 | 62.78 | 0.61 | 72.70 | 70.73 | 0.68 |
| Fused | 68.02 | 73.04 | 0.63 | 73.94 | 72.38 | 0.69 |

The highest accuracy is found for the reference data set and the second highest for the fused data set

*OA* overall accuracy, *AA* average accuracy

### 2.2.2 Mineral Mapping

The Haib River Cu-Mo deposit in the lower Orange River region represents a unique test site for the demonstration of synergistic applications in geological remote sensing. Using simulations from HyMAP data, Mielke et al. (2014a) showed that it is possible to link data from state-of-the-art MS systems such as Sentinel-2 and Landsat-8 to results from hyperspectral systems such as EnMAP via the IFD iron index introduced in Sect. 2.1.2. This absorption in the 900-nm region is caused by iron-bearing minerals that may be produced by the weathering of metal sulfides such as pyrite and chalcopyrite (Chavez 2000). This process forms gossan surfaces that may be targeted with, e.g., Sentinel-2, calculating the normalized iron feature depth from Sentinel-2 L1C data as shown in Fig. 8. The IFD is a simple three-point band depth index for MS sensor systems that proxies the band depth of the iron feature near 900 nm using the two spectral bands which are closest to the shoulders of the 900-nm iron absorption feature (Mielke et al. 2014a). These two shoulder bands encompass the absorption band, which is closest to the 900-nm iron absorption feature (Mielke et al. 2014a). The feature depth is found by an interpolation of the aforementioned iron absorption feature band with the shoulder bands. The difference between interpolated and measured iron feature absorption band yields the IFD (Mielke et al. 2014b). This may be used in mineral exploration to highlight gossan zones that may indicate the presence of sulfide ore deposits (Taylor 2011). If this concept is expanded to other sensors, for example ASTER SWIR measurements, it is possible to derive a false color composite of the normalized feature depths, which highlight the dominant material mixture at a specific location, shown for the Haib River area in Fig. 8b.

However, hyperspectral data from spaceborne sensors such as EnMAP or PRISMA are necessary for a more detailed mineral mapping using, e.g., expert systems such as the EnMAP Geologic Mapper (EnGeoMAP) Base, which is a fully automated system for the detection of mineralogical surface cover types over mineral deposit areas (Mielke et al. 2016). IS may be used for a detailed view on the local dominant minerals in one area, as shown in Fig. 9. Here the gossan zones, which have been identified in Fig. 8 via the normalized iron feature depth, may be subdivided into hematite-, goethite- and jarosite-dominated gossans. Only IS data with its superior spectral resolution are able to correctly highlight and discriminate the most prominent minerals in the shortwave infrared from 2000 to 2500 nm. This shows the potential synergies in mineral exploration between large, multispectral, global mapping missions, such as Sentinel-2, and regional scale hyperspectral instruments such as EnMAP. The Global Mappers identify and highlight interesting anomalies for scientists working in mineral exploration, while IS data

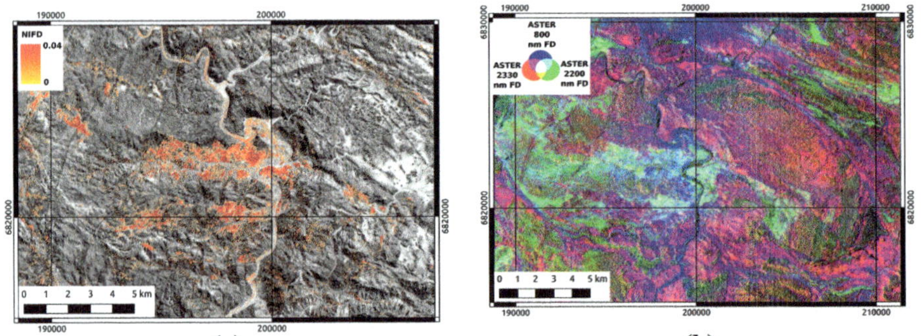


(a) (b)

**Fig. 8** Indirect mapping of mineral types from multispectral remote sensing data over the Haib River Quartz Feldspar, Porphyry. The Sentinel-2 normalized iron feature depth data, which highlights the main ore-bearing unit, appears as a large oval-shaped anomaly in the central part of the image in (**a**). ASTER normalized feature depth composite image is shown in (**b**). It highlights mineral mixtures that are dominated by absorption features near 2330 nm (red), e.g., epidote, chlorite and carbonates. Illite-, alunite- and muscovite-dominated areas are colored in green. Areas with material that shows prominent iron absorption features are colored blue

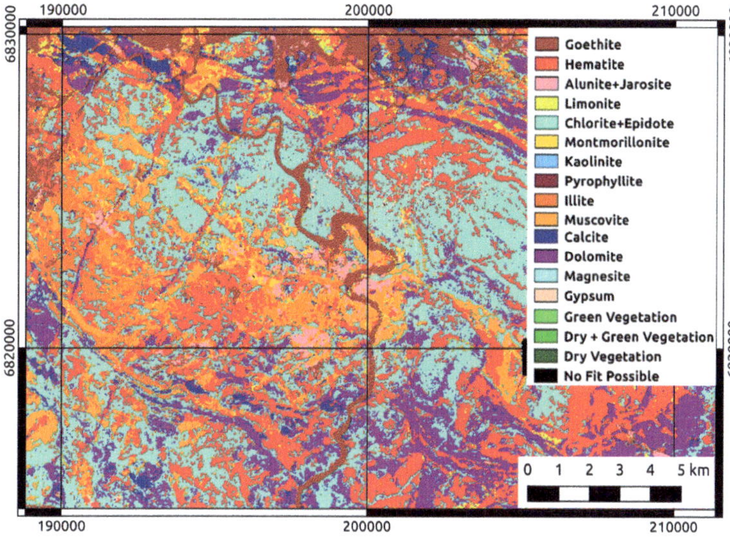


**Fig. 9** EnGeoMAP Base classification result from simulated EnMAP data. The areas dominated by chlorite, epidote and carbonates correspond well to the areas colored in magenta and red in Fig. 8b

offer the capability to characterize these anomalies in much more detail using spectral geology tools such as EnGeoMAP for material identification.

### 2.2.3 Retrieval of Vegetation Parameters Through Model Inversion

We can use the physical modeling introduced earlier to theoretically understand the limitations of different sensors for the retrieval of vegetation parameters, and how combining observations from different sensors might benefit parameter retrieval. To this end, we simulate a set of spectral acquisitions for EnMAP, as well as simultaneous observations from Sentinel-2/MSI. The simulations are done using the PROSPECT-D leaf RT model (Féret et al. 2017) and the 4SAIL model (Verhoef 1984) for leaf and canopy levels, respectively. The atmospheric effects are simulated by the 6S model (Vermote et al. 1997), taking into account the multiple interactions between land and atmosphere. The simulation thus presents a mapping from surface and atmospheric composition parameters to at-sensor reflectances. With some indication of the uncertainty in the observations at the sensor level, we can study the uncertainty of the retrieved parameters, under the assumption that the "true state" of the combined surface and atmospheric system can be retrieved. This uncertainty is encoded in the posterior covariance matrix, which we approximate by a linearization of the Hessian as in Lewis et al. (2012). We consider an scenario where we have a thick vegetation canopy (high leaf area index, LAI), and a moderate atmospheric loading. We can see the uncertainty associated with the retrieval of different parameters in the left-hand side column of Fig. 10, which depicts the posterior correlation matrix, where a perfect retrieval would be indicated by an identity matrix. We can see that for a single EnMAP observation, a substantial number of off-diagonal elements are present, suggesting parameters that result in changes to the measured reflectance in the same spectral window cannot be individually differentiated with a single observation. This is unsurprising, and that is the reason that many retrieval schemes prescribe some of these parameters (e.g., the leaf structure

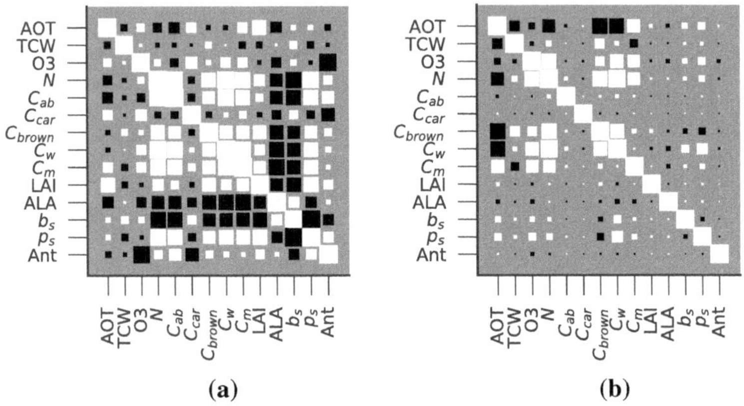


**Fig. 10** Posterior correlation matrices for the combined land and atmosphere RT models PROSAIL and 6S used to invert a single observation from the EnMAP sensor (**a**) and the equivalent matrices when the EnMAP observation is supplemented with a contemporary Sentinel-2/MSI observation (**b**). The variables in the axis correspond to 6S and PROSAIL input parameters. The white (... black) squares indicate positive (... negative) correlation, and the size of the square is proportional to its absolute value. Elements along the main diagonal have a correlation of one

parameter *N*, the parameter(s) controlling the leaf angle distribution, *ALA*, or prescribing a soil response). The left-hand side of Fig. 10 is in essence a depiction of the ill-posedness of the inversion problem. It is important to note that this example is contrived, as no prior information has been used at all, which would not be the case on any practical scenario.

On the right-hand side panel of Fig. 10, we show the posterior correlation matrix where the EnMAP observations have been optimally combined with the information retrieved from a Sentinel-2/MSI observation. We have assumed that these occur at different times, but close enough for the land surface to only have experienced a small change, but no extra information is gained on the atmospheric composition, and we have also ignored the 1375-nm band in Sentinel-2 MSI. It is clear that the posterior correlation matrices are much closer to a diagonal matrix, suggesting that some parameters such as leaf chlorophyll content, carotenoid content and anthocyanins might be well resolved. Other important parameters, such as LAI, can be retrieved for high canopy cover, but the multiple scattering between the canopy and the atmosphere for low LAI results in a strong compensatory effect with aerosol optical thickness (AOT). Note that we cannot show the posterior covariance matrix for the Sentinel-2 MSI observation in this example: We are approximating the problem as a linear problem, in which we try to infer fourteen surface and atmosphere parameters, and with Sentinel-2 MSI we only have twelve bands, which results in an undetermined linear system.

Although in this example we have not assessed whether the solution can be found from the data, only what shape the uncertainty would take, the method used forms the basis of any Bayesian update, being the fundamental basis of techniques like Kalman or particle filters and smoothers (Gómez-Dans et al. 2016). Extended with a state vector dynamic model, the system would not only just provide inferences at the time of the acquisitions, but also be able to optimally interpolate the state vector and provide uncertainty-quantified inferences. The probabilistic basis of the Bayesian combination method rests on the two observations being interpreted by the same physical model with the same parameters, the assumption that the two sensors are accurately calibrated to a common standard, and that the different spatial resolutions can be bridged (by, e.g., modeling the individual IFOV of the individual sensors). Provided these conditions are maintained, the method can be extended to other sensors.

### 2.2.4 Monitoring of Inland and Coastal Waters

Regular observations of physical and biogeochemical components in inland and coastal waters provide essential information in the form of maps of water quality, bottom properties and bathymetry as needed for science and resource management (e.g., Palmer et al. 2015; Olmanson et al. 2008; Dekker et al. 2011; Mouw et al. 2015; Tyler et al. 2016). Depending on the scale of observations, the developments in water quality and biophysical parameter retrieval algorithms are driven by airborne IS (e.g., AVIRIS, APEX), ocean color (OC) radiometry (e.g., MODIS, MERIS) and MS sensors (e.g., Landsat, Sentinel-2). In particular, the Sentinel-3/OLCI MS instrument offers an improved mapping potential for its specific capacities to resolve turbid, productive waters and for having daily revisit with a 300-m GSD, whereas the Sentinel-2/MSI can provide data at a higher spatial resolution every 5 days. When combined with Landsat, a fine-scale global mapping at a temporal resolution close to that of OC missions is also feasible. Finally, the previously mentioned IS satellite missions (e.g., EnMAP, PRISMA, HyspIRI) are anticipated to provide greatly enhanced capability to effectively enable wider applications for coastal and inland waters that, so far, have been mostly based on Hyperion (e.g., Kutser 2004), HICO (e.g., Garcia et al. 2014) and CHRIS-Proba (e.g., Casal et al. 2011).

IS data in fact allow both to increase the estimation accuracy of inland and coastal water variables currently observed by OC and MS sensors and to access to new variables of interest (e.g., identification and quantification of particulate and dissolved matter: type and size of suspended particles, types of pigments, organic matter composition, cyanobacteria) for multiple applications (Hestir et al. 2015; Giardino et al. 2018). In such a context, a prime example regards phytoplankton, a key parameter for water managers and of considerable interest to scientists who are for instance interested to freshwater ecology. As a proxy of phytoplankton biomass, the chlorophyll-a concentration (chl-a) was mapped in lakes already in 1974 from aircraft and satellite (Strong 1974). It also represents a primary parameter quantitatively derived from OC (Mishra et al. 2017). Then, IS provides further insights for detecting the accessory pigments of phytoplankton such as phycocyanin and phycoerythrin pigments, which are often associated with harmful algal blooms. As an example, in occasion of a red ciliate blooms in coastal waters, Dierssen et al. (2015) used OC MODIS to map the chl-a concentration and IS HICO to further distinguish phycoerythrin pigments. An additional application in which IS provided enhanced mapping capability occurs in shallow waters (those where the bottom is visible from the water surface and measurably influences the remote sensing reflectance). The patchy structure typical of these environments hinders the use of OC sensors so that monitoring of bottom types and benthic communities (e.g., mud, sand-mud mixture, coral sands, coral reefs, seagrass, macrophytes) has been commonly achieved through MS sensors (e.g., Dekker et al. 2005). With a 10–30-m GSD, these sensors are ideal for most of the application scales, but are limited to identify species with similar spectral characteristics or to assess particular processes such as the state health of coral reefs (Botha et al. 2013). This is especially true for fine tracking of biodiversity and ecosystem functioning (identification of invasive and resident species). To this aim, airborne IS simultaneously providing high spatial and high spectral resolutions has been extensively used to make large-scale inventories of benthic photosynthetic organisms, such as macrophytes, seagrasses and corals (Phinn et al. 2008).

To summarize, the optical complexity of inland and coastal waters, which also usually show a fast degree of change and a patchy distribution of both water components and benthic habitats, makes crucial synergic applications of IS and MS. Moreover, dealing with water optics, the availability of OC sensors has to be naturally included. As an example, we present the Lake Garda (Italy) test site, where at the beginning of the 1990s, Zilioli et al. (1994) started to study the lake color from Landsat. The lake is characterized by clear yet optically complex deep waters, with occasional cyanobacterial blooms and optically shallow areas with important submerged macrophyte beds, which make it challenging to develop robust retrieval algorithms. Nevertheless, the lake relevance (it is visited by more than 20 million tourists every year and stores about 50 $km^3$ of water, used for both recreational purposes and water supply) is demanding a series of applications that only synergistic use of IS, MS and OC is able to provide. Some of these applications (e.g., the support to the EU Water Framework Directive to report on both chl-a concentration and extension of submerged vegetation beds) are qualitatively shown in Fig. 11. For the sake of brevity we can only mention that most of these applications (namely A, B, C and E) were developed based on Hyperion data that were used as a benchmark for establishing a sensor-independent physically based approach (Giardino et al. 2007). The use of HICO was instead useful to recognize the spectral feature due to the phycocyanin (Fig. 11D) according to Kutser (2004), while the neural network C2R operationally provided the MERIS-derived chl-a time series (Fig. 11E). A complete description of the applications can be found in Bresciani et al. (2011, 2012).

## 3 Potential Synergies of IS with Non-optical Remote Sensing Data

### 3.1 Synergies of IS with TIR Data

TIR measurements hold a strong synergistic potential with optical data in general and with IS in particular. The exploitation of synergies between VSWIR IS and multispectral TIR measurements is actually at the core of the NASA HyspIRI mission concept, which is intended to address a number of science questions focused on world ecosystems and natural hazards (Lee et al. 2015). HyspIRI is currently awaiting decision for implementation.

Based on the review by Lee et al. (2015), synergies between optical IS and multispectral TIR data can be important for, at least, the following applications:

- Canopy biochemistry: Optical IS has demonstrated its potential to retrieve important leaf photosynthetic pigments, such as chlorophylls, carotenoids and anthocyanins, as well as leaf and canopy liquid water content (e.g., Ustin et al. 2004). This capability is well complemented by the ability of TIR measurements to measure other vegetation parameters such as cellulose, hemicellulose, cutin and other biochemicals with absorption features in the 8–14 μm region (Ribeiro da Luz and Crowley 2007).
- Plant functioning: Simultaneous measurements of vegetation biophysical and biochemical properties and surface temperature can help monitor plant physiological functioning and potential stress situations as well as to estimate evapotranspiration (Anderson and Kustas 2008), which is important to, e.g., agricultural applications, water use practices and mitigation strategies in response to drought.

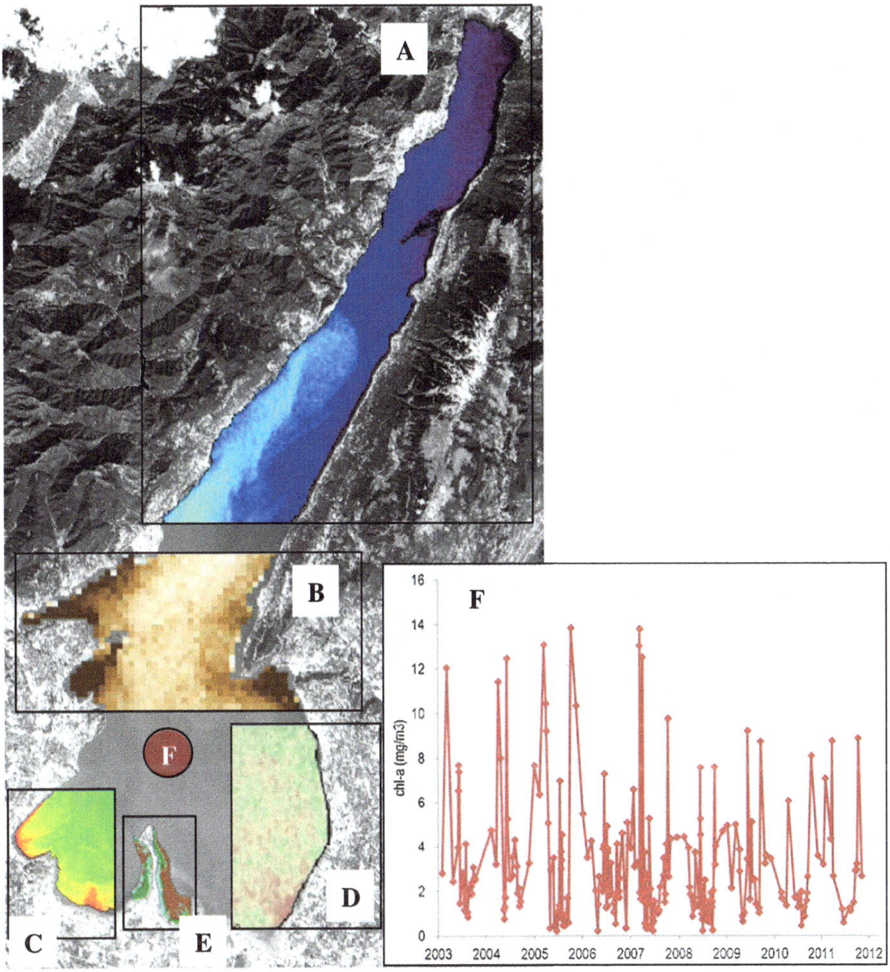


**Fig. 11** An overview of satellite products developed for Lake Garda. A: total suspended matter (SPM) to trace water dynamics from Sentinel-2 (17 August 2016, from blue to green the SPM ranges from 0.1 to 2 g/m$^3$); B: colored dissolved organic matter (CDOM) as a result of primary producers degradation from MERIS (11 October 2006, from light to dark brown CDOM ranges from 0.01 to 0.1 m-1); C: fine-scale mapping of chlorophyll-a concentrations from Sentinel-2 (17 August 2016, from green to red chl-a ranges from 2 to 5 mg/m$^3$); D: map of cyanobacterial bloom from HICO (23 August 2012, from green–yellow–red the cyanobacterial index, as a proxy of its biomass, increases); E: substrate type from airborne IS data (15 July 2005, in brown nude substrates, from cyan to light green to dark green: submerged vegetation beds with increasing vegetation density cover); F: time series of chl-a from a pelagic station from MERIS

- Earth surface composition and change: The composition of exposed rock and soils can benefit from synergies between optical IS and TIR measurements, as the combination of spectral reflectance and emissivity measurements has been shown to be very helpful in identifying rocks, minerals and soils (e.g., Calvin et al. 2015; Eisele et al. 2015) which is of especial importance for geological applications of remote sensing (van der Meer et al. 2012). This is due to the fact that the spectral features of, e.g., silicates, clay minerals, iron oxides and hydroxides from the VSWIR and

TIR range, complement each other perfectly to material discrimination and quantification.

- Wildfires: Optical IS and TIR measurements yield complementary capabilities to understand wildfire processes, and in particular the coupling between fires, vegetation and associated trace gas emissions. TIR measurements can be used to calculate fire radiative power and temperature, whereas spectroscopic measurements in the SWIR can be used to distinguish small hotter fires from large cooler fires (Matheson and Dennison 2012) as well as to evaluate fire severity and vegetation recovery (Veraverbeke et al. 2012).
- Volcanoes: In addition to the capabilities of TIR remote sensing to monitor changes in temperature and gases indicating volcanic activity, the combination with optical IS measurements can help predict lava flows through the characterization of effusion rate and temperature.
- Urban environments: Although limited by the high spatial resolution ($< 5$ m) typically required by urban applications, the combination of optical IS and TIR has been shown to be very useful to characterize urban environments. The IS provides a high power of discrimination of manmade materials through the different spectral signatures, whereas TIR measurements can be used to measure temperatures and characterize the associated urban heat island effect (Roberts et al. 2012).

In addition to HyspIRI, concurrent IS and TIR measurements will also be available from the ISS through the coordinated operation of ECOSTRESS and HISUI (Stavros et al. 2017). Such combined measurements can be used for a preliminary assessment of some of HyspIRI's scientific questions.

### 3.2 Synergies of IS with LiDAR Data

LiDAR is an active remote sensing technology, using the time of flight of a laser pulse to compute a distance between the instrument and the backscattering (reflecting) object. With a precise estimate of the location and orientation of the measurement platform, a 3D coordinate of the reflecting object can be computed. LiDAR can be applied across a range of scales, from ground-based instruments to space-borne designs, serving very different purposes, as, e.g., surveying forest stands (Danson et al. 2007) or measuring the decline of the polar ice caps (Zwally et al. 2002). The most common LiDAR implementation, however, is in airborne laser scanners (ALS), where LiDARs are combined with a scanner to cover an across flight-track swath. Very high scanning frequencies (modern systems send several hundred thousand laser pulses per second) provide detailed 3D point clouds of the Earth's surface. These 3D data are of largest benefit for surface types that have inherent 3D features, e.g., urban areas and forests. For the latter, IS suffers from structural effects (Hilker et al. 2008), and consequently, the fusion of ALS and IS has the largest potential for forested areas. So far, most studies were focused on airborne instruments, as there were only few spaceborne IS instruments and LiDARs, and the latter only in very large footprints not well suited for vegetation studies (e.g., ICESAT GLAS). The fusion of IS and ALS can be performed at various levels of the processing chain, but accurate coregistration of the data sets is key to all approaches. One of the straightforward ways of synergistic use of IS and ALS is to use the ALS-derived digital surface model (DSM) to geo-locate the data of an IS system (typically push-broom sensors). Additional to the elevation model, more advanced approaches use both the IS and ALS intensity information in an overlapping wavelength domain (Fig. 12a) (Brell et al. 2016, 2017). This strategy enhances the synergistic use especially for the accurate coregistration of the two sensors (Brell et al. 2016).

In addition to the data level, fusion can as well be carried out at the product level, e.g., when combining two separate land-cover classifications based on separate IS and ALS data sets. Another aspect to classify fusion approaches is the choice of method. So far, most studies have used empirical frameworks for the fusion, e.g., a classifier or a regression model, but only few have used physical models of the radiative transfer to potentially improve results. A recent review of IS and ALS fusion approaches and their categorization by fusion level, approach and application can be found in Torabzadeh et al. (2014b). An important consideration is the notion of scale. IS and ALS data need to be similar in spatial extent and resolution for a genuine fusion of data sets. Larger mismatches in these two categories will result in methods being either up- or down-scaling or point-based cross-validation, e.g., in the case of a space-borne sampling LiDAR design (e.g., GEDI on the ISS) with a wall-to-wall IS instrument.

*Empirical approaches* Up to now, the majority of studies on IS and ALS fusion have used empirical approaches for land-cover classification (Torabzadeh et al. 2014b). In an early example, Koetz et al. (2008) fused IS and ALS data layers in a support vector machine framework to classify land-cover types including fuel types in a wildland–urban interface to assess and mitigate forest fire risk. ALS and IS were highly complementary, having a much higher accuracy when combined, with the height information of ALS being particularly helpful for vegetation canopy height classes. Regarding tree species classification, accuracy for a temperate mixed forest comprising 8 species was increased from 75%

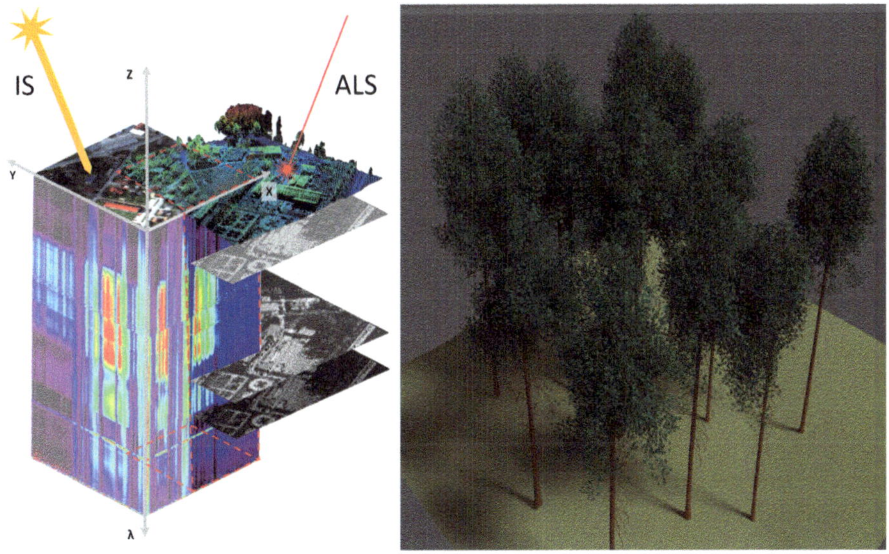


(a) (b)

**Fig. 12** Combination of IS and LiDAR (airborne laser scanner, ALS) data. A schematic view of the combination is shown in (**a**), and a 3D model of a pine stand (Tharandt, Germany) reconstructed within the 3D vegetation laboratory project using laser scanning is displayed in (**b**). Laser scanning was at three different levels to model the pine stand: laboratory (for the shoot structure), ground (stems and branch structure) and airborne (tree locations and dimensions) (see Eysn et al. (2013) for details) when using either ALS or IS alone to 90% when combining the data sets (Torabzadeh et al. 2014a). Both ALS and IS features were aggregated to individual tree crowns in this study. The high complementarity of ALS and IS for vegetation studies is as well confirmed by the large-scale campaigns and results of the Carnegie Airborne Observatory, flying IS and ALS sensors simultaneously, forming an ideal tool for 3D ecosystem assessment (Asner et al. 2012).

*Physical approaches* A potential tool to maximize the complementary exploitation of IS and ALS data is radiative transfer models, which add a physical layer in understanding and exploiting the signals recorded by both ALS and IS. A first of its kind, Koetz et al. (2007), used two different RTMs, one for IS, one for LiDAR, in a fused look-up table inversion retrieval of biochemical and biophysical variables such as LAI and fractional cover. However, results were mixed, likely hindered by the two models having distinct physical realities and differing parameterizations. For increased understanding of IS signals over vegetation, a possible solution is to derive an ALS-based parameterization of the vegetation canopy, applied in an RTM to forward simulate the spectral response. This was successfully done with the ESA STSE "3D Vegetation Laboratory" project (see Fig. 12b for an example 3D model), and such modeling tools will contribute to better fusion approaches for future missions (Schneider et al. 2017).

Besides the modeling approaches, physically based synergies between inflight IS and ALS data are being investigated. Based on radiative transfer modeling and ray-tracing approaches, the highly complementary sensor responses are physically adopted, and the active-passive dualism can be used to acquire more reliable and comparable hyperspectral data (Brell et al. 2017). This intensity-based cross-calibration between the two sensors is a first physical-based step to exploit active-passive synergies, with the specified goal of combining structural and spectral information for a comprehensive surface object description.

*Level-0 fusion (multispectral LiDAR)* The ultimate fusion of IS and ALS would be the development of a hyper-spectral LiDAR and implement that in an airborne design, remedying many issues of passive optical systems. However, up to now, only laboratory designs are truly hyperspectral due to their use as a supercontinuum light source (Junttila et al. 2015), while the commercial ALS systems are multispectral only (3 wavelengths maximum in the Optech Titan system). In these implementations, however, not the small number of bands is the largest limitation, but the missing spatial and temporal synchronization of the wavelengths. This will lead to large errors for band ratios with only small signals (e.g., the photochemical reflectance index), as reflectance and structural differences are mixed between the different wavelengths. Only a design where all wavelengths sample exactly the same footprint (preferably at the same time) will provide accurate band ratios, which can be linked to foliage biochemistry (Morsdorf et al. 2009; Woodhouse et al. 2011).

Technology and system characteristic of spaceborne LiDAR instruments differ significantly from airborne systems. However, airborne-based insights can be used for the development of synergies between various spectral and spatial sparse overlaps of spaceborne IS and LiDAR instruments. Upcoming systems, e.g., GEDI, ICESat-2 ATLAS (Abdalati et al. 2010) and the Methane Remote Sensing LiDAR Mission (MERLIN) (Ehret et al. 2017), demonstrate the rapid advances in spaceborne LiDAR technology. This indicates that, in the future, large footprints and the lack of spatial coverage will be overcome.

It must be stated that synergistic applications can also be found for the combination of IS with radar, although to a lesser extent than with LiDAR in part due to the non-overlapping wavelength domains. Demonstrated synergies between IS and radar are relatively similar to those between IS and LiDAR in the sense that radar provides the information on object structure to complete the spectral information from the IS image. Examples in the literature show the potential of synergies between IS and radar for the retrieval of vegetation properties (Treuhaft et al. 2002), urban mapping and land use classification (Hu et al. 2017) and oil spill mapping (Dabbiru et al. 2015).

## 4 Summary and Conclusions

This contribution has discussed potential avenues for the synergistic use of IS and other sources of remote sensing data, with a focus on synergies with optical MS satellite missions because of their coexistence with spaceborne IS missions in the coming years after the launch of EnMAP and PRISMA.

Optical IS and MS satellite missions can benefit from each other in two directions. On the one hand, IS acquisitions can be spatially enhanced through fusion with higher spatial resolution MS data sets, as it would be the case of the combination of EnMAP and PRISMA with Sentinel-2 10–20-m GSD data. Such spatial sharpening of IS data has been shown to have a strong potential for LULC applications, and in especial for those dealing with urban environments. On the other hand, the spatio-temporal monitoring potential of wide-swath MS systems can be complemented by the rich spectral information in IS data. For example, IS data can be used to interpret and refine information retrieved from the MS data over a wider area through the addition of complementary information (e.g., for multisensor monitoring of the composition of land surfaces and coastal and inland waters). The combination with IS data also allows to improve retrievals by MS systems through the extra spectral information provided by the IS data (e.g., endmembers derived from the IS data can be used as input for spectral unmixing techniques applied to the MS data). In the particular case of vegetation, for which relatively accurate physical radiative transfer models exist to link spectral reflectance with leaf and canopy parameters, model inversion and data assimilation techniques have a strong potential for parameter retrieval and the consistent merging of time series of MS and IS data.

Two other important aspects regarding optical IS and MS synergies have not been addressed in the text. First, the high spectral resolution and coverage of spaceborne IS instruments can be used to support calibration/validation activities of MS instruments in vicarious calibration exercises. Second, a wide range of optical MS missions with very high spatial resolution are being deployed by the private sector. These very high spatial resolution data hold an even bigger potential for spatial enhancement of IS data than Sentinel-2. This is especially true for the Worldview-3 mission (DigitalGlobe Inc.) which includes several spectral channels in the SWIR which can be expected to substantially improve the fusion results with respect to Sentinel-2.

In addition to optical MS data, TIR data yield strong synergistic potential with optical IS data for a number of application domains including the monitoring of vegetation functioning, natural hazards and surface composition. Such a synergy of colocated IS-TIR observations is the basis of the HyspIRI mission concept currently under development by NASA (Lee et al. 2015). The potential of merging IS and LiDAR data for the characterization of, e.g., vegetation covers and urban objects, has been proven by several studies based on airborne data. Synergies of IS with both TIR and LiDAR data can be further tested through the joint operation of the HISUI spectrometer, the GEDI LiDAR and the ECOSTRESS multispectral TIR instrument to coexist onboard the ISS (Stavros et al. 2017). However, the exploitation of such data set will be restricted to particular scientific studies because of the limited temporal and spatial overlap of the four instruments. In this regard, only the combination of optical IS and MS data can be considered for regular application in the next years thanks to the availability of Sentinel-2, Landsat and at least EnMAP and PRISMA in the 2018–2020 time frame. The development of unsupervised algorithms for the automatic colocation and synergistic exploitation of the two sources of data, either through the spatial enhancement of the IS data or the improvement of information extraction from the MS, can thus be considered as an important field of research in the next years. If such combined IS-MS data exploitation could become quasi-operational, it might have an impact on the definition of future spaceborne IS missions, as some observational requirements for the most demanding applications (e.g., spatial resolution for urban mapping) could be relaxed.

**Acknowledgements** This paper is an outcome of a workshop on requirements capabilities and directions in spaceborne imaging spectroscopy held at the International Space Science Institute (ISSI) in Bern, Switzerland, in November 2016. LG, KS, MB and CM were partly funded by the EnMAP scientific preparation program (FKZ: 50EE1617).

## References

Abdalati W, Zwally HJ, Bindschadler R, Csatho B, Farrell SL, Fricker HA, Harding D, Kwok R, Lefsky M, Markus T, Marshak A, Neumann T, Palm S, Schutz B, Smith B, Spinhirne J, Webb C (2010) The ICESat-2 laser altimetry mission. Proc IEEE 98(5):735–751

Anderson M, Kustas W (2008) Thermal remote sensing of drought and evapotranspiration. EOS Trans Am Geophys Union 89(26):233–234

Asner GP, Knapp DE, Boardman J, Green RO, Kennedy-Bowdoin T, Eastwood M, Martin RE, Anderson C, Field CB (2012) Carnegie airborne observatory-2: increasing science data dimensionality via high-fidelity multi-sensor fusion. Remote Sens Environ 124(Supplement C):454–465

Asner GP, Brodrick PG, Anderson CB, Vaughn N, Knapp DE, Martin RE (2016) Progressive forest canopy water loss during the 2012–2015 California drought. Proc Natl Acad Sci 113(2):E249–E255

Barnsley MJ, Settle JJ, Cutter M, Lobb D, Teston F (2004) The PROBA/CHRIS mission: a low-cost smallsat for hyperspectral, multi-angle, observations of the Earth surface and atmosphere. IEEE Trans Geosci Remote Sens 42:1512–1520

Bishop CA, Liu JG, Mason PJ (2011) Hyperspectral remote sensing for mineral exploration in Pulang, Yunnan Province, China. Int J Remote Sens 32(9):2409–2426

Botha EJ, Brando VE, Anstee JM, Dekker AG, Sagar S (2013) Increased spectral resolution enhances coral detection under varying water conditions. Remote Sens Environ 131(Supplement C):247–261

Brell M, Rogass C, Segl K, Bookhagen B, Guanter L (2016) Improving sensor fusion: a parametric method for the geometric coalignment of airborne hyperspectral and lidar data. IEEE Trans Geosci Remote Sens 54(6):3460–3474

Brell M, Segl K, Guanter L, Bookhagen B (2017) Hyperspectral and lidar intensity data fusion: a framework for the rigorous correction of illumination, anisotropic effects, and cross calibration. IEEE Trans Geosci Remote Sens 55(5):2799–2810

Bresciani M, Stroppiana D, Odermatt D, Morabito G, Giardino C (2011) Assessing remotely sensed chlorophyll-a for the implementation of the water framework directive in European perialpine lakes. Sci Total Environ 409(17):3083–3091

Bresciani M, Bolpagni R, Braga F, Oggioni A, Giardino C (2012) Retrospective assessment of macrophytic communities in southern Lake Garda (Italy) from in situ and mivis (multispectral infrared and visible imaging spectrometer) data. J Limnol 71(1):19

Bush A, Sollmann R, Wilting A et al (2017) Connecting Earth observation to high-throughput biodiversity data. Nat Ecol Evol 1:0176

Calvin WMF, Littlefield E, Kratt C (2015) Remote sensing of geothermal-related minerals for resource exploration in Nevada. Geothermics 53:517–526

Candela L, Formaro R, Guarini R, Loizzo R, Longo F, Varacalli G (2016) The PRISMA mission. In: 2016 IEEE international geoscience and remote sensing symposium (IGARSS), pp 253–256

Casal G, Kutser T, Domínguez-Gómez J, Sánchez-Carnero N, Freire J (2011) Mapping benthic macroalgal communities in the coastal zone using CHRIS-PROBA mode 2 images. Estuar Coast Shelf Sci 94(3):281–290

Chan JC-W, Yokoya N (2016) Mapping land covers of Brussels capital region using spatially enhanced hyperspectral images. In: WHISPERS 2016, pp 1–5. IEEE Xplore

Chan JC-W, Paelinckx D (2008) Evaluation of random forest and adaboost tree-based ensemble classification and spectral band selection for ecotope mapping using airborne hyperspectral imagery. Remote Sens Environ 112(6):2999–3011

Chan JCW, Ma J, de Voorde TV, Canters F (2011) Preliminary results of superresolution-enhanced angular hyperspectral (CHRIS/PROBA) images for land-cover classification. IEEE Geosci Remote Sens Lett 8(6):1011–1015

Chavez W Jr (2000) Supergene oxidation of copper deposits: zoning and distribution of copper oxide minerals. SEG Newsl 41:9–21

Chen Z, Pu H, Wang B, Jiang GM (2014) Fusion of hyperspectral and multispectral images: a novel framework based on generalization of pan-sharpening methods. IEEE Geosci Remote Sens Lett 11(8):1418–1422

Clark RN, Swayze GA, Livo KE, Kokaly RF, Sutley SJ, Dalton JB, McDougal RR, Gent CA (2003) Imaging spectroscopy: Earth and planetary remote sensing with the usgs tetracorder and expert systems. J Geophys Res Planets 108(E12):5131

Dabbiru L, Samiappan S, Nobrega RAA, Aanstoos JA, Younan NH, Moorhead RJ (2015) Fusion of synthetic aperture radar and hyperspectral imagery to detect impacts of oil spill in gulf of mexico. In: 2015 IEEE international geoscience and remote sensing symposium (IGARSS), pp 1901–1904

Danson FM, Hetherington D, Morsdorf F, Koetz B, Allgower B (2007) Forest canopy gap fraction from terrestrial laser scanning. IEEE Geosci Remote Sens Lett 4(1):157–160

Dekker AG, Brando VE, Anstee JM (2005) Retrospective seagrass change detection in a shallow coastal tidal Australian lake. Remote Sens Environ 97(4):415–433

Dekker AG, Phinn SR, Anstee J, Bissett P, Brando VE, Casey B, Fearns P, Hedley J, Klonowski W, Lee ZP, Lynch M, Lyons M, Mobley C, Roelfsema C (2011) Intercomparison of shallow water bathymetry, hydro-optics, and benthos mapping techniques in Australian and caribbean coastal environments. Limnol Oceanogr Methods 9(9):396–425

Demarchi L, Chan JC-W, Ma J, Canters F (2012) Mapping impervious surfaces from superresolution enhanced CHRIS/PROBA imagery using multiple endmember unmixing. ISPRS J Photogramm Remote Sens 72(Supplement C):99–112

Dierssen H, Mcmanus G, Chlus A, Qiu D, Gao B-C, Lin S (2015) Space station image captures a red tide ciliate bloom at high spectral and spatial resolution. Proc Natl Acad Sci USA 112:14783–14787

Drusch M, Bello UD, Carlier S, Colin O, Fernandez V, Gascon F, Hoersch B, Isola C, Laberinti P, Martimort P, Meygret A, Spoto F, Sy O, Marchese F, Bargellini P (2012) Sentinel-2: ESA's optical high-resolution mission for GMES operational services. Remote Sens Environ 120:25–36

Drusch M, Moreno J, Bello UD, Franco R, Goulas Y, Huth A, Kraft S, Middleton EM, Miglietta F, Mohammed G, Nedbal L, Rascher U, Schüttemeyer D, Verhoef W (2017) The fluorescence explorer mission concept—ESA's Earth explorer 8. IEEE Trans Geosci Remote Sens 55(3):1273–1284

Ehret G, Bousquet P, Pierangelo C, Alpers M, Millet B, Abshire JB, Bovensmann H, Burrows JP, Chevallier F, Ciais P, Crevoisier C, Fix A, Flamant P, Frankenberg C, Gibert F, Heim B, Heimann M, Houweling S, Hubberten HW, Jöckel P, Law K, Löw A, Marshall J, Agusti-Panareda A, Payan S, Prigent C, Rairoux P, Sachs T, Scholze M, Wirth M (2017) Merlin: a French–German space lidar mission dedicated to atmospheric methane. Remote Sens 9(10):1052

Eisele A, Chabrillat S, Hecker C, Hewson R, Lau IC, Rogass C, Segl K, Cudahy TJ, Udelhoven T, Hostert P, Kaufmann H (2015) Advantages using the thermal infrared (TIR) to detect and quantify semi-arid soil properties. Remote Sens Environ 163(Supplement C):296–311

Eysn L, Pfeifer N, Ressl C, Hollaus M, Grafl A, Morsdorf F (2013) A practical approach for extracting tree models in forest environments based on equirectangular projections of terrestrial laser scans. Remote Sens 5(11):5424–5448

Féret JB, Gitelson AA, Noble SD et al (2017) PROSPECT-D: towards modeling leaf optical properties through a complete lifecycle. Remote Sens Environ 193:204–215

Foody GM (2002) Status of land cover classification accuracy assessment. Remote Sens Environ 80(1):185–201

Garcia RA, Fearns PR, McKinna LI (2014) Detecting trend and seasonal changes in bathymetry derived from hico imagery: a case study of Shark Bay, western Australia. Remote Sens Environ 147(Supplement C):186–205

Giardino C, Brando VE, Dekker AG, Strömbeck N, Candiani G (2007) Assessment of water quality in Lake Garda (Italy) using hyperion. Remote Sens Environ 109(2):183–195

Giardino C, Brando VE, Gege P, Pinnel N, Hochberg E, Knaeps E, Reusen I, Doerffer R, Bresciani M, Braga F, Foerster S, Champollion N, Dekker A (2018) Imaging spectrometry of inland and coastal waters: state of the art, achievements and perspectives. Surv Geophys. https://doi.org/10.1007/s10712-018-9476-0

Goetz AFH, Vane G, Salomon JE, Rock BN (1985) Imaging spectroscopy for Earth remote sensing. Science 228:1147–1153

Gómez-Dans J, Lewis P, Disney M (2016) Efficient emulation of radiative transfer codes using Gaussian processes and application to land surface parameter inferences. Remote Sens 8(2):119

Green RO, Eastwood M, Sarture C, Chrien T, Aronsson M, Chippendale B, Faust J, Pavri B, Chovit C, Solis M, Olah M, Williams O (1998) Imaging spectroscopy and the airborne visible/infrared imaging spectrometer (AVIRIS). Remote Sens Environ 65:227–248

Grohnfeldt C, Zhu XX, Bamler R (2013) Jointly sparse fusion of hyperspectral and multispectral imagery. In: 2013 IEEE international geoscience and remote sensing symposium—IGARSS, pp 4090–4093

Guanter L, Kaufmann H, Segl K, Foerster S, Rogass C, Chabrillat S, Kuester T, Hollstein A, Rossner G, Chlebek C, Straif C, Fischer S, Schrader S, Storch T, Heiden U, Mueller A, Bachmann M, Mühle H, Müller R, Habermeyer M, Ohndorf A, Hill J, Buddenbaum H, Hostert P, van der Linden S, Leitão PJ, Rabe A, Doerffer R, Krasemann H, Xi H, Mauser W, Hank T, Locherer M, Rast M, Staenz K, Sang B (2015) The enmap spaceborne imaging spectroscopy mission for earth observation. Remote Sens 7(7):8830–8857

Herold M, Roberts DA, Gardner ME, Dennison PE (2004) Spectrometry for urban area remote sensing-development and analysis of a spectral library from 350 to 2400 nm. Remote Sens Environ 91(3):304–319

Hestir EL, Brando VE, Bresciani M, Giardino C, Matta E, Villa P, Dekker AG (2015) Measuring freshwater aquatic ecosystems: the need for a hyperspectral global mapping satellite mission. Remote Sens Environ 167(Supplement C):181–195

Hilker T, Coops NC, Hall FG, Black TA, Wulder MA, Nesic Z, Krishnan P (2008) Separating physiologically and directionally induced changes in PRI using BRDF models. Remote Sens Environ 112(6):2777–2788

Hu J, Mou L, Schmitt A, Zhu XX (2017) Fusionet: a two-stream convolutional neural network for urban scene classification using polsar and hyperspectral data. In: 2017 Joint urban remote sensing event (JURSE), pp 1–4

Hubbard B, Crowley JK (2005) Mineral mapping on the Chilean–Bolivian altiplano using co-orbital ali, aster and hyperion imagery: data dimensionality issues and solutions. Remote Sens Environ 99:173–186

Hubbard BE, Crowley JK, Zimbelman DR (2003) Comparative alteration mineral mapping using visible to shortwave infrared (0.4–2.4 mu m) hyperion, ali, and aster imagery. IEEE Trans Geosci Remote Sens 41(6):1401–1410

Junttila S, Kaasalainen S, Vastaranta M, Hakala T, Nevalainen O, Holopainen M (2015) Investigating bi-temporal hyperspectral lidar measurements from declined trees-experiences from laboratory test. Remote Sens 7(10):13863–13877

Koetz B, Sun G, Morsdorf F, Ranson K, Kneubühler M, Itten K, Allgöwer B (2007) Fusion of imaging spectrometer and lidar data over combined radiative transfer models for forest canopy characterization. Remote Sens Environ 106(4):449–459

Koetz B, Morsdorf F, van der Linden S, Curt T, Allgöwer B (2008) Multi-source land cover classification for forest fire management based on imaging spectrometry and lidar data. For Ecol Manag 256(3):263–271. Impacts of forest ecosystem management on greenhouse gas budgets

Kutser T (2004) Quantitative detection of chlorophyll in cyanobacterial blooms by satellite remote sensing. Limnol Oceanogr 49(6):2179–2189

Lanaras C, Baltsavias E, Schindler K (2015) Hyperspectral super-resolution by coupled spectral unmixing. In: The IEEE international conference on computer vision (ICCV)

Lee CM, Cable ML, Hook SJ, Green RO, Ustin SL, Mandl DJ, Middleton EM (2015) An introduction to the NASA hyperspectral infrared imager (hyspiri) mission and preparatory activities. Remote Sens Environ 167(Supplement C):6–19

Lewis P, Gómez-Dans J, Kaminski T, Settle J, Quaife T, Gobron N, Styles J, Berger M (2012) An Earth observation land data assimilation system (EO-LDAS). Remote Sens Environ 120:219–235

Lucke RL, Corson M, McGlothlin NR, Butcher SD, Wood DL, Korwan DR, Li RR, Snyder WA, Davis CO, Chen DT (2011) Hyperspectral imager for the coastal ocean: instrument description and first images. Appl Opt 50(11):1501–1516

Matheson DS, Dennison PE (2012) Evaluating the effects of spatial resolution on hyperspectral fire detection and temperature retrieval. Remote Sens Environ 124(Supplement C):780–792

Mielke C, Boesche NK, Rogass C, Kaufmann H, Gauert C, de Wit M (2014a) Spaceborne mine waste mineralogy monitoring in South Africa, applications for modern push-broom missions: hyperion/OLI and EnMAP/Sentinel-2. Remote Sens 6(8):6790–6816
Mielke CK, Boesche N, Rogaß C, Segl K, Gauert C, Kaufmann H (2014b) Potential applications of the Sentinel-2 multispectral sensor and the EnMAP hyperspectral sensor in mineral exploration. In: EARSeL eProceedings, vol 13, p 93
Mielke C, Rogass C, Boesche N, Segl K, Altenberger U (2016) EnGeoMAP 2.0—automated hyperspectral mineral identification for the German EnMAP space mission. Remote Sens 8(2):127
Milewski R, Chabrillat S, Behling R (2017) Analyses of recent sediment surface dynamic of a Namibian Kalahari salt pan based on multitemporal landsat and hyperspectral hyperion data. Remote Sens 9(2):170
Mishra D, Ogashawara I, Gitelson A (2017) Bio-optical modeling and remote sensing of inland waters. Elsevier, Amsterdam, p 332
Morsdorf F, Nichol C, Malthus T, Woodhouse IH (2009) Assessing forest structural and physiological information content of multi-spectral lidar waveforms by radiative transfer modelling. Remote Sens Environ 113(10):2152–2163
Mouw CB, Greb S, Aurin D, DiGiacomo PM, Lee Z, Twardowski M, Binding C, Hu C, Ma R, Moore T, Moses W, Craig SE (2015) Aquatic color radiometry remote sensing of coastal and inland waters: challenges and recommendations for future satellite missions. Remote Sens Environ 160(Supplement C):15–30
Olmanson LG, Bauer ME, Brezonik PL (2008) A 20-year landsat water clarity census of minnesota's 10,000 lakes. Remote Sens Environ 112(11):4086–4097
Palmer SC, Kutser T, Hunter PD (2015) Remote sensing of inland waters: challenges, progress and future directions. Remote Sens Environ 157(Supplement C):1–8
Palsson F, Sveinsson JR, Ulfarsson MO, Benediktsson JA (2016) Quantitative quality evaluation of pansharpened imagery: consistency versus synthesis. IEEE Trans Geosci Remote Sens 54(3):1247–1259
Phinn S, Roelfsema C, Dekker A, Brando V, Anstee J (2008) Mapping seagrass species, cover and biomass in shallow waters: an assessment of satellite multi-spectral and airborne hyper-spectral imaging systems in moreton bay (Australia). Remote Sens Environ 112(8):3413–3425
Rainforth T, Wood F (2015) Canonical correlation forests. ArXiv e-prints
Ribeiro da Luz B, Crowley JK (2007) Spectral reflectance and emissivity features of broad leaf plants: prospects for remote sensing in the thermal infrared (8.0–14.0 μm). Remote Sens Environ 109(4):393–405
Roberts DA, Quattrochi DA, Hulley GC, Hook SJ, Green RO (2012) Synergies between VSWIR and TIR data for the urban environment: an evaluation of the potential for the hyperspectral infrared imager (hyspiri) decadal survey mission. Remote Sens Environ 117(Supplement C):83–101
Rodriguez JJ, Kuncheva LI, Alonso CJ (2006) Rotation forest: a new classifier ensemble method. IEEE Trans Pattern Anal Mach Intell 28(10):1619–1630
Roy D, Wulder M, Loveland T, Woodcock CE, Allen R, Anderson M, Helder D, Irons J, Johnson D, Kennedy R, Scambos T, Schaaf C, Schott J, Sheng Y, Vermote E, Belward A, Bindschadler R, Cohen W, Gao F, Hipple J, Hostert P, Huntington J, Justice C, Kilic A, Kovalskyy V, Lee Z, Lymburner L, Masek J, McCorkel J, Shuai Y, Trezza R, Vogelmann J, Wynne R, Zhu Z (2014) Landsat-8: science and product vision for terrestrial global change research. Remote Sens Environ 145:154–172
Schmid T, Koch M, Gumuzzio J (2005) Multisensor approach to determine changes of wetland characteristics in semiarid environments (central Spain). IEEE Trans Geosci Remote Sens 43(11):2516–2525
Schneider FD, Morsdorf F, Schmid B, Petchey OL, Hueni A, Schimel DS, Schaepman ME (2017) Mapping functional diversity from remotely sensed morphological and physiological forest traits. Nat Commun 8:1441
Segl K, Guanter L, Rogass C, Kuester T, Roessner S, Kaufmann H, Sang B, Mogulsky V, Hofer S (2012) Eetes—the enmap end-to-end simulation tool. IEEE J Sel Top Appl Earth Obs Remote Sens 5(2):522–530
Segl K, Guanter L, Gascon F, Kuester T, Rogass C, Mielke C (2015) S2etes: an end-to-end modeling tool for the simulation of Sentinel-2 image products. IEEE Trans Geosci Remote Sens 53(10):5560–5571
Selva M, Aiazzi B, Butera F, Chiarantini L, Baronti S (2015) Hyper-sharpening: a first approach on SIM-GA data. IEEE J Sel Top Appl Earth Obs Remote Sens 8(6):3008–3024
Stavros EN, Schimel D, Pavlick R, Serbin S, Swann A, Duncanson L, Fisher JB, Fassnacht F, Ustin S, Dubayah R, Schweiger A, Wennberg P (2017) ISS observations offer insights into plant function. Nat Ecol Evol 1:0194
Strong AE (1974) Remote sensing of algal blooms by aircraft and satellite in Lake Erie and Utah Lake. Remote Sens Environ 3(2):99–107
Taylor R (2011) Gossans and leached cappings—field assessment. Springer, Berlin, p 146

Thompson DR, Boardman JW, Eastwood ML, Green RO (2017) A large airborne survey of earth's visible-infrared spectral dimensionality. Opt Express 25(8):9186–9195

Torabzadeh H, Morsdorf F, Leiterer R, Schaepman M (2014a) Fusing imaging spectrometry and airborne laser scanning data for tree species discrimination. In: Geoscience and remote sensing symposium (IGARSS), 2014 IEEE international, pp 1253–1256

Torabzadeh H, Morsdorf F, Schaepman ME (2014b) Fusion of imaging spectroscopy and airborne laser scanning data for characterization of forest ecosystems—a review. ISPRS J Photogramm Remote Sens 97(Supplement C):25–35

Treuhaft RN, Asner GP, Law BE, Van Tuyl S (2002) Forest leaf area density profiles from the quantitative fusion of radar and hyperspectral data. J Geophys Res Atmos 107(D21):ACL 7-1–ACL 7-13

Tyler AN, Hunter PD, Spyrakos E, Groom S, Constantinescu AM, Kitchen J (2016) Developments in earth observation for the assessment and monitoring of inland, transitional, coastal and shelf-sea waters. Sci Total Environ 572(Supplement C):1307–1321

Ungar SG, Pearlman JS, Mendenhall JA, Reuter D (2003) Overview of the Earth Observing One (EO-1) mission. IEEE Trans Geosci Remote Sens 41:1149–1159

Ustin SL, Roberts DA, Gamon JA, Asner GP, Green RO (2004) Using imaging spectroscopy to study ecosystem processes and properties. Bioscience 54(6):523–534

van der Meer FD, van der Werff HM, van Ruitenbeek FJ, Hecker CA, Bakker WH, Noomen MF, van der Meijde M, Carranza EJM, de Smeth JB, Woldai T (2012) Multi- and hyperspectral geologic remote sensing: a review. Int J Appl Earth Obs Geoinf 14(1):112–128

Veraverbeke S, Hook S, Harris S (2012) Synergy of vswir (0.4–2.5 μm) and MTIR (3.5–12.5 μm) data for post-fire assessments. Remote Sens Environ 124(Supplement C):771–779

Verhoef W (1984) Light scattering by leaf layers with application to canopy reflectance modeling: the SAIL model. Remote Sens Environ 16(2):125–141

Vermote EF, Tanre D, Deuze J-L, Herman M, Morcette J-J (1997) Second simulation of the satellite signal in the solar spectrum, 6S: an overview. IEEE Trans Geosci Remote Sens Publ IEEE Geosci Remote Sens Soc 35(3):675–686

Wei Q, Dobigeon N, Tourneret JY (2015) Bayesian fusion of multi-band images. IEEE J Sel Top Signal Process 9(6):1117–1127

Woodhouse IH, Nichol C, Sinclair P, Jack J, Morsdorf F, Malthus TJ, Patenaude G (2011) A multispectral canopy lidar demonstrator project. IEEE Geosci Remote Sens Lett 8(5):839–843

Xu B, Gong P (2007) Land-use/land-cover classification with multispectral and hyperspectral EO-1 data. Photogramm Eng Remote Sens 73:955–965

Yokoya N, Yairi T, Iwasaki A (2012) Coupled nonnegative matrix factorization unmixing for hyperspectral and multispectral data fusion. IEEE Trans Geosci Remote Sens 50(2):528–537

Yokoya N, Chan JC-W, Segl K (2016) Potential of resolution-enhanced hyperspectral data for mineral mapping using simulated enmap and Sentinel-2 images. Remote Sens 8(3):172

Yokoya N, Grohnfeldt C, Chanussot J (2017) Hyperspectral and multispectral data fusion: a comparative review of the recent literature. IEEE Geosci Remote Sens Mag 5(2):29–56

Zilioli E, Brivio P, Gomarasca M (1994) A correlation between optical properties from satellite data and some indicators of eutrophication in Lake Garda (Italy). Sci Total Environ 158(Supplement C):127–133

Zwally H, Schutz B, Abdalati W, Abshire J, Bentley C, Brenner A, Bufton J, Dezio J, Hancock D, Harding D, Herring T, Minster B, Quinn K, Palm S, Spinhirne J, Thomas R (2002) Icesat's laser measurements of polar ice, atmosphere, ocean, and land. J Geodyn 34(3):405–445


## Affiliations


**Luis Guanter[1] · Maximilian Brell[1] · Jonathan C.-W. Chan[2] · Claudia Giardino[3] · Jose Gomez-Dans[4] · Christian Mielke[1] · Felix Morsdorf[5] · Karl Segl[1] · Naoto Yokoya[6]**

[1] Helmholtz Centre Potsdam German Centre for Geosciences (GFZ), Telegrafenberg A17, 14473 Potsdam, Germany

[2] Department of Electronics and Informatics, Vrije Universiteit Brussel, 1050 Brussels, Belgium

[3] National Research Council–Institute for Electromagnetic Sensing of the Environment (CNR-IREA), Via Bassini 15, 20133 Milan, Italy

[4] Department of Geography, University College London and National Centre for Earth Observation, Pearson Building, Gower Street, London WC1E 6BT, UK

[5] Remote Sensing Laboratories, Department of Geography, University of Zurich, Winterthurerstrasse 190, 8057 Zurich, Switzerland

[6] RIKEN Center for Advanced Intelligence Project, 1-4-1 Nihonbashi, Chuo-ku, Tokyo 103-0027, Japan

Printed by Printforce, the Netherlands